AF509463

New Trends in Photo(Electro)catalysis: From Wastewater Treatment to Energy Production

New Trends in Photo(Electro)catalysis: From Wastewater Treatment to Energy Production

Editor

Simonetta Palmas

MDPI • Basel • Beijing • Wuhan • Barcelona • Belgrade • Manchester • Tokyo • Cluj • Tianjin

Editor
Simonetta Palmas
Mechanical, Chemical and
Materials Engineering
Università degli studi di
Cagliari
Italy

Editorial Office
MDPI
St. Alban-Anlage 66
4052 Basel, Switzerland

This is a reprint of articles from the Special Issue published online in the open access journal *Catalysts* (ISSN 2073-4344) (available at: www.mdpi.com/journal/catalysts/special_issues/photoelectrocatal_wastewater).

For citation purposes, cite each article independently as indicated on the article page online and as indicated below:

LastName, A.A.; LastName, B.B.; LastName, C.C. Article Title. *Journal Name* **Year**, *Volume Number*, Page Range.

ISBN 978-3-0365-4098-6 (Hbk)
ISBN 978-3-0365-4097-9 (PDF)

Contents

About the Editor

Simonetta Palmas

Simonetta Palmas is a Chemical Engineer, PhD, Full Professor, at the Dept of Mechanical Chemical and Materials Engineering of the University of Cagliari in the scientific sector of Industrial Chemistry and Technology (SSD ING/IND27). She also attained the National academic qualification for full professor in Chemical Technology Fundamentals (SSD CHIM/07). Her scientific and didactics activity is specifically addressed to Electrochemical Engineering and applications related to environmental remediation and energy conversion and storage.

Her contribution to these topics is also relevant at the international level: she is the Italian Delegate in the sitting members of the Working European Party on Electrochemical Engineering, in the European Federation of Chemical Engineering (EFCE). Moreover, she was the vice chair of the ISE (International Society of Electrochemistry) Division 5 (Electrochemical Process Engineering and Technology) from 2018 to 2021.

Editorial

Special Issue: New Trends in Photo (Electro)catalysis: From Wastewater Treatment to Energy Production

Simonetta Palmas

Chimica e dei Materiali, Dipartimento di Ingegneria Meccanica, Università degli studi di Cagliari, Via Marengo 2, 09123 Cagliari, Italy; simonetta.palmas@unica.it

Citation: Palmas, S. Special Issue: New Trends in Photo (Electro)catalysis: From Wastewater Treatment to Energy Production. *Catalysts* **2021**, *11*, 586. https://doi.org/10.3390/catal11050586

Received: 8 March 2021
Accepted: 23 March 2021
Published: 30 April 2021

Publisher's Note: MDPI stays neutral with regard to jurisdictional claims in published maps and institutional affiliations.

This Special Issue aimed at focusing on photo- and photo-electrocatalytic processes specifically devoted to present both new catalytic materials and possible applications in environmental and energetic fields.

In fact, solar-driven photoelectrocatalytic (PEC) processes could be considered as one of the focal points to which the research should be addressed in the next future to achieve optimal results in environmental recovery and energy production. However, due to the low efficiency and selectivity of photocatalytic processes under solar energy, a major challenge exists to improve the performance of photocatalytic materials and increase the effectiveness of single and combined processes.

In this context, I would like to sincerely thank all the authors who accepted the challenge and collaborated with their excellent contributions to this Special Issue, which includes ten articles and five review papers.

Summarizing, the papers present both state-of-the-art and new trends on wastewater treatment and sustainable methods for organic degradation and energy production.

Among the different proposals TiO_2-based materials remain one of the protagonists in most of the papers, which used titania as single electrode material or in combination with either metal oxides or noble metal nanoparticles [1,2]. The effectiveness of the prepared electrodes, as well as the effect of the physical characteristics of the synthesized materials on the final PEC performances, are generally assessed during waste water treatment processes (WWTP) for the removal of organic pollutants, with special attention on persistent organic pollutants, drugs, or dyes which represent the most problematic classes of substances in traditional WWTP.

Analogous processes have been examined at different catalysts such as spinel $ZnFe_2O_4$ (ZFO) [3] ZnO/Ag [4], or Niobium-based metal oxides [5], which are generally proposed as photoanode materials for organic degradation in different range of wavelength of the irradiating light.

The synthesis of cathodic materials has also been investigated. Electrochemical dealloying has been proposed [6] where starting from Ni-Cu co-deposits, highly porous Ni electrodes were obtained, which demonstrated greater activity towards hydrogen evolution reactions, in comparison with commercial smooth Ni electrodes.

Cathodic materials were also the focus of some papers in which WWTP were coupled with the electrical energy production in microbial fuel cells [7,8]

The effectiveness of PEC processes has been assessed also in combined processes with sono-electrolysis for the degradation of drugs [9]: in that case boron-doped diamond electrodes were used, which are widely known in the electrocatalysis field.

Moreover, the development of photoactive and durable floating devices has been proposed for the treatment of water basins [10]. The major challenge in this field is the development of effective devices which allow an easier retrieval of the photocatalyst than in the case of powder catalysts. This allows for a more efficient light usage, since light, especially UV, attenuates rapidly in water.

Five review papers completed the picture, by presenting a wide panoramic on the recent progress in the development of novel photocatalysts for H2 production by water splitting or for WWTP [11,12]. Graphitic carbon nitride materials [13] and metal sulphide composite nanomaterials [14] have been proposed as photocatalysts, and numerous strategies have been developed for their preparation. Finally, a brief summary of the current challenges and an outlook for the development of composite photocatalysts in the area of wastewater treatment were provided [15].

The guest editor would like to express her gratitude to both the editorial team for their professional assistance and all the authors for their valuable scientific contributions.

Conflicts of Interest: The authors declare no conflict of interest.

References

1. Akel, S.; Dillert, R.; Balayeva, N.O.; Boughaled, R.; Koch, J.; El Azzouzi, M.; Bahnemann, D.W. Ag/Ag$_2$O as a Co-Catalyst in TiO$_2$ Photocatalysis: Effect of the Co-Catalyst/Photocatalyst Mass Ratio. *Catalysts* **2018**, *8*, 647. [CrossRef]
2. Matarrese, R.; Mascia, M.; Vacca, A.; Mais, L.; Usai, E.M.; Ghidelli, M.; Mascaretti, L.; Bricchi, B.R.; Russo, V.; Casari, C.S.; et al. Integrated Au/TiO$_2$ Nanostructured Photoanodes for Photoelectrochemical Organics Degradation. *Catalysts* **2019**, *9*, 340. [CrossRef]
3. Granone, L.I.; Nikitin, K.; Emeline, A.; Dillert, R.; Bahnemann, D.W. Effect of the Degree of Inversion on the Photoelectrochemical Activity of Spinel ZnFe$_2$O$_4$. *Catalysts* **2019**, *9*, 434. [CrossRef]
4. González-Rodríguez, J.; Fernández, L.; Bava, Y.B.; Buceta, D.; Vázquez-Vázquez, C.; López-Quintela, M.A.; Feijoo, G.; Moreira, M.T. Enhanced Photocatalytic Activity of Semiconductor Nanocomposites Doped with Ag Nanoclusters under UV and Visible Light. *Catalysts* **2019**, *10*, 31. [CrossRef]
5. Dos Santos, A.J.; Batista, L.M.B.; Martínez-Huitle, C.A.; Alves, A.P.D.M.; Garcia-Segura, S. Niobium Oxide Catalysts as Emerging Material for Textile Wastewater Reuse: Photocatalytic Decolorization of Azo Dyes. *Catalysts* **2019**, *9*, 1070. [CrossRef]
6. Mais, L.; Palmas, S.; Mascia, M.; Sechi, E.; Casula, M.F.; Rodriguez, J.; Vacca, A. Porous Ni Photocathodes Obtained by Selective Corrosion of Ni-Cu Films: Synthesis and Photoelectrochemical Characterization. *Catalysts* **2019**, *9*, 453. [CrossRef]
7. Włodarczyk, P.P.; Włodarczyk, B. Preparation and Analysis of Ni–Co Catalyst Use for Electricity Production and COD Reduction in Microbial Fuel Cells. *Catalysts* **2019**, *9*, 1042. [CrossRef]
8. Włodarczyk, P.P.; Włodarczyk, B. Wastewater Treatment and Electricity Production in a Microbial Fuel Cell with Cu–B Alloy as the Cathode Catalyst. *Catalysts* **2019**, *9*, 572. [CrossRef]
9. Silva, F.L.; Saez, C.; Lanza, M.R.; Cañizares, P.; Rodrigo, M.A. The Role of Mediated Oxidation on the Electro-irradiated Treatment of Amoxicillin and Ampicillin Polluted Wastewater. *Catalysts* **2018**, *9*, 9. [CrossRef]
10. Sabatini, V.; Rimoldi, L.; Tripaldi, L.; Meroni, D.; Farina, H.; Ortenzi, M.A.; Ardizzone, S. TiO$_2$-SiO$_2$-PMMA Terpolymer Floating Device for the Photocatalytic Remediation of Water and Gas Phase Pollutants. *Catalysts* **2018**, *8*, 568. [CrossRef]
11. Al Jitan, S.; Palmisano, G.; Garlisi, C. Synthesis and Surface Modification of TiO$_2$-Based Photocatalysts for the Conversion of CO$_2$. *Catalysts* **2020**, *10*, 227. [CrossRef]
12. Wang, Y.-H.; Rahman, K.H.; Wu, C.-C.; Chen, K.-C. A Review on the Pathways of the Improved Structural Characteristics and Photocatalytic Performance of Titanium Dioxide (TiO$_2$) Thin Films Fabricated by the Magnetron-Sputtering Technique. *Catalysts* **2020**, *10*, 598. [CrossRef]
13. Mun, S.J.; Park, S.-J. Graphitic Carbon Nitride Materials for Photocatalytic Hydrogen Production via Water Splitting: A Short Review. *Catalysts* **2019**, *9*, 805. [CrossRef]
14. Lee, S.L.; Chang, C.-J. Recent Progress on Metal Sulfide Composite Nanomaterials for Photocatalytic Hydrogen Production. *Catalysts* **2019**, *9*, 457. [CrossRef]
15. Ge, J.; Zhang, Y.; Heo, Y.-J.; Park, S.-J. Advanced Design and Synthesis of Composite Photocatalysts for the Remediation of Wastewater: A Review. *Catalysts* **2019**, *9*, 122. [CrossRef]

 *catalysts* 

Review

A Review on the Pathways of the Improved Structural Characteristics and Photocatalytic Performance of Titanium Dioxide (TiO$_2$) Thin Films Fabricated by the Magnetron-Sputtering Technique

Yu-Hsiang Wang [1], Kazi Hasibur Rahman [2], Chih-Chao Wu [3] and Kuan-Chung Chen [1,4,*]

[1] Department of Environmental Science and Engineering, National Pingtung University of Science and Technology,1 Shuehfu Rd., Neipu, Pingtung 91201, Taiwan; p9831002@mail.npust.edu.tw

[2] Department of Physics, Indian Institute of Technology (Indian School of Mines) Dhanbad, Jharkhand 826004, India; via.kazi786@gmail.com

[3] Department of Environmental Engineering and Science, Feng Chia University, 100Wenhwa Rd., Seatwen, Taichung 40724, Taiwan; ccwu@fcu.edu.tw

[4] Emerging Compounds Research Center (ECOREC), National Pingtung University of Science and Technology, 1 Shuehfu Rd., Neipu, Pingtung 91201, Taiwan

* Correspondence: kcchen@mail.npust.edu.tw; Tel.: +886-8-7703202 (ext. 7039)

Received: 26 April 2020; Accepted: 25 May 2020; Published: 27 May 2020

Abstract: Titanium dioxide (TiO$_2$) thin films are used for a broad range of applications such as wastewater treatment, photocatalytic degradation activity, water splitting, antibacterial and also in biomedical applications. There is a wide range of synthesis techniques for the deposition of TiO$_2$ thin films, such as chemical vapor deposition (CVD) and physical vapor deposition (PVD), both of which are well known deposition methods. Layer by layer deposition with good homogeneity, even thickness and good adhesive nature is possible by using the PVD technique, with the products being used for photocatalytic applications. This review studies the effects of magnetron sputtering conditions on TiO$_2$ films. This innovative technique can enhance the photocatalytic activity by increasing the thickness of the film higher than any other methods. The main purpose of this article is to review the effects of DC and RF magnetron sputtering conditions on the preparation of TiO$_2$ thin films for photocatalysis. The characteristics of TiO$_2$ films (i.e., structure, composition, and crystallinity) are affected significantly by the substrate type, the sputtering power, the distance between substrate and target, working pressure, argon/oxygen ratio, deposition time, substrate temperature, dopant types, and finally the annealing treatment. The photocatalytic activity and optical properties, including the degree of crystallinity, band gap (E_g), refractive index (n), transmittance (T), and extinction coefficient (k), of TiO$_2$ films are dependent on the above- mentioned film characteristics. Optimal TiO$_2$ films should have a small particle size, a strong degree of crystallinity, a low band gap, a low contact angle, a high refractive index, transmittance, and extinction coefficient. Finally, metallic and nonmetallic dopants can be added to enhance the photocatalytic activity of TiO$_2$ films by narrowing the band gap.

Keywords: magnetron sputtering; titanium dioxide (TiO$_2$) film; photocatalytic activity; metal and non-metal doping; optical properties

---

1. Introduction

Titanium dioxide (TiO$_2$) is a low-cost non-toxic oxide semiconductor that is extensively employed in various industries due to its optical, electronic, and photocatalytic properties. It is mostly used in thin film form, which enhances the quantum efficiency. Because TiO$_2$ has high transparency and a high refractive index, it can be employed for optical coatings such as those for dielectric interference filters [1], multilayer mirrors, and anti-reflection coatings [2]. Its high dielectric constant and lower resistivity ($\sim 10^{-7} \Omega \cdot cm^{-1}$) give TiO_2 the potential for use in the fabrication of capacitors for microelectronic devices [3,4]. In addition, the high chemical stability and modest band gap of TiO_2 make it suitable for fabricating dye-sensitized solar cells [5] and photocatalysts. Photocatalysis using TiO_2 films has been used to clean up environmental problems [6,7]. TiO_2 films have gradually replaced conventional TiO_2 powder, which requires stirring during the reaction process and is difficult to separate after the end of reaction [8].

According to electrochemical properties, TiO_2 is typically characterized as an n-type semiconductor [9]. TiO_2 can exist as an amorphous layer or one of three crystalline phases: anatase (tetragonal, $a = 0.3785$ nm, $c = 0.9514$ nm), rutile (tetragonal, $a = 0.4594$ nm, $c = 0.2958$ nm), and brookite (orthorhombic, $a = 0.9184$ nm, $b = 0.5447$ nm, $c = 0.5145$ nm) (Figure 1). In all crystalline forms of TiO_2, titanium atoms surrounded by six oxygen atoms form TiO_6 [10]. In the anatase phase, corner (vertice)-sharing octahedra form (001) planes (Figure 1a) and result in a tetragonal structure. In the rutile phase, the octahedral share edges at (001) planes, therefore, forming a tetragonal structure (Figure 1b). In the brookite phase, both edges and corners are shared, which create an orthorhombic structure (Figure 1c) [11].

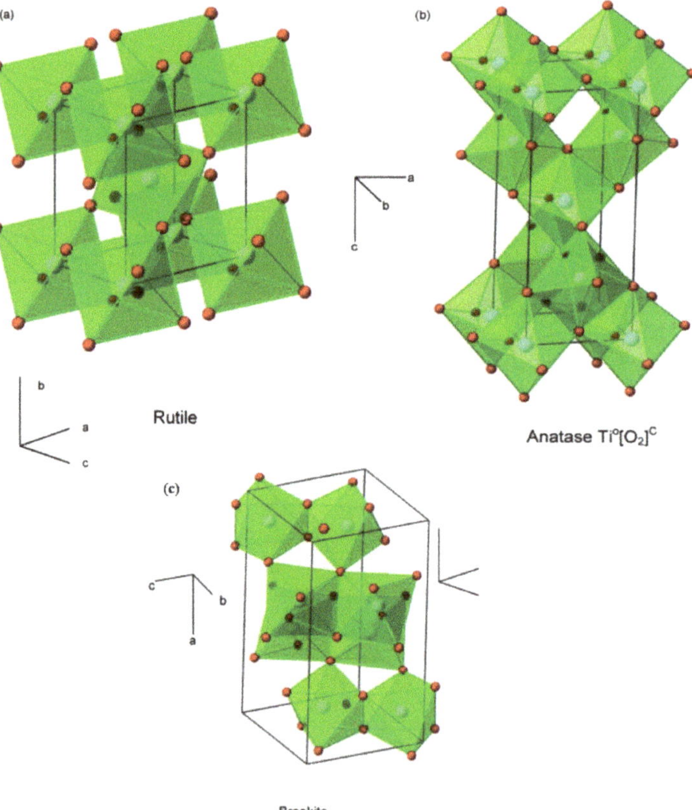

Figure 1. Building-block representation of TiO_2 (**a**) rutile phase, (**b**) anatase phase, and (**c**) brookite phase. Reprinted with permission from [11], Copyright 2009, Elsevier.

The thermal stability of crystallineTiO$_2$ is affected by the film structure and particle size [10]. Rutile is a thermodynamically stable phase, whereas anatase and brookite are metastable. It has been reported that for pure TiO$_2$ films, the metastable anatase phase can easily transform into the more stable rutile phase at high temperature [12,13]. Each crystalline form has practical applications. Rutile is desirable for optical applications, whereas anatase has suitable photocatalytic properties [14,15]. The bandgaps of anatase, rutile, and brookite are 3.2, 3.0, and ~3.2 eV, respectively [16]. In photocatalysis, TiO$_2$ is excited by photons with energy equal to or higher than their band gap energy level and electrons on the TiO$_2$ surface are excited to the conduction band (e_{CB}^-, CB), generating positive holes in the valence band (h_{VB}^+, VB) (Equation (1)). Moreover, the CB and VB can further react with water and molecular oxygen, respectively, resulting in the formation of the hydroxyl radical ($\cdot$OH) (Equation (2)) and the superoxide radical anion ($O_2^{\cdot-}$) (Equation (4)). $O_2^{\cdot-}$ subsequently reacts with H$^+$ to generate the hydroperoxyl radical ($\cdot$OOH) (Equation (6)), and then it may further react with itself to produce H$_2$O$_2$ (Equation (8)). The radicals produced by photocatalysis are powerful oxidants that can efficiently oxidize organic species, with mineralization producing mineral salts, CO$_2$, and H$_2$O (Equations (3), (5) and (7)). For TiO$_2$, the photocatalysis reactions (1) to (8) are listed below [16]:

$$\text{TiO}_2 \xrightarrow{h\upsilon} e_{CB}^- + h_{VB}^+ \tag{1}$$

$$\text{H}_2\text{O} + h_{VB}^+ \rightarrow \cdot\text{OH} + \text{H}^+ \tag{2}$$

$$\cdot\text{OH} + \text{pollutant} \rightarrow \text{H}_2\text{O} + \text{CO}_2 \tag{3}$$

$$\text{O}_2 + e_{CB}^- \rightarrow \text{O}_2^{\cdot-} \tag{4}$$

$$\text{O}_2^{\cdot-} + \text{pollutant} \rightarrow \text{H}_2\text{O} + \text{CO}_2 \tag{5}$$

$$\text{O}_2^{\cdot-} + \text{H}^+ \rightarrow \cdot\text{OOH} \tag{6}$$

$$\cdot\text{OOH} + \text{pollutant} \rightarrow \text{H}_2\text{O}_2 + \text{CO}_2 \tag{7}$$

$$\cdot\text{OOH} + \cdot\text{OOH} \rightarrow \text{H}_2\text{O}_2 + \text{O}_2 \tag{8}$$

TiO$_2$ films have been modified to enhance their photoelectric properties and make them excitable by visible light (narrowing the band gap) [17]. Many studies have demonstrated that the addition of dopants (metal and non-metal) onto TiO$_2$ films can enhance the effect of photocatalytic activity [16].

Metal dopants can act as electron traps at the semiconductor interface that prevent recombination (see Figure 2) [18,19], and non-metal dopants substitute the lattice O atom in the structure of TiO$_2$ to increase the absorption of light in the visible spectrum (see Figure 3) [20,21].

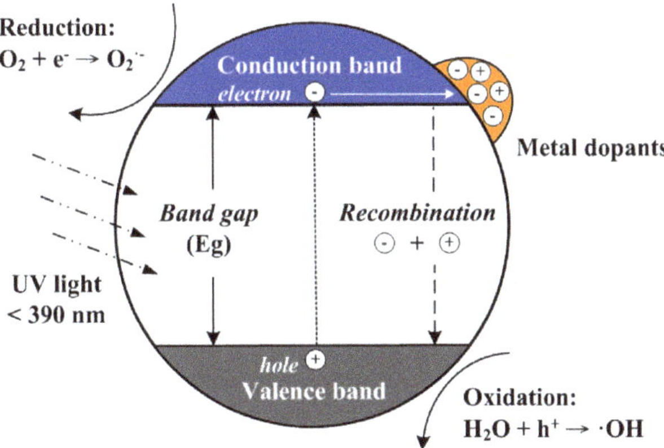

Figure 2. Schematic of metal-TiO$_2$ photocatalytic mechanism.

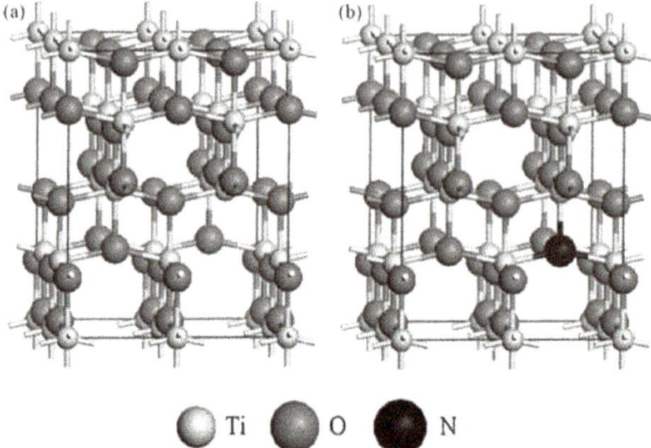

Figure 3. Schematic models in a 2 × 2 × 1 supercell for (**a**) oxygen vacancy and (**b**) substitutional N-doping. Reprinted with permission from [21], copyright 2009, Elsevier.

Moreover, non-metal dopants can also increase the thermal stability of anatase [22]. There are many important parameters which control and enhance the photocatalytic performance of the TiO_2 thin films. The parameters are listed as follows: different pathways during reaction process, i.e., nucleation growth, formation of different phases like anatase or rutile, surface area, at the interface of the reactants. TiO_2 thin films can be deposited by some advances techniques like physical vapor deposition (PVD), chemical vapor deposition (CVD), and wet chemical methods including dip coating, spin coating, spray pyrolysis, etc. Moreover, thermal evaporation, reactive sputtering, ion or electron beam evaporation methods are widely used. These mentioned applications are low cost, and easy to access which is preferable for outdoor applications. The disadvantages of these techniques are that they must be followed by secondary process which include drying and annealing of the thin films to increase the stability and crystallinity of the films, which are some of the essential criteria for the desired enhanced photocatalytic activities. Generally, vapor phase methods present various advantages such as well controlled homogeneity and thickness over a large area and good adhesive nature [23]. The CVD method requires higher temperatures (400–900 °C) compared to PVD processes [24]. This disadvantage of requirement of excessive heat can be incompatible with some substrate materials. Some titanium precursors and their byproducts are highly corrosive in nature, which leads to various material handling and storage problems [25]. However, the PVD processes have garnered great interest since they are not limited to deposition only at thermodynamically equilibrium and they run at much lower costs in comparison to CVD processes [26]. There are many PVD methods that are being applied for the deposition of thin films such as oxidation, deposition, DC magnetron sputtering, RF magnetron sputtering, thermal evaporation, pulsed laser deposition (PLD), etc.

Highly homogenous and high-density films are obtained in these processes due to bombardment of energetic particles, thus these energetic hot particles take part in the photocatalysis process followed by surface plasmon resonance mechanism [27]. This method yields highly porous and dense films at relative low temperatures. TiO_2 thin films prepared in this technique can exhibit diverse photocatalytic activities, including high surface area, morphology, defect density, and crystallization pathways. In comparison with TiO_2 bulk nanoparticles, TiO_2 thin films exhibit a limited surface area, which hinders the efficiency of the photocatalysis for the decomposition of the organic compounds [28]. Thus, various strategies are reported to enhance the surface area of thin films for higher photocatalytic dye degradation studies, where novel sputtering methods are established to produce highly porous thin films for better adsorption of the dye molecules on the surface of the photocatalyst TiO_2 thin films. Suzuki et al. [28] have reported a sculptured TiO_2 thin film providing a higher photocatalytic efficiency.

To achieve a high surface area, nowadays another effective process is established, namely to control nano-crack network formation within sputter deposited TiO_2 thin films [29,30].

This review is focused on the effect of magnetron sputtering conditions on TiO_2 films. Magnetron sputtering allows the structure and composition of TiO_2 films to be easily controlled, yields uniform films, and is suitable for large-scale industrial production. Thus, the optimal operating conditions for magnetron sputtering are of interest. The considered operating conditions, which influence the quality of photocatalytic thin films (PTFs) during magnetron sputtering, are the substrate type, sputtering power, distance between substrate and target, working pressure, argon/oxygen ratio, deposition time, substrate temperature, dopant type, and annealing treatment.

2. Working Principle of Magnetron Sputtering Technique

DC magnetron sputtering techniques are used for depositing thin films over large areas. The continuous current glow discharge that originates the energetic particles that support sputtering process can be obtained by applying a potential difference between two electrodes in the presence of a gas (usually argon) kept at low pressure inside a vacuum chamber. The potential difference can be delivered by a continuous current power supply that can provide tens of thousands of volts, depending on the equipment configuration. An electric field is formed between the electrodes separated by a distance d and having a potential difference V due to the presence of electrons formed by processes such as ionization, thermionic emission and collisions between particles. Here ionization is caused by cosmic rays, but it cannot be maintained. A schematic diagram is shown in Figure 4a.

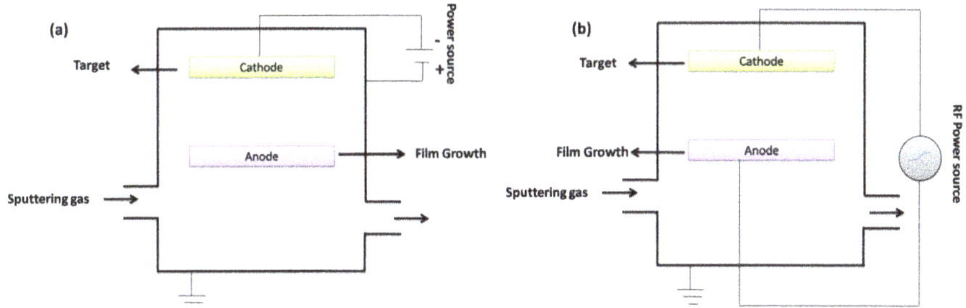

Figure 4. Schematic diagram of the working principle of (**a**) DC and, (**b**) RF magnetron.

The gas pressure inside the chamber should not be very low in order to allow a collision between electrons and some atoms before the electrons hit the anode. If the gas pressure is very high, the electrons will neither reach enough velocity nor achieve enough energy in order to form new ions or excited species in the collisions. Under the electrical field, positive ions are accelerated towards the cathode while electrons are accelerated towards the anode. Once electrons travel a long enough distance before the collision, they will acquire enough kinetic energy to promote a new ionization. The main difference between the RF and DC magnetron sputtering is that they work in AC source and DC source mode, respectively. The main advantage of using RF sputtering techniques is that they work well with insulating targets. Moreover, the electric field inside the plasma chamber changes with RF frequency, which avoids any charge up effects. A schematic diagram of RF magnetron sputtering is shown in Figure 4b.

3. Effect of Deposition Conditions on Photocatalytic Thin Films

3.1. Substrate Type

Many materials have been used as substrates in magnetron sputtering, including organic materials (cotton fabrics [31] and polycarbonate [18,32]), inorganic materials (glass [16,33,34] and quartz [35,36]), metals (wafers of alumina (Al_2O_3) [37] zinc oxide (ZnO) [38], and stainless steel [39]), and minerals (sapphire [5]). Since the transmittance of PTFs affects the photocatalytic efficiency, materials such as glass and quartz are widely used as substrates due to their high transmittance.

Applying TiO_2 coatings on a glass surface is the most commonly used method of fabricating PTFs. Sun et al. [40] reported that TiO_2 coated on glass has good antibacterial, disinfectant, antifogging, and self-cleaning properties. PTFs have been deposited using chemical vapor deposition, spin coating [40], the impregnation method [41], electrodeposition [42], the sol-gel process [43], evaporation [44], and other sputtering methods [45,46]. Using magnetron sputtering for fabricating PTFs has several advantages, including a high deposition rate, low substrate temperature, good surface flatness, and high density of the deposited layer. However, many researchers have demonstrated that the sodium ions (Na^+) from the glass can diffuse into the TiO_2 film at a high substrate temperature [47,48] and this Na^+ ion diffusion from the glass can reduce the photocatalytic actively of TiO_2 films. In order to restrain Na^+ ion diffusion, many researchers have applied a silicon (Si) pre-coating on the glass [49]. Meng et al. [50] and Nair et al. [36] used a quartz substrate to prevent light refraction and Na^+ ion diffusion. Many researchers have found that a metal oxide coating on the surface of glass (e.g., indium tin oxide (ITO) [51–53] and fluorine-doped tin oxide (FTO) [32,54]) can increase the electrical conductivity, however, the photocatalytic activity of TiO_2 films mainly depends on the crystalline structure of the surface, not the substrate type [32].

3.2. Sputtering Power

Sputtering power is an important parameter for fabricating TiO_2 films via RF [33,55–57] and DC [58] sputtering. The crystal phase of TiO_2, which affects photocatalytic activity, strongly depends on the sputtering power used [34,58–60].

Chen et al. [32] investigated the effect of RF power on the photodecomposition of the dye methylene blue (MB). The results showed that an increase in RF power during TiO_2 film fabrication decreases subsequent MB photodecomposition. This result is consistent with that reported by Huang et al. [59] that RF power has a significant influence on MB photodecomposition. This can be attributed to changes in the crystalline phase, surface morphology, and optical properties caused by RF power.

TiO_2 exists in an amorphous phase or one of three crystal polymorphs, namely anatase (tetragonal), rutile (tetragonal), and brookite (orthorhombic). These crystalline phases can be prepared using magnetron sputtering [12]. Huang et al. [59] fabricated TiO_2 films at increasing RF powers (50–250 W) and observed the conversion of the crystalline phase using X-ray diffraction (XRD). Their results showed that the rutile phase is the favored structure during deposition. With an increase in RF power during deposition, the amorphous, rutile, and a combination of rutile and anatase phases are obtained in sequence. Nevertheless, the anatase phase content levels are similar (in the range of 34% to 37%) at all RF power levels. Sputtering power also affects the elemental composition of the deposited films. The bonding condition of titanium on the surface of TiO_2 films can be investigated by X-ray photoelectron spectroscopy (XPS). Lin [58] and Lin and Wu [61] found that Ti^{4+} forms compared favorably to other types of titanium ion and that the deposited films become non-stoichiometric with increasing sputtering power density, as determined using XPS. The components with the highest and lowest energies on TiO_2 films were Ti^{4+} and Ti^0, respectively. The intermediate components (Ti^{3+}, Ti^{2+}, and Ti^+) have an energy shift, with resulting energies that fall between those of Ti^{4+} and Ti^0. Substantial contributions from lower oxidation states (Ti^{3+} or Ti^{2+} and Ti^+) are their large oxygen deficiency. The photocatalytic activity of TiO_2 mainly depends on the crystallinity and density of surface defect sites [61]. Liu et al. [62] reported that the dominant defects on TiO_2 surfaces are Ti^{3+} defects and oxygen vacancies. Ti^{3+} is considered to be an important reactive agent for many adsorbates. Hence, some studies have suggested that Ti^{3+} sites might play an essential role in the photocatalytic process over a TiO_2 photocatalyst [61,63]. Su et al. [60] studied the orientation transformation of crystalline TiO_2 film for various RF powers. The sputtering power was set in the range of 100 to 300 W. The results showed that the schematic phase diagram of the crystalline phases can be divided into four regions, as shown in Figure 5.

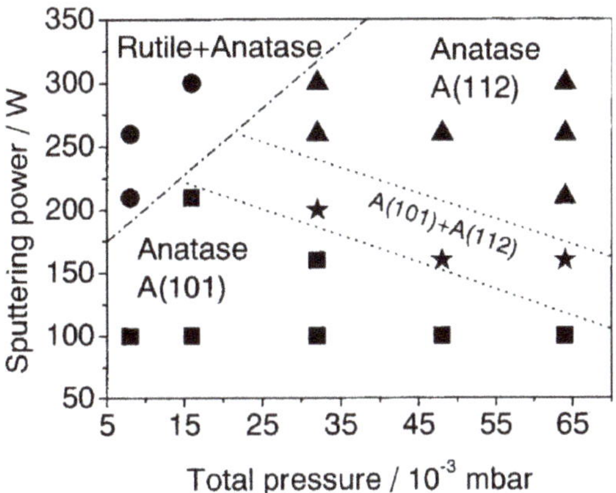

Figure 5. Schematic phase diagram of TiO$_2$ films defined by sputtering power. The diagram is compiled from experimental data obtained from XRD: (■) (101)-preferred orientation anatase, (▲) (112)-preferred orientation anantase, (★) anatase with almost equivalent (101) and (112) peaks, and (●) rutile and anatase mixture.Reprited with permission from [60], Copyright 2004, IOP Publishing.

An anatase-preferred orientation transformation between (101) and (112) is observed. At low total pressure conditions (<11.5 mTorr), anatase (101) exists at low sputtering power (200 W), whereas mixed anatase and rutile phases formed at high RF power (200 to 300 W). Anatase (112) is obtained only at high levels of RF power and high total pressure. These findings are confirmed by using a high RF power, resulting in a mixture of highly crystalline anatase and rutile phases [55]. Rutile is a thermodynamically stable phase, whereas anatase and brookite are metastable. Therefore, the rutile phases of TiO$_2$ films are formed with high sputtering power. In addition, these results imply that the crystalline phase of TiO$_2$ in turn becomes amorphous, anatase, and rutile as the sputtering power is increased. The efficiency of photodecomposition of anatase phase in TiO$_2$ films is found to be better than those of amorphous, rutile, brookite, and anatase/rutile phases [64]. Bombardment on the substrate at higher power may cause film damage and decrease compactness, which can be an obstacle to rutile growth [55]. This trend has also been reported by Chow et al. [65], who found a similar effect of negative-bias-induced bombardment.

Film density is also directly affected by sputtering power. It has been well established that the refractive index is closely correlated with film density [66]. Many researchers found that the refractive index increases with increasing sputtering power [34,35,67]. Theoretically, for a well-balanced magnetron, the power per sputtered adatom increases with increasing total sputtering power, delivering more energy to the growing film and thus improving densification. Therefore, a higher refractive index (n) means a denser film [34] and a finer film nano/micro-structure [37,68,69]. In addition, the refractive index of a transparent thin film is directly proportional to its electronic polarization, which is in turn inversely proportional to the inter-atomic separation [1]. The refractive index of TiO$_2$ films also depends on porosity (volume of pores per volume of film) of the film layer and can be calculated using the Swanepoel method [1,35,68,70] (Equations (9) and (10)). The porosity of the deposited film was calculated from the refractive index obtained from previous studies [70–72] (Equations (10) and (11)). From the refractive index, the porosity of each sputtered film can be derived using Equation (12) [72].

$$N' = \frac{1}{2}\left(1 + n_s^2\right) + \frac{2n_s\left(T_{max} - T_{min}\right)}{T_{max} \cdot T_{min}} \tag{9}$$

$$n = \sqrt{N\prime + \sqrt{N\prime^2 - n_s^2}} \tag{10}$$

$$\frac{\rho_{film}}{\rho} = \frac{n^2 + 1}{n^2 - 1}\frac{n_{film}^2 - 1}{n_{film}^2 + 2} \tag{11}$$

$$Porosity(\%) = \left(1 - \frac{\left(n_p^2 - 1\right)}{\left(n^2 - 1\right)}\right) \times 100 \tag{12}$$

where T_{max} and T_{min} are the corresponding transmittance maximum and minimum, respectively, at a certain wavelength λ in the optical transmittance spectrum (for determination of T_{max} and T_{min}, refer to [69]). n_s is the refractive index of fused quartz. ρ_{film} is the film density, ρ is the bulk anatase TiO_2 density (3.9 g/cm^3) or rutile TiO_2 density (4.23 g/cm^3), n_P is the refractivity of porous thin films, and n is the refractivity of bulk TiO_2. As is well known, the refractivity of bulk anatase TiO_2 is $n_a = 2.52$ and that of rutile TiO_2 is $n_r = 2.92$ at 500 nm [72].

Nair et al. [36] attributed the increase in the refractive index to the densification of layers treated at a high RF power. Since the densification of layers is correlated with the thickness of the films, an increase in the sputter power increases the deposition rate at the target [57,73]. Yang et al. [35] found that a high sputtering power can increase the surface roughness of the film, which results in decreased transmittance [57,74]. The surface roughness affects the contact angle of films; the contact angle increases with roughness [75]. Many researchers have demonstrated that the rutile phase is more hydrophilic than the anatase phase [74,76,77]. Therefore, the sputtering power is an importer factor when the magnetron sputtering process is used to fabricate TiO_2 films. It also directly influences the crystalline phase (including elemental composition), optical properties, and physical characteristics of deposited films.

3.3. Distance between Substrate and Target

Relatively little research has been conducted on the effect of the distance between substrate and target (S-T) on the crystalline phase of TiO_2 films during the magnetron sputtering. Ogawa et al. [15] have investigated the crystalline phase of TiO_2 films obtained with various S-T distances during RF magnetron sputtering with an MgO substrate. According to XRD results, the crystalline structure of TiO_2 changes with S-T distance. When the S-T distance is under 90 mm, the rutile phase of TiO_2 film is dominant. At an S-T distance of 100 nm, mixed rutile and anatase phases are observed. At S-T distances of 110 and 120 mm, the anatase phase is dominant, with epitaxial growth with its α-axis perpendicular to the MgO substrate. Shibata et al. [78] reported that the anatase phase grows from the reaction between neutral Ti and neutral O_2 or O_2^-, and that the rutile phase results from the reaction between decelerated Ti^+ or activated Ti and O_2^-. Hence, more anatase TiO_2 film is created at low kinetic energy of sputtering compared to rutile TiO_2 film [59]. This result implies that the crystal structure shift is due to the reaction of sputtering particles between the substrate and the Ti target. When the substrate bias voltage is increased, the electrons in the plasma bombarded Ti atoms, creating Ti^+ or activated Ti, resulting in increased formation of rutile TiO_2 [79]. In the magnetron sputtering process, a magnetic field configured parallel to the target surface can constrain secondary electron motion to the vicinity of the target [80]. Therefore, a decrease of S-T distance (increase in magnetic field range) increases the kinetic energy of sputtering, and thus rutile TiO_2 film is formed under this sputtering condition. While the S-T distance increased resulting in the decrease of magnetic field strength, secondary electrons can easily recombine in the plasma, resulting in a decrease in the kinetic energy of sputtering and the formation of anatase TiO_2 film.

The S-T distance also affects the deposition rate and particle size. Sundaram and Khan [81] found that a 2–7 cm distance between S-T during the magnetron sputtering process decreases deposition rate due to a large number of negative and neutral ions in a small space, which led to the occurrence of resputtering on the surface of the growing film. Regarding particle size, Barnes et al. [82] indicated that the clusters nucleate close to the target and migrate to the substrate via convection. A small S-T distance limits convection, leading to an almost uniform cluster size and small grain size. An increase in S-T distance increases convection space, and thus leads to an increased size distribution of clusters and grain size on TiO$_2$ films. This phenomenon has been observed in metal evaporation systems [83]. The grain size influences the photocatalytic activity of TiO$_2$ films [18,84]. Using the Scherrer Equation, the crystalline grain size can be calculated from an XRD pattern as:

$$D = \frac{0.9 \times \lambda}{B \cos \theta} \tag{13}$$

where D is the mean grain size for crystalline planes, B is the full width at half maximum (FWHM) intensity in radians, and λ is the wavelength of the Cu K$_\alpha$ radiation source. Many researchers have reported that smaller particles enhance photocatalytic activity [18,85,86] due to the decrease of distance between the crystal interface and the surface of the substrate [87]. Smaller particles induce the quantum size effect, which influences the optical absorption edge of the films, resulting in an increase of the band gap and the absorption onset blue-shift. Therefore, photocatalytic activity increases with increasing oxidation rate. For large particles, the e^-/h^+ volume recombination is the dominant process. An appropriate S-T distance is thus important when fabricating TiO$_2$ films using magnetron sputtering.

3.4. Working Pressure

The deposition rate, surface morphology, and crystalline phase of TiO$_2$ films are affected by the working pressure during the sputtering process [14,87,88]. The working pressure directly affects the photocatalytic activity of TiO$_2$ films since it affects the gas collisions, which affect the characteristics of deposited films. Many researchers have also reported that the photocatalytic activity of TiO$_2$ films is significantly correlated with the working pressure [8,89]. For example, TiO$_2$ films with high photocatalytic activity can be prepared using a relatively low-working pressure [90]. Thus, specific parameters of deposited films are influenced by the working pressure, namely the deposition rate (thickness), surface morphology, and crystal phase of TiO$_2$ films.

The effect of working pressure in the sputtering process on the deposition rate has been investigated by many researchers [14,87,91]. Zeman and Takabayashi [14] found that the increasing working pressure (0.18—1.50 Pa) significantly decreases the deposition rate during magnetron sputtering (Figure 5). Zhu et al. [91] obtained different experimental results when using ZnO and Al to fabricate films. They measured the deposition rate for fabricating ZnO-Al films at various working pressures. The experimental results showed that the deposition rate is low at low working pressure (0.5 Pa) and then it increases with the working pressure up to 1.0 Pa. Above 1.0 Pa, the deposition rate is decreased with increasing working pressure. They also found that the decrease in deposition rate with higher working pressure (above 1.0 Pa) can be fitted using the Keller-Simmons model [92], which suggested a decay of the deposition rate at high pressures is caused by shielding effects of gas particles between the target and substrate. Nevertheless, the low deposition rate at low working pressure can be explained by the resputtering of ZnO by highly energetic particles, which are more prevalent at lower pressure. Chaoumead et al. [87] used RF magnetron sputtering to deposit ITO thin film on glass substrates. The trend of the deposition rate of the ITO film was similar to that reported by Zhu et al. [91]. Decreases in deposition rate of ITO film is observed when the working pressure is increased from 5 to 15 mTorr (low working pressure). When the working pressure is further increased to 20 mTorr yielding decreases in the deposition rate. Finally, it is clear that the surface morphology is affected by the working pressure.

Scanning electron microscopy (SEM) [14,92] and atomic force microscopy (AFM) [89,93] can be applied to observe the surface morphology of PTFs. In PTFs, the particle size influences surface morphology. Many researchers have demonstrated that particle size depends on sputtering pressure [88,89,94]. It has been suggested that the pressure influences the energy transfer from the applied electric field to the species present in the discharge, and thus increasing the sputtering pressure results in an increase in the number of gas-phase collisions and a reduced fraction of particles with sufficient energy reaching the TiO$_2$ film [95]. In other words, higher sputtering pressure leads to lower ion kinetic energy, which decreases the particle size [96]. Jin et al. [88] indicated that the surface morphology is strongly influenced by working pressure. Spherical TiO_2 particles are observed on the surface of TiO_2 films via field-emission SEM. When the working pressure is increased from 5 to 10 mTorr, the grain size of TiO_2 particles decreased from 35 to 25 nm. The working pressure thus affects the grain size of TiO_2 particles; hence, increasing working pressure increases surface roughness [97] during the sputtering process. Zeman and Takabayashi [14] also found that working pressure affects surface morphology. A significant change in film density is observed when the working pressure is increased from 1.35 to 15.3 mTorr. Three types of surface morphology are observed. (i) A dense surface with only a few distinct symmetrical particles corresponds to rutile film (working pressure: 1.35 mTorr) [14]. (ii) The anatase surface (working pressure: 4.8 to 9.75 mTorr) is characterized by relatively densely packed surface particles resembling an array of different polygons. These particles have a symmetrical or elongated shape with a size of 10 to 50 nm. (iii) When the working pressure exceeded approximately 1.30 Pa, the surface morphology changed again [14]. The surface obtained at a higher working pressure is less dense, which implies that voids occupy more space and the surface roughness increases. Zhang et al. [93] have confirmed these findings by using AFM (Figure 6)

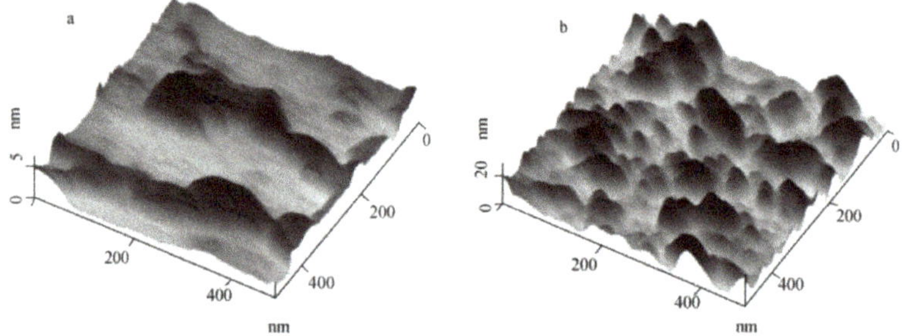

Figure 6. AFM morphological images of TiO_2 films deposited at (**a**) 0.4 Pa and (**b**) 1.4 Pa. Reprinted with permission from [93], copyrught 2009, Elsevier.

The surface of TiO_2 film deposited at 0.4 Pa is much smoother than that of film deposited at 1.4 Pa. El Akkad et al. [98] found similar results, where rough surfaces are obtained at high working pressures. The change in surface morphology is due to the transition of the crystal phase at high working pressure [14].

Many researchers have found that the anatase phase of TiO_2 films is dominant at high working pressure, whereas the rutile phase is dominant at low working pressure [14,87]. Wang et al. [89] showed that the band gap increases with increasing working pressure. This is due to the different content levels of anatase and rutile phases. For wavelengths below 400 nm, the transparency of films sharply decreases in the ultraviolet (UV) region; it is almost constant for wavelengths above 400 nm. This decrease is caused by the fundamental absorption of light [99]. This phenomenon is due to the quantum size effect when the particle size is very small [100]. Nair et al. [36] found that the increase in porosity with increasing sputtering pressure results in a decrease of the refractive index. This suggests that sputtered particles undergo more scattering and that fewer atoms reach the substrate, resulting in lower crystallinity at high working pressure.

The absorption coefficient (α) can be expressed as [36,90,100]:

$$\alpha = d^{-1} \ln\left(\frac{1}{T}\right) \tag{14}$$

where T is the transmittance and d is the thickness of the film.

The band gap energy (E_g) can be calculated based on the absorption spectra using [36,94]:

$$\alpha h\upsilon = A\left(h\upsilon - E_g\right)^{n/2} \tag{15}$$

where α, hv, E_g, and A are the absorption coefficient, light frequency, band gap energy, and a constant, respectively. The value of n depends on the type of optical transition of the semiconductor ($n = 1$ for direct transition and $n = 4$ indirect transition). TiO_2 has been demonstrated to have both directly forbidden and indirectly allowed transitions. The indirectly allowed transition dominates the optical absorption because the directly forbidden transition is weak. The value of the direct optical band gap is obtained by plotting (αhv^2) versus hv (Taucplot) in the high-absorbance region. Extrapolating the linear portion of the graph to the X axis yields E_g, which is the sum of terms correlated to carrier concentration [17], structural disorder [1], and thermal contribution [101].

An increase in the working pressure can effectively increase the density of gas particles in the chamber and decrease the cathode potential, which may further influence the probability of collisions and the acceleration of particles, resulting in changes in the crystalline structures formed. Šícha et al. [97] obtained the following results: (i) For the formation of amorphous TiO_2 film, the energy E delivered to the deposited film decreases when the working pressure is below 2 Pa, and is thus insufficient to stimulate film crystallization. (ii) For the formation of rutile TiO_2 film, crystalline phases are formed at low values of P_T (total pressure) and P_{O2} (oxygen pressure) when the energy E (incident and condensing atoms) is sufficient for the crystallization and is delivered to the deposited film. The energy E increases with decreasing P_T in consequence of a decrease of collisions between particles. In addition, the condition of low P_{O2} can change the chemical process, contributing to the formation of the rutile phase. (iii) For the formation of anatase TiO_2 film, crystalline phases are formed when the working pressure is higher than 2 Pa. The increase of sputtering time also ensures that a sufficient amount of energy is delivered to the deposited films and stimulates the crystallization of the anatase phase. In addition, Zeman and Takabayashi [14] found that the boundary between the metallic and reactive modes shifts when the working pressure and the ratio of P_{O2}/P_T are increased.

The change in crystalline phase of TiO_2 films is not only affected by the working pressure but also the argon/oxygen ratio in magnetron sputtering.

3.5. Argon/Oxygen Ratio

During magnetron sputtering, argon and oxygen serve as the plasma gas and the reactive gas, respectively. Thus, the oxygen partial pressure may influence the discharge parameters, such as plasma potential, discharge voltage, deposition rate, and ion composition of the discharge, and thus the characteristics of TiO_2 films in the magnetron sputtering process [14]. Although numerous attempts have been made to study the relationship between the argon/oxygen ratio and photocatalytic activity, no clear consensus has been reached. Zhang et al. [101] found that TiO_2 film deposited at a high argon flow rate absorbs lighter irradiation, which results in more electron-hole pairs generated in the TiO_2 film and thus enhances photocatalytic activity. Huang et al. [66] reported that the photocatalytic activity of TiO_2 films increases with increasing oxygen flow rate. These results differ from those reported by Chiou et al. [33], who found that the argon/oxygen ratio and the photocatalytic activity of TiO_2 films are not significantly related. Since the photocatalytic activity of TiO_2 depends on the strength of the crystalline phase and the characteristics of the deposited films, the effect of the argon/oxygen ratio on the characteristics of TiO_2 films needs further study.

The effects of the argon/oxygen ratio on the photocatalytic activity of TiO_2 films have been extensively investigated. Many researchers have found that a lot of properties of TiO_2 film are affected by the argon/oxygen ratio in magnetron sputtering, including the deposition rate [62,74,82,102,103], grain size and surface roughness [37,76,87,104], crystallinity and surface chemical composition [33,86], and optical properties [1,37,62,105]. The dissociation energy of argon is about 15.76 eV and that of oxygen is about 48.77 eV. Thus, argon gas is more easily dissociated [33].

The deposition rate decreases slowly with increasing proportion of oxygen (Figure 7) [62,74,102,103]. Pradhan et al. [105] and Liu et al. [62] suggested that this decrease is due to the transition from metallic to reactive sputtering modes. Tomaszewski et al. [102] investigated the transition mechanism on the surface during magnetron sputtering. They found that increasing the oxygen flow rate led to a significant decrease in the sputtering voltage. In the low-energy state, Ti and O-related species sputtered from a TiO_{2-x} target easily reach the substrate, forming stoichiometric TiO_2 films. However, the oxygen of the sputter gas can be decomposed and changed to O^- ions in the plasma, which are accelerated by the electric field around the target, increasing the opportunity for ion-induced secondary electron emission and thus decreasing the deposition rate. This explanation is consistent with those results reported by Zhang et al. [104].

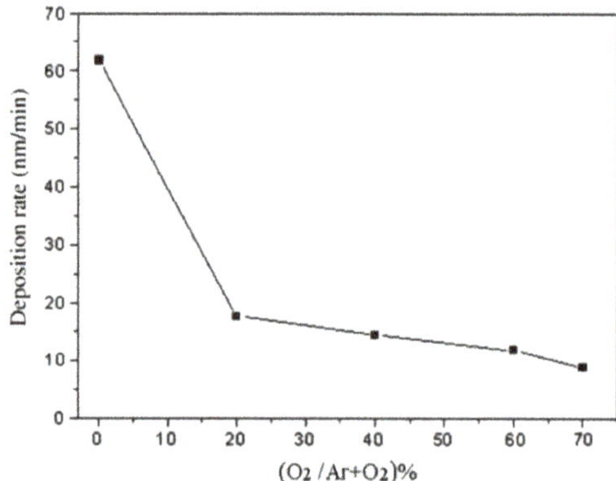

Figure 7. Deposition rate of TiO_2 films as a function of oxygen partial pressure at a constant sputtering pressure (10 mTorr). Reprinted with permission from [74], copyright 2010, Elsevier.

The grain size and root-mean-square (RMS) surface roughness of TiO_2 films are affected by the argon/oxygen ratio in magnetron sputtering [37,76,87,104,106]. Argon is the main sputtering gas that imparts kinetic energy to the sputtered atoms. The partial pressure of oxygen can affect particle collisions and the mobility and diffusion of the species approaching the substrate. Studies have found three different phenomena: (i) The grain size increases with increasing oxygen partial pressure. Laha et al. [37] suggested that the increase in the oxygen flow rate decreases the mobility of the sputtered ions, resulting in increased particle size. (ii) The grain size decreases with increasing oxygen partial pressure. Lin et al. [74] and Huang and Hsu [103] reported that the grain size decreases as the percentage of O_2 increases. Huang and Hsu [103] found that high oxygen gas content in the sputtering process increases the resputtering effect, which results in insufficient energy for grain growth. (iii) The grain size initially decreases and then increases with further increases in the oxygen partial pressure [86,87]. Regarding the relation of surface roughness and the argon/oxygen ratio, many researchers have demonstrated that the surface roughness decreases when oxygen pressure increases [2,37,76,87,106]. Jeong et al. [2] found that Ti and Si atoms decreased with increasing proportion of oxygen. Laha et al. [37] found that the surface of the films become smooth and flat as a result of increased grain size. However, Xu and Shen [86] found that the RMS surface roughness of TiO_2 films decreased with increasing oxygen gas pressure, and that the RMS surface roughness is not related to the grain size. Toku et al. [107] gave a different interpretation of the influence of oxygen pressure on the surface roughness of deposited films. They indicated that increasing the percentage of O_2 makes the potential drop across plasma sheath decreases due to the abundant negative oxygen ions formed in the plasma; therefore, the impact of energetic ions on the surface is reduced. This impact may be the reason for the surface is becoming uniform in the absence of impurities. However, Huang and Hsu [103] reported an increase in RMS surface roughness (from 12.3 to 18.1 nm) when the percentage of oxygen is increased from 0% to 20% (argon% + oxygen% = 100%). Water and Chu [106] found a simialr trend, with an increase in the oxygen content there is increasing in the RMS surface roughness of ZnO films. Many researchers agree that both the grain size and surface roughness of TiO_2 films are important factors of photocatalytic performance. A small grain size can decrease the recombination of electron-hole pairs [18,86] and a low surface roughness can prevent light scattering [108,109] and thus increase the hydrophilicity of deposited films [76]. However, studies on the effects of the argon/oxygen ratio on the grain size and roughness of TiO$_2$ films have reported mixed results. The optimal argon/oxygen ratio for photocatalytic activity is thus difficult to determine.

The effects of the argon/oxygen ratio on the crystallinity and surface chemical composition of TiO$_2$ films have been investigated in many studies [79,83,105,106,110–112]. Oxygen plays an important role in the preparation of TiO$_2$ films via magnetron sputtering. Many researchers believe that the oxygen concentration during magnetron sputtering affects oxygen vacancies, which cause structural defects in the deposited films [37]. Zhang et al. [101] found that the discharge voltage increases with increasing oxygen flow rate up to a certain threshold. With increasing oxygen flow rate, the secondary electron emission in the reaction chamber is also increased. A mixture of metallic Ti and titanium oxides is deposited on the substrate before the threshold is reached. Only after the oxygen flow rate reaches the threshold can a uniform transparent TiO$_2$ thin film form on the substrate

A large secondary electron emission may thus help convert the target into titanium oxides. In addition, the oxygen in the chamber can promote the substitution of Ti in the TiO$_2$ crystal lattice, change the crystallinity of TiO$_2$ films [113], and implant anion species on TiO$_2$ films, leading to changes in the surface chemical composition [79]. Toku et al. [107] deposited TiO$_2$ films at various oxygen concentrations and found that the degree of crystallinity (anatase 101) of TiO$_2$ films is greatest at an oxygen content of 30%, and then decreases with increasing oxygen content (see Figure 8).

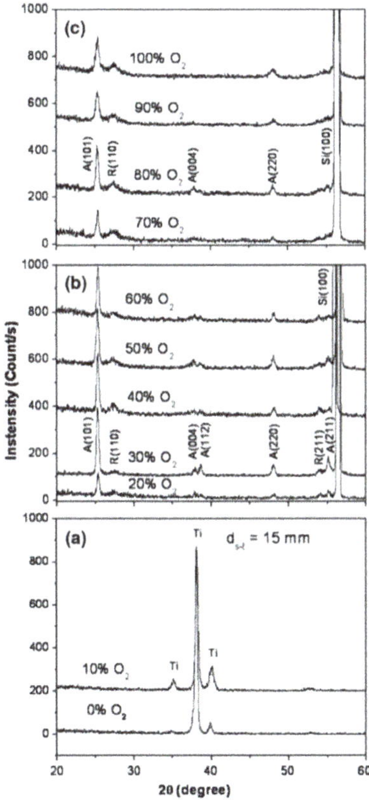

Figure 8. XRD patterns of TiO$_2$ films deposited on silicon (100) at various O$_2$ concentrations in Ar/O$_2$ gas mixture (**a–c**). Reprinted with permission from [107], copyright 2008, Elsevier.

The crystal phases of TiO_2 films are also affected by the oxygen concentration in magnetron sputtering. When the oxygen concentration is increased, the crystallinity of TiO_2 films slightly changes from the anatase phase to a mixture of anatase and rutile phases [14]. However, the authors concluded that working pressure has the most influence on the crystalline phase of deposited films. Tavares et al. [109] also found that the oxygen flow rate influences the crystalline phase of TiO_2 films. At flow rates below the 2.4 sccm threshold, the texture is very weak and only a slightly pronounced anatase (101) peak is observed. With increasing oxygen flow rate, other anatase peaks, namely (004), (200), (105), (211), (204), and (106), are enhanced at threshold conditions. When the oxygen flow rate is above the threshold, the crystalline phase of TiO_2 films gradually changes from anatase to rutile because the latter is more stable. Above 3.6 sccm, the crystal phase becomes amorphous due to the high pressure and inherent reduced adatom mobility. Xiong et al. [110] reported similar results, with TiO_2 films showing a phase transition from anatase to rutile when the oxygen flow rate is increased. Kinetic energy (sputtering power or working pressure) is thus not the only factor in determining the crystal phase. Consequently, they inferred that the species sputtered from the target and the surface reactions on the substrates have a large impact on crystal nucleation and growth.

Several researchers have found that the argon/oxygen ratio affects the surface chemical composition. Ti^{4+} and Ti^{3+} are dominant species of the elemental compositions. Peaks at $2p_{3/2}$ (458.5 eV, TiO_2) and $2p_{1/2}$ (464.1 eV, TiO_2) are observed, representing the existence of $Ti_4{}^+$, when XPS is used to examine TiO_2 films fabricated at high oxygen pressure in the magnetron sputtering process. If a high argon pressure is used to fabricate TiO_2 films, Ti^{3+} will be the dominant species and peaks at $2p_{3/2}$ (457.5 eV, Ti_2O_3) and $2p_{1/2}$ (463.3 eV, TiO_2) will appear in the XPS spectrum [2,62,106,114]. However, the $Ti_3{}^+$ state of TiO_2 films may decrease the probability of electron-hole recombination, leading to enhanced photocatalytic activity. Vancoppenolle et al. [115] investigated the reactive species (Ti^+, TiO^+, $TiO_2{}^+$) are generated on TiO_2 films at different argon/oxygen ratios. Ti^+ decreases and TiO^+ and $TiO_2{}^+$ increases when the oxygen content was increased from 0% to 20%. In addition, O_2 is incorporated practically and completely into the growing films at a low flow rate. Barnes et al. [82] provided a schematic of the theory of charged clusters in terms of nanocrystal and thin film growth that is depicted in Figure 9.

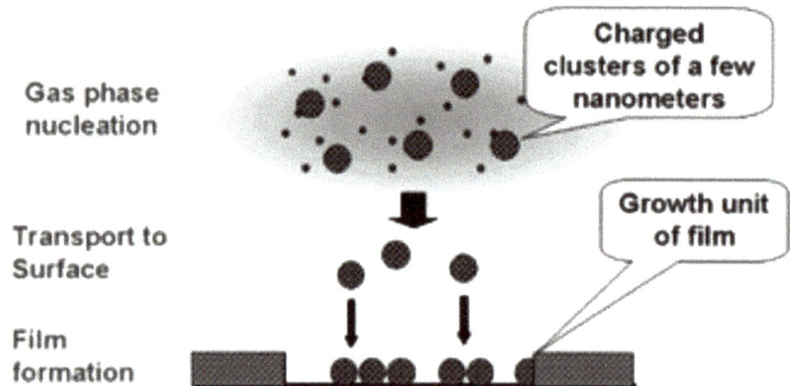

Figure 9. Schematic of the theory of charged clusters in terms of nanocrystal and thin film growth. Reprinted with permission from [82], copyright 2004, Elsevier.

This figure shows the nucleation of TiO_2 clusters, which is a simulated reaction between Ti and O atoms during the magnetron sputtering process. The clusters are the growth unit for the thin film. Shibata [78] reported that the anatase phase is the result of the reaction between metallic Ti with neutral O_2 or $O_2{}^-$ and that the rutile phase results from the reaction between ionic Ti^+ or activated Ti with $O_2{}^-$.

Toku et al. [107] used mass spectra to measure the variation of Ti and TiO$_2$ particles on TiO$_2$ films with oxygen content (Figure 10).

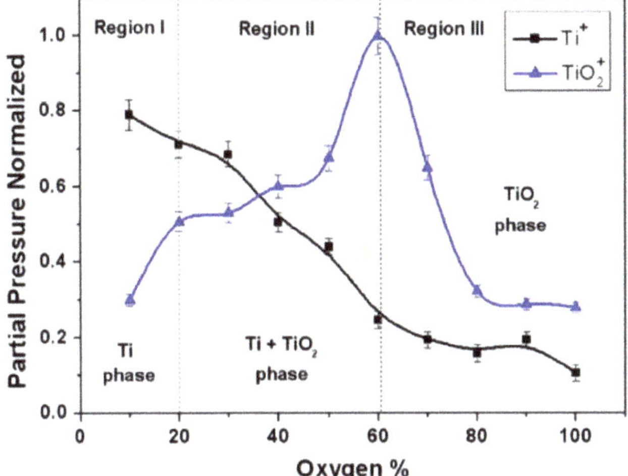

Figure 10. Variation of intensity of Ti$^+$ and TiO$_2^+$ species as a function of O$_2$%. Partial pressures are normalized to 1 at maximum value. Reprinted with permission from [107], copyright 2008, Elsevier.

They divided the elemental composition of TiO$_2$ films into three categories: metallic, transition, and oxide. The results show that the three regions can be distinguished, since each one describing a different process occurring during discharge. At the condition of pure argon and low oxygen (Region I), the peak of Ti is predominantly due to the bombardment by argon ions. When the oxygen content is increased, an increase in the TiO$_2^+$ peak is observed (Region II). With oxygen content higher than 60% (Region III), an increase in oxygen content causes a decrease in the intensity of the anatase phase. As the oxygen is increased, the target becomes gradually oxidized, and consequently fewer Ti atoms and more TiO$_2$ molecules are ejected from the target. TiO$_2$ molecules in magnetron sputtering can originate from either the sputtering of the oxidized target or from the reaction of Ti atoms and O$_2$ molecules being charged. Ti atoms cannot be deposited on the substrate without argon gas in the magnetron sputtering process. Some researchers claim that the overdose of oxygen content in the magnetron sputtering might be responsible for the growth of low-crystallinity films [107].

The optical properties of PTFs is depending on the crystal phase and surface characteristics. The crystal phase of PTFs is affected by the argon/oxygen ratio in the sputtering process. Jeong et al. [2] found that the structural characteristics of both TiO$_2$ and SiO$_2$ are significantly improved with oxygen gas addition in the sputtering ambient compared to those of films grown without oxygen gas. Choi et al. [52] obtained similar results, finding that low mobility could be caused by high oxygen content. Their study showed that the decrease of collisions of Ar/O$_2$ gases during surface diffusion may result in the deficiency of crystallinity of TiO$_2$ films. The band gaps of the rutile and anatase phases are 3.0 and 3.2 eV, respectively. Structural defects in deposited films may also cause a variation in the band gap. The structure of deposited films consists of the UV emission band and the visible emission broad band. UV emission causes exciton recombination, and visible emission is attributed to structural defects, which are related to deep-level emissions [37]. Laha et al. [37] and Tomaszewski et al. [102] found that the argon/oxygen ratio during magnetron sputtering influences the level of oxygen vacancies in deposited films. Therefore, the level of oxygen vacancies is related to the band gap of the deposited films.

Transmittance (*T*) and the refractive index (*n*) are also affected by the argon/oxygen ratio in magnetron sputtering. High transparency in the visible range and a proper band absorption edge are very important for self-cleaning applications and photocatalysis. Xiong et al. [110] have investigated the effects of the argon/oxygen ratio in RF sputtering on the optical properties of TiO_2 films. They found that light transmittance in the visible range (400–800 nm) is slightly increased when the oxygen content is increased. At wavelengths below 350 nm, a significant drop in transmittance is discovered, which is attributed to the absorption of light caused by the excitation of electrons from the VB to the CB. Laha et al. [37] found that the transmittances of TiO_2/Al_2O_3 films in the spectral range (200–1100 nm) is increased with increasing argon/oxygen ratio. Moreover, the maximum transmittances are red-shifted with increasing argon/oxygen ratio, which can be explained by the particle size effect. Regarding the refractive index (*n*), Tomaszewski et al. [102] found that it is reduced with increasing oxygen content. They also indicated that the refractive index is correlated with the deposition rate. Kumar et al. [1] found a similar trend for $MgTiO_3$ films fabricated by RF magnetron sputtering at various oxygen content levels. The change in the refractive index is attributed to the packing density and crystallinity of the deposited films. The fabrication of thick TiO_2 films requires a long sputtering time and sufficient energy, which results in high packing density. Kumar et al. [1] found that the amorphous nature of deposited films is highly disordered and thus the films have a low refractive index. This is attributed to the low film density and low adatom mobility when the energy applied during the magnetron sputtering is insufficient. The oxygen content affects the crystal phase of TiO_2 films and its optical properties. The thickness of TiO_2 films also affects these properties.

3.6. Deposition Time

The deposition time is directly proportional to the thickness of deposited films. The influence of thickness on photocatalytic activity is still being debated [8,46,66]. Some researchers hypothesize that photocatalytic activity mainly depends on the crystalline phase and optical properties of PTFs, not thickness [46,66]. Huang et al. [66] found that sputtering power is a more important factor than sputtering time in enhancing photocatalytic activity of fabricated PTF films to decompose MB. Chen et al. [32] revealed that the sputtering time of PTF film fabrication has no effect on the photodecomposition of MB. Jamuna-Thevi et al. [39] used Ag-doped TiO_2 coatings on stainless steel to obtain antibacterial properties against Staphylococcus (S.) aureus. The results showed that the level of silver ions affected the reduction of S. aureus, not the deposition time. Nevertheless, Zheng et al. [8] and Eufinger et al. [67] found that film thickness plays an important role in the photocatalytic activity of TiO_2 films. The thickness (*d*) of films can be calculated as [1]:

$$d = \frac{\lambda_1 \lambda_2}{2(\lambda_1 n_2 - \lambda_2 n_1)} \tag{16}$$

where n_1 and n_2 are the refractive indices of two adjacent maximum and minimum at wavelengths λ_1 and λ_2, respectively.

When the film thickness is below a critical value, the photocatalytic activity of TiO_2 films increases with increasing thickness. Once the film thickness exceeds the critical thickness, it has little influence on the photocatalytic activity of TiO_2 films. Yang et al. [46] indicated that the deposition time had a dominant effect on MB absorbance. Zhang et al. [104] observed the optical properties of PTFs obtained with deposition times of 2, 3, and 5 h. The results showed that the three kinds of TiO_2 film exhibits quantum size effects and that the blue-shift of their band edges is slightly increased with decreasing deposition time. The authors attributed this to increases in the film thickness and grain size, which resulted in complete crystallinity. Rahman et al. [111] shows the deposition of TiO_2 homogeneous highly porous thin films by hydrothermal method and enhanced photocatalytic degradation efficiency with increase in film thickness with the increase in titania precursor concentration. They also reported that the thickness can be determined from the transmittance spectra, which shows that the transmittance of the films decreases with film thickness. This phenomenon occurs due to the increase of the atom on the surface because of the increase in deposition time or concentration of the precursor. In another article revealed that on increasing the deposition time of the magnetron sputtering the thickness, crystalline size and resistivity of the films are increased gradually [112]. In addition the structural and electrical properties of the ZnS film has been studied with the variation of film thickness have been investigated. It was observed that with the increasing the deposition time, the crystalline quality could be improved with developing better conducting behavior according to the increase in film thickness [113,116]. The better conducting behavior can increase the generation of the charge carriers on the surface of the material, which is the key reason for the enhanced photocatalytic activity. With the increase in film thickness, the quality of the crystallinity is improved as well as grain size value is increased, with increase in film thickness [117]. All characteristics are enhanced with the increase in deposition time. The roughness of prepared films is acquired from AFM measurements, and their evolutions with deposition time are increased by increasing the thickness, where it increases from 2 nm to a maximum value of 7 nm, when film thickness is from 70 nm to 170 nm, respectively. These results were confirmed by Abdalah et al. [118]. The cross section of the films, investigated from the SEM characterization, can be increased with deposition time and reaction rate. In sputtering technique, the parameters like base pressure, substrate temperature, deposition time and distance between the substrate and target are vital for better films properties. To obtain better optical properties and photo response properties in photocatalysis reactions doping elements, substrate, as well as deposition time can be changed for enhanced results. Abbas et al. [119] has reported in his research that the crystallization of the PbS thin films can be improved by increasing the deposition time and annealing temperature to increase thickness of film. The low value of transmittance in the UV region and increases gradually as the wavelength shifts to the NIR region. It can also be emphasized that the transparency window decreases with the deposition time, due to increasing thickness with time. Thickness of the film increases with increased deposition time or different nature of the substrate. Deposition time can affect the surface roughness by RF sputtering process [115]. The results show the highest roughness of the sample after depositing the films for 10 minutes, but it decreases with a deposition time of 15 minutes [120]. By lowering the amount of surface roughness the rate of surface reactions will increase.

Recent studies [36,117] have found that increased thickness increases the refractive index. This infers that higher thickness leads to higher photocatalytic activity. Šícha et al. [97] found that the deposition time (relative to the thickness of deposited films) in the magnetron sputtering process could be adjusted to form different crystalline phases of TiO_2 films. When the thickness was above 1000 nm, the deposited films have an anatase (101) phase structure. With a thickness of 330 nm, the deposited films have a poor crystallinity and exhibited a mixture of anatase and rutile phases with very low intensity of the rutile phase. When the thickness of the deposited films was below 140 nm, the crystalline phase became an amorphous phase. In addition, Zhang et al. [101] and Šícha et al. [97] both found that the substrate temperature increases with increasing deposition time. The final substrate temperature was approximately 150 °C. The energy of nucleation sites is a function of temperature [118]. Therefore, increasing deposition time might cause the substrate temperature to increase, which could further promote crystallite growth and increase grain size. This implies that substrate temperature also influences the optical properties of deposited films.

3.7. Substrate Temperature

The substrate temperature during magnetron sputtering can influence the characteristics of deposited films, including photocatalytic activity [32,46,55,121], crystallinity [48,56,72,122–124], refractive index [2,58,70], surface roughness [48,58,119,122], and optical properties [58,72,119,122]. Photocatalytic activity is usually tested using MB solution. When a substrate temperature higher than 400 °C is applied for depositing films, higher decomposition effect of MB solution can be achieved [46,55]. Substrate temperatures lower than 400 °C do not significantly influence the decomposition of MB [32]. This phenomenon can be attributed to the change of crystallinity with substrate temperature. Xu et al. [86] reported that a monotonic increase of the rutile/anatase phase ratio is observed with increasing substrate temperature. They suggested that the nucleation site energy increases with increasing substrate temperature. For substrate temperatures higher than 400 °C, nucleation sites have enough energy to transform metastable anatase into the thermally stable rutile phase.

It is generally recognized that deposition on unheated substrates usually results in amorphous films [101]. Increasing the substrate temperature promotes particle diffusion and provides the extra energy needed to grow the crystalline phase of deposited films. Haseeb et al. [48] prepared TiO_2 thin films using the RF sputtering process at various substrate temperatures. They found that an increase in substrate temperature promoted the growth of the crystalline phases of TiO_2 films. XRD patterns showed peaks of the main anatase phase, namely those for (101), (200), and (211), of TiO_2 films at room temperature. Other anatase planes, namely (103), (004), (105), and (204), are formed when the substrate temperature was increased. However, anatase planes (101), (200), and (211) decreased significantly with increasing the substrate temperature. Ananthakumar et al. [122] used the DC sputtering process to prepare TiO_2 films and obtained similar results. The degree of crystallinity of the deposited films is increased with increasing substrate temperature (in the range of 100 to 500 °C). They found that the polycrystalline tetragonal structures of anatase phase (101), (004), and (105) are obtained for TiO_2 films when the substrate temperature was 400 °C. Random growth orientations are found for the deposited films due to the amorphous nature of the substrates. When the substrate temperature is higher than 300 °C, only the (004) peak is observed. When the substrate temperature is increased to 500 °C, the (004) peak decreases. This means that the formation of the (004) peak plays a major role in the decreasing c-axis orientation. Zheng and Li [125] studied the properties of Nb-doped TiO_2 films deposited on $LaAlO_3$ by RF sputtering. Their results show that the anatase (004) phase in the deposited films at temperatures of 650 to 800 °C became the main reflection, which indicates that the films had a good c-axis texture (growth direction). The diffraction peak of the anatase (004) phase in TiO_2 only appears when magnetron sputtering is used [126]. The FWHM values of the (004) peak were used to estimate the crystallite size along the c-axis using Scherrer's relation (see Equation (13)) [127]. According to Zheng and Li [125], the (004) peak becomes the main reflection for films deposited by RF sputtering. Moreover, the FWHM values of the (004) peak decreased with increasing substrate temperature (from 600 to 800 °C), while the intensity of the (004) diffraction increased with increasing substrate temperature. This indicates that better c-axis textured anatase films with larger grain sizes are likely to form at high substrate temperatures. Many researchers have found that better c-axis textured anatase ITO films with larger grain sizes are likely to form at high substrate temperatures [128,129]. The increase in particle size with increasing temperature can thus explain the diffraction patterns. XRD was used to analyze the effect of substrate temperature in magnetron sputtering on the crystalline phase of films. Except for the anatase (004) phase, the intensity of other anatase phases in PTFs decrease with increasing substrate temperature [72,126,130]. This means that the anatase (004) phase is relatively stable. During magnetron sputtering, the crystalline phases of TiO_2 film change from amorphous to anatase and then to rutile with increasing substrate temperature [2,56,122,130]. A high substrate temperature may thus provide sufficient energy to generate crystalline phases.

The particle size increases with increasing substrate temperature. Because the surface roughness depends on the grain size, substrate temperature affects surface roughness. Takeuchi et al. [131] and Haseeb et al. [48] both found that surface roughness increased with increasing substrate temperature. They attributed this to the substrate giving extra energy, resulting in an increase in particle collisions. The surface roughness of films affects optical properties; a high surface roughness causes a lot of light scattering [58].

It has been demonstrated that the refractive index of films can be increased with increasing substrate temperature [2,58,72], which implies that the use of high substrate temperature during RF sputtering will lead to high packing density, crystallization, and oxygen deficiency. The transmittance of deposited films is affected by the substrate temperature during magnetron sputtering. Parameters such as stoichiometry, crystallinity, and thickness may also influence the transmittance of deposited films. The crystallinity of TiO_2 films is easily affected by substrate temperature. Lin [58] used simultaneous RF and DC sputtering to deposit Ti-doped TiO_2 nano ceramic films. The visible light transmittance is decreased following the increasing in substrate temperature. Takeuchi et al. [131] used RF sputtering with a high substrate temperatures ranging from 100 to 700 °C for the preparation of TiO_2 films have confirmed these results; the anatase phase of TiO_2 has higher visible transmittance than those of other crystalline phases [90]. However, the substrate temperature is a key factor for the transmittance of deposited films. Hasan et al. [72] used various substrate temperatures (room temperature, 200 °C, and 300 °C) to prepare TiO_2 films via the RF sputtering method. They observed that substrate temperatures of 200 and 300 °C do not affect transmittance because the crystallinity is not altered significantly and there is no phase transformation to the anatase phase in TiO_2 films. A change of crystalline phase in deposited films requires sufficient extra energy from the heated substrate. Moreover, the structure of the crystalline phases in deposited films may alter the transmittance. Ananthakumar et al. [122] used DC sputtering to prepare TiO_2 films at various substrate temperatures and found that the degree of crystallinity of the deposited films are significantly increased with an increasing in the substrate temperature. The authors have postulated that multiple factors, such as the increase in structural homogeneity [122], defects in the oxygen structure, and impurities in plasma gas [132] are involved in this process. Moreover, a substrate temperature in the sputtering process may increase the band gap [122,127] because the substrate temperature affects the crystallinity of deposited films. Lin [58] used the co-sputtering process (RF and DC sputtering) to deposit TiO_2 films and found that the band gap is decreased with increasing substrate temperature due to a possible Moss-Burstein shift; the change of the energy gap is the result of a large decrease in the free carrier concentration, and there is a corresponding downward shift of the Fermi level to below the band edge. In conclusion, a higher substrate temperature can provide extra energy to promote crystalline growth in deposited films, changing the optical properties.

4. Metal Oxide Doping

Many researchers have found that doping metals into TiO_2 films enhances their photocatalytic activity and extends the absorption edge into the visible light region [107,133]. Metals can be doped into TiO_2 films using layer-by-layer [45,134–136] and co-sputtering [57,137] methods. The co-sputtering method allows easy composition control and the layer-by-layer method yields higher quality thin films [136].

TiO_2 is an important material for transparent conducting oxide (TCO) layers in thin films because it has high optical transmittance (>80%) in the visible range and low resistivity ($10^{-7} \Omega \cdot cm^{-1}$) [4]. Optical materials based on ITO [51,117] and FTO [52,133] are the most widely used TCO films due to their low resistivity ($< 10^{-4} \Omega \cdot cm^{-1}$) [120]. TCO thin films can be used as substrates for the deposition of TiO_2 using sputtering to enhance the effect of electron transfer [51]. However, due to the scarcity and high cost of indium, ITO may not be able to satisfy anticipated demand in the future. Moreover, TiO_2 films usually demonstrate poor responses under visible light due to their wide band gap (E_g = 3.20 eV). In order to solve this problem, many researchers have found that metal doping effectively enhances the photoelectric properties [57,94,138] and lowers the band gap of TiO_2 films [139]. When the two materials are connected electrically, electrons migrate from the semiconductor to the metal until the two Fermi levels are aligned [17]. Therefore, several dopants, such as Ag [94,125], Al [61], Au [57], CdS, CeO [109], CeO_2 [109], Cr, Cu [135], Fe [18], Mo [132], Nb [140,141], Pt [134], and Si [113], have been used for TiO_2 thin film.

Metals such as Cr, Co, and Fe that are deposited on a TiO_2 surface can react with oxygen, resulting in the reduction of TiO_2 and the oxidation of the metals [136]. These metals can produce a series change of structure, which affects photocatalytic activity. Liu et al. [62] found that the electron transfer between TiO_2 and CeO_2 networks via Ti–O–Ce structural units plays an important role in photocatalysis. The results revealed that the $CeO_{1.6}2TiO_2$ phase is the original cause of visible light photocatalysis [121]. The holes drifting from $CeO_{1.6}2TiO_2$ to TiO_2 under an inner electric field can inject photo-induced holes into TiO_2 and endow the TiO_2-CeO_2 composite films with photocatalytic properties. Therefore, the TiO_2-CeO_2 films have better activity than that of pure TiO_2 films under UV illumination [121]. Matsuoka et al. [135] indicated that Pt loading in TiO_2 films enhances photocatalytic activity; the optimum Pt loading is determined to be 21 μgcm^{-2}. Carneiro et al. [18] found that the absorption edges of Fe-doped TiO_2 films shifted to the visible region with increasing concentrations of iron. Fe-doped TiO_2 films prepared at a low iron concentration have a better photocatalytic activity than that of pure TiO_2 films. In TiO_2, the CB consists of Ti 3d atomic orbital. For example, in Fe-doped TiO_2, the mixing of Fe 2p state with O 2p state shifts the CB downward and narrows the band-gap energy of the photocatalyst [139]. Fe with different ion forms (Fe^{2+}, Fe^{3+} and Fe^{4+}) can act as traps for electron hole pairs and consequently inhibit their recombination. However, increasing metal concentration may increase the recombination rate of photo-generated electron-hole due to the metal ion species acting as both electron and hole traps, and thus forming multiple trap sites. If there are too many traps of charge carriers in the bulk of catalyst or on its way to the surface, its apparent mobility may become low and it will likely recombine with its mobile counterpart generated by subsequent photons before it can reach the surface [142]. The schematic diagram of the mechanism is shown in Figure 11 below.

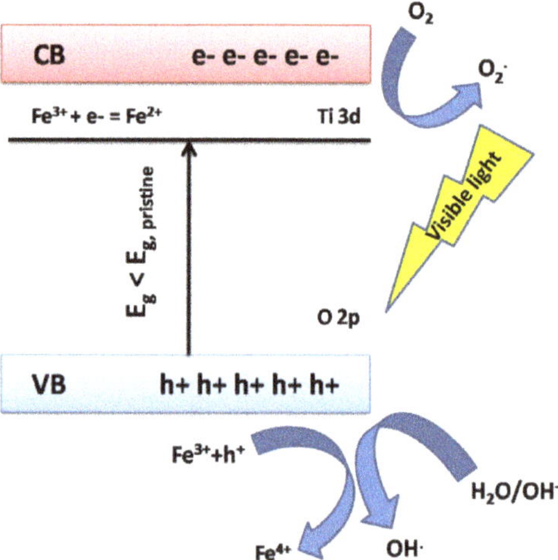

Figure 11. Schematic mechanism of visible light driven photocatalysis process Fe doped TiO_2 according to tailor the band gap.

Zhang et al. [104] reported that the absorption edges of Cu-doped samples shifted to the longer wavelength region; however, the optical transmittance of these films decreases abruptly with increasing copper concentration. Subrahmanyam et al. [19] further tested the photocatalytic activity of Ag-modified TiO_2 under UV and sunlight. The silver content on the TiO_2 surface is between 0.89 and 5.74 at.%. These results demostrated that the maximum photodegradations of rhodamine red molecules are 3.77 at.% and 0.89 at.% with Ag-doped TiO_2 film for irradiation at 254 and 352 nm, respectively. This result suggests that the optimum Ag content in TiO_2 films, which depends on the irradiation wavelength, is below 5.74 at.%. Similarly, the results of Jung et al. [57] confirmed that increasing the concentrations of Au in TiO_2 films can decrease photocatalytic activity. In addition, Seong et al. [143] found that the type of doping metal in TiO_2 films is an important factor; the photocatalytic activity of TiO_2/SiO_x double-layer samples is superior to that of pure TiO_2 thin films and the TiO_2/SiO_2 double layer. Metal dopants can act as recombination centers for photogenerated holes. Therefore, excess mental dopant decreases the photocatalytic activity of TiO_2.

The functional mechanism of photocatalytic activity on metal-doped TiO_2 has been investigated [133,144]. Lee and Lee [49] employed the RF magnetron sputtering method to prepare

TiO$_2$/SiO$_2$/glass thin films for their investigation of the effect of the SiO$_2$ interlayer on the film properties. It is found that the SiO$_2$ interlayer enhances crystallinity and induces a very fine columnar structure that prevents Na$^+$ ion diffusion. For understanding the photochemical process for metal-doped TiO$_2$ films, Zuo [94] has proposed that the surface plasmon resonance (SPR) of the metal dopant increases the absorption wavelength of the films. For instance, the SPR of Ag nanoparticles was centered at around 490 nm for a cluster-like film [145]. The electrons are trapped in individual clusters or islands, and cannot move freely. Sufficient energy (i.e., from irradiation) can excite surface plasmons, leading to absorption. Hence, the increase of the absorption edge on TiO$_2$:Ag film is predictable. Generally, the mechanism of photocatalytic process is initiated when light with energy equal to or greater than the bandgap, incident on the TiO$_2$ valence band that can lead to the generating positive (holes) and negative (electrons) charge carriers on the surface of TiO$_2$ nanoparticle. These photogenerated electron hole pairs causes recombination, become it gets trapped in metastable states between the CB and VB, or react with the organic pollutant adsorbed on the surface of TiO$_2$. These charge carriers initiate reduction and oxidation reactions of electrons and holes, respectively, at the surface of catalyst which in turns leading to the generation of highly reactive agents, such as superoxide and ·OH hydroxyl radicals, by reacting with dissolved molecular oxygen, surface hydroxyl groups, and adsorbed water molecules. In the case of metal doping, defect levels are created near the CB, while nonmetals induce additional defect levels above the VB, which in turn narrows the bandgap and thus contributes to visible-light photoactivity. In co-doped and multi doped TiO$_2$, electrons can jump either from these defect levels or from VB to the defect impurity level of the metal or to the topmost level of CB of TiO$_2$. Metal ions existing in variable oxidation states can act as trapping centers of electrons, which results in the enhancement in the lifetime of charge carriers, thereby increasing the photocatalytic activity of TiO$_2$.

A surface plasmon is a collection of electrons of the conduction band which oscillates in the interface between conductor and nonconductor or semiconductor and metal oxide or metal, or semiconductor oxide as well [136,146]. Warren et al. [145] has reported in his work that when an electric field (here we consider solar light, visible light or UV light source) is irradiated the metal's surface plasmon, the density of electron is decreased on one side of the metal while is increased on the another side. This redistribution of the charge carriers inside the surface plasmon's, creates an electric field inside and outside the metal. The metal is in opposite direction. A series of charge density oscillations will take place with this displacement of electron density which is caused by the coulombic restoring force. This phenomenon of oscillation of charge density and electric field creation are called either surface plasmon resonance (SPR) or localized surface plasmon resonance (LSPR) [147]. According to the mechanism of hot electron and hole present in the SPR, the excitation of the electron from the plasmonic metal because of the SPR effect, which further comes back to the conduction band of TiO$_2$ for reactions of photocatalysis. The study of Tian et al. with Au-doped TiO$_2$ nanocomposites in an electrochemical cell may be used to explain this mechanism [148]. In most of the studies Au, Ag and Cu have been used as plasmonic metals which elicit hot electrons to transfer from the metal to the semiconductor under solar light illumination. The difference stands for the surface plasmon on Au and Ag is that Au resonates under solar light, whereas Ag resonates in the near-UV light region. Electrons can be excited from the noble metals via the SPR mechanism [149–151]. Furube et al. [149] discovered that the flow of excited electrons uses the SPR effect from gold nanoparticles to TiO$_2$ nanoparticles using femtosecond transient absorption spectroscopy. Chen et al. [152] reported the number of electrons that obtained sufficient energy to overcome the Schottky barrier via the Fowler Theory. Water oxidation assisted by plasmon using Au-NR–TiO$_2$ electrodes was studied by Nishijima et al. [151] who confirmed that the evolution of O$_2$ and H$_2$O$_2$ from the electrodes in the solution based on the finding of the induction of electron–hole pairs during water oxidation with a stable plasmon. By using a photoelectrochemical cell Chen et al. [152] have also evaluated the mechanism of water splitting over Au-ZnO photoelectrodes under visible light. The possible mechanism of the photocatalytic process and the transfer of hot electrons in Au- doped TiO$_2$ are shown in Figure 12 [153].

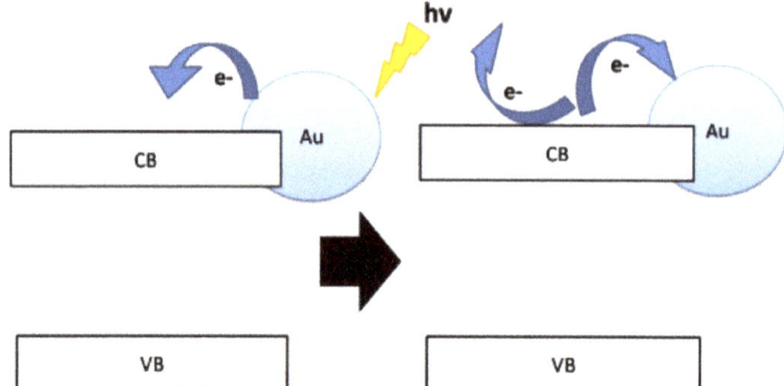

Figure 12. Proposed possible mechanism of the pathway of the hot electrons transferred in Au doped TiO$_2$.

Yan et al. [128] hypothesized that the mixing of metals with TiO$_2$ implanted under a lightly doped layer may lead to: (i) the introduction of an upward self-built electric field which could promote charge separation; (ii) an increase in more photogenerated carriers under visible light as the energy gap is curtailed; (iii) the inhibition of charge separation because of a large number of defects; (iv) the intensify of the Schottky barrier between TiO$_2$ and the metal substrate, and thus hindering a charge transfer.

György and Pérez del Pino [154] found that the dopant incorporation leads to the decrease of the crystallites' average dimensions. Carneiro et al. [18] and Subrahmanyam et al. [19] reported that metals doped into TiO$_2$ films can act as electron traps at the semiconductor interface. Kamisaka et al. [138] suggested that electron-trapping character of metal doped TiO$_2$ films is affected by the crystal phase of TiO$_2$ and the formal charged state of metals. The trapping of charge carriers can decrease the volume recombination rate of (e$^-$/h$^+$) pairs and thus increase the lifetime of charge carriers. The process of charge trapping is depicted as follows [84]:

$$M^{n+} + e_{cb}^- \rightarrow M^{(n-1)+} \tag{17}$$

where M^{n+} is the metal ion dopant. The energy level of M^{n+}/M$^{(n-1)+}$ lies below the CB edge. Thus, the energy level of transition metal ions affects the trapping efficiency.

Seong et al. [143] illustrated the pathway of charged electrons (e$^-$) and holes (h$^+$) in TiO$_2$/SiO$_x$ thin films (Figure 13).

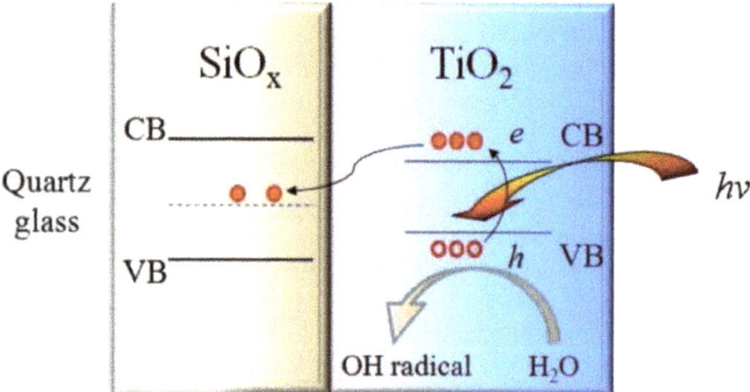

Figure 13. Schematic illustration of photocatalytic process for TiO$_2$/SiO$_x$ double-layer photocatalyst. Reprinted with permission from [143], copyright 2009, Elsevier.

The CB and VB edges of TiO_2 are located within the SiO_x bandgap. Under UV irradiation, the electron-hole pairs are initially generated in the TiO_2 layer. The generated electrons are accumulated in the CB of TiO_2 that are attracted to the trap level of SiO_x due to a difference in the work functions of TiO_2 and SiO_x and thus preventing e^-/h^+ recombination. This may explain why the photoactivity of a dual-layer photocatalyst is the highest among the prepared samples, even though its light absorbance is not the largest [45]. Jung et al. [57] used the co-sputtering method to prepare Au/TiO_2 thin films and found that some of the electrons reacted with lattice metal ions (Ti^{4+}) to form Ti^{3+} defective sites. The formation processes of defective sites on the TiO_2 surface can be expressed as:

$$TiO_2 + hv \rightarrow h^+_{vb} + e^-_{cb} \tag{18}$$

$$e^-_{cb} + Au \rightarrow Au^- \tag{19}$$

$$Au^- + Ti^{4+} \rightarrow Ti^{3+} + Au \tag{20}$$

$$H_2O_{(abs)} + h^+_{vb} \rightarrow OH\cdot + H^+ \tag{21}$$

$$Au^- + h^+ \rightarrow Au \tag{22}$$

The Ti^{3+} sites arise from the Ti^{4+} sites at which the photogenerated electrons are trapped [144]. A certain number of Ti^{3+} ions reduce the electron-hole recombination rate and thus enhance photocatalytic activity [113]. Liu et al. [155] proposed that the doped Ag^- ions convert Ti^{4+} to Ti^{3+} by charge compensation. These phenomena can be explained by the defective sites on the TiO_2 surface. The electrons react with adsorbed oxygen molecules or surface Ti^{4+} can generate reactive species $O_2^{\bullet-}$ and reactive center surface Ti^{3+}, respectively. The number of recombination centers of inner Ti^{3+} thus decreases accordingly. In addition, a loaded silver particle can either transfer one electron to an adsorbed oxygen molecule to form $O_2^{\bullet-}$ or move to the TiO_2 surface Ti^{4+} to form surface Ti^{3+}. These photochemical reactions explain why the photocatalytic activity of metal-doped TiO_2 films is higher than that of regular TiO_2 films.

The photocatalytic efficiency of metal-doped TiO_2 films seems to depend on the metal particle size, type of metal dopant, fabrication technology, and the hydrophilicity. Many researchers have proved that a large number of small epitaxial deposits on a semiconductor substrate is energetically capable of trapping photoelectrons which might decrease the space distance of charge carriers and increase the probability of the recombination of electrons and holes [147,149,156]. Thus, smaller particles have better photocatalytic activity because the transportation length of the e^-/h^+ pair from the crystal interface to the surface is shorter; and small grains (<40 nm) are more easily achieve charge separation as compared to large grains [147]. Lin and Wu [61] doped Al into TiO_2 films using RF sputtering and resulting in a smaller grain size, lower porosity, a higher linear refractive index, a lower stress-optical coefficient, and a higher visible-infrared transmission. However, Zuo [94] reported that Ag doping in TiO_2 films increases the FWHM due to large Ag clusters and broader size distribution. Ag nanoparticles are aggregated and form island-like structures. The large number of irregular nanoparticles and the interaction among the densely packed Ag nanoparticles resulted in broader peaks and red-shifts of absorption curves. The type of metal dopant can also influence photocatalytic activity [57,144]. Anpo and Takeuchi [136] have doped various metals into TiO_2 films and then examined their effect on the level of red-shift. It is demonstrated that the order of the red-shift efficiency of doped metals is found to be V > Cr >Mn> Fe > Ni. Such a shift allows metal-ion-implanted TiO_2 to use the solar irradiation more effectively in the range of 20–30%. Maeda and Watanabe [13] compared the photocatalytic activity of the plasma-enhanced chemical vapor deposition (PECVD) with those of the sol-gel method for fabricating TiO_2 films. Experimental results revealed that the grain size of TiO_2 films obtained using the PECVD method is almost uniform after annealing at 900 °C, and that these films have a higher photocatalytic activity than that of those obtained using the sol-gel method. It is inferred that the sputtering process can be used to fabricate optimal photocatalytic thin films. Additional studies have demonstrated that the surface of TiO_2 becomes highly hydrophilic under UV light irradiation [55,74]. This is attributed to the structural change in the TiO$_2$ surface. Electrons can reduce the Ti^{4+} cations to the Ti^{3+} state, and the holes can oxidize the O$_2^-$ anions, creating oxygen vacancies. Water molecules can occupy these oxygen vacancies, producing adsorbed OH groups, and making the surface hydrophilic [156]. It has been shown that more OH groups can improve the photocatalytic activity and thus the hydrophilicity may improve further the photocatalysis [147,149,150]. Nevertheless, the putative photocatalytic effects of doping various metals into TiO$_2$ films are still controversial [49,141,148]. Moreover, it has been reported that doping suitable amounts of Si and Ag in TiO$_2$ films significantly enhance hydrophilicity [49,147,148]. However, Jun and Lee [141] found that Cr doping is accompanied by the formation of not only the rutile TiO$_2$ phase but also the Cr$_2$O$_3$ phase, all of which degraded hydrophilicity [151–153]. The hydrophilicity of films has been evaluated by examining photographs and measuring the contact angle of a water droplet. The water contact angle (θ) can be calculated as [50]:

$$\theta \ = \ \arctan \frac{4HL}{L^2 - 4H^2} \tag{23}$$

where L and H are the diameter and height of the spherical crown, respectively.

5. Non-Metal Oxide Doping

Doping materials in the TiO$_2$ thin films is a very common approach for enhancement of the TiO$_2$ photocatalytic activity [154,155,157–160]. The main strategy of the doping is the tailoring of the band gap extending the absorption at longer wavelength. The dopants include transition, noble metals, non-metals and oxide materials as well. Noble metals like Ag, Au, Pd, and Pt exhibits better absorption properties in incorporating with TiO$_2$ followed by the mechanism of surface plasmon resonance (SPR) [161]. TiO$_2$ thin film incorporated with Au and Ag exhibited red-shifted wavelength and boost plasmon resonance, which can improve the overall photocatalytic organic decomposition [162,163]. However, metallic NPs can also enhance the photocatalytic performance of TiO$_2$ thin films through a non-plasmonic mechanism. But they have high cost which limits their usage in doping science with a large scale applications, whereas, transition metals offer more cost-effective processes [164]. On the other hand, non-metals are heavily preferred in the doping of TiO$_2$ thin films. Some of commonly used non-metals dopants are N, B, and S. [165,166]. Other non-metals such as, H, F, and I have also been incorporated into TiO$_2$ thin films during the preparation process to lower the band gap of the composite films to drives it to provide visible light photocatalytic activity. Valentin et al. have employed the density functional theory (DFT) to investigate the effects on the electronic structure of replacing lattice O atoms with B, C, N, or F dopants, or the inclusion of the same atoms in interstitial positions. The energy level of the bands introduced by doping increases with decreasing electronegativity. Fluorine is very electronegative, which introduces states below the O 2p valence band and leads to the formation of Ti^{3+} ions due to the charge compensation. In case of doping with B and C, this leads to the combination of energy levels above and below the valence band. For further improvement of visible light-driven photocatalysis, binary and ternary co-doping of two nonmetal elements such as N-C and N-B has also been studied. In many cases the synergistic effect is observed for the co-doping, but the parameters used in co-doping are very complex so there is no clear cut conclusion on doping effects [167–172].

Non-metal doping into TiO$_2$ thin films has many effects on the microstructure of the material that is depending on the reaction route in spite of band gap narrowing and visible light-driven photocatalysis. B-doped TiO$_2$ shows reduced size with a resulting higher surface area and suppressed phase transformation, and improved photogenerated charge separation, which may contribute to the photocatalysis-enhanced performance of titania [173–176]. The general mechanism of the doped TiO$_2$ with metal or nonmetal is visualized below (Figure 14):

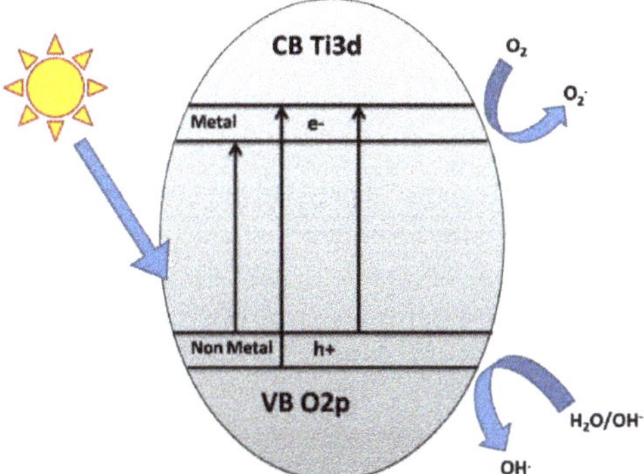

Figure 14. Schematic diagram for general photocatalysis mechanism for metal and nonmetal doped TiO$_2$ nanocomposite.

The doping of nonmetals into TiO$_2$ films leads to great photocatalytic activity [16]. Non-metals used as dopants include nitrogen (N) [171,172], carbon (C) [171], sulfur (S) [144], and fluorine (F) [22]. Many researchers have found that non-metals introduced into TiO$_2$ films can narrow the band gap [156], which is believed to be responsible for increasing the photocatalytic activity at long wavelengths ($\lambda > 500$ nm) [171].

Nitrogen can be easily introduced into the TiO$_2$ structure due to its comparable atomic size with that of oxygen, small ionization energy, and high stability. Therefore, nitrogen is the most promising non-metal dopant. Pelaez et al. [16] used various nitrogen-containing organic compounds in a modified sol-gel method to synthesize N-doped TiO$_2$. The obtained samples showed a better photocatalytic activity compared to those of other metal-ion-doped TiO$_2$ samples and Evonik P25-TiO$_2$. Bersani et al. [20] demonstrated that a significantly high substitution of the lattice O atoms of TiO$_2$ with N atoms narrows the band gap of TiO$_2$ thin films, enabling them to absorb and operate under visible light irradiation as a highly reactive, effective photocatalyst. Prabakar et al. [177] prepared visible-light-active nitrogen-doped TiO$_2$ films using DC-reactive magnetron sputtering with a Ti target in an Ar + O$_2$/N gas mixture. These results indicated that increasing the amount of N decreased the particle size of the anatase phase. Moreover, the absorption edge tends to be red-shifted as the N concentrations are increased. Parker and Siegel [178] used Raman spectra to analyze photocatalysts discovered that oxygen deficiency is the origin of all these effects. A clear decrease in the band gap and the nitrogen 2p states on the top of the VB on N-doped TiO$_2$ (compared to those of pure TiO$_2$) has found by Sathish et al. [179]. Di Valentin et al. [180] reported that the Ti–N bond lengths (1.964 and 2.081 Å) are similar to the Ti–O bond lengths (1.942 and 2.002 Å) in the matrix. This infers that N can easily substitute the O sites in the TiO$_2$ structure, and that the change may not degrade the crystallinity. Moreover, Lindgren et al. [181] found that a small concentration of N promotes the growth of anatase. Li and Shang [182] reported that N-doped TiO$_2$ photocatalyst is an anatase-rutile mixed phase which leads to a better electron-hole separation and enhances the visible light photocatalytic activity.

Both carbon and sulfur dopants have shown positive results for visible light activity in TiO$_2$, since they narrow the band gap [16,174]. Hamal and Klabunde [183] indicated that carbon embedded in the TiO$_2$ structure increases visible light absorption, and that carbon can act as a trap site within the CB and the VB, increasing the lifetime of photogenerated charge carriers. Khan et al. [153] prepared carbon-modified rutile TiO$_2$ using flame spray pyrolysis and observed that carbon replaces some of the lattice oxygen sites. Hsu et al. [184] has obtained a similar result, finding that carbon in TiO$_2$ substitutes the lattice oxygen atoms and forms Ti–C bonds. XPS spectra of C-doped TiO$_2$ has revealed a C 1s peak at ~281.8 eV and a free graphitic peak (C–C) at 284.6 eV [182]. The former is regarded as a response to visible light and the latter is served as a photosensitizer [184,185]. Furthermore, carbon doping of TiO_2 increases the absorption edge, which can be attributed to the visible light sensitivity originating from the localized C 2p formed in the band gap. Periyat et al. [176] used sulfuric acid to modify TiO_2 and found that the anatase phase can resist high temperatures ($\geq$800 °C) and that the presence of sulfur increases visible light photocatalytic activity. The effect of fluorine dopant is similar to that of sulfur dopant in TiO_2; it increases the anatase to rutile phase transformation temperature [22]. Fluorine dopant is unable to shift the TiO_2 band gap; nevertheless, it improves the surface acidity and causes the formation of reduced Ti^{3+} ions due to the charge compensation between F^- and Ti^{4+} [16]. The present review has identified that nitrogen and carbon are the most suitable non-metals for doping TiO_2 films in the sputtering process.

6. Annealing Treatment

It has been demonstrated that annealing treatment provides extra energy for the growth of the crystalline phase of TiO_2 films which is similar to the effect of substrate temperature all of which affects the crystalline phases [50,71]. XRD patterns show that the anatase (211) peak increases with an increasing annealing temperature on TiO_2 films (Figure 15) [186].

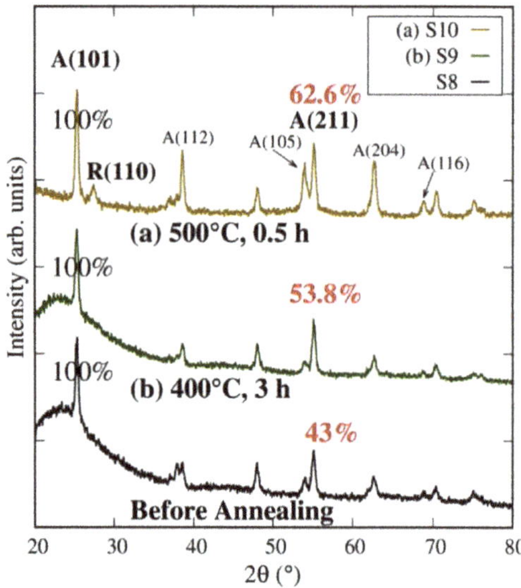

Figure 15. XRD pattern of anatase TiO_2 thin film annealed at (a) 500 °C 0.5 h and (b) 400 °C 3 h. Each noted percentage is the ratio of XRD intensity of A(211) to that of A(101). 'A' and 'R' represent anatase and rutile phase, respectively. Reprinted with permission from [186], copyright 2017, Elsevier.

Zhang et al. [104] found that the anatase phase appears when the annealing temperature is higher than 300 °C. Others observed that the rutile phase is first found at annealing temperatures of 600 to 800 °C [12,56,71]. The anatase phase changes into the rutile phase completely at 1000 °C [50], while the brookite phase initially forms at annealing temperatures of above 1000 °C [50].

Zhang et al. [104] also reported that the photocatalytic activity decreases with an increasing in the annealing temperature from 200 to 900 °C. Ye et al. [71] annealed as-deposited TiO_2 films at 500 °C and obtained the same result, which is attributed to the reduction of oxygen vacancies on the surface of TiO_2; this has implied that a change of elemental composition (Ti^{3+} changed to Ti^{4+}) on the surface of TiO_2 films occurres when the deposited films are annealed, which can decrease the photocatalytic activity on the surface of TiO_2 films by increasing the electron-hole recombination rate. However, Meng and Lu [187] have obtained a different result. They found that the photocatalytic activity of TiO$_2$ films annealed at 600 °C is the lowest, and then increases with increasing annealing temperature. The content of anatase TiO$_2$ is gradually decreased while the rutile TiO$_2$ is rather increased. This suggests that the annealing temperature affects the crystalline phase of TiO$_2$ and thus influences its photocatalytic activity.

Annealing treatment also influences the optical properties of TiO$_2$ films. Hasan et al. [72] found that the band gap of annealed films increases with increasing temperature (from 300 to 600 °C). Other optical properties (the extinction coefficient (see Equation (24)), refractive index, and transmittance) are decreased with increasing temperature. Zhang et al. [104] also found that annealing can cause opacity and that transmittance decreases with increasing annealing temperature (200 to 900 °C). Kumar et al. [1] found that annealed films exhibit an enhanced refractive index and a larger band gap, which is ascribed to the improvement in packing density, crystallinity, and decrease in the porosity ratio. These results are indirect evidence that anatase TiO$_2$ transformed into rutile TiO$_2$. Such degraded optical properties after annealing treatment, which may be attributed to higher packing density within the film and a slight increase in crystallinity [1,72].

The extinction coefficient (k) can be derived [37,69] as:

$$\alpha = \frac{4\pi\kappa}{\lambda}$$

(24)

where α is the absorption coefficient and λ is the absorption wavelength. A low extinction coefficient implies low absorption [37].

Many studies have demonstrated that the grain size is increased with increasing annealing temperature [15,50,185]. Ye et al. [71] found that increased grain size results in increased density of TiO$_2$ thin films results in increased refractivity, but a decreased porosity ratio ($P\%$). Ye et al. [71] and Hasan et al. [72] reported that $P\%$ is decreased with increasing annealing temperature (Figure 16).

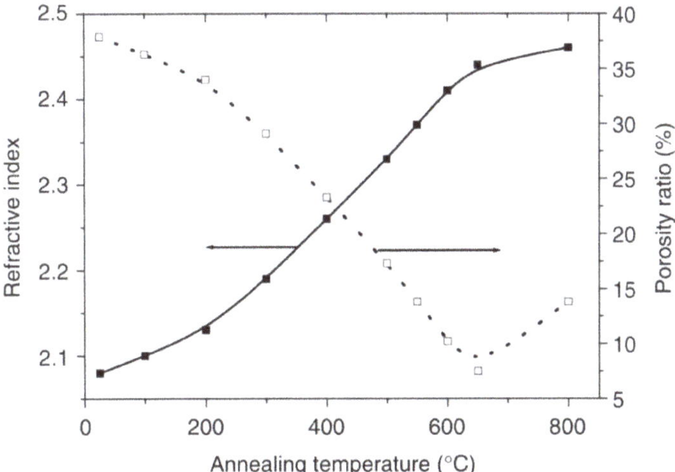

Figure 16. Dependence of refractive index and porosity of TiO$_2$ films on annealing temperature. Reprinted with permission from [71], copyright 2007, Elsevier.

A correlation between the photocatalytic activity and porosity has been confirmed by a study that obtained the highest photocatalytic activity is associated with a high-porosity photocatalyst [188].

Regarding hydrophilicity, Ye et al. [71] found that the decrease in water contact angle is associated with increases annealing temperature in the range of 200 to 600 °C, and then is further increased for a higher temperatures above 800 °C (Figure 17).

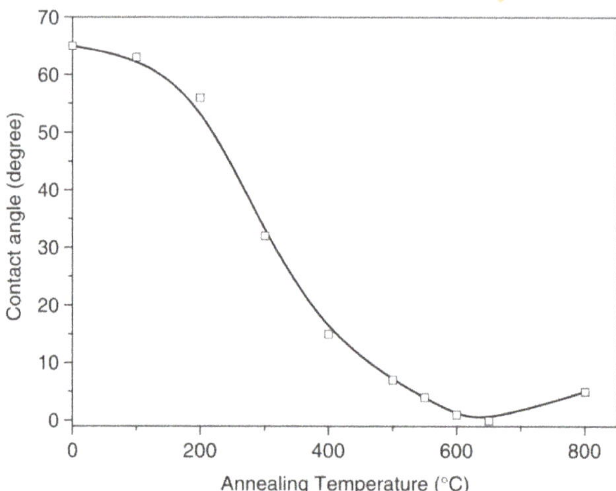

Figure 17. Relationship between water contact angle of TiO_2 films and annealing temperature. Reprinted with permission from [71], copyright 2007, Elsevier.

This result is consistent with those reported by Liu et al. [155] and Meng et al. [50], who concluded that amorphous TiO_2 has poor hydrophilicity, anatase TiO_2 has excellent hydrophilicity, and rutile TiO_2 has good hydrophilicity. The hydrophilicity effect of anatase TiO_2 is superior to that of rutile TiO_2 which is due to the surface roughness [155,189]. Khadar and Shanid [190] has demonstrated that the surface roughness of TiO_2 films is increased with increasing annealing temperature; this phenomenon is due mainly to the ballistic agglomeration vertically on the film.

7. Conclusions

This review demonstrated that TiO_2 is a promising semiconductor and photocatalyst due to its physical, structural, and optical properties under UV light. The working principle of the magnetron sputtering is discussed in detail, from which we can examine the process of electron transfer from the cathode to the target substrate material and the effect on sputtering power, pressure, etc. In order to prepare optimal TiO_2 films, the sputtering power, working pressure, argon/oxygen ratio, substrate temperature, dopant type, and annealing treatment must be carefully selected. Due to a high surface area, TiO_2 nanoparticles exhibit a high photocatalytic activity, nevertheless, their direct use in water treatment application is limited since their recovery from such aqueous medium needs advanced technologies which is very costly. Therefore, there is a need of TiO_2 thin films for high photocatalytic activity for both air and water treatments. By tailoring the surface morphology and properties, the photocatalytic activity of TiO_2 thin film can be improved in spite of the worse performance than TiO_2 nanoparticles. Thus, reactive sputtering systems and evaporation techniques are critical strategies for improving the morphological as well as other properties which can enhance the efficiency of the TiO_2 thin films. Optimal TiO_2 films must have a small particle size, a strong degree of crystallinity, a low band gap, a low contact angle (hydrophilic), and a high refractive index, transmittance, and extinction coefficient. Metal dopants in TiO_2 films can act as electron traps, preventing recombination. The photocatalytic activity of metal-doped TiO_2 films is thus higher than that of pure TiO_2 films. Non-metal dopants, in particular nitrogen, can substitute sites in the TiO_2 lattice. Other non-metal dopants, including carbon, fluorine, and sulfur, have been shown to increase visible light photo-induced activity by retaining the anatase phase of TiO_2 films at high temperature. However, only nitrogen and carbon are suitable for the magnetron sputtering process.

Author Contributions: Conceptualization, Y.-H.W., C.-C.W., K.-C.C.; investigation, Y.-H.W., C.-C.W., K.-C.C.; resources, K.-C.C.; writing—original draft preparation, Y.-H.W.; writing—review and editing, Y.-H.W., K.H.R., K.-C.C.; supervision, K.-C.C.; project administration, K.-C.C.; funding acquisition, K.-C.C. All authors have read and agreed to the published version of the manuscript.

Funding: The authors would like to thank the Ministry of Science and Technology (MOST) of Taiwan for financial support under Grant MOST 108-2637-E-020-007.

Conflicts of Interest: The authors declare no conflict of interest.

References

1. Kumar, T.S.; Bhuyan, R.K.; Pamu, D. Effect of post annealing on structural, optical and dielectric properties of $MgTiO_3$ thin films deposited by RF magnetron sputtering. *Appl. Surf. Sci.* **2013**, *264*, 184–190. [CrossRef]

2. Jeong, S.H.; Kim, J.K.; Kim, B.S.; Shim, S.H.; Lee, B.T. Characterization of SiO_2 and TiO_2 films prepared using RF magnetron sputtering and their application to anti-reflection coating. *Vacuum* **2004**, *76*, 507–515. [CrossRef]

3. Hu, W.; Li, L.; Li, G.; Tang, C.; Sun, L. High-quality brookite TiO_2 flowers: Synthesis, characterization, and dielectric performance. *Cryst. Growth Des.* **2009**, *9*, 3676–3682. [CrossRef]

4. Tighineanu, A.; Ruff, T.; Albu, S.; Hahn, R.; Schmuki, P. Conductivity of TiO_2 nanotubes: Influence of annealing time and temperature. *Chem. Phys. Lett.* **2010**, *494*, 260–263. [CrossRef]

5. Lee, C.H.; Kim, K.H.; Choi, H.W. Enhancing Efficiency of Dye-Sensitized Solar Cells Using TiO_2 Composite Films and RF-Sputtered Passivating Layer. *Mol. Cryst. Liq. Cryst.* **2012**, *567*, 9–18. [CrossRef]

6. Pansila, P.; Witit-Anun, N.; Chaiyakun, S. Influence of sputtering power on structure and photocatalyst properties of DC magnetron sputtered TiO_2 thin film. *Procedia Eng.* **2012**, *32*, 862–867. [CrossRef]

7. Chen, C.; Ma, W.; Zhao, J. Semiconductor-mediated photodegradation of pollutants under visible-light irradiation. *Chem. Soc. Rev.* **2010**, *39*, 4206–4219. [CrossRef]

8. Zheng, S.K.; Xiang, G.; Wang, T.M.; Pan, F.; Wang, C.; Hao, W.C. Photocatalytic activity studies of TiO_2 thin films prepared by RF magnetron reactive sputtering. *Vacuum* **2003**, *72*, 79–84. [CrossRef]

9. Wang, Q.; Zhu, K.; Neale, N.R.; Frank, A.J. Constructing ordered sensitized heterojunctions: Bottom-up electrochemical synthesis of p-type semiconductors in oriented n-TiO_2 nanotube arrays. *Nano Lett.* **2009**, *9*, 806–813. [CrossRef]

10. Hanaor, D.A.; Sorrell, C.C. Review of the anatase to rutile phase transformation. *J. Mater. Sci.* **2011**, *46*, 855–874. [CrossRef]

11. Yang, Z.; Choi, D.; Kerisit, S.; Rosso, K.M.; Wang, D.; Zhang, J.; Graff, G.; Liu, J. Nanostructures and lithium electrochemical reactivity of lithium titanites and titanium oxides: A review. *J. Power Sour.* **2009**, *192*, 588–598. [CrossRef]

12. Yoo, D.; Kim, I.; Kim, S.; Hahn, C.H.; Lee, C.; Cho, S. Effects of annealing temperature and method on structural and optical properties of TiO_2 films prepared by RF magnetron sputtering at room temperature. *Appl. Surf. Sci.* **2007**, *253*, 3888–3892. [CrossRef]

13. Maeda, M.; Watanabe, T. Effects of crystallinity and grain size on photocatalytic activity of titania films. *Surf. Coat. Technol.* **2007**, *201*, 9309–9312. [CrossRef]

14. Takabayashi, S.; Zeman, P. Effect of total and oxygen partial pressures on structure of photocatalytic TiO_2 films sputtered on unheated substrates. *Surf. Coat. Technol.* **2002**, *153*, 93–99.

15. Ogawa, H.; Higuchi, T.; Nakamura, A.; Tokita, S.; Miyazaki, D.; Hattori, T.; Tsukamoto, T. Growth of TiO_2 thin film by reactive RF magnetron sputtering using oxygen radical. *J. Alloys Compd.* **2008**, *449*, 375–378. [CrossRef]

16. Pelaez, M.; Nolan, N.T.; Pillai, S.C.; Seery, M.K.; Falaras, P.; Kontos, A.G.; Dunlop, P.S.; Hamilton, J.W.; Byrne, J.A.O.; Shea, K.; et al. A review on the visible light active titanium dioxide photocatalysts for environmental applications. *Appl. Catal. B* **2012**, *125*, 331–349. [CrossRef]

17. Herrmann, J.M. Heterogeneous photocatalysis: State of the art and present applications In honor of Pr. RL Burwell Jr. (1912–2003), Former Head of Ipatieff Laboratories, Northwestern University, Evanston (Ill). *Top. Catal.* **2005**, *34*, 49–65. [CrossRef]

18. Carneiro, J.O.; Teixeira, V.; Portinha, A.; Dupak, L.; Magalhaes, A.; Coutinho, P. Study of the deposition parameters and Fe-dopant effect in the photocatalytic activity of TiO_2 films prepared by dc reactive magnetron sputtering. *Vacuum* **2005**, *78*, 37–46. [CrossRef]

19. Subrahmanyam, A.; Biju, K.P.; Rajesh, P.; Kumar, K.J.; Kiran, M.R. Surface modification of sol gel TiO_2 surface with sputtered metallic silver for Sun light photocatalytic activity: Initial studies. *Sol. Energy Mater. Sol. Cell* **2012**, *101*, 241–248. [CrossRef]

20. Bersani, D.; Lottici, P.P.; Ding, X.Z. Phonon confinement effects in the Raman scattering by TiO_2 nanocrystals. *Appl. Phys. Lett.* **1998**, *72*, 73–75. [CrossRef]

21. Lee, S.H.; Yamasue, E.; Okumura, H.; Ishihara, K.N. Effect of oxygen and nitrogen concentration of nitrogen doped TiOx film as photocatalyst prepared by reactive sputtering. *Appl. Catal. A-Gen.* **2009**, *371*, 179–190. [CrossRef]

22. Padmanabhan, S.C.; Pillai, S.C.; Colreavy, J.; Balakrishnan, S.; McCormack, D.E.; Perova, T.S.; Gun'ko, Y.; Hinder, S.J.; Kelly, J.M. A simple sol– gel processing for the development of high-temperature stable photoactive anatase titania. *Chem. Mater.* **2007**, *19*, 4474–4481. [CrossRef]

23. Wiatrowski, A.; Mazur, M.; Obstarczyk, A.; Wojcieszak, D.; Kaczmarek, D.; Morgiel, J.; Gibson, D. Comparison of the physicochemical properties of TiO_2 thin films obtained by magnetron sputtering with continuous and pulsed gas flow. *Coatings* **2018**, *8*, 412. [CrossRef]

24. Veith, M.; Lee, J.; Miró, M.M.; Akkan, C.K.; Dufloux, C.; Aktas, O.C. Bi-phasic nanostructures for functional applications. *Chem. Soc. Rev.* **2012**, *41*, 5117–5130. [CrossRef] [PubMed]

25. Su, J.; Boichot, R.; Blanquet, E.; Mercier, F.; Pons, M. Chemical vapor deposition of titanium nitride thin films: Kinetics and experiments. *Cryst. Eng. Comm.* **2019**, *21*, 3974–3981. [CrossRef]

26. Henkel, B.; Neubert, T.; Zabel, S.; Lamprecht, C.; Selhuber-Unkel, C.; Raetzke, K.; Strunskus, T.; Vergöhl, M.; Faupel, F. Photocatalytic properties of titania thin films prepared by sputtering versus evaporation and aging of induced oxygen vacancy defects. *Appl. Catal. B* **2016**, *180*, 362–371. [CrossRef]

27. Rodríguez, J.; Gómez, M.; Lu, J.; Olsson, E.; Granqvist, C.G. Reactively Sputter-Deposited Titanium Oxide Coatings with Parallel Penniform Microstructure. *Adv. Mater.* **2000**, *12*, 341–343. [CrossRef]

28. Suzuki, M.; Ito, T.; Taga, Y. Photocatalysis of sculptured thin films of TiO_2. *Appl. Phys. Lett.* **2001**, *78*, 3968–3970. [CrossRef]

29. Merenda, A.; Kong, L.; Zhu, B.; Duke, M.C.; Gray, S.R.; Dumée, L.F. Functional nanoporous titanium dioxide for separation applications: Synthesis routes and properties to performance analysis. In *Water Scarcity and Ways to Reduce the Impact*; Springer: Cham, Switzerland, 2019; pp. 151–186.

30. Henkel, B.; Vahl, A.; Aktas, O.C.; Strunskus, T.; Faupel, F. Self-organized nanocrack networks: A pathway to enlarge catalytic surface area in sputtered ceramic thin films, showcased for photocatalytic TiO_2. *Nanotechnology* **2017**, *29*, 035703. [CrossRef]

31. Mejia, M.I.; Marin, J.M.; Restrepo, G.; Pulgarin, C.; Mielczarski, E.; Mielczarski, J.; Arroyo, Y.; Lavanchy, J.C.; Kiwi, J. Self-cleaning modified TiO_2–cotton pretreated by UVC-light (185 nm) and RF-plasma in vacuum and also under atmospheric pressure. *Appl. Catal. B* **2009**, *91*, 481–488. [CrossRef]

32. Chen, D.Y.; Tsao, C.C.; Hsu, C.Y. Photocatalytic TiO_2 thin films deposited on flexible substrates by radio frequency (RF) reactive magnetron sputtering. *Curr. Appl. Phys.* **2012**, *12*, 179–183. [CrossRef]

33. Chiou, A.H.; Kuo, C.G.; Huang, C.H.; Wu, W.F.; Chou, C.P.; Hsu, C.Y. Influence of oxygen flow rate on photocatalytic TiO_2 films deposited by rf magnetron sputtering. *J. Mater. Sci.-Mater. Electron.* **2012**, *23*, 589–594. [CrossRef]

34. Wei, C.H.; Chang, C.M. Polycrystalline TiO_2 thin films with different thicknesses deposited on unheated substrates using RF magnetron sputtering. *Mater. Trans.* **2011**. [CrossRef]

35. Yang, W.; Liu, Z.; Peng, D.L.; Zhang, F.; Huang, H.; Xie, Y.; Wu, Z. Room-temperature deposition of transparent conducting Al-doped ZnO films by RF magnetron sputtering method. *Appl. Surf. Sci.* **2009**, *255*, 5669–5673. [CrossRef]

36. Nair, P.B.; Justinvictor, V.B.; Daniel, G.P.; Joy, K.; Ramakrishnan, V.; Thomas, P.V. Effect of RF power and sputtering pressure on the structural and optical properties of TiO_2 thin films prepared by RF magnetron sputtering. *Appl. Surf. Sci.* **2011**, *257*, 10869–10875. [CrossRef]

37. Laha, P.; Panda, A.B.; Mahapatra, S.K.; Barhai, P.K.; Das, A.K.; Banerjee, I. Development of RF plasma sputtered Al_2O_3–TiO_2 multilayer broad band antireflecting coatings and its correlation with plasma parameters. *Appl. Surf. Sci.* **2012**, *258*, 2275–2282. [CrossRef]

38. Shet, S.; Ahn, K.S.; Nuggehalli, R.; Yan, Y.; Turner, J.; Al-Jassim, M. Effects of substrate temperature and RF power on the formation of aligned nanorods in ZnO thin films. *JOM-US* **2010**, *62*, 25–30. [CrossRef]

39. Jamuna-Thevi, K.; Bakar, S.A.; Ibrahim, S.; Shahab, N.; Toff, M.R. Quantification of silver ion release, in vitro cytotoxicity and antibacterial properties of nanostuctured Ag doped TiO$_2$ coatings on stainless steel deposited by RF magnetron sputtering. *Vacuum* **2011**, *86*, 235–241. [CrossRef]

40. Sun, H.; Wang, C.; Pang, S.; Li, X.; Tao, Y.; Tang, H.; Liu, M. Photocatalytic TiO$_2$ films prepared by chemical vapor deposition at atmosphere pressure. *J. Non-Cryst. Solids* **2008**, *354*, 1440–1443. [CrossRef]

41. Amadelli, R.; Samiolo, L.; Maldotti, A.; Molinari, A.; Valigi, M.; Gazzoli, D. Preparation, characterisation, and photocatalytic behaviour of Co-TiO$_2$ with visible light response. *Int. J. Photoenergy* **2008**, *2008*, 853753. [CrossRef]

42. Lei, Y.; Zhang, L.D.; Fan, J.C. Fabrication, characterization and Raman study of TiO$_2$ nanowire arrays prepared by anodic oxidative hydrolysis of TiCl$_3$. *Chem. Phys. Lett.* **2001**, *338*, 231–236. [CrossRef]

43. Arconada, N.; Castro, Y.; Durán, A. Photocatalytic properties in aqueous solution of porous TiO$_2$-anatase films prepared by sol–gel process. *Appl. Catal. A-Gen.* **2010**, *385*, 101–107. [CrossRef]

44. Khoa, N.T.; Pyun, M.W.; Yoo, D.H.; Kim, S.W.; Leem, J.Y.; Kim, E.J.; Hahn, S.H. Photodecomposition effects of graphene oxide coated on TiO$_2$ thin film prepared by electron-beam evaporation method. *Thin Solid Films* **2012**, *520*, 5417–5420. [CrossRef]

45. Liao, C.H.; Huang, C.W.; Wu, J.C. Novel dual-layer photoelectrode prepared by RF magnetron sputtering for photocatalytic water splitting. *Int. J. Hydrog. Energy* **2012**, *37*, 11632–11639. [CrossRef]

46. Yang, W.J.; Hsu, C.Y.; Liu, Y.W.; Hsu, R.Q.; Lu, T.W.; Hu, C.C. The structure and photocatalytic activity of TiO$_2$ thin films deposited by dc magnetron sputtering. *Superlattice Microstruct.* **2012**, *52*, 1131–1142. [CrossRef]

47. Garlisi, C.; Scandura, G.; Palmisano, G.; Chiesa, M.; Lai, C.Y. Integrated nano-and macroscale investigation of photoinduced hydrophilicity in TiO$_2$ thin films. *Langmuir* **2016**, *32*, 11813–11818. [CrossRef]

48. Haseeb, A.S.; Hasan, M.M.; Masjuki, H.H. Structural and mechanical properties of nanostructured TiO$_2$ thin films deposited by RF sputtering. *Surf. Coat. Technol.* **2010**, *205*, 338–344. [CrossRef]

49. Lee, K.S.; Lee, S.H. Influence of SiO$_2$ interlayer on the hydrophilicity of TiO$_2$/SiO$_2$/glass produced by RF-magnetron sputtering. *Mater. Lett.* **2007**, *61*, 3516–3518. [CrossRef]

50. Meng, F.; Xiao, L.; Sun, Z. Thermo-induced hydrophilicity of nano-TiO$_2$ thin films prepared by RF magnetron sputtering. *J. Alloys Compd.* **2009**, *485*, 848–852. [CrossRef]

51. Ohsaki, H.; Andou, R.; Kinbara, A.; Watanabe, T. Crystallization of ITO and TiO$_2$ by RF plasma treatment. *Vacuum* **2013**, *87*, 145–149. [CrossRef]

52. Choi, W.J.; Kwak, D.J.; Park, C.S.; Sung, Y.M. Characterization of transparent conductive ITO, ITiO, and FTO Films for application in photoelectrochemical cells. *J. Nanosci. Nanotechnol.* **2012**, *12*, 3394–3397. [CrossRef] [PubMed]

53. Song, S.; Yang, T.; Li, Y.; Pang, Z.; Lin, L.; Lv, M.; Han, S. Structural, electrical and optical properties of ITO films with a thin TiO$_2$ seed layer prepared by RF magnetron sputtering. *Vacuum* **2009**, *83*, 1091–1094. [CrossRef]

54. Rhee, S.W.; Kim, K.H.; Choi, H.W. Effect of ZnO Passivating Layer Using RF-Sputtered for Dye-Sensitized Solar Cells. *Mol. Cryst. Liq. Cryst.* **2012**, *565*, 131–137. [CrossRef]

55. Chen, C.C.; Yang, W.J.; Hsu, C.Y. Investigation into the effects of deposition parameters on TiO$_2$ photocatalyst thin films by rf magnetron sputtering. *Superlattice Microstruct.* **2009**, *46*, 461–468. [CrossRef]

56. Wenbin, X.; Shurong, D.; Demiao, W.; Gaochao, R. Investigation of microstructure evolution in Pt-doped TiO$_2$ thin films deposited by RF magnetron sputtering. *Physica B* **2008**, *403*, 2698–2701. [CrossRef]

57. Jung, J.M.; Wang, M.; Kim, E.J.; Hahn, S.H. Photocatalytic properties of Au/TiO$_2$ thin films prepared by RF magnetron co-sputtering. *Vacuum* **2008**, *82*, 827–832. [CrossRef]

58. Lin, S.S. The optical properties of Ti-doped TiO$_2$ nanoceramic films deposited by simultaneous RF and DC magnetron sputtering. *Ceram. Int.* **2012**, *38*, 3129–3134. [CrossRef]

59. Huang, H.H.; Huang, C.C.; Huang, P.C.; Yang, C.F.; Hsu, C.Y. Preparation of rutile and anatase phases titanium oxide film by RF sputtering. *J. Nanosci. Nanotechnol.* **2008**, *8*, 2659–2664. [CrossRef]

60. Su, Y.F.; Chou, T.C.; Ling, T.R.; Sun, C.C. Photocurrent performance and nanostructure analysis of TiO$_2$/ITO electrodes prepared using reactive sputtering. *J. Electrochem. Soc.* **2004**, *151*, A1375–A1382. [CrossRef]

61. Lin, S.S.; Wu, D.K. Effect of RF deposition power on the properties of Al-doped TiO$_2$ thin films. *Surf. Coat. Technol.* **2010**, *204*, 2202–2207. [CrossRef]

62. Liu, B.; Zhao, X.; Zhao, Q.; Li, C.; He, X. The effect of O_2 partial pressure on the structure and photocatalytic property of TiO_2 films prepared by sputtering. *Mater. Chem. Phys.* **2005**, *90*, 207–212. [CrossRef]
63. Magyari-Köpe, B.; Park, S.G.; Lee, H.D.; Nishi, Y. First principles calculations of oxygen vacancy-ordering effects in resistance change memory materials incorporating binary transition metal oxides. *J. Mater. Sci.* **2012**, *47*, 7498–7514. [CrossRef]
64. Sirisuk, A.; Klansorn, E.; Praserthdam, P. Effects of reaction medium and crystallite size on Ti_3^+ surface defects in titanium dioxide nanoparticles prepared by solvothermal method. *Catal. Commun.* **2008**, *9*, 1810–1814. [CrossRef]
65. Chow, L.L.; Yuen, M.M.; Chan, P.C.; Cheung, A.T. Reactive sputtered TiO_2 thin film humidity sensor with negative substrate bias. *Sens. Actuators B-Chem.* **2001**, *76*, 310–315. [CrossRef]
66. Huang, C.H.; Tsao, C.C.; Hsu, C.Y. Study on the photocatalytic activities of TiO_2 films prepared by reactive RF sputtering. *Ceram. Int.* **2011**, *37*, 2781–2788. [CrossRef]
67. Eufinger, K.; Poelman, D.; Poelman, H.; De Gryse, R.; Marin, G.B. Photocatalytic activity of dc magnetron sputter deposited amorphous TiO_2 thin films. *Appl. Surf. Sci.* **2007**, *254*, 148–152. [CrossRef]
68. Hou, Y.Q.; Zhuang, D.M.; Zhang, G.; Zhao, M.; Wu, M.S. Influence of annealing temperature on the properties of titanium oxide thin film. *Appl. Surf. Sci.* **2003**, *218*, 98–106. [CrossRef]
69. Miao, L.; Jin, P.; Kaneko, K.; Terai, A.; Nabatova-Gabain, N.; Tanemura, S. Preparation and characterization of polycrystalline anatase and rutile TiO_2 thin films by rf magnetron sputtering. *Appl. Surf. Sci.* **2003**, *212*, 255–263. [CrossRef]
70. Zhang, T.J.; Li, S.Z.; Zhang, B.S.; Pan, R.K.; Jiang, J.; Ma, Z.J. Optical properties of RF-sputtered $Ba_{0.65}Sr_{0.35}TiO_3$ thin films. *J. Electroceram.* **2008**, *21*, 174–177. [CrossRef]
71. Ye, Q.; Liu, P.Y.; Tang, Z.F.; Zhai, L. Hydrophilic properties of nano-TiO_2 thin films deposited by RF magnetron sputtering. *Vacuum* **2007**, *81*, 627–631. [CrossRef]
72. Hasan, M.M.; Haseeb, A.S.; Saidur, R.; Masjuki, H.H.; Hamdi, M. Influence of substrate and annealing temperatures on optical properties of RF-sputtered TiO_2 thin films. *Opt. Mater.* **2010**, *32*, 690–695. [CrossRef]
73. Heo, C.H.; Lee, S.B.; Boo, J.H. Deposition of TiO_2 thin films using RF magnetron sputtering method and study of their surface characteristics. *Thin Solid Films* **2005**, *475*, 183–188. [CrossRef]
74. Lin, W.S.; Huang, C.H.; Yang, W.J.; Hsu, C.Y.; Hou, K.H. Photocatalytic TiO_2 films deposited by RF magnetron sputtering at different oxygen partial pressure. *Curr. Appl. Phys.* **2010**, *10*, 1461–1466. [CrossRef]
75. Kang, S.H.; Lim, J.W.; Kim, H.S.; Kim, J.Y.; Chung, Y.H.; Sung, Y.E. Photo and electrochemical characteristics dependent on the phase ratio of nanocolumnar structured TiO_2 films by RF magnetron sputtering technique. *Chem. Mater.* **2009**, *21*, 2777–2788. [CrossRef]
76. Zhao, X.T.; Sakka, K.; Kihara, N.; Takata, Y.; Arita, M.; Masuda, M. Hydrophilicity of TiO_2 thin films obtained by RF magnetron sputtering deposition. *Curr. Appl. Phys.* **2006**, *6*, 931–933. [CrossRef]
77. Song, P.K.; Yamagishi, M.; Odaka, H.; Shigesato, Y. Photoinduced hydrophilicity of epitaxially grown TiO_2 films by RF magnetron sputtering. *Jpn. J. Appl. Phys.* **2003**, *42*, L1529. [CrossRef]
78. Shibata, A.; Okimura, K.; Yamamoto, Y.; Matubara, K. Effect of heating probe on reactively sputtered TiO_2 film growth. *Jpn. J. Appl. Phys.* **1993**, *32*, 5666. [CrossRef]
79. Kasemanankul, P.; Witit-Anan, N.; Chaiyakun, S.; Limsuwan, P.; Boonamnuayvitaya, V. Low-temperature deposition of (1 1 0) and (1 0 1) rutile TiO_2 thin films using dual cathode DC unbalanced magnetron sputtering for inducing hydroxyapatite. *Mater. Chem. Phys.* **2009**, *117*, 288–293. [CrossRef]
80. Ngaruiya, J.M.; Kappertz, O.; Mohamed, S.H.; Wuttig, M. Structure formation upon reactive direct current magnetron sputteringof transition metal oxide films. *Appl. Phys. Lett.* **2004**, *85*, 748–750. [CrossRef]
81. Sundaram, K.B.; Khan, A. Characterization and optimization of zinc oxide films by RF magnetron sputtering. *Thin Solid Films* **1997**, *295*, 87–91. [CrossRef]
82. Barnes, M.C.; Gerson, A.R.; Kumar, S.; Hwang, N.M. The mechanism of TiO_2 deposition by direct current magnetron reactive sputtering. *Thin Solid Films* **2004**, *446*, 29–36. [CrossRef]
83. Barnes, M.C.; Kim, D.Y.; Ahn, H.S.; Lee, C.O.; Hwang, N.M. Deposition mechanism of gold by thermal evaporation: Approach by charged cluster model. *J. Cryst. Growth* **2000**, *213*, 83–92. [CrossRef]
84. Shah, S.I.; Li, W.; Huang, C.P.; Jung, O.; Ni, C. Study of Nd_3^+, Pd_2^+, Pt_4^+, and Fe_3^+ dopant effect on photoreactivity of TiO_2 nanoparticles. *Proc. Natl. Acad. Sci. USA* **2002**, *99* (Suppl. 2), 6482–6486. [CrossRef]
85. Lin, W.S.; Kao, L.M.; Li, W.P.; Hsu, C.Y.; Hou, K.H. Fabricating TiO_2 photocatalysts by RF reactive magnetron sputtering at varied oxygen partial pressures. *J. Mater. Eng. Perform.* **2011**, *20*, 1063–1067. [CrossRef]

86. Xu, Y.; Shen, M. Fabrication of anatase-type TiO$_2$ films by reactive pulsed laser deposition for photocatalyst application. *J. Mater. Process. Technol.* **2008**, *202*, 301–306. [CrossRef]

87. Chaoumead, A.; Joo, B.H.; Kwak, D.J.; Sung, Y.M. Structural and electrical properties of sputtering power and gas pressure on Ti-dope In$_2$O$_3$ transparent conductive films by RF magnetron sputtering. *Appl. Surf. Sci.* **2013**, *275*, 227–232. [CrossRef]

88. Jin, Y.S.; Kim, K.H.; Kim, W.J.; Jang, K.U.; Choi, H.W. The effect of RF-sputtered TiO$_2$ passivating layer on the performance of dye sensitized solar cells. *Ceram. Int.* **2012**, *38*. S505–S509. [CrossRef]

89. Wang, T.M.; Zheng, S.K.; Hao, W.C.; Wang, C. Studies on photocatalytic activity and transmittance spectra of TiO$_2$ thin films prepared by rf magnetron sputtering method. *Surf. Coat. Technol.* **2002**, *155*, 141–145. [CrossRef]

90. Choi, Y.L.; Kim, S.H.; Song, Y.S.; Lee, D.Y. Photodecomposition and bactericidal effects of TiO$_2$ thin films prepared by a magnetron sputtering. *J. Mater. Sci.* **2004**, *39*, 5695–5699. [CrossRef]

91. Zhu, H.; Hüpkes, J.; Bunte, E.; Gerber, A.; Huang, S.M. Influence of working pressure on ZnO: Al films from tube targets for silicon thin film solar cells. *Thin Solid Films* **2010**, *518*, 4997–5002. [CrossRef]

92. Drüsedau, T.P.; Löhmann, M.; Garke, B. Decay length of the pressure dependent deposition rate for magnetron sputtering. *J. Vac. Sci. Technol. A* **1998**, *16*, 2728–2732. [CrossRef]

93. Zhang, C.; Ding, W.; Wang, H.; Chai, W.; Ju, D. Influences of working pressure on properties for TiO$_2$ films deposited by DC pulse magnetron sputtering. *J. Environ. Sci.* **2009**, *21*, 741–744. [CrossRef]

94. Zuo, J. Deposition of Ag nanostructures on TiO$_2$ thin films by RF magnetron sputtering. *Appl. Surf. Sci.* **2010**, *256*, 7096–7101. [CrossRef]

95. Barreca, D.; Gasparotto, A.; Tondello, E.; Bruno, G.; Losurdo, M. Influence of process parameters on the morphology of Au/SiO$_2$ nanocomposites synthesized by radio-frequency sputtering. *J. Appl. Phys.* **2004**, *96*, 1655–1665. [CrossRef]

96. Okimura, K.; Shibata, A. Mass and Energy Analyses of Substrate-incident Ions in TiO$_2$ Deposition by RF Magnetron Sputtering. *Jpn. J. Appl. Phys.* **1997**, *36*, 313. [CrossRef]

97. Šícha, J.; Musil, J.; Meissner, M.; Čerstvý, R. Nanostructure of photocatalytic TiO$_2$ films sputtered at temperatures below 200 C. *Appl. Surf. Sci.* **2008**, *254*, 3793–3800. [CrossRef]

98. El Akkad, F.; Paulose, T.A. Optical transitions and point defects in F: SnO$_2$ films: Effect of annealing. *Appl. Surf. Sci.* **2014**, *295*, 8–17. [CrossRef]

99. Mardare, D.; Tasca, M.; Delibas, M.; Rusu, G.I. On the structural properties and optical transmittance of TiO$_2$ RF sputtered thin films. *Appl. Surf. Sci.* **2000**, *156*, 200–206. [CrossRef]

100. Vorontsov, A.V.; Savinov, E.N.; Zhensheng, J. Influence of the form of photodeposited platinum on titania upon its photocatalytic activity in CO and acetone oxidation. *J. Photochem. Photobiol. A* **1999**, *125*, 113–117. [CrossRef]

101. Zhang, W.; Li, Y.; Zhu, S.; Wang, F. Influence of argon flow rate on TiO$_2$ photocatalyst film deposited by DC reactive magnetron sputtering. *Surf. Coat. Technol.* **2004**, *182*, 192–198. [CrossRef]

102. Tomaszewski, H.; Poelman, H.; Depla, D.; Poelman, D.; De Gryse, R.; Fiermans, L.; Reyniers, M.F.; Heynderickx, G.; Marin, G.B. TiO$_2$ films prepared by DC magnetron sputtering from ceramic targets. *Vacuum* **2002**, *68*, 31–38. [CrossRef]

103. Huang, C.L.; Hsu, C.H. Properties of reactively radio frequency-magnetron sputtered (Zr, Sn) TiO$_4$ dielectric films. *J. Appl. Phys.* **2004**, *96*, 1186–1191. [CrossRef]

104. Zhang, W.J.; Zhu, S.L.; Li, Y.; Wang, F.H. Photocatalytic property of TiO$_2$ films deposited by pulsed DC magnetron sputtering. *J. Mater. Sci. Technol.* **2004**, *20*, 31–34.

105. Pradhan, S.S.; Pradhan, S.K.; Bhavanasi, V.; Sahoo, S.; Sarangi, S.N.; Anwar, S.; Barhai, P.K. Low temperature stabilized rutile phase TiO$_2$ films grown by sputtering. *Thin Solid Films* **2012**, *520*, 1809–1813. [CrossRef]

106. Water, W.; Chu, S.Y. Physical and structural properties of ZnO sputtered films. *Mater. Lett.* **2002**, *55*, 67–72. [CrossRef]

107. Toku, H.; Pessoa, R.S.; Maciel, H.S.; Massi, M.; Mengui, U.A. The effect of oxygen concentration on the low temperature deposition of TiO$_2$ thin films. *Surf. Coat. Technol.* **2008**, *202*, 2126–2131. [CrossRef]

108. Jeong, S.H.; Kim, B.S.; Lee, B.T.; Kim, J.K.; Park, H.R. Structural and optical properties of TiO$_2$ films prepared using reactive RF magnetron sputtering. *J. Korean Phys. Soc.* **2002**, *41*, 67–71.

109. Tavares, C.J.; Vieira, J.; Rebouta, L.; Hungerford, G.; Coutinho, P.; Teixeira, V.; Carneiro, J.O.; Fernandes, A.J. Reactive sputtering deposition of photocatalytic TiO_2 thin films on glass substrates. *Mater. Sci. Eng. B* **2007**, *138*, 139–143. [CrossRef]

110. Xiong, J.; Das, S.N.; Kim, S.; Lim, J.; Choi, H.; Myoung, J.M. Photo-induced hydrophilic properties of reactive RF magnetron sputtered TiO_2 thin films. *Surf. Coat. Technol.* **2010**, *204*, 3436–3442. [CrossRef]

111. Rahman, K.H.; Kar, A.K. Effect of precursor concentration of microstructured titanium-di-oxide (TiO_2) thin films and their photocatalytic activity. *Mater. Res. Express* **2019**, *6*, 096436. [CrossRef]

112. Tondare, R.S.; Shivaraj, B.W.; Narasimhamurthy, H.N.; Krishna, M.; Subramanyam, T.K. Effect of deposition time on structural, electrical and optical properties of Aluminium doped ZnO thin films by RF magnetron sputtering. *Mater. Today-Proc.* **2018**, *5*, 2710–2715. [CrossRef]

113. Abdallah, B.; Jazmati, A.K.; Refaai, R. Oxygen effect on structural and optical properties of ZnO thin films deposited by RF magnetron sputtering. *Mater. Res.* **2017**, *20*, 607–612. [CrossRef]

114. Nakajima, A.; Hashimoto, K.; Watanabe, T.; Takai, K.; Yamauchi, G.; Fujishima, A. Transparent superhydrophobic thin films with self-cleaning properties. *Langmuir* **2000**, *16*, 7044–7047. [CrossRef]

115. Vancoppenolle, V.; Jouan, P.Y.; Wautelet, M.; Dauchot, J.P.; Hecq, M. Dc magnetron sputtering deposition of TiO_2 films in argon–oxygen gas mixtures: Theory and experiments. *Surf. Coat. Technol.* **1999**, *116*, 933–937. [CrossRef]

116. Alnama, K.; Abdallah, B.; Kanaan, S. Deposition of ZnS thin film by ultrasonic spray pyrolysis: Effect of thickness on the crystallographic and electrical properties. *Compos. Interface* **2017**, *24*, 499–513. [CrossRef]

117. Abdallah, B.; Ismail, A.; Kashoua, H.; Zetoun, W. Effects of deposition time on the morphology, structure, and optical properties of PbS thin films prepared by chemical bath deposition. *J. Nanomater.* **2018**, *2018*, 1826959. [CrossRef]

118. Abdallah, B.; Kakhia, M.; Abou Shaker, S. Deposition of Na_2WO_4 films by ultrasonic spray pyrolysis: Effect of thickness on the crystallographic and sensing properties. *Compos. Interface* **2016**, *23*, 663–674. [CrossRef]

119. Abbas, M.M.; Shehab, A.A.; Al-Samuraee, A.K.; Hassan, N.A. Effect of deposition time on the optical characteristics of chemically deposited nanostructure PbS thin films. *Energy Procedia* **2011**, *6*, 241–250. [CrossRef]

120. Daghrir, R.; Drogui, P.; Robert, D. Modified TiO_2 for environmental photocatalytic applications: A review. *Ind. Eng. Chem. Res.* **2013**, *52*, 3581–3599. [CrossRef]

121. Liu, B.; Zhao, X.; Zhang, N.; Zhao, Q.; He, X.; Feng, J. Photocatalytic mechanism of TiO_2–CeO_2 films prepared by magnetron sputtering under UV and visible light. *Surf. Sci.* **2005**, *595*, 203–211. [CrossRef]

122. Ananthakumar, R.; Subramanian, B.; Yugeswaran, S.; Jayachandran, M. Effect of substrate temperature on structural, morphological and optical properties of crystalline titanium dioxide films prepared by DC reactive magnetron sputtering. *J. Mater. Sci.-Mater. Electron.* **2012**, *23*, 1898–1904. [CrossRef]

123. Metin, H.; Esen, R. Kimyasal Depolama Yöntemi Teknigiyle Büyütülen cds Filimlerinde Fotoiletkenlik Çalismalari. *Erciyes Üniversitesi Fen Bilimleri Enstitüsü Fen Bilimleri Dergisi* **2003**, *19*, 96–102.

124. Purniawan, A.; Hermastuti, R.; Purwaningsih, H.; Atmono, T.M. Effect of deposition time of sputtering Ag-Cu thin film on mechanical and antimicrobial properties. In *AIP Conference Proceedings 2018*; AIP Publishing LLC: Melville, NY, USA, 2018; Volume 1945, p. 020008.

125. Zheng, X.W.; Li, Z.Q. Structural and electrical transport properties of Nb-doped TiO_2 films deposited on $LaAlO_3$ by rf sputtering. *Appl. Surf. Sci.* **2009**, *255*, 8104–8109. [CrossRef]

126. Singh, S.; Srinivasa, R.S.; Major, S.S. Effect of substrate temperature on the structure and optical properties of ZnO thin films deposited by reactive rf magnetron sputtering. *Thin solid films.* **2007**, *515*, 8718–8722. [CrossRef]

127. Lin, S.S.; Huang, J.L. Effect of thickness on the structural and optical properties of ZnO films by rf magnetron sputtering. *Surf. Coat. Technol.* **2004**, *185*, 222–227. [CrossRef]

128. Yan, B.X.; Luo, S.Y.; Mao, X.G.; Shen, J.; Zhou, Q.F. Unusual photoelectric behaviors of Mo-doped TiO_2 multilayer thin films prepared by RF magnetron co-sputtering: Effect of barrier tunneling on internal charge transfer. *Appl. Phys. A* **2013**, *110*, 129–135. [CrossRef]

129. Zuo, J.; Keil, P.; Grundmeier, G. Synthesis and characterization of photochromic Ag-embedded TiO_2 nanocomposite thin films by non-reactive RF-magnetron sputter deposition. *Appl. Surf. Sci.* **2012**, *258*, 7231–7237. [CrossRef]

130. Sato, Y.; Akizuki, H.; Kamiyama, T.; Shigesato, Y. Transparent conductive Nb-doped TiO$_2$ films deposited by direct-current magnetron sputtering using a TiO$_2$− x target. *Thin Solid Films* **2008**, *516*, 5758–5762. [CrossRef]

131. Takeuchi, M.; Sakai, S.; Matsuoka, M.; Anpo, M. Preparation of the visible light responsive TiO$_2$ thin film photocatalysts by the RF magnetron sputtering deposition method. *Res. Chem. Intermed.* **2009**, *35*, 973. [CrossRef]

132. Ievtushenko, A.I.; Karpyna, V.A.; Lazorenko, V.I.; Lashkarev, G.V.; Khranovskyy, V.D.; Baturin, V.A.; Karpenko, O.Y.; Lunika, M.M.; Avramenko, K.A.; Strelchuk, V.V.; et al. High quality ZnO films deposited by radio-frequency magnetron sputtering using layer by layer growth method. *Thin Solid Films* **2010**, *518*, 4529–4532. [CrossRef]

133. Zhang, W.; Li, Y.; Zhu, S.; Wang, F. Copper doping in titanium oxide catalyst film prepared by dc reactive magnetron sputtering. *Catal. Today* **2004**, *93*, 589–594. [CrossRef]

134. Laird, R.; Belkind, A. Cosputtered films of mixed TiO$_2$/SiO$_2$. *J. Vac. Sci. Technol. A* **1992**, *10*, 1908–1912. [CrossRef]

135. Matsuoka, M.; Kitano, M.; Takeuchi, M.; Anpo, M.; Thomas, J.M. Photocatalytic water splitting on visible light-responsive TiO$_2$ thin films prepared by a RF magnetron sputtering deposition method. *Top. Catal.* **2005**, *35*, 305–310. [CrossRef]

136. Anpo, M.; Takeuchi, M. The design and development of highly reactive titanium oxide photocatalysts operating under visible light irradiation. *J. Catal.* **2003**, *216*, 505–516. [CrossRef]

137. Trupina, L.; Miclea, C.; Amarande, L.; Cioangher, M. Growth of highly oriented iridium oxide bottom electrode for Pb (Zr, Ti) O$_3$ thin films using titanium oxide seed layer. *J. Mater. Sci.* **2011**, *46*, 6830–6834. [CrossRef]

138. Kamisaka, H.; Mizuguchi, N.; Yamashita, K. Electron trapping at the lattice Ti atoms adjacent to the Nb dopant in Nb-doped rutile TiO$_2$. *J. Mater. Sci.* **2012**, *47*, 7522–7529. [CrossRef]

139. Diebold, U. The surface science of titanium dioxide. *Surf. Sci. Rep.* **2003**, *48*, 53–229. [CrossRef]

140. Gillispie, M.A.; van Hest, M.F.; Dabney, M.S.; Perkins, J.D.; Ginley, D.S. rf magnetron sputter deposition of transparent conducting Nb-doped TiO$_2$ films on SrTiO$_3$. *J. Appl. Phys.* **2007**, *101*, 033125. [CrossRef]

141. Jun, T.H.; Lee, K.S. Cr-doped TiO$_2$ thin films deposited by RF-sputtering. *Mater. Lett.* **2010**, *64*, 2287–2289. [CrossRef]

142. Zhu, J.; Chen, F.; Zhang, J.; Chen, H.; Anpo, M. Fe$_3$$^+$-TiO$_2$ photocatalysts prepared by combining sol–gel method with hydrothermal treatment and their characterization. *J. Photochem. Photobiol. A* **2006**, *180*, 196–204. [CrossRef]

143. Seong, S.G.; Kim, E.J.; Kim, Y.S.; Lee, K.E.; Hahn, S.H. Influence of deposition atmosphere on photocatalytic activity of TiO$_2$/SiOx double-layers prepared by RF magnetron sputtering. *Appl. Surf. Sci.* **2009**, *256*, 1–5. [CrossRef]

144. Ohko, Y.; Tatsuma, T.; Fujii, T.; Naoi, K.; Niwa, C.; Kubota, Y.; Fujishima, A. Multicolour photochromism of TiO$_2$ films loaded with silver nanoparticles. *Nat. Mater.* **2003**, *2*, 29–31. [CrossRef] [PubMed]

145. Warren, S.C.; Thimsen, E. Plasmonic solar water splitting. *Energy Environ. Sci.* **2012**, *5*, 5133–5146. [CrossRef]

146. Dionne, J.A.; Atwater, H.A. Plasmonics: Metal-worthy methods and materials in nanophotonics. *Mrs Bull.* **2012**, *37*, 717–724. [CrossRef]

147. Lee, M.K.; Kim, T.G.; Kim, W.; Sung, Y.M. Surface plasmon resonance (SPR) electron and energy transfer in noble metal− zinc oxide composite nanocrystals. *J. Phys. Chem. C* **2008**, *112*, 10079–10082. [CrossRef]

148. Tian, Y.; Tatsuma, T. Mechanisms and applications of plasmon-induced charge separation at TiO$_2$ films loaded with gold nanoparticles. *J. Am. Chem. Soc.* **2005**, *127*, 7632–7637. [CrossRef] [PubMed]

149. Furube, A.; Du, L.; Hara, K.; Katoh, R.; Tachiya, M. Ultrafast plasmon-induced electron transfer from gold nanodots into TiO$_2$ nanoparticles. *J. Am. Chem. Soc.* **2007**, *129*, 14852–14853. [CrossRef]

150. Yin, P.; Chen, L.; Wang, Z.; Qu, R.; Liu, X.; Ren, S. Production of biodiesel by esterification of oleic acid with ethanol over organophosphonic acid-functionalized silica. *Bioresour Technol.* **2012**, *110*, 258–263. [CrossRef]

151. Nishijima, Y.; Ueno, K.; Kotake, Y.; Murakoshi, K.; Inoue, H.; Misawa, H. Near-infrared plasmon-assisted water oxidation. *J. Phys. Chem. Lett.* **2012**, *3*, 1248–1252. [CrossRef]

152. Chen, M.H.; Chen, C.K.; Chen, C.J.; Cheng, L.C.; Wu, C.P.; Cheng, B.H.; Ho, Z.Y.; Tseng, L.M.; Hsu, Y.Y.; Chan, T.S. Plasmon inducing effects for enhanced photoelectrochemical water splitting X-ray absortpion approach to electronic structrures. *ACS Nano* **2012**, *6*, 7362–7372. [CrossRef]

153. Khan, M.R.; Chuan, T.W.; Yousuf, A.; Chowdhury, M.N.; Cheng, C.K. Schottky barrier and surface plasmonic resonance phenomena towards the photocatalytic reaction: Study of their mechanisms to enhance photocatalytic activity. *Catal. Sci. Technol.* **2015**, *5*, 2522–2531. [CrossRef]

154. György, E.; del Pino, A.P. Tunable optical and nano-scale electrical properties of WO_3 and $Ag-WO_3$ nanocomposite thin films. *J. Mater. Sci.* **2011**, *46*, 3560–3567. [CrossRef]

155. Liu, S.X.; Qu, Z.P.; Han, X.W.; Sun, C.L. A mechanism for enhanced photocatalytic activity of silver-loaded titanium dioxide. *Catal. Today* **2004**, *93*, 877–884. [CrossRef]

156. Nooke, A.; Beck, U.; Hertwig, A.; Krause, A.; Krüger, H.; Lohse, V.; Negendank, D.; Steinbach, J. On the application of gold based SPR sensors for the detection of hazardous gases. *Sensor Actuators B-Chem.* **2010**, *149*, 194–198. [CrossRef]

157. Atabaev, T.S.; Hossain, M.A.; Lee, D.; Kim, H.K.; Hwang, Y.H. Pt-coated TiO_2 nanorods for photoelectrochemical water splitting applications. *Results Phys.* **2016**, *6*, 373–376. [CrossRef]

158. Sadeghi, M.; Liu, W.; Zhang, T.G.; Stavropoulos, P.; Levy, B. Role of photoinduced charge carrier separation distance in heterogeneous photocatalysis: Oxidative degradation of CH_3OH vapor in contact with Pt/TiO_2 and cofumed $TiO_2-Fe_2O_3$. *J. Phys. Chem.-US* **1996**, *100*, 19466–19474. [CrossRef]

159. Meng, F.; Sun, Z. Enhanced photocatalytic activity of silver nanoparticles modified TiO_2 thin films prepared by RF magnetron sputtering. *Mater. Chem. Phys.* **2009**, *118*, 349–353. [CrossRef]

160. Singh, J.; Khan, S.A.; Shah, J.; Kotnala, R.K.; Mohapatra, S. Nanostructured TiO_2 thin films prepared by RF magnetron sputtering for photocatalytic applications. *Appl. Surf. Sci.* **2017**, *422*, 953–961. [CrossRef]

161. Li, H.; Li, Z.; Yu, Y.; Ma, Y.; Yang, W.; Wang, F.; Yin, X.; Wang, X. Surface-plasmon-resonance-enhanced photoelectrochemical water splitting from Au-nanoparticle-decorated 3D TiO_2 nanorod architectures. *J. Phys. Chem. C* **2017**, *121*, 12071–12079. [CrossRef]

162. Vahl, A.; Veziroglu, S.; Henkel, B.; Strunskus, T.; Polonskyi, O.; Aktas, O.C.; Faupel, F. Pathways to Tailor Photocatalytic Performance of TiO_2 Thin Films Deposited by Reactive Magnetron Sputtering. *Materials* **2019**, *12*, 2840. [CrossRef]

163. Chen, W.F.; Koshy, P.; Adler, L.; Sorrell, C.C. Photocatalytic activity of V-doped TiO_2 thin films for the degradation of methylene blue and rhodamine B dye solutions. *J. Aust. Ceram. Soc.* **2017**, *53*, 569–576. [CrossRef]

164. Bensouici, F.; Bououdina, M.; Dakhel, A.A.; Tala-Ighil, R.; Tounane, M.; Iratni, A.; Souier, T.; Liu, S.; Cai, W. Optical, structural and photocatalysis properties of Cu-doped TiO_2 thin films. *Appl. Surf. Sci.* **2017**, *395*, 110–116. [CrossRef]

165. Gomes, J.; Lincho, J.; Domingues, E.; Quinta-Ferreira, R.M.; Martins, R.C. N–TiO_2 photocatalysts: A review of their characteristics and capacity for emerging contaminants removal. *Water* **2019**, *11*, 373. [CrossRef]

166. Banerjee, S.; Pillai, S.C.; Falaras, P.; O'shea, K.E.; Byrne, J.A.; Dionysiou, D.D. New insights into the mechanism of visible light photocatalysis. *J. Phys. Chem. Lett.* **2014**, *5*, 2543–2554. [CrossRef]

167. El-Sheikh, S.M.; Zhang, G.; El-Hosainy, H.M.; Ismail, A.A.; O'Shea, K.E.; Falaras, P.; Kontos, A.G.; Dionysiou, D.D. High performance sulfur, nitrogen and carbon doped mesoporous anatase–brookite TiO_2 photocatalyst for the removal of microcystin-LR under visible light irradiation. *J. Hazard. Mater.* **2014**, *280*, 723–733. [CrossRef] [PubMed]

168. Liu, G.; Zhao, Y.; Sun, C.; Li, F.; Lu, G.Q.; Cheng, H.M. Synergistic effects of B/N doping on the visible-light photocatalytic activity of mesoporous TiO_2. *Angew. Chem. Int. Ed.* **2008**, *47*, 4516–4520. [CrossRef]

169. Xu, J.H.; Li, J.; Dai, W.L.; Cao, Y.; Li, H.; Fan, K. Simple fabrication of twist-like helix N, S-codoped titania photocatalyst with visible-light response. *Appl. Catal. B-Environ.* **2008**, *79*, 72–80. [CrossRef]

170. Liu, G.; Han, C.; Pelaez, M.; Zhu, D.; Liao, S.; Likodimos, V.; Kontos, A.G.; Falaras, P.; Dionysiou, D.D. Enhanced visible light photocatalytic activity of CN-codoped TiO_2 films for the degradation of microcystin-LR. *J. Mol. Catal. A-Chem.* **2013**, *372*, 58–65. [CrossRef]

171. Katsanaki, A.V.; Kontos, A.G.; Maggos, T.; Pelaez, M.; Likodimos, V.; Pavlatou, E.A.; Dionysiou, D.D.; Falaras, P. Photocatalytic oxidation of nitrogen oxides on NF-doped titania thin films. *Appl. Catal. B-Environ.* **2013**, *140*, 619–625. [CrossRef]

172. Cavalcante, R.P.; Dantas, R.F.; Bayarri, B.; González, O.; Giménez, J.; Esplugas, S.; Junior, A.M. Synthesis and characterization of B-doped TiO_2 and their performance for the degradation of metoprolol. *Catal. Today* **2015**, *252*, 27–34. [CrossRef]

173. Wieslander, G.; Norbäck, D.; Björnsson, E.; Janson, C.; Boman, G. Asthma and the indoor environment: The significance of emission of formaldehyde and volatile organic compounds from newly painted indoor surfaces. *Int. Arch. Occup. Environ. Health* **1996**, *69*, 115–124. [CrossRef]

174. Wu, K.R.; Hung, C.H. Characterization of N, C-codoped TiO_2 films prepared by reactive DC magnetron sputtering. *Appl. Surf. Sci.* **2009**, *256*, 1595–1603. [CrossRef]

175. Fujishima, A.; Zhang, X.; Tryk, D.A. TiO_2 photocatalysis and related surface phenomena. *Surf. Sci. Rep.* **2008**, *63*, 515–582. [CrossRef]

176. Periyat, P.; Pillai, S.C.; McCormack, D.E.; Colreavy, J.; Hinder, S.J. Improved high-temperature stability and sun-light-driven photocatalytic activity of sulfur-doped anatase TiO_2. *J. Phys. Chem. C* **2008**, *112*, 7644–7652. [CrossRef]

177. Prabakar, K.; Takahashi, T.; Nezuka, T.; Takahashi, K.; Nakashima, T.; Kubota, Y.; Fujishima, A. Visible light-active nitrogen-doped TiO_2 thin films prepared by DC magnetron sputtering used as a photocatalyst. *Renew. Energy* **2008**, *33*, 277–281. [CrossRef]

178. Parker, J.C.; Siegel, R.W. Raman microprobe study of nanophase TiO_2 and oxidation-induced spectral changes. *J. Mater. Res.* **1990**, *5*, 1246–1252. [CrossRef]

179. Sathish, M.; Viswanathan, B.; Viswanath, R.P.; Gopinath, C.S. Synthesis, characterization, electronic structure, and photocatalytic activity of nitrogen-doped TiO_2 nanocatalyst. *Chem. Mater.* **2005**, *17*, 6349–6353. [CrossRef]

180. Di-Valentin, C.; Pacchioni, G.; Selloni, A. Origin of the different photoactivity of N-doped anatase and rutile TiO_2. *Phys. Rev. B* **2004**, *70*, 1–4. [CrossRef]

181. Lindgren, T.; Mwabora, J.M.; Avendano, E.; Jonsson, J.; Hoel, A.; Granqvist, C.G.; Lindquist, S.E. Photoelectrochemical and optical properties of nitrogen doped titanium dioxide films prepared by reactive DC magnetron sputtering. *J. Phys. Chem. B* **2003**, *107*, 5709–5716. [CrossRef]

182. Li, Q.; Shang, J.K. Heavily Nitrogen-Doped Dual-Phase Titanium Oxide Thin Films by Reactive Sputtering and Rapid Thermal Annealing. *J. Am. Ceram. Soc.* **2008**, *91*, 3167–3172. [CrossRef]

183. Hamal, D.B.; Klabunde, K.J. Synthesis, characterization, and visible light activity of new nanoparticle photocatalysts based on silver, carbon, and sulfur-doped TiO_2. *J. Colloid Interface Sci.* **2007**, *311*, 514–522. [CrossRef] [PubMed]

184. Hsu, S.W.; Yang, T.S.; Chen, T.K.; Wong, M.S. Ion-assisted electron-beam evaporation of carbon-doped titanium oxide films as visible-light photocatalyst. *Thin Solid Films* **2007**, *515*, 3521–3526. [CrossRef]

185. Matsunaga, T.; Inagaki, M. Carbon-coated anatase for oxidation of methylene blue and NO. *Appl. Catal. B-Environ.* **2006**, *64*, 9–12. [CrossRef]

186. Lee, M.K.; Park, Y.C. Super-hydrophilic anatase TiO_2 thin film in-situ deposited by DC magnetron sputtering. *Thin Solid Films* **2017**, *638*, 9–16. [CrossRef]

187. Meng, F.; Lu, F. Characterization and photocatalytic activity of TiO_2 thin films prepared by RF magnetron sputtering. *Vacuum.* **2010**, *85*, 84–88. [CrossRef]

188. Sakatani, Y.; Grosso, D.; Nicole, L.; Boissière, C.; de AASoler-Illia, G.J.; Sanchez, C. Optimised photocatalytic activity of grid-like mesoporous TiO_2 films: Effect of crystallinity, pore size distribution, and pore accessibility. *J. Mater. Chem.* **2006**, *16*, 77–82. [CrossRef]

189. Liu, Q.; Wu, X.; Wang, B.; Liu, Q. Preparation and super-hydrophilic properties of TiO2/SnO2 composite thin films. *Mater. Res. Bull.* **2002**, *37*, 2255–2262. [CrossRef]

190. Khadar, M.A.; Shanid, N.M. Nanoscale fine-structure evaluation of RF magnetron sputtered anatase films using HRTEM, AFM, micro-Raman spectroscopy and fractal analysis. *Surf. Coat. Technol.* **2010**, *204*, 1366–1374. [CrossRef]

Review

Graphitic Carbon Nitride Materials for Photocatalytic Hydrogen Production via Water Splitting: A Short Review

Seong Jun Mun and Soo-Jin Park *

Department of Chemistry, Inha University, 100 Inharo, Incheon 22212, Korea; dajoasj@gmail.com
* Correspondence: sjpark@inha.ac.kr; Tel.: +82-32-876-7234

Received: 19 August 2019; Accepted: 23 September 2019; Published: 25 September 2019

Abstract: The generation of photocatalytic hydrogen via water splitting under light irradiation is attracting much attention as an alternative to solve such problems as global warming and to increase interest in clean energy. However, due to the low efficiency and selectivity of photocatalytic hydrogen production under solar energy, a major challenge persists to improve the performance of photocatalytic hydrogen production through water splitting. In recent years, graphitic carbon nitride (g-C_3N_4), a non-metal photocatalyst, has emerged as an attractive material for photocatalytic hydrogen production. However, the fast recombination of photoexcited electron–hole pairs limits the rate of hydrogen evolution and various methods such as modification, heterojunctions with semiconductors, and metal and non-metal doping have been applied to solve this problem. In this review, we cover the rational design of g-C_3N_4-based photocatalysts achieved using methods such as modification, metal and non-metal doping, and heterojunctions, and we summarize recent achievements in their application as hydrogen production photocatalysts. In addition, future research and prospects of hydrogen-producing photocatalysts are also reviewed.

Keywords: graphitic carbon nitride; photocatalysis; H_2 generation; water splitting

1. Introduction

As interest in the fossil fuel depletion and environmental pollution has increased, the development of clean energy has also recently attracted increased attention. It is important to find new alternative energy sources because of the increased use of energy, depletion of fossil fuels, and the need for sustainable energy development [1]. Among the many alternative energy sources, hydrogen-based energy systems are considered candidates for future energy because they are nonpolluting, inexhaustible, efficient, and can provide high-quality energy services in a wide range of applications [2,3]. However, most hydrogen production processes are based on natural gas [4], coal [5], crude oil [6], or the electrolysis of water [7], and unfortunately, the application of most of these processes is limited because heat and electrical energy are required. Thus, photocatalytic hydrogen production using solar energy, a clean energy resource for the foreseeable future, is considered to be an attractive way of solving the global energy issue and environmental pollution [8,9].

The overall water splitting by a photocatalyst under sunlight irradiation enables the production of environmentally friendly molecular hydrogen and does not use fossil fuel [10]. A photocatalytic system should consider the following prerequisites. First, to absorb as many photons as possible, the photocatalyst must have a narrow band-gap; to generate hydrogen from water splitting, the bottom of the conduction band (CB) must be more negative than the reduction potential of H^+/H_2 and the top of the valence band (VB) must be more positive than the oxidation potential of H_2O/O_2 [11]. Second, efficient charge separation and fast charge transport that simultaneously avoid bulk and surface charge recombination are essential to transfer the photogenerated charge to the surface reaction site [12].

Third, because the charge carriers at the interface lack the capacity to boost the transportation process, the charge carriers mostly move via a random path and require a surface chemical reaction that is active between the charge carrier and the water or other molecules [13]. A variety of semiconductor materials such as TiO_2, ZnO, CdS, and WO_3 have been extensively studied for hydrogen generation via photocatalytic water splitting [14–17]. Among them, WO_3 absorbs visible light but has a problem in that the CB is not useful for hydrogen production because it is lower than the H reduction potential [18,19]. In addition, hydrogen evolution through photocatalytic water splitting has been extensively studied for metal oxides, quantum dots, and metal–organic frameworks, etc. However, some methods are difficult to use due to their low efficiency under visible light and the fast recombination rate of the electron–hole pairs [20–25]. Therefore, it is a major challenge to develop photocatalysts that exhibit stable water-splitting performance under visible-light irradiation for the efficient use of solar energy.

Recently, graphitic carbon nitride (g-C_3N_4) has attracted attention as a hydrogen-generating photocatalyst via water splitting. g-C_3N_4 is synthesized by the thermal condensation of nitrogen-rich precursors with a tri-s-triazine ring structure such as cyanamide, dicyandiamide, urea, or thiourea, resulting in a graphene-like structure after exfoliation (Figure 1) [26]. In addition, it has a band gap of ~2.7 eV corresponding to 460 nm in the visible range and high thermal and chemical stability [27].

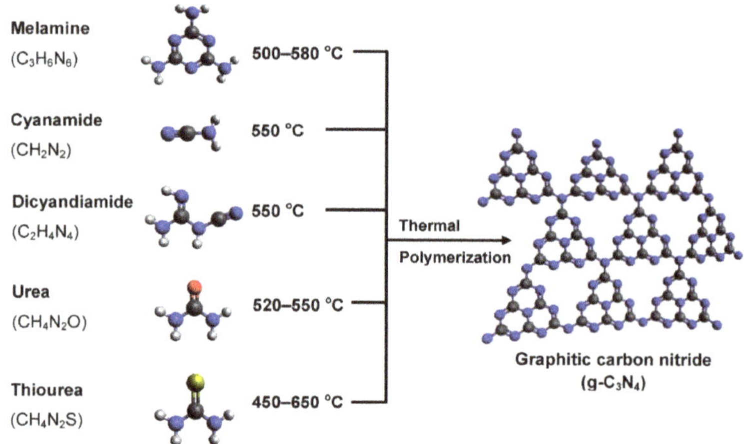

Figure 1. Schematic illustration of the synthesis process from the possible precursors of g-C_3N_4. Reproduced with permission from [26]; copyright (2016), the American Chemical Society.

However, there are some drawbacks to using g-C_3N_4 as a water-splitting photocatalyst. The relatively large band-gap and low charge-carrier mobility limit the electron and hole separation and transport and thus limit the effective use of visible light [28]. Thus, increasing hydrogen production during photocatalytic water splitting under visible-light irradiation is necessary through a variety of methods such as creating heterojunctions with semiconductors and doping with other elements [29–33]. As a result, the focus of this review is on summarizing the current and prospective advances in photocatalysis research based on g-C_3N_4 that make it effective even under visible-light irradiation.

2. The Principles of H_2 Generation via Water Splitting

Photocatalytic reactions can be divided into three parts. The first step is to obtain photons with energies that exceed the photocatalyst's band gap of the electron–hole pairs, the second step is the separation of the carrier in the photocatalyst by transfer, and the third step is the reaction between the carrier and H_2O. In addition, the electron–hole pairs are concurrently combined with each other. The photocatalyst is involved in the production of hydrogen, but the lowest position of the CB should be lower than the reduction position of H_2O/H_2 and the position of the VB should be higher than the potential of H_2O/O_2 [34–40].

Figure 2 shows the band gap and band edge positions of various semiconductor photocatalysts [41]. A variety of these, such as ZnO, TiO_2, and WO_3 have been studied for solar hydrogen production and degradation of organic pollutants [42–45]. However, although there are exceptions for some semiconductor photocatalysts, most of the semiconductor photocatalysts have low efficiency under visible-light irradiation. Therefore, it is a major challenge to develop photocatalysts that efficiently exploit solar energy.

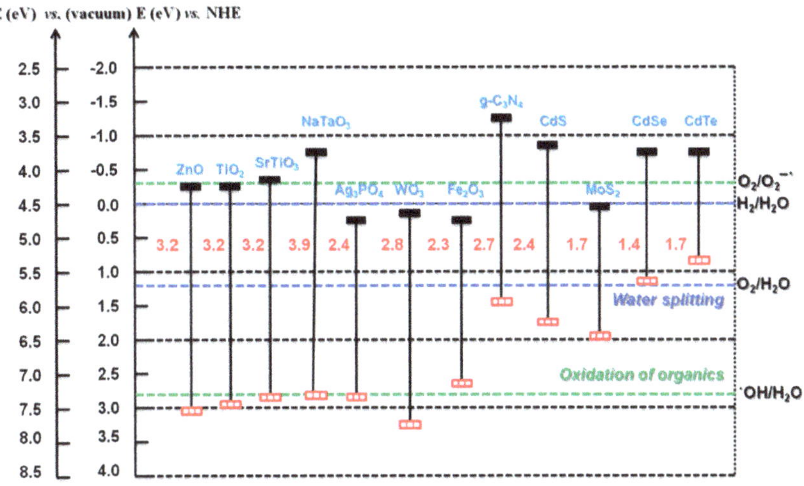

Figure 2. A schematic illustration of the band-gap energy of several typical semiconductor photocatalysts. Reproduced with permission from [41].

Recently, g-C_3N_4, which has a unique electron band structure for photo-oxidation and reduction, has been confirmed by several researchers as an efficient photocatalyst for visible-light activation for photochemical reactions [46]. This achieves the photoexcited state when creating electron–hole pairs where photogenerated electrons are involved in the reduction process while the holes are consumed in the oxidation process [47]. The excited electrons and holes act as reactive species that are highly oxidizing and reducing. The excited electrons and holes travel to the active sites on the surface, thereby splitting the water into oxygen and hydrogen (Figure 3) [48]. However, despite its excellent electron and optical properties, g-C_3N_4 has low efficiency for visible-light utilization and a high recombination speed of photoelectric carrier, resulting in the poor formation of radical species causing redox reaction during the photocatalytic reaction [49]. It has a low specific surface area, provides fewer reactive sites, and reduces light harvesting. In addition, the low bandgap (2.7 eV) of g-C_3N_4 is still quite large for efficient visible-light harvesting and has limited use, leaving much of the visible-light spectrum unexploited.

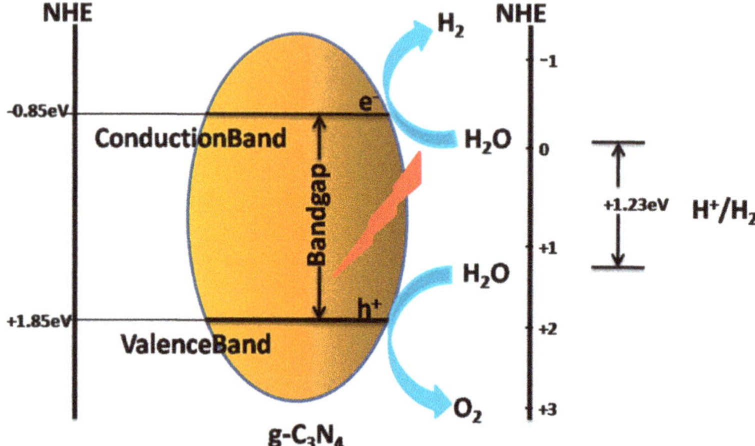

Figure 3. Schematic illustration of the charge-transfer mechanism of neat g-C$_3$N$_4$ as a photocatalyst. Reproduced with permission from [48]; copyright (2016), Royal Society of Chemistry Advances.

3. Hydrogen Generation of g-C$_3$N$_4$-Based Photocatalysts

In recent years, a review of the technological improvements of the photocatalytic efficiency of g-C$_3$N$_4$-based materials has been published, mostly focusing on contaminant removal, the reaction mechanisms, principles of photocatalysis, and the effects of operational parameters [50,51]. Therefore, the aim of this review is to summarize recent trends in the improvement of hydrogen production by photocatalysts using various methods for the purpose of improving g-C$_3$N$_4$-based photocatalytic hydrogen production: (1) modification of g-C$_3$N$_4$; (2) heterojunctions from g-C$_3$N$_4$/semiconductors; and (3) metal- and non-metal-doped g-C$_3$N$_4$.

3.1. Modification of g-C$_3$N$_4$

Improving the photocatalytic activity of g-C$_3$N$_4$ by introducing various nanostructures such as nanoparticles, nanosheets, nanorods, and nanowires has recently been studied [52–56]. Surface modification of the catalytic structure and morphology has the potential to promote charge separation and narrow the band gap due to increased surface area and efficient charge-carrier separation [57,58].

In 2016, Han et al. [59] reported an atomically thin mesoporous nanomesh of g-C$_3$N$_4$ for hydrogen evolution by highly efficient photocatalysts (Figure 4a) fabricated via the solvothermal exfoliation of mesoporous g-C$_3$N$_4$ prepared by the thermal polymerization of freeze-dried nanostructured precursors. The delamination of the layer material to provide the two-dimensional single-atom sheet has led to unique physical properties such as a large surface area, a very high unique carrier mobility, and a significant change in the energy band structure [60]. The mesoporous g-C$_3$N$_4$ nanomesh shows inherent structural advantages, electron transfer capability, and efficient light harvesting. Figure 4b shows the electronic band structure of the monolayer mesoporous g-C$_3$N$_4$ nanomesh and bulk counterparts. The band gap is 2.75 eV for the monolayer mesoporous g-C$_3$N$_4$ nanomesh and 2.59 eV for the bulk counterpart, as determined from optical absorption spectra. The VB of the monolayer mesoporous g-C$_3$N$_4$ nanomesh (2.41 eV) identified via X-ray photoelectron spectroscopy is also 0.35 eV higher than the bulk counterparts (2.06 eV). The CB is upshifted by 0.51 eV when considering the 0.16 eV increase in the VB and a negative shift of 0.35 eV. The monolayer mesoporous g-C$_3$N$_4$ nanomesh exhibits significantly improved the light-harvesting ability mainly due to the multiple scattering effect and the presence of defect sites associated with the mesoporous surface. A 30 h reaction was performed with intermittent evacuation every 5 h to confirm the hydrogen production ability of mesoporous g-C$_3$N$_4$

nanomesh under visible-light irradiation (Figure 4c). As a result, the 2.6 mmol H_2 gas (59 mL) produced by the atomically thin mesoporous g-C_3N_4 nanomesh was not visibly deactivated and the H_2 gas was generated continuously. Wavelength-dependent H_2 evolution shows the optical absorption spectrum of monolayer g-C_3N_4 nanomesh, indicating that the H_2 generation is driven by photoinduced electrons in g-C_3N_4 (Figure 4d). In conclusion, the mesoporous g-C_3N_4 nanomesh produces an atomically thin mesoporous layer during the freeze-dried assembly and solvothermal exfoliation. Its good application benefits from structural advantages for light harvesting, electron transport, and accessible reaction sites [61]. This new type of mesoporous g-C_3N_4 nanomesh could be applied to photocatalytic and various engineering fields.

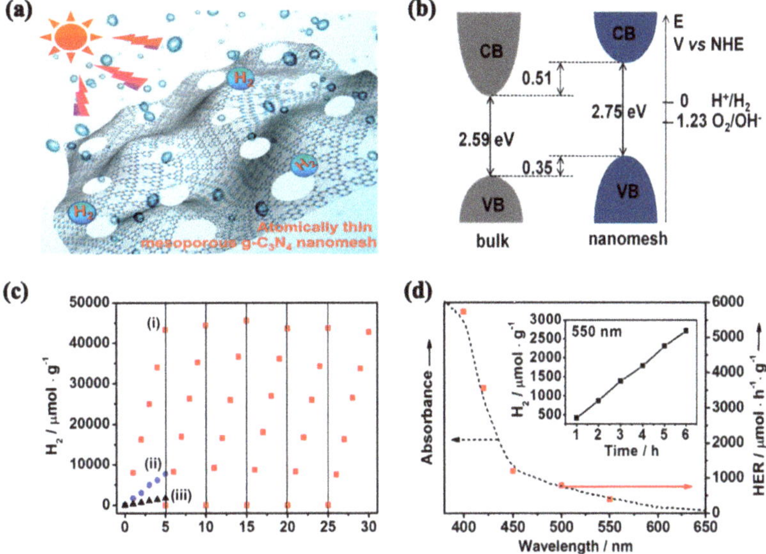

Figure 4. (**a**) Schematic illustration of atomically thin mesoporous g-C_3N_4 nanomesh photocatalyst and (**b**) a band gap schematic of the monolayer mesoporous g-C_3N_4 nanomesh and bulk counterparts. (**c**) Hydrogen production rate of the monolayer mesoporous g-C_3N_4 nanomesh, the bulk counterpart, and the traditional g-C_3N_4 bulk under visible-light irradiation. (**d**) H_2 evolution rate on the monolayer mesoporous g-C_3N_4 nanomesh with wavelength dependence. Reproduced with permission from [59]; copyright (2016), American Chemical Society.

In 2018, Zhao et al. [62] reported the fabrication of a mesoporous g-C_3N_4 consisting of hollow nanospheres (MCNHN) via a simple vapor-deposition method that improved hydrogen production under visible-light irradiation. Figure 5a shows the photocatalytic hydrogen evolution by MCNHN under visible-light irradiation. Both MCNHN and bulk g-C_3N_4 achieved a stable average rate of hydrogen production within 4 h, but the hydrogen evolution of MCNHN was 659.8 μmol g^{-1} h^{-1}, which is 22.3 times greater than bulk g-C_3N_4 (29.6 μmol g^{-1} h^{-1}). The excellent hydrogen production activity of MCNHN is due to its well-defined structure. The increased surface area provides more active sites in the photocatalytic reaction, thereby allowing more light to be harvested. Moreover, the planarized unit layer and the decreased interlayer space of g-C_3N_4 crystals facilitate the transfer and separation of photoinduced charge carriers in MCNHN. As a result, photocatalytic hydrogen generation is significantly improved due to the large surface area and decreased interlayer space of g-C_3N_4. Figure 5b shows the proposed photocatalytic mechanism of H_2 evolution for MCNHN based on the aforementioned results and the literature. The active site of MCNHN absorbs visible light. Electrons in the VB are excited to the CB by absorption of photons, and are then transferred to the Pt nanoparticles loaded on the surface of MCNHN; the corresponding photoexcited holes remain in the VB. The electron-rich Pt nanoparticles become active sites where water can be split into hydrogen. In addition, multiple reflections of visible light in the MCNHN with Pt nanoparticles improves light absorption.

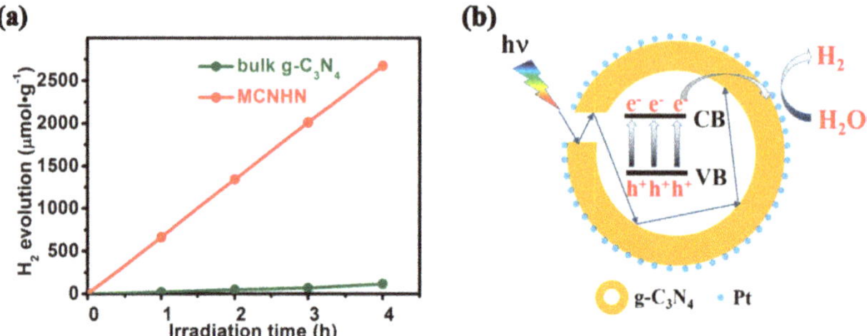

Figure 5. (**a**) Time course of H₂ evolution and (**b**) a schematic mechanism for photocatalytic H₂ evolution on MCNHN. Reproduced with permission from [62]; copyright (2018), Elsevier.

3.2. Heterojunctions and Photocatalysis

Electron–hole charge pairs formed by the photocatalytic hydrogen evolution reaction are transferred to the surface of the photocatalyst or else recombine with each other. To better understand this point, let us illustrate it by reviewing the presentation in [63]: a comparison of the influence of gravitational force on a man jumping off the ground and electrons jumping from the VB to the CB (Figure 6a,b, respectively). If a man (electron) jumps from the ground (VB) into the sky (CB), it will return to the floor quickly (recombine with the hole) due to gravitational force. However, a stool (semiconductor B) can be provided to get the man off the ground (separate the photogenerated electron–hole pair), as illustrated in Figure 6c,d, respectively. Subsequently, the aforementioned man will land again on the stool rather than the ground (the electron–hole pair recombination will be inhibited). Preventing electron–hole recombination is an urgent issue, but it can be achieved by the proper design of materials [64–66]. Many methods have been proposed to achieve better separation of the photogenerated electron–hole pairs in semiconductor photocatalysts, such as element combining, metal and non-metal doping, and heterojunctions [67–72]. Among these strategies, heterojunctions in photocatalysts have proved to be one of the most promising methods for efficient photocatalyst preparation due to their improved separation of electron–hole pairs [73].

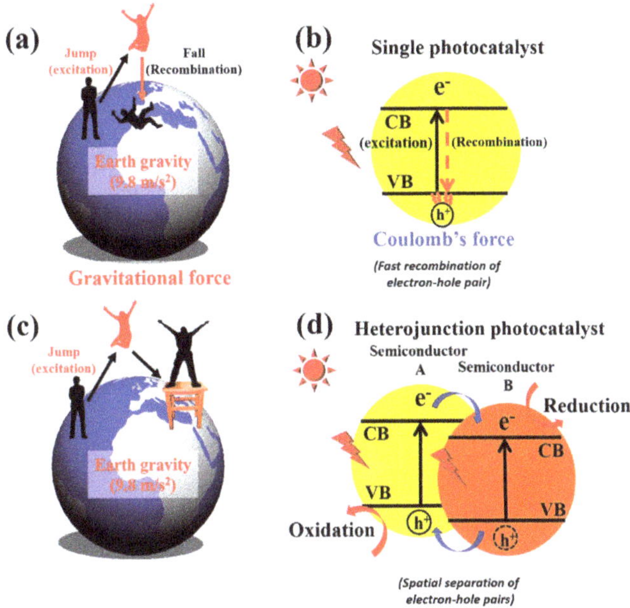

Figure 6. Schematics of (**a**) the influence of gravitational force on a man jumping; (**b**) recombination of photocatalyst electron–hole pair; (**c**) using stool to keep a man from returning to the ground; and (**d**) electron–hole pairs separated in a heterojunction photocatalyst. Reproduced with permission from [63]; copyright (2017), John Wiley & Sons, Inc.

3.2.1. Semiconductor Heterojunction Photocatalysts

Suppressing the electron–hole recombination rate is the most important solution to increase photocatalytic efficiency. Bulk g-C_3N_4 has low ability to collect visible light, low charge-transport properties, and small surface area, so there have been many studies to make it an efficient photocatalyst [74]. Various strategies have been proposed to achieve better electron–hole pair separation such as element combining, metal doping, and creating heterojunctions. Among these strategies, g-C_3N_4/semiconductor heterojunctions have shown the improved separation capability of electron–hole pairs; the charge carrier is transferred through the heterostructure interface to inhibit recombination, thereby improving the photocatalytic performance [75–77]. In addition, a g-C_3N_4/semiconductor heterostructure can be formed by combining a visible-light excited photocatalyst semiconductor material having a narrow band-gap and a photoexcited photocatalyst having a large band-gap in a coupling process; the connection between the two different kinds of photocatalyst having different band structures induces a new band arrangement [78,79].

In 2017, Zhang et al. [80] reported the in situ synthesis of a g-C_3N_4/TiO_2 heterostructure photocatalyst which greatly improved the hydrogen evolution performance under visible light. The g-C_3N_4 nanosheets were synthesized by calcining urea at 550 °C for 4 h. Two hundred milligrams of the as-prepared g-C_3N_4 nanosheets were dispersed in 20 mL ethanol and sonicated for 1 hour to obtain a homogeneous suspension. Under continuous stirring, 40 mL of ammonia solution (~28 wt%) and tetrabutyl titanate (TBT) (0, 100, 200, 300 and 400 μL) were added and stirred for 12 h to achieve the in situ synthesis of amorphous TiO_2. The obtained products were expressed as CNTO-*x* (*x* = 0–4) according to the TBT content. As shown in Figure 7a, the shape of the CNTO-2 sample seen in a transmission electron microscopy (TEM) image shows that the TiO_2 nanoparticles are uniformly distributed in the g-C_3N_4 nanosheets. As a result, there is uniform interfacial contact between the

TiO_2 phase and the g-C_3N_4 phase. Figure 7b shows the average rate of hydrogen production within 3 h. Pure TiO_2 does not react with visible light and produces negligible H_2, while CNTO-0 exhibits a low hydrogen production rate of 15 μmol h^{-1} due to the fast recombination of photogenerated charge carriers. In contrast, the CNTO-2 sample exhibits significantly improved hydrogen production performance at 40 μmol h^{-1}. However, as the amount of TiO_2 is further increased, TiO_2 occupies the surface of g-C_3N_4 resulting in less active sites for H_2 evolution. The proposed mechanism of heterostructure composites is also shown in Figure 7c. According to previous reports, the CB and VB potentials of g-C_3N_4 and TiO_2 are −1.12 and +1.58 V, and −0.29 and +2.91 V, respectively. Under visible light irradiation, only g-C_3N_4 can absorb light to generate electron–hole pairs. However, in pure g-C_3N_4, photogenerated electrons and holes recombine rapidly, and only a few of the electrons participate in the reaction, resulting in low reactivity. When g-C_3N_4 is modified by TiO_2 to form a heterojunction structure, the CB edge of g-C_3N_4 is more negative than TiO_2, so that electrons excited in the CB of g-C_3N_4 can be injected directly into the CB of TiO_2. Consequently, Pt^{2+} adsorbed on the surface is reduced by electrons transferred from the CB of TiO_2, and newly formed Pt nanoparticles are deposited on the surface of TiO_2 as an efficient cocatalyst for hydrogen production. The electrons then accumulate in Pt nanoparticles and participate in hydrogen evolution. Therefore, the photocatalytic activity of the g-C_3N_4/TiO_2 composite with Pt nanoparticles as a cocatalyst is significantly improved.

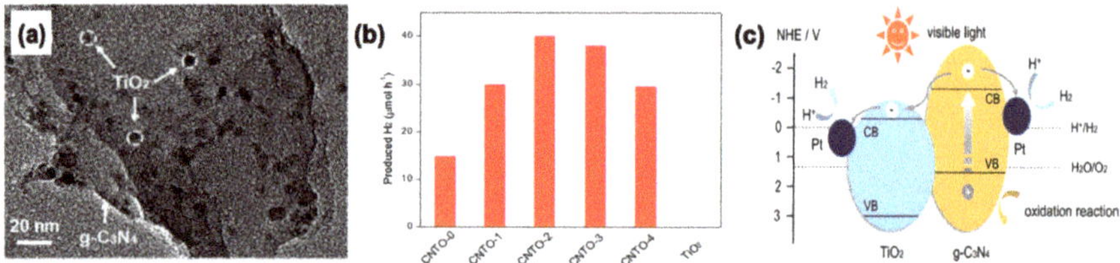

Figure 7. (**a**) A TEM image of CNTO-2, (**b**) H_2 evolution rates of the CNTO-*x* samples under visible light ($\lambda \geq 420$ nm), and (**c**) an illustration of the g-C_3N_4/TiO_2 heterojunction system. Reproduced with permission from [80]; copyright (2017), The Royal Society of Chemistry.

3.2.2. Z-Scheme Heterojunction Photocatalysts

In 2017, Lu et al. [81] reported a Z-scheme photocatalyst that improved the photocatalytic hydrogen production of g-C_3N_4 nanosheets by loading porous silicon (PSi). The Z-scheme heterostructure improved the photocatalytic H_2 evolution performance by loading PSi onto the g-C_3N_4 photocatalyst. g-C_3N_4/PSi composites were prepared by the facile polycondensation reaction of PSi with urea at various PSi content ratios and included pure g-C_3N_4 that was not PSi loaded for comparison. The photocatalytic performance of the g-C_3N_4/PSi composites and pure g-C_3N_4 in Figure 8a was evaluated by H_2 evolution from water under visible-light irradiation. For composite materials loaded with PSi on g-C_3N_4 nanosheets, the rate of H_2 evolution was better than that of pure g-C_3N_4 (427.28 μmol g^{-1} h^{-1}). In particular, the g-C_3N_4/2.50 wt% composite exhibited the highest photocatalytic activity with a hydrogen evolution rate of 870.58 μmol g^{-1} h^{-1}, which is around twice as high as that of pure g-C_3N_4. However, in the case of the Si-based photocatalyst, a passive oxide film was formed on the Si surface, and thus the stability suffered. When the PSi content was larger than 2.50 wt%, the H_2 generation activity was reduced. Figure 8b depicts an energy band diagram of g-C_3N_4/PSi with the redox potential of the photocatalytic reaction. The Z-scheme heterostructure system is recognized as the photocatalytic mechanism for the g-C_3N_4/PSi composite, and the electrons excited from the CB of PSi in the photocatalyst system can be transferred to the VB of g-C_3N_4. In addition, the holes generated in g-C_3N_4 can move to the CB of PSi through the interface formed between g-C_3N_4 and PSi. The recombination at the interface between the electrons and the holes accumulates a large number of bonds and acts as a recombination center for the electron–hole pairs [82,83]. As a result, the efficiency of the photogenerated electron–hole pairs is improved, thereby improving photocatalytic hydrogen production under visible-light irradiation.

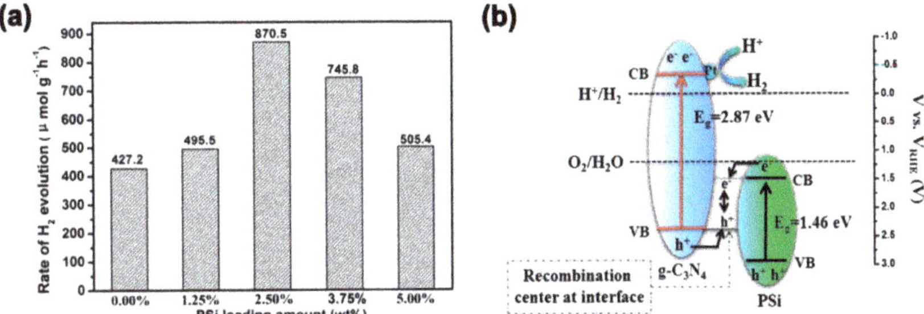

Figure 8. (**a**) Photocatalytic H_2 evolution with 100 mg pure g-C_3N_4 and g-C_3N_4/PSi composite photocatalysts under visible light (400 nm) and (**b**) a schematic diagram of the g-C_3N_4/TiO_2 heterojunction system. Reproduced with permission from [81]; copyright (2017), Elsevier.

3.3. Metal- and Non-Metal-Doped g-C_3N_4

Among the strategies for making g-C_3N_4 as a photocatalyst capable of effective hydrogen production, sufficient doping with metallic and nonmetallic elements is known to enhance the photocatalytic activity of g-C_3N_4. Metal doping is an effective strategy to adjust the electronic structure of g-C_3N_4 and promotes surface kinetics to accelerate photogenerated electron transfer and provide active sites for better photocatalytic hydrogen production. In addition, the light transmittance can be maximized since the spatial distribution and the particle size of the metal can be finely controlled to provide a sufficient active size.

In 2016, Li et al. [84] reported water splitting by Cu- and Fe-doped g-C_3N_4 visible-light-activated photocatalysts. Figure 9a shows the mechanism of water splitting by light-driven catalysis with Fe- and Cu-doped g-C_3N_4. Under visible-light irradiation, water is converted to H_2 and H_2O_2, and then H_2O_2 is further converted to O_2 and H_2O via the photocatalytic imbalance path. After absorbing visible light, g-C_3N_4 forms excited electrons and holes by electron catalysis, and the electrons move from the energy potential difference between g-C_3N_4 and Fe or Cu to the metal Fe or Cu sites. The potential of these electrons is around -0.25 eV and has enough force to induce H_2O_2 disproportionation to form $\cdot$OH and OH^-. In the hole catalytic process (HCP), OH^- and H_2O_2 could form the $\cdot O_2^-$ and H_2O species reaction with the holes. Finally, O_2^- and OH can recombine to form O_2. Electron catalysis is an energy-consuming process whereas HCP and recombination processes can be viewed as energy-releasing processes. Figure 9b shows the oxygen and hydrogen evolution rates of Fe/C_3N_4 (0.37 wt%) and Cu/C_3N_4 (0.42 wt%) under visible-light irradiation ($\lambda \geq 420$ nm) for 12 h. In this case, the production of hydrogen and oxygen by the Cu/C_3N_4 and Fe/C_3N_4 photocatalysts were 1.4 and 0.5 μmol, and 2.1 and 0.8 μmol, respectively. In addition, the potential of the Fe/g-C_3N_4 photocatalyst is obviously lower than those of the g-C_3N_4 and Cu/g-C_3N_4 photocatalysts, which leads to the O_2 and H_2 evolution activity over the Fe/g-C_3N_4 photocatalyst being clearly higher than that over the g-C_3N_4 and Cu/g-C_3N_4 photocatalysts. The findings of this study give new insight into the designing of efficient catalysts for overall water splitting.

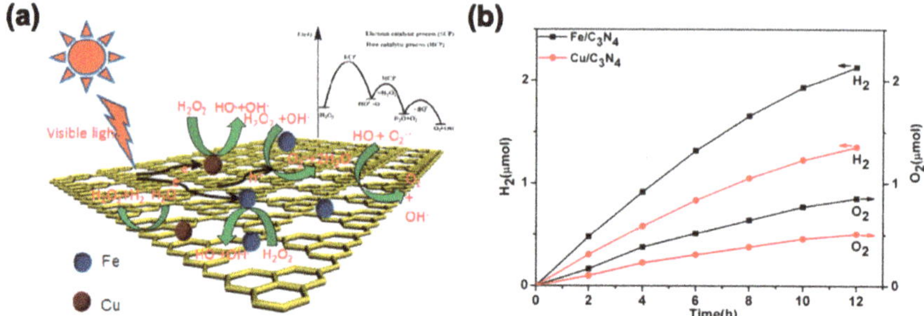

Figure 9. (a) A schematic diagram of the water splitting mechanism by Fe/C$_3$N$_4$ and Cu/C$_3$N$_4$ photocatalysts under visible-light irradiation. (b) Production of H$_2$ and O$_2$ by water splitting by the Fe/C$_3$N$_4$ and Cu/C$_3$N$_4$ photocatalysts under visible-light irradiation for 12 h. Reproduced with permission from [84]; copyright (2015), American Chemical Society.

Non-metal doping is a useful strategy to adjust the electronic structure of g-C$_3$N$_4$ and to increase the photocatalytic effect by promoting the reaction surface. When the non-metal elements B, N, O, P, and S are used to dope g-C$_3$N$_4$, the photocatalyst is efficiently optimized by lowering the charge recombination rate due to optical absorption and accelerated charge mobility, and thus the amount of H$_2$ produced can be increased [85,86]. Consequently, the potential of the Fe/g-C$_3$N$_4$ photocatalyst is obviously lower than those of the g-C$_3$N$_4$ and Cu/g-C$_3$N$_4$ photocatalysts. This indicates that the Fe/g-C$_3$N$_4$ photocatalyst has higher activity on photocatalytic hydrogen evolution than the g-C$_3$N$_4$ and Cu/g-C$_3$N$_4$ photocatalysts. The findings of this study give new insights into designing efficient photocatalytic hydrogen generation and catalysts through overall water splitting.

In 2018, Feng et al. [87] reported P nanostructures with P-doped g-C$_3$N$_4$ as light photocatalysts for H$_2$ evolution. P nanostructures and P-doped g-C$_3$N$_4$ (P@P-g-C$_3$N$_4$) were synthesized via a solid reaction, and P@P-g-C$_3$N$_4$ showed increased optical absorption, high-efficiency transmission, and efficient separation of photogenerated electron–hole pairs. When C atoms are replaced with P atoms (the gray and red balls in Figure 10a, respectively) in the base frame of g-C$_3$N$_4$, the extra electrons are decentralized into a π-conjugated triazine ring and generate a positive-charge P$^+$ center, thereby facilitating rapid separation of the photogenerated excited electrons. Furthermore, efficient band gap transfers between the P and P-doped g-C$_3$N$_4$ leads to a significant improvement in photoactivity (Figure 10b). P-doped g-C$_3$N$_4$ photoexcited electrons can be delivered to phosphorus via intimate contact because the CB edge of g-C$_3$N$_4$ (-1.2 V vs. normal hydrogen electrode (NHE)) is more negative than P (-0.25 V vs. NHE) which provides an interface under the buildup of the internal electric field. Thus, the extra electrons superimposed on the P surface can easily be captured by the oxygen molecules in the solution and react with $\cdot$O^{2-} and $\cdot$OH. Figure 10c,d shows the hydrogen evolution yield and the improvement in hydrogen production ability of the photocatalysts prepared at different weight ratios of P/g-C$_3$N$_4$. P@P-g-C$_3$N$_4$-15 showed the highest hydrogen production rate (941.80 μmol h^{-1} g^{-1}), which is around four times that of conventional g-C$_3$N$_4$.

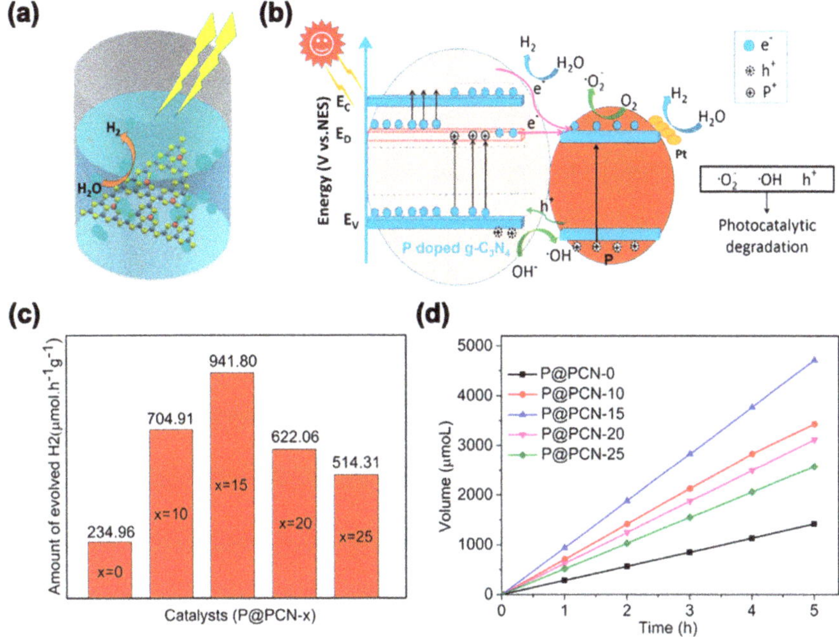

Figure 10. (**a**,**b**) A schematic of the mechanism of H_2 evolution by the P@P-g-C_3N_4 catalyst; (**c**) comparison of the evolution rates of H_2; and (**d**) H_2 evolution rate of the P@P-g-C_3N_4 composites. Reproduced with permission from [87]; copyright (2018), American Chemical Society.

4. Summary and Perspectives

Photocatalytic action is a key factor for the future of environmental pollution and hydrogen generation due to water splitting. Over the past several years, photocatalytic reactions have emerged as a promising method to generate hydrogen, and interest in the photocatalyst g-C_3N_4 has received attention in a variety of scientific disciplines. However, a major problem that limits the rate of production of H_2 by g-C_3N_4-based photocatalysis is the fast recombination of photoexcited electron–hole pairs. This problem can be solved in a variety of ways, including modification, heterojunctions, and metal and non-metal doping. Table 1 summarizes the literature on the photocatalytic H_2 generation of g-C_3N_4-based materials. We reviewed the rational design of photocatalysts for efficient H_2 generation though a variety of methods. Furthermore, the improvement of g-C_3N_4-based photocatalysts will likely result from advances in science. Herein, we have covered the recent progress of g-C_3N_4-based materials involved in hydrogen production in improving their overall photocatalytic activity and have characterized their performance and importance. We hope that this report will support further research efforts related to photocatalytic development.

Table 1. Photocatalytic H_2 generation of $g-C_3N_4$-based materials.

Entry	Type	Mass Fraction of $g-C_3N_4$	Mass of Photocatalyst	Reactant Solution	Light Source	H_2 Generation Rate (μmol h^{-1})	Reference
Figure 4	Monolayer mesoporous $g-C_3N_4$ nanomesh	100 wt%	0.01 g	100 mL of 10 vol% triethanolamine aqueous solution; 3 wt% Pt as a cocatalyst	300 W Xe lamp (>420 nm)	85.10	[59]
Figure 5	Mesoporous $g-C_3N_4$ comprising hollow nanospheres	100 wt%	0.1 g	100 mL of 10 vol.% triethanolamine aqueous solution; 3 wt% Pt as a cocatalyst	300 W Xe lamp (>420 nm)	65.98	[62]
Figure 7	$g-C_3N_4$ nanosheets/TiO_2	50 wt%	0.05 g	100 mL of 10 vol% triethanolamine aqueous solution; 3 wt.% Pt as a cocatalyst	300 W Xe lamp (>420 nm)	40	[80]
Figure 8	Porous Si-loaded $g-C_3N_4$	97.50 wt%	0.1 g	100 mL of 10 vol% triethanolamine aqueous solution; 3 wt% Pt as a cocatalyst	300 W Xe lamp (>400 nm)	87.05	[81]
Figure 9	Fe-doped $g-C_3N_4$ / Cu-doped $g-C_3N_4$	99.63 wt% / 99.58 wt%	0.01 g	Pure water; without other cocatalyst	300 W Xe lamp (>420 nm)	0.175	[84]
Figure 10	P@P-doped $g-C_3N_4$	75 wt%	0.1 g	100 mL of 10 vol% triethanolamine aqueous solution; 1 wt% Pt as a cocatalyst	300 W Xe lamp (>420 nm)	94.18	[87]

Funding: This research was supported by the Technology Innovation Program (or Industrial Strategic Technology Development Program) (10080293), Development of carbon-based non phenolic electrode materials with 3000 m^2g^{-1} grade surface area for energy storage device funded by the Ministry of Trade, Industry and Energy (MOTIE, Korea) and the Commercialization Promotion Agency for R&D Outcomes (COMPA) funded by the Ministry of Science and ICT (MSIT) 2018_RND_002_0064, Development of 800 mA·h·g^{-1} pitch carbon coating.

Conflicts of Interest: The authors declare no conflict of interest.

References

1. Armaroli, N.; Balzani, V. The future of energy supply: Challenges and opportunities. *Chem.-Int. Ed.* **2007**, *46*, 52–66. [CrossRef] [PubMed]
2. Kudo, A.; Miseki, Y. Heterogeneous photocatalyst materials for water splitting. *Chem. Soc. Rev.* **2009**, *38*, 253–278. [CrossRef] [PubMed]
3. Zhou, H.; Fan, T.; Zhang, D. Biotemplated materials for sustainable energy and environment: Current status and challenges. *ChemSusChem* **2011**, *4*, 1344–1387. [CrossRef] [PubMed]
4. Dicks, A.L. Hydrogen generation from natural gas for the fuel cell systems of tomorrow. *J. Power Sources* **1996**, *61*, 113–124. [CrossRef]
5. Lin, S.; Harada, M.; Suzuki, Y.; Hatano, H. Hydrogen production from coal by separating carbon dioxide during gasification. *Fuel* **2002**, *81*, 2079–2085. [CrossRef]
6. Rana, M.S.; Sámano, V.; Ancheyta, J.; Diaz, J.A.I. A review of recent advances on process technologies for upgrading of heavy oils and residua. *Fuel* **2007**, *86*, 1216–1231. [CrossRef]
7. Zeng, K.; Zhang, D. Recent progress in alkaline water electrolysis for hydrogen production and applications. *Prog. Energy Combust. Sci.* **2010**, *36*, 307–326. [CrossRef]
8. Acar, C.; Dincer, I.; Naterer, G.F. Review of photocatalytic water-splitting methods for sustainable hydrogen production. *Int. J. Energy Res.* **2016**, *40*, 1449–1473. [CrossRef]
9. Tong, H.; Ouyang, S.; Bi, Y.; Umezawa, N.; Oshikiri, M.; Ye, J. Nano-photocatalytic materials: Possibilities and challenges. *Adv. Mater.* **2012**, *24*, 229–251. [CrossRef]
10. Lubitz, W.; Tumas, W. Hydrogen: An overview. *Chem. Rev.* **2007**, *107*, 3900–3903. [CrossRef]
11. Christoforidis, K.C.; Fornasiero, P. Photocatalytic hydrogen production: A rift into the future energy supply. *ChemCatChem* **2017**, *9*, 1523–1544. [CrossRef]
12. Maeda, K.; Domen, K. Photocatalytic Water Splitting: Recent Progress and Future Challenges. *J. Phys. Chem. Lett.* **2010**, *1*, 2655–2661. [CrossRef]
13. Wei, P.; Liu, J.; Li, Z. Effect of Pt loading and calcination temperature on the photocatalytic hydrogen production activity of TiO_2 microspheres. *Ceram. Int.* **2013**, *39*, 5387–5391. [CrossRef]
14. Choi, S.K.; Yang, H.S.; Kim, J.H.; Park, H. Organic dye-sensitized TiO_2 as a versatile photocatalyst for solar hydrogen and environmental remediation. *Appl. Catal. B Environ.* **2012**, *121*, 206–213. [CrossRef]
15. Liu, D.; Zhang, S.; Wang, J.; Peng, T.; Li, R. Direct Z-Scheme 2D/2D Photocatalyst Based on Ultrathin g-C_3N_4 and WO_3 Nanosheets for Efficient Visible-Light-Driven H_2 Generation. *ACS Appl. Mater. Interfaces* **2019**, *11*, 27913–27923. [CrossRef] [PubMed]
16. Kudo, A. Photocatalyst materials for water splitting. *Catal. Surv. Asia* **2003**, *7*, 31–38. [CrossRef]
17. Yuan, L.; Han, C.; Yang, M.-Q.; Xu, Y.-J. Photocatalytic water splitting for solar hydrogen generation: Fundamentals and recent advancements. *Int. Rev. Phys. Chem.* **2016**, *35*, 1–36. [CrossRef]
18. Huang, L.; Xu, H.; Li, Y.; Li, H.; Cheng, X.; Xia, J.; Cai, G. Visible-light-induced WO_3/g-C_3N_4 composites with enhanced photocatalytic activity. *Dalton Trans.* **2013**, *42*, 8606–8616. [CrossRef]
19. Wang, G.; Ling, Y.; Wang, H.; Yang, X.; Wang, C.; Zhang, J.Z.; Li, Y. Hydrogen-treated WO_3 nanoflakes show enhanced photostability. *Energy Environ. Sci.* **2012**, *5*, 6180–6187. [CrossRef]
20. Wang, C.C.; Yi, X.H.; Wang, P. Powerful combination of MOFs and C_3N_4 for enhanced photocatalytic performance. *Appl. Catal. B Environ.* **2019**, *247*, 24–48. [CrossRef]
21. Pant, B.; Park, M.; Kim, H.Y.; Park, S.J. CdS-TiO_2 NPs decorated carbonized eggshell membrane for effective removal of organic pollutants: A novel strategy to use a waste material for environmental remediation. *J. Alloy. Compd.* **2017**, *699*, 73–78. [CrossRef]

22. Pullen, S.; Fei, H.; Orthaber, A.; Cohen, S.M.; Ott, S. Enhanced photochemical hydrogen production by a molecular diiron catalyst incorporated into a metal–organic framework. *J. Am. Chem. Soc.* **2013**, *135*, 16997–17003. [CrossRef]

23. Cao, S.W.; Yuan, Y.P.; Fang, J.; Shahjamali, M.M.; Boey, F.Y.; Barber, J.; Xue, C. In-situ growth of CdS quantum dots on g-C$_3$N$_4$ nanosheets for highly efficient photocatalytic hydrogen generation under visible light irradiation. *Int. J. Hydrog. Energy* **2013**, *38*, 1258–1266. [CrossRef]

24. Wang, W.; Jimmy, C.Y.; Shen, Z.; Chan, D.K.; Gu, T. g-C$_3$N$_4$ quantum dots: Direct synthesis, upconversion properties and photocatalytic application. *Chem. Commun.* **2014**, *50*, 10148–10150. [CrossRef] [PubMed]

25. Zhou, L.; Tian, Y.; Lei, J.; Wang, L.; Liu, Y.; Zhang, J. Self-modification of g-C$_3$N$_4$ with its quantum dots for enhanced photocatalytic activity. *Catal. Sci. Technol.* **2018**, *8*, 2617–2623. [CrossRef]

26. Ong, W.J.; Tan, L.L.; Ng, Y.H.; Yong, S.T.; Chai, S.P. Graphitic carbon nitride (g-C$_3$N$_4$)-based photocatalysts for artificial photosynthesis and environmental remediation: Are we a step closer to achieving sustainability? *Chem. Rev.* **2016**, *116*, 7159–7329. [CrossRef] [PubMed]

27. Naseri, A.; Samadi, M.; Pourjavadi, A.; Moshfegh, A.Z.; Ramakrishna, S. Graphitic carbon nitride (g-C$_3$N$_4$)-based photocatalysts for solar hydrogen generation: Recent advances and future development directions. *J. Mater. Chem. A* **2017**, *5*, 23406–23433. [CrossRef]

28. Liao, G.; Chen, S.; Quan, X.; Yu, H.; Zhao, H. Graphene oxide modified g-C$_3$N$_4$ hybrid with enhanced photocatalytic capability under visible light irradiation. *J. Mater. Chem.* **2012**, *22*, 2721–2726. [CrossRef]

29. Bhunia, M.K.; Yamauchi, K.; Takanabe, K. Harvesting solar light with crystalline carbon nitrides for efficient photocatalytic hydrogen evolution. *Angew. Chem. Int. Ed. Engl.* **2014**, *53*, 11001–11005. [CrossRef]

30. Hen, S.; Wang, C.; Bunes, B.R.; Li, Y.; Wang, C.; Zang, L. Enhancement of visible-light-driven photocatalytic H$_2$ evolution from water over g-C$_3$N$_4$ through combination with perylene diimide aggregates. *Appl. Catal. A Gen.* **2015**, *498*, 63–68.

31. Jiang, L.; Yuan, X.; Pan, Y.; Liang, J.; Zeng, G.; Wu, Z.; Wang, H. Doping of graphitic carbon nitride for photocatalysis: A reveiw. *Appl. Catal. B Environ.* **2017**, *217*, 388–406. [CrossRef]

32. Zhu, Y.-P.; Ren, T.-Z.; Yuan, Z.-Y. Mesoporous phosphorus-doped g-C$_3$N$_4$ nanostructured flowers with superior photocatalytic hydrogen evolution performance. *ACS Appl. Mater. Interfaces* **2015**, *7*, 16850–16856. [CrossRef] [PubMed]

33. Han, C.; Ge, L.; Chen, C.; Li, Y.; Xiao, X.; Zhang, Y.; Guo, L. Novel visible light induced Co$_3$O$_4$-g-C$_3$N$_4$ heterojunction photocatalysts for efficient degradation of methyl orange. *Appl. Catal. B Environ.* **2014**, *147*, 546–553. [CrossRef]

34. Li, R.G.; Weng, Y.X.; Zhou, X.; Wang, X.L.; Mi, Y.; Chong, R.F.; Han, H.X.; Li, C. Achieving overall water splitting using titanium dioxide-based photocatalysts of different phases. *Energy Environ. Sci.* **2015**, *8*, 2377–2382. [CrossRef]

35. Miyoshi, A.; Nishioka, S.; Maeda, K. Water splitting on rutile TiO$_2$-based photocatalysts. *Chem. Eur. J.* **2018**, *24*, 18204–18219. [CrossRef] [PubMed]

36. Ibhadon, A.O.; Fitzpatrick, P. Heterogeneous photocatalysis: Recent advances and applications. *Catalysts* **2013**, *3*, 189–218. [CrossRef]

37. Ohtani, B. Titania photocatalysis beyond recombination: A critical review. *Catalysts* **2013**, *3*, 942–953. [CrossRef]

38. Kato, H.; Kudo, A. Water splitting into H$_2$ and O$_2$ on alkali tantalate photocatalysts ATaO$_3$ (A = Li, Na, and K). *J. Phys. Chem. B* **2001**, *105*, 4285–4292. [CrossRef]

39. Pan, C.; Jia, J.; Hu, X.; Fan, J.; Liu, E. *In situ* construction of g-C$_3$N$_4$/TiO$_2$ heterojunction films with enhanced photocatalytic activity over magnetic-driven rotating frame. *Appl. Surf. Sci.* **2018**, *430*, 283–292. [CrossRef]

40. Zhang, Y.; Heo, Y.J.; Lee, J.W.; Lee, J.H.; Bajgai, J.; Lee, K.J.; Park, S.J. Photocatalytic hydrogen evolution via water splitting: A short review. *Catalysts* **2018**, *8*, 655. [CrossRef]

41. Kumar, S.; Karthikeyan, S.; Lee, A. g-C$_3$N$_4$-based nanomaterials for visible light-driven photocatalysis. *Catalysts* **2018**, *8*, 74. [CrossRef]

42. Pant, B.; Park, M.; Park, S.J.; Kim, H.Y. One-pot synthesis of CdS sensitized TiO$_2$ decorated reduced graphene oxide nanosheets for the hydrolysis of ammonia-borane and the effective removal of organic pollutant from water. *Ceram. Int.* **2016**, *42*, 15247–15252. [CrossRef]

43. Ullah, R.; Dutta, J. Photocatalytic degradation of organic dyes with manganese-doped ZnO nanoparticles. *J. Hazard. Mater.* **2008**, *156*, 194–200. [CrossRef] [PubMed]

44. Yu, W.; Chen, J.; Shang, T.; Chen, L.; Gu, L.; Peng, T. Direct Z-scheme g-C$_3$N$_4$/WO$_3$ photocatalyst with atomically defined junction for H$_2$ production. *Appl. Catal. B Environ.* **2017**, *219*, 693–704. [CrossRef]
45. Seong, D.B.; Son, Y.R.; Park, S.-J. A study of reduced graphene oxide/leaf-shaped TiO$_2$ nanofibers for enhanced photocatalytic performance via electrospinning. *J. Solid State Chem.* **2018**, *266*, 196–204. [CrossRef]
46. Nasir, M.S.; Yang, G.; Ayub, I.; Wang, S.; Wang, L.; Wang, X.; Ramakarishna, S. Recent development in graphitic carbon nitride based photocatalysis for hydrogen generation. *Appl. Catal. B Environ.* **2019**, *257*, 117855. [CrossRef]
47. Mamba, G.; Mishra, A.K. Graphitic carbon nitride (g-C$_3$N$_4$) nanocomposites: A new and exciting generation of visible light driven photocatalysts for environmental pollution remediation. *Appl. Catal. B Environ.* **2016**, *198*, 347–377. [CrossRef]
48. Patnaik, S.; Martha, S.; Parida, K.M. An overview of the structural, textural and morphological modulations of g-C$_3$N$_4$ towards photocatalytic hydrogen production. *RSC Adv.* **2016**, *6*, 46929–46951. [CrossRef]
49. Zhang, Y.; Liu, J.; Wu, G.; Chen, W. Porous graphitic carbon nitride synthesized via direct polymerization of urea for efficient sunlight-driven photocatalytic hydrogen production. *Nanoscale* **2012**, *4*, 5300–5303. [CrossRef]
50. Fajrina, N.; Tahir, M. A critical review in strategies to improve photocatalytic water splitting towards hydrogen production. *Int. J. Hydrog. Energy* **2018**, *44*, 540–577. [CrossRef]
51. Tay, Q.; Kanhere, P.; Ng, C.F.; Chen, S.; Chakraborty, S.; Huan, A.C.H.; Chen, Z. Defect engineered g-C$_3$N$_4$ for efficient visible light photocatalytic hydrogen production. *Chem. Mater.* **2015**, *27*, 4930–4933. [CrossRef]
52. Chen, Y.; Lin, B.; Wang, H.; Yang, Y.; Zhu, H.; Yu, W.; Basset, J.M. Surface modification of g-C$_3$N$_4$ by hydrazine: Simple way for noble-metal free hydrogen evolution catalysts. *Chem. Eng. J.* **2016**, *286*, 339–346. [CrossRef]
53. Wang, L.; Wang, C.; Hu, X.; Xue, H.; Pang, H. Metal/graphitic carbon nitride composites: Synthesis, structures, and applications. *Chem.-Asian J.* **2016**, *11*, 3305–3328. [CrossRef] [PubMed]
54. Martha, S.; Nashim, A.; Parida, K.M. Facile synthesis of highly active g-C$_3$N$_4$ for efficient hydrogen production under visible light. *J. Mater. Chem. A* **2013**, *1*, 7816–7824. [CrossRef]
55. Reddy, K.R.; Reddy, C.V.; Nadagouda, M.N.; Shetti, N.P.; Jaesool, S.; Aminabhavi, T.M. Polymeric graphitic carbon nitride (g-C$_3$N$_4$)-based semiconducting nanostructured materials: Synthesis methods, properties and photocatalytic applications. *J. Environ. Manag.* **2019**, *238*, 25–40. [CrossRef] [PubMed]
56. Zhou, L.; Zhang, H.; Sun, H.; Liu, S.; Tade, M.O.; Wang, S.; Jin, W. Recent advances in non-metal modification of graphitic carbon nitride for photocatalysis: A historic review. *Catal. Sci. Technol.* **2016**, *6*, 7002–7023. [CrossRef]
57. Xiao, M.; Luo, B.; Wang, S.; Wang, L. Solar energy conversion on g-C$_3$N$_4$ photocatalyst: Light harvesting, charge separation, and surface kinetics. *J. Energy Chem.* **2018**, *27*, 1111–1123. [CrossRef]
58. Yang, H.M.; Park, S.J. Influence of mesopore distribution on photocatalytic behaviors of anatase TiO$_2$ spherical nanostructures. *J. Ind. Eng. Chem.* **2016**, *41*, 33–39. [CrossRef]
59. Han, Q.; Wang, B.; Gao, J.; Cheng, Z.; Zhao, Y.; Zhang, Z.; Qu, L. Atomically thin mesoporous nanomesh of graphitic C$_3$N$_4$ for high-efficiency photocatalytic hydrogen evolution. *ACS Nano* **2016**, *10*, 2745–2751. [CrossRef]
60. Zhang, J.; Chen, Y.; Wang, X. Two-dimensional covalent carbon nitride nanosheets: Synthesis, functionalization, and applications. *Energy Environ. Sci.* **2015**, *8*, 3092–3108. [CrossRef]
61. Jing, L.; Ong, W.J.; Zhang, R.; Pickwell-MacPherson, E.; Jimmy, C.Y. Graphitic carbon nitride nanosheet wrapped mesoporous titanium dioxide for enhanced photoelectrocatalytic water splitting. *Catal. Today* **2018**, *315*, 103–109. [CrossRef]
62. Zhao, Z.; Wang, X.; Shu, Z.; Zhou, J.; Li, T.; Wang, W.; Tan, Y. Facile preparation of hollow-nanosphere based mesoporous g-C$_3$N$_4$ for highly enhanced visible-light-driven photocatalytic hydrogen evolution. *Appl. Surf. Sci.* **2018**, *455*, 591–598. [CrossRef]
63. Low, J.; Yu, J.; Jaroniec, M.; Wageh, S.; Al-Ghamdi, A.A. Heterojunction photocatalysts. *Adv. Mater.* **2017**, *29*, 1601694. [CrossRef] [PubMed]
64. Dong, F.; Zhao, Z.; Xiong, T.; Ni, Z.; Zhang, W.; Sun, Y.; Ho, W.K. In situ construction of g-C$_3$N$_4$/g-C$_3$N$_4$ metal-free heterojunction for enhanced visible-light photocatalysis. *ACS Appl. Mater. Interfaces* **2013**, *5*, 11392–11401. [CrossRef] [PubMed]

65. Kim, T.W.; Park, M.; Kim, H.Y.; Park, S.J. Preparation of flower-like TiO_2 sphere/reduced graphene oxide composites for photocatalytic degradation of organic pollutants. *J. Solid State Chem.* **2016**, *239*, 91–98. [CrossRef]

66. Chai, B.; Peng, T.; Mao, J.; Li, K.; Zan, L. Graphitic carbon nitride (g-C_3N_4)–Pt-TiO_2 nanocomposite as an efficient photocatalyst for hydrogen production under visible light irradiation. *Phys. Chem. Chem. Phys.* **2012**, *14*, 16745–16752. [CrossRef] [PubMed]

67. Ismail, A.A.; Bahnemann, D.W. Photochemical splitting of water for hydrogen production by photocatalysis: A review. *Sol. Energy Mater. Sol. Cells* **2014**, *128*, 85–101. [CrossRef]

68. Villa, K.; Domènech, X.; Malato, S.; Maldonado, M.I.; Peral, J. Heterogeneous photocatalytic hydrogen generation in a solar pilot plant. *Int. J. Hydrog. Energy* **2013**, *38*, 12718–12724. [CrossRef]

69. Pant, B.; Park, M.; Kim, H.Y.; Park, S.J. Ag-ZnO photocatalyst anchored on carbon nanofibers: Synthesis, characterization, and photocatalytic activities. *Synth. Met.* **2016**, *220*, 533–537. [CrossRef]

70. Gholipour, M.R.; Dinh, C.-T.; Béland, F.; Do, T.-O. Nanocomposite heterojunctions as sunlight-driven photocatalysts for hydrogen production from water splitting. *Nanoscale* **2015**, *7*, 8187–8208. [CrossRef]

71. Xu, J.; Huo, F.; Zhao, Y.; Liu, Y.; Yang, Q.; Cheng, Y.; Min, S.; Jin, Z.; Xiang, Z. *In-situ* La doped Co_3O_4 as highly efficient photocatalyst for solar hydrogen generation. *Int. J. Hydrog. Energy* **2018**, *43*, 8674–8682. [CrossRef]

72. Samsudin, E.M.; Hamid, S.B.A. Effect of band gap engineering in anionic-doped TiO_2 photocatalyst. *Appl. Surf. Sci.* **2017**, *391*, 326–336. [CrossRef]

73. Zhou, L.; Wang, L.; Zhang, J.; Lei, J.; Liu, Y. The preparation, and applications of g-C_3N_4/TiO_2 heterojunction catalysts—A review. *Res. Chem. Intermed.* **2017**, *43*, 2081–2101. [CrossRef]

74. Cao, S.; Jiang, J.; Zhu, B.; Yu, J. Shape-dependent photocatalytic hydrogen evolution activity over a Pt nanoparticle coupled g-C_3N_4 photocatalyst. *Phys. Chem. Chem. Phys.* **2016**, *18*, 19457–19463. [CrossRef] [PubMed]

75. He, Y.; Wang, Y.; Zhang, L.; Teng, B.; Fan, M. High-efficiency conversion of CO_2 to fuel over ZnO/g-C_3N_4 photocatalyst. *Appl. Catal. B Environ.* **2015**, *168*, 1–8.

76. Zhang, Z.; Liu, K.; Feng, Z.; Bao, Y.; Dong, B. Hierarchical sheet-on-sheet $ZnIn_2S_4$/g-C_3N_4 heterostructure with highly efficient photocatalytic H_2 production based on photoinduced interfacial charge transfer. *Sci. Rep.* **2016**, *6*, 19221. [CrossRef] [PubMed]

77. Yuan, J.; Wen, J.; Zhong, Y.; Li, X.; Fang, Y.; Zhang, S.; Liu, W. Enhanced photocatalytic H_2 evolution over noble-metal-free NiS cocatalyst modified CdS nanorods/g-C_3N_4 heterojunctions. *J. Mater. Chem. A* **2015**, *3*, 18244–18255. [CrossRef]

78. Zhang, J.; Ma, Z. $Ag_6Mo_{10}O_{33}$/g-C_3N_4 1D-2D hybridized heterojunction as an efficient visible-light-driven photocatalyst. *Mol. Catal.* **2017**, *432*, 285–291. [CrossRef]

79. Tan, Y.; Shu, Z.; Zhou, J.; Li, T.; Wang, W.; Zhao, Z. One-step synthesis of nanostructured g-C_3N_4/TiO_2 composite for highly enhanced visible-light photocatalytic H_2 evolution. *Appl. Catal. B Environ.* **2018**, *230*, 260–268. [CrossRef]

80. Zhang, H.; Liu, F.; Wu, H.; Cao, X.; Sun, J.; Lei, W. *In situ* synthesis of g-C_3N_4/TiO_2 heterostructures with enhanced photocatalytic hydrogen evolution under visible light. *RSC Adv.* **2017**, *7*, 40327–40333. [CrossRef]

81. Shi, Y.; Chen, J.; Mao, Z.; Fahlman, B.D.; Wang, D. Construction of Z-scheme heterostructure with enhanced photocatalytic H_2 evolution for g-C_3N_4 nanosheets via loading porous silicon. *J. Catal.* **2017**, *356*, 22–31. [CrossRef]

82. Lam, S.M.; Sin, J.C.; Mohamed, A.R. A review on photocatalytic application of g-C_3N_4/semiconductor (CNS) nanocomposites towards the erasure of dyeing wastewater. *Mater. Sci. Semicond. Process.* **2016**, *47*, 62–84. [CrossRef]

83. Ren, Y.; Dong, Y.; Feng, Y.; Xu, J. Compositing two-dimensional materials with TiO_2 for photocatalysis. *Catalysts* **2018**, *8*, 590. [CrossRef]

84. Li, Z.; Kong, C.; Lu, G. Visible photocatalytic water splitting and photocatalytic two-electron oxygen formation over Cu-and Fe-doped g-C_3N_4. *J. Phys. Chem. C* **2015**, *120*, 56–63. [CrossRef]

85. Iqbal, W.; Yang, B.; Zhao, X.; Rauf, M.; Waqas, M.; Gong, Y.; Mao, Y. Controllable synthesis of graphitic carbon nitride nanomaterials for solar energy conversion and environmental remediation: The road travelled and the way forward. *Catal. Sci. Technol.* **2018**, *8*, 4576–4599. [CrossRef]

86. Guo, Q.; Zhang, Y.; Qiu, J.; Dong, G. Engineering the electronic structure and optical properties of g-C$_3$N$_4$ by non-metal ion doping. *J. Mater. Chem. C* **2016**, *4*, 6839–6847. [CrossRef]

87. Feng, J.; Zhang, D.; Zhou, H.; Pi, M.; Wang, X.; Chen, S. Coupling P nanostructures with P-doped g-C$_3$N$_4$ as efficient visible light photocatalysts for H$_2$ evolution and RhB degradation. *ACS Sustain. Chem. Eng.* **2018**, *6*, 6342–6349. [CrossRef]

 *catalysts*

Review

Recent Progress on Metal Sulfide Composite Nanomaterials for Photocatalytic Hydrogen Production

Sher Ling Lee and Chi-Jung Chang *

Department of Chemical Engineering, Feng Chia University, 100, Wenhwa Road, Seatwen, Taichung 40724, Taiwan; sherlinglee0209@gmail.com
* Correspondence: changcj@fcu.edu.tw

Received: 25 April 2019; Accepted: 13 May 2019; Published: 17 May 2019

Abstract: Metal sulfide-based photocatalysts have gained much attention due to their outstanding photocatalytic properties. This review paper discusses recent developments on metal sulfide-based nanomaterials for H_2 production, acting as either photocatalysts or cocatalysts, especially in the last decade. Recent progress on key experimental parameters, in-situ characterization methods, and the performance of the metal sulfide photocatalysts are systematically discussed, including the forms of heterogeneous composite photocatalysts, immobilized photocatalysts, and magnetically separable photocatalysts. Some methods have been studied to solve the problem of rapid recombination of photoinduced carriers. The electronic density of photocatalysts can be investigated by in-situ C *K*-edge near edge X-ray absorption fine structure (NEXAFS) spectra to study the mechanism of the photocatalytic process. The effects of crystal properties, nanostructure, cocatalyst, sacrificial agent, electrically conductive materials, doping, calcination, crystal size, and pH on the performance of composite photocatalysts are presented. Moreover, the facet effect and light trapping (or light harvesting) effect, which can improve the photocatalytic activity, are also discussed.

Keywords: metal sulfides; H_2 production; photocatalyst; facet effect; light trapping; crystal size

---

1. Introduction

Hydrogen, which has high energy yield and less greenhouse gas emissions after combustion, is an environmentally friendly and attractive fuel. Hydrogen production and hydrogen economy, together with potential applications in fuel cell hydrogen electric cars and hydrogen-diesel fuel co-combustion, are hot topics in research related to hydrogen energy [1–8]. Hydrogen is an alternative energy source for fossil fuels. Research on generating electricity from solar energy has also gained increasing attention. Environmentally friendly photocatalytic water-splitting by semiconductor nanomaterials is very important for the development of a hydrogen economy. The photoexcited electron–hole pairs play an important role in the water splitting process. Photocatalysts should be able to absorb radiation and efficiently reduce protons to hydrogen molecules with photogenerated electrons.

Metal sulfide-based photocatalysts are widely investigated because of their unique physical and chemical properties [9–11]. In comparison with their metal oxide analogs, sulfide semiconductors with narrower band gaps hold more promise as photocatalysts for H_2 production [12]. However, the H_2 production performance of metal sulfide-based composite photocatalysts is still limited because of the fast recombination of photoexcited carriers [13]. In addition, photocorrosion is also a major problem for metal sulfide-based photocatalysts. Therefore, many researchers have focused on solving the problems of photocorrosion and carrier recombination, examining the formation of a heterojunction by decorating cocatalysts, introducing noble metal nanoparticles, or using porous conductive substrate. It has been reported that the formation of a heterojunction interface with other semiconductors or coupling with other cocatalysts can overcome the carrier recombination and photocorrosion problems [14]. The formation of a junction helps to promote the transport of photoexcited electrons. In addition, it is reported that the performance of photocatalysts can also be enhanced further by doping, modifying the surface texture, and tuning the crystal structure [15,16]. Furthermore, current research on metal sulfide-based photocatalysts is focused on two hot topics: the search for novel metal-sulfide nanomaterials that do not contain acutely toxic metals such as cadmium and lead, and the search for cocatalyst nanomaterials that do not contain noble metals.

In this review, we discuss recent developments in the use of metal sulfide-based nanomaterials as either photocatalysts or cocatalysts for H_2 production, especially those in the last 10 years. Recent progress on the fabrication and performance of metal sulfide-based composite photocatalysts is reviewed, including the form of heterogeneous composite photocatalysts, immobilized photocatalysts, and magnetically separable photocatalysts. In addition, the effects of experimental parameters—including noble metal loading, transition metal doping, anion doping, calcination, the pH level of solution, sacrificial agent, structure of photocatalysts, facet effect, light trapping effect, crystal size of photocatalysts, and fabrication methods—on photocatalytic performance are also systematically reviewed.

2. Composite Photocatalysts

2.1. Semiconductor-Based Composite Photocatalysts

2.1.1. Cadmium Sulfide

CdS is a widely-investigated metal sulfide-based photocatalyst that has a narrow band gap and high hydrogen production activity. The synergetic effects of this coupled composite photocatalyst on hydrogen evolution can be studied by investigating the decay behavior of photoexcited carriers. The synergetic effect is important for the enhancement of hydrogen production by sulfides of transition metals. Wang et al. [17] reported that $(ZnO)_1/(CdS)_{0.2}$ composite showed noticeably slower decay kinetics as compared to bare ZnO and CdS when observed with time-resolved fluorescence emission decay spectra. This finding indicates that the photocatalyst $(ZnO)_1/(CdS)_{0.2}$ is able to produce the highest H_2 evolution among different ZnO/CdS heterostructures due to the transfer of photoexcited carriers between CdS and ZnO, which may hinder recombination. A similar trend of this synergetic effect was also observed in other studies, such as those incorporating g-C_3N_4 on the outer shell of a CdS core [18], g-C_3N_4/Ni(OH)$_2$-CdS [19], g-C_3N_4Ni/CdS [20], co-loading of MoS_2 and graphene to CdS nanorods [21], Ni_3N/CdS [22], and MoS_2/CdS [23]. Transition metal chalcogenides are appropriate candidates for composite photocatalysts due to their conduction band positions, which are appropriate for the reduction reaction of water to form hydrogen [24], and their excellent luminescence and photochemical properties. Table 1 summarizes the photocatalytic performances of CdS-based photocatalysts.

Table 1. Photocatalytic performances of CdS-based photocatalysts.

Photocatalyst	Morphology	Synthetic Method	Sacrificial Agent	Activity (μmol h^{-1}g^{-1})	Ref. (Year)
CdS/ZnO	Heterostructure	Two-step precipitation	Na_2S, Na_2SO_3	1805	[17] (2009)
g-C_3N_4/CdS	Core–shell	Solvothermal & chemisorption	Na_2S, Na_2SO_3	4200	[18] (2013)
Ni(OH)$_2$-CdS/g-C_3N_4	Core–shell	Mixture	Na_2S, Na_2SO_3	115.2	[19] (2016)
Ni/CdS/g-C_3N_4	Hybrid system	$NaBH_4$ reduction method	triethanolamine	1258.7	[20] (2016)
MoS$_2$-graphene/CdS	Nanoparticle/nanorods	Hydrothermal	Lactic acid	2320 μmol h^{-1}	[21] (2014)
Ni$_3$N/CdS	Nanorods	Two step in-situ growth method	Na_2S, Na_2SO_3	~62	[22] (2016)
MoS$_2$/CdS	Heterostructure	Precipitation	Lactic acid	~540	[23] (2008)

2.1.2. Copper Sulfide

CuS/ZnS composites are efficient photocatalysts for hydrogen evolution. Wang et al. reported that CuS/ZnS nanomaterials exhibit high visible light-induced H_2 generation activity. The H_2 generation rate increases with increasing Cu^{2+} ions. However, as with other cocatalysts, when the maximum amount of Cu^{2+} is reached (above 7 mol %), the hydrogen evolution rate decreases significantly. This is due to light shielding by excess CuS, which reduces the number of active sites on the surface [25]. In addition to the solvothermal method for the fabrication of CuS cocatalysts, the growth of CuS/g-C_3N_4 by an in-situ method has been investigated by other authors. That study found that CuS nanoparticles were uniformly distributed on g-C_3N_4 nanosheets [26].

Chang et al. found that CuS-ZnS$_{1-x}$O$_x$/g-C_3N_4 heterostructured photocatalyst had a high photocatalytic H_2 generation property. That study demonstrated that the decoration of CuS on the surface helps to enhance the absorption of the heterostructured photocatalyst [27]. In addition, CuS is used to decorate free-standing ZnS-carbon nanotube films because CuS can form heterojunctions with ZnS to improve the separation of photoexcited charge carriers, resulting in higher rates of hydrogen production [28]. In another work, Markovskaya et al. suggested the important role of CuS in enhancing the hydrogen evolution rate of the photocatalyst $Cd_{0.3}Zn_{0.7}S$. The optimized performance (3520 μmol h^{-1}g^{-1}) was obtained with 1 mol % CuS/$Cd_{0.3}Zn_{0.7}S$ [29]. Similar contributions were also found for CuS/TiO_2 nanocomposites [30,31]. Table 2 lists the photocatalytic activity of CuS-based composite photocatalysts.

Table 2. Photocatalytic activity of copper sulfide as the cocatalyst.

Photocatalyst	Structure	Synthetic Method	Sacrificial Agent	Activity (μmol h^{-1}g^{-1})	Ref. (Year)
CuS/ZnS	Hexagonal plates	Solvothermal	Na_2S, Na_2SO_3	1233.5	[25] (2015)
CuS/g-C_3N_4	Nanocomposites	In-situ growth method	triethanolamine	17.2	[26] (2017)
CuS-ZnS$_{1-x}$O$_x$/g-C_3N_4	Heterostructure	Thermal decomposition and hydrothermal	Na_2S, Na_2SO_3, NaCl	10,900	[27] (2017)
CuS-ZnS/CNTF	Nanocomposite	Hydrothermal	Na_2S, Na_2SO_3, NaCl	1213.5	[28] (2018)
CuS/$Cd_{0.3}Zn_{0.7}S$	Nanoparticles	Two-step technique	Na_2S, Na_2SO_3	3520	[29] (2015)
CuS/TiO_2	Nanocomposite	Hydrothermal	Na_2S, Na_2SO_3	1262	[30] (2018)
CuS/TiO_2	Nanocomposite	Hydrothermal	Methanol	570 μmol h^{-1}	[31] (2013)

2.1.3. Silver Sulfide

Recently, silver sulfide was used as a cocatalyst of ZnS photocatalyst to enhance the hydrogen evolution rate. For instance, Hsu et al. reported the use of Ag_2S-coupled ZnO@ZnS core–shell nanorods to achieve efficient H_2 production. The maximum hydrogen production rate is reached when the $AgNO_3$ concentration is 2 mM; further increasing the concentration only decreases the hydrogen production rate [32]. Figure 1 demonstrates the morphology of and a possible mechanism for the photocatalytic reaction. Because of the growth of Ag_2S on ZnO@ZnS nanorods, photoexcited electrons can effectively transfer from ZnS to Ag_2S or migrate to the conductive wire mesh substrate. The reaction of electrons and H^+ can produce H_2. Similarly, Yue et al. fabricated a novel Ag_2S/ZnS/carbon nanofiber ternary composite to increase the hydrogen production rate well above that of the reported ZnS composite photocatalyst. In addition, the synergetic effect of Ag_2S and CNF is also important for inhibiting the recombination of charge carriers [33]. Moreover, nanosheets of ZnS:Ag_2S also exhibit a trend of enhancement similar to that of the above mentioned nanostructure. As reported by Yang et al., porous ZnS:Ag_2S nanosheets were synthesized such that the porous nanostructure provided a large surface area for intimate contact with the sacrificial solution [34]. Table 3 shows the H_2 generation performances of photocatalysts loaded with silver sulfide.

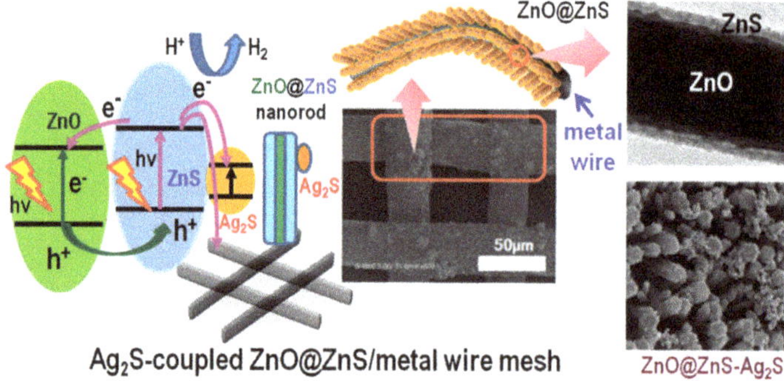

Figure 1. Morphology and a proposed mechanism of the photocatalytic H_2 production by metal wire mesh based immobilized photocatalysts with Ag_2S-coupled ZnO@ZnS core–shell nanorods. Figure adapted from [32].

Table 3. Photocatalytic H_2 generation property of photocatalysts loaded with silver sulfide.

Photocatalyst	Morphology	Synthetic Method	Sacrificial Agent	Activity (μmol h^{-1}g^{-1})	Ref. (Year)
Ag_2S-ZnO@ZnS core–shell	Nanorods	Hydrothermal	Na_2S, Na_2SO_3, NaCl	6406	[32] (2016)
Ag_2S/ZnS/carbon nanofiber	Nanofibers	Solid-state process and cation-exchange	Na_2S, Na_2SO_3	224.9 μmol h^{-1}	[33] (2016)
ZnS:Ag_2S	Porous nanosheets	Thermal decomposition	Na_2S, Na_2SO_3	104.9	[34] (2014)

2.1.4. Zinc Sulfide

ZnS is an excellent photocatalyst for photocatalytic water splitting, for it can produce high negative potentials of photoexcited electrons. Xin et al. reported that ZnS@CdS-Te photocatalysts with a p–n heterostructure exhibited improved H_2 generation rates. Based on the possible mechanism, after the loading of ZnS on the CdS-Te nanostructure, more surface active sites can be produced, leading to increased hydrogen generation activity [35]. Figure 2 presents a possible mechanism of ZnS@CdS-Te photocatalysts. Moreover, a similar heterojunction between ZnS/g-C$_3$N$_4$ has been studied by Hao et al. In that study, the close contact between ZnS and g-C$_3$N$_4$ increased the capacity of light harvesting and efficiency of charge separation. The key factor that enhanced the photocatalytic hydrogen was the two-photo excitation of ZnS [36].

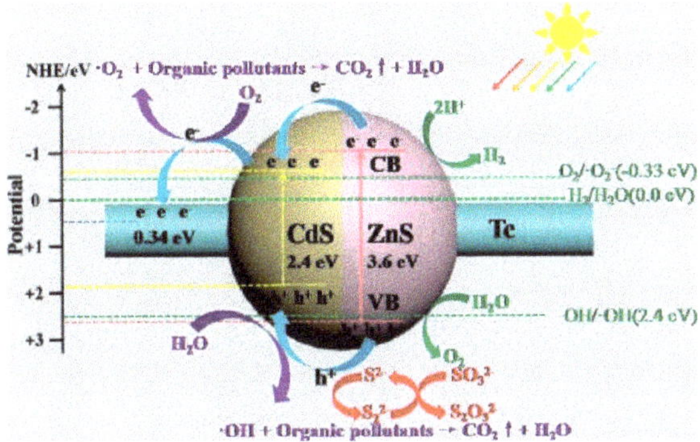

Figure 2. Possible mechanism of ZnS@CdS-Te composite photocatalysts. Figure adapted from reference [35].

In addition to nanostructures, the chloroplast-like structure of Bi$_2$S$_3$/ZnS also possesses high photocatalytic activity due to its band gap structure. Based on the reported reaction mechanism, photogenerated electrons migrate from Bi$_2$S$_3$ to ZnS to generate H$_2$, while holes transfer from ZnS to Bi$_2$S$_3$. This efficient charge separation improves the H$_2$ generation rate in this chloroplast-like structure [37].

Efficient H$_2$ generation has been achieved by CdS-ZnS photocatalyst without facilitation by a cocatalyst. This includes the study done by Jiang et al. on a CdS nanorod/ZnS nanoparticle photocatalyst. The highly efficient hydrogen production likely resulted from the rapid transport of carriers in the core–shell nanorod structure [38]. Table 4 presents the H$_2$ generation performances of ZnS-based photocatalysts.

Table 4. Photocatalytic H$_2$ generation performances of ZnS-based photocatalysts.

Photocatalyst	Morphology	Synthetic Method	Sacrificial Agent	Activity (μmol h^{-1}g^{-1})	Ref. (Year)
ZnS@CdS-Te	p–n heterostructure	Microwave hydrothermal	Na$_2$S·H$_2$O, Na$_2$SO$_3$	592.5	[35] (2018)
ZnS/g-C$_3$N$_4$	Nanocomposite	One-pot hydrothermal	Na$_2$S, Na$_2$SO$_3$	713.68	[36] (2018)
Bi$_2$S$_3$/ZnS	Chloroplast-like structure	Solvothermal	Na$_2$S, Na$_2$SO$_3$	176.24	[37] (2017)
CdS/ZnS	Core–shell nanoparticle composite	Solvothermal	Na$_2$S, Na$_2$SO$_3$	239 μmol h^{-1}mg^{-1}	[38] (2015)

2.2. Electrically Conductive Materials (Non-Noble Metal)-Based Composite Photocatalysts

2.2.1. Graphene

In addition to decoration with noble and non-noble metals, combining a photocatalyst with graphene or carbon dots is also a useful way to improve activity [39–44]. Graphene can be incorporated solely or co-loaded with other compounds. Loading graphene provides several benefits, such as a large surface area and enhanced separation of photoexcited electron–hole pairs. In other words, the exceptional electron transfer capability of graphene and intimate contact between the photocatalyst and graphene can help to transport photoexcited electrons efficiently, thus improving the activity for photocatalytic hydrogen generation. As shown in work done by Azarang et al. on nitrogen-doped graphene-supported ZnS nanorods, the photocatalytic activity of ZnS was multiplied by as much as 6 times when solely graphene was loaded and by 14 times when loaded with NG-ZnS [45].

Chang et al. [46] found that the incorporation of graphene with ZnO-ZnS nanoparticles improved the rate of H_2 generation from glycerol. The irradiated and dark states of in-situ C *K*-edge NEXAFS spectra were monitored to investigate the electronic properties of the photocatalyst. In-situ NEXAFS spectra revealed that photoexcited electrons can be transported from ZnO-ZnS nanomaterials to graphene. NEXAFS spectra were used to investigate the interfacial electronic states of AgI/BiOI/graphene (A10B/G10) samples [47]. Figure 3a–c presents the intensity change (ΔA) between the irradiated and dark states of the Ag L_3-edge, C *K*-edge, and Bi lL_3-edge of BiOI/graphene photocatalyst (BG10), and A10B/G10 photocatalyst. In comparison with BG10, A10B/G10 presents more positive ΔA values of NEXAFS for the Bi L_3-edge and the Ag L_3-edge under in situ light exposure, revealing the increased amounts of the unoccupied density of states (DOS) of the Ag L-edge and Bi L-edge after light exposure due to donating photoexcited electrons for BiOI and AgI. In contrast, compared with A10B/G10, B/G10 exhibits more negative ΔA values of NEXAFS for the C *K*-edge under light irradiation. The reduced C *K*-edge indicates the decrease of unoccupied DOS of graphene after light irradiation due to receiving the photoexcited electrons. These results imply the migration of a charge from AgI to BiOI and then to graphene under light exposure. (Figure 3d). The activity can be improved by the formation of graphene/BiOI and BiOI/AgI heterostructures.

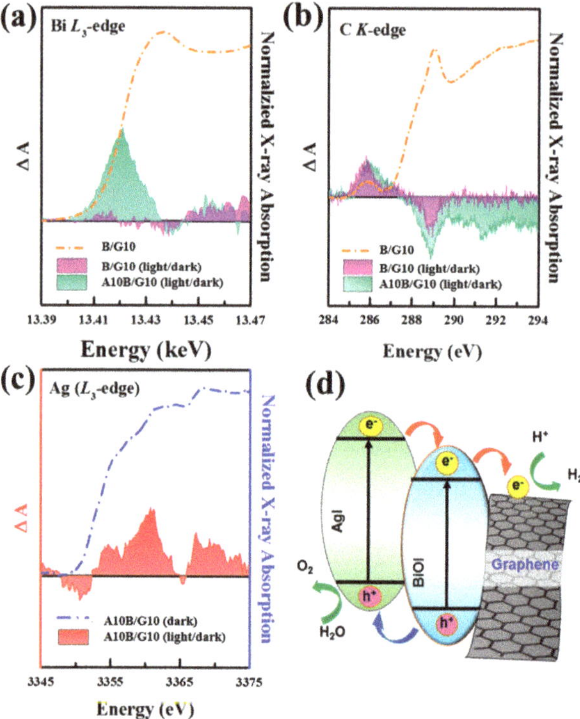

Figure 3. Intensity change between the irradiated and dark states of (**a**) Bi lL_3-edge, (**b**) C *K*-edge, and (**c**) Ag L_3-edge. ($\Delta A = A_{light} - A_{dark}$) of B/G10 and A10B/G10 photocatalysts, (**d**) a proposed mechanism for the photocatalytic H_2 generation reaction by A10B/G10 [47].

The incorporation of graphene provides not only the above-mentioned advantages but also good water dispersity of the composite photocatalyst. This advantage was confirmed by Zhang et al. based on the results of CdSe/CdS-Au (QD-Au) core–satellite heteronanocrystal assembled on graphene nanosheets [48]. In addition to the common graphene nanosheets, Chang et al. also reported using flower-like graphene with a 3D porous structure for application in photocatalytic activity. The application of flower-like graphene enables the photocatalyst to have a higher BET surface area, which leads to enhanced photocatalytic hydrogen production [49]. Table 5 shows the hydrogen generation properties of graphene-loaded composite photocatalysts.

Table 5. Photocatalytic H_2 production properties of graphene-loaded composite photocatalysts.

Photocatalyst	Morphology	Synthetic Method	Sacrificial Agent	Activity (μmol h^{-1}g^{-1})	Ref. (Year)
MoS_2/graphene-CdS	Nanocomposite	Solution-chemistry	Lactic acid	1800 μmol h^{-1}	[44] (2014)
Nitrogen-doped graphene/ZnS	Nanorods	Thermal annealing of G-ZnS	Na_2S, Na_2SO_3, NaCl	1755.7	[45] (2018)
Graphene/ZnO-ZnS	Particle-on-sheet	Two-step heating	Glycerol	1070	[46] (2018)
Graphene/CdS	Nanocomposite	Solvothermal	Lactic acid	1120 μmol h^{-1}	[47] (2011)
CdSe/CdS-Au-graphene	Nanocrystals	SILAR technique	Na_2S, Na_2SO_3	3113	[48] (2014)
Graphene/ZnS	Nanocomposite	Hydrothermal	Na_2S, Na_2SO_3, NaCl	11,600	[49] (2017)

2.2.2. Reduced Graphene Oxide and Graphene Oxide

Since the discovery of graphene (G) and/or graphene oxide (GO) [50], they have been widely studied [51–53]. GO consists of graphene nanosheets with epoxy or hydroxyl group-modified basal planes and carbonyl/carboxylic acid-modified edges [54–56]. In contrast to graphite, GO can be exfoliated easily and dispersed in aqueous solution because of these hydrophilic groups on the surface [57–59]. Hence, a few studies have used graphene oxide as the supporting material. For instance, in a recent study done by Peng et al., GO-cadmium sulfide was immobilized and well dispersed on a graphene oxide sheet. Increasing the GO-loading up to 5 wt % promoted the hydrogen generation rate to a maximum of 314 μmol h^{-1} [60]. Hou et al. found that the introduction of GO can extend the lifetime of the photoexcited carriers because they can act as both electron transporter and electron acceptor [61].

A number of authors have also driven the further development of reduced graphene oxide. For instance, Zhang et al. enhanced solar photocatalytic hydrogen production by introducing reduced graphene oxide nanosheets and $Zn_xCd_{1-x}S$. That research provided a green method for using reduced graphene oxide (RGO) as a support material to enhance the performance of $Zn_xCd_{1-x}S$ photocatalyst, for the hydrogen production of RGO-$Zn_{0.8}Cd_{0.2}S$ was 450% higher than that of pristine $Zn_{0.8}Cd_{0.2}S$ [62]. Similar effects after the incorporation of reduced graphene oxide have been evidenced in other works, such as studies of ZnO-CdS/RGO [63] and ternary NiS/$Zn_xCd_{1-x}S$/RGO nanocomposites [64]. Table 6 presents the photocatalytic H_2 production properties of graphene oxide loaded with composite photocatalysts.

Table 6. Photocatalytic H_2 production for graphene oxide loaded with composite photocatalysts.

Photocatalyst	Morphology	Synthetic Method	Sacrificial Agent	Activity ($\mu mol\ h^{-1}g^{-1}$)	Ref. (Year)
GO-CdS	Nanocomposite	Precipitation	Na_2S, Na_2SO_3	314 $\mu mol\ h^{-1}$	[60] (2012)
GO-CdS@TaON	Hybrid composites	Hydrothermal	Na_2S, Na_2SO_3	633 $\mu mol\ h^{-1}$	[61] (2012)
RGO-$Zn_xCd_{1-x}S$	Nanocomposite	Coprecipitation-hydrothermal	Na_2S, Na_2SO_3	1824	[62] (2012)
ZnO-CdS/RGO	Heterostructure	Light irradiation-induced reduction	Na_2S, Na_2SO_3	510 $\mu mol\ h^{-1}$	[63] (2014)
NiS/$Zn_xCd_{1-x}S$/RGO	Ternary nanocomposite	Coprecipitation-hydrothermal	Na_2, Na_2SO_3	205.9 $\mu mol\ h^{-1}$	[64] (2014)

2.2.3. Conductive Polymer

Many papers have reported on the improvement of photocatalytic activity due to the loading of conductive polymers, such as polythiophene, polypyrrole, PEDOT, and PSS. Conductive polymers are capable of inducing charge separation in the composite photocatalysts [65]. A recent study showed that polyaniline (PANI)/ZnS synthesized by solvothermal method increases the hydrogen evolution rate up to 6750 $\mu mol\ h^{-1}g^{-1}$ because PANI has unique electron and hole transporting properties [66]. PANI-ZnS composite photocatalysts exhibit improved dispersibility, light harvesting, and photocurrent response. Wang et al. [67] also reported on conducting polymers (such as polypyrrole, poly-3,4-ethylenedioxythiophene (PEDOT), and PANI) on the surface of CdS nanorods. It was found that the rate of hydrogen production of polyaniline@CdS was almost 5 times that of PEDOT@CdS and 3 times that of polypyrrole@CdS. That study showed that polyaniline is an efficient conducting material for modifying CdS nanomaterials, for it enables better light penetration than does a polypyrrole shell. Zielińska et al. [68] reported that the hydrogen production of PANI/$NaTaO_3$ photocatalyst under UV light irradiation was about twice that of pristine $NaTaO_3$ photocatalyst. PANI/$NaTaO_3$ exhibits a lower PL spectrum than that of $NaTaO_3$, indicating a slower recombination of photoexcited charge carriers. The photoluminescence (PL) spectrum confirmed that the enhancement resulted from the efficient charge separation process. Sasikala et al. studied the photocatalytic hydrogen production performance of MoS_2-PANI-CdS photocatalysts [69]. The incorporation of MoS_2 and PANI improved the visible light absorption ability and improved the lifetime of photoexcited electron–hole pairs of the composite photocatalysts. To achieve high photocatalytic activity, 4% MoS_2 and 5% PANI are the optimum amounts for the composite photocatalysts. The enhanced light absorption and lifetime of the photoexcited charge carriers result in improved activity of the photocatalysts. Figure 4 presents a possible mechanism for the electron transfer of MoS_2-PANI-CdS photocatalysts. Incorporating polyaniline helps to separate photoinduced charges across the ZnS-polyaniline interfaces.

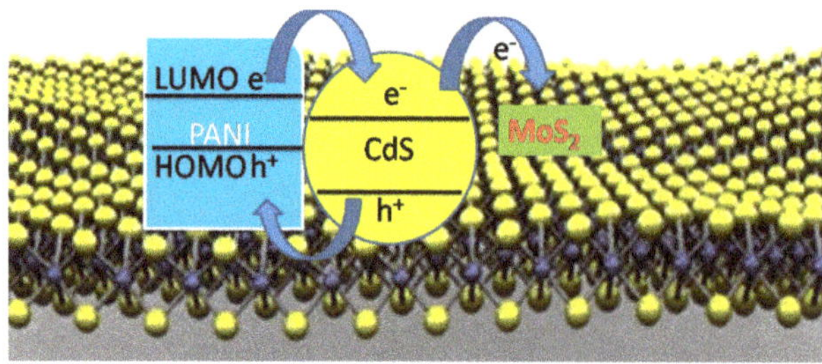

Figure 4. A proposed mechanism for the electron transfer of MoS_2-PANI-CdS photocatalysts when illuminated. Figure adapted from [69].

2.2.4. Conductive Substrate

Chang et al. studied the performance of Ni-doped ZnS/NiO/Ni foam photocatalysts [70]. Ni foam was decorated with doped ZnS to prepare the porous immobilized photocatalysts. The photocatalysts had an optimized activity of 2500 μmol/g^{-1} h^{-1}, resulting from their matched band structure, porous microstructure, and conductive Ni foam as substrate. The surface turned from hydrophobic to superhydrophilic after Ni-doped ZnS was grown on the surface of the Ni foam. In addition, the porous microstructure facilitated the transport of reactant and generated a large amount of surface active sites, and the conductive Ni foam aided in the separation of photoexcited carriers. Figure 5 shows a proposed mechanism, illustrating the band structure of the photocatalysts and the transport of photoexcited electrons. When the composite photocatalysts were irradiated with light, the photoexcited electrons were effectively separated by their transportation from ZnS to NiO. In addition, stainless-steel wire mesh is also a good candidate for the preparation of porous immobilized photocatalysts [71].

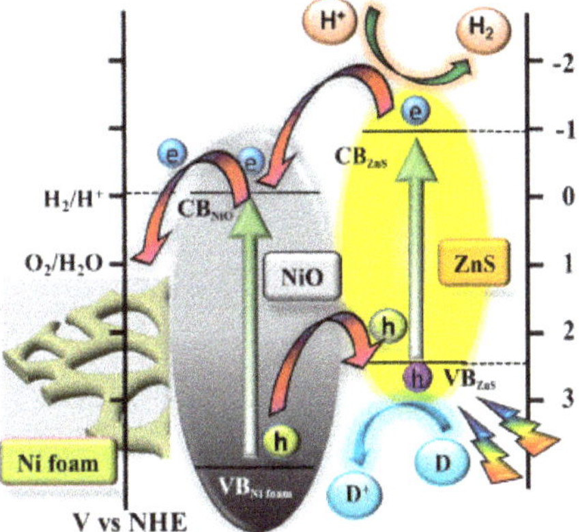

Figure 5. A possible mechanism for the photoexcitation and carrier transporting process of Ni-doped ZnS/NiO/Ni foam photocatalysts. Figure adapted from [70].

2.3. Magnetic Materials-Based Composite Photocatalysts

Magnetic nanomaterials have been used as the core for the preparation of magnetically separable photocatalysts. To improve the activity, photocatalyst nanomaterials should be highly dispersible in solution. Nonetheless, it will become more difficult to separate and reuse the nanoparticles of photocatalysts by centrifugation for repeated operation if such good dispersal stability is achieved. Hence, the introduction of magnetic nanomaterials in photocatalysts will enable efficient separation for the repeated use of photocatalysts. For instance, in studies conducted on NiCo$_2$O$_4$@ZnS and Fe$_3$O$_4$@ZnS core–shell nanoparticles, the recycled photocatalyst exhibited a good hydrogen evolution rate even after being recycled three times [72]. The ZnS shell deposited on a magnetic core decreases the magnetic saturation of core–shell microspheres. CoFe$_2$O$_4$@ZnS nanoparticles exhibit superparamagnetic properties where no residual magnetism is left after repeated use of the photocatalyst [73]. Figure 6 shows the dispersion and magnetic separation of calcinated CoFe$_2$O$_4$@ZnS-0.5h photocatalyst. The superparamagnetic property is quite important for photocatalytic H$_2$ generation in practical operations. After the magnet is removed, there should be no residual magnetism so as to prevent the aggregation of recycled photocatalysts.

Figure 6. (**a**) Dispersion of CoFe$_2$O$_4$@ZnS-0.5h photocatalysts, (**b**) the separation of CoFe$_2$O$_4$@ZnS-0.5h photocatalyst by an external magnet. Figure adapted from [73].

3. Experimental Parameters for Enhancing Photocatalytic Activity

3.1. Loading with Metal

The majority of the prior research to improve the activity of photocatalysts involved the loading of noble and non-noble metals and the coupling of semiconductors. Then the photogenerated electrons could be transported to the noble metals or delocalized and transported between photocatalysts.

3.1.1. Noble Metal Loading

The photocatalytic activity for the production of hydrogen is affected by the noble metal loaded on the photocatalyst. Noble metal co-catalysts can accelerate the transport of photoexcited charge and thus create H$_2$ desorption sites. This will eventually lead to higher H$_2$ generation activity [74]. Although the rate of H$_2$ generation will increase with increasing noble metal loads, the loading amount will reach a maximum, above which further increases of a noble metal will not decrease the photocatalytic activity. The reduction is probably due to two factors: (i) excess noble metal will lead to shielding of the incident light; and (ii) a higher amount of noble metal decorated on the photocatalyst surface will causes light scattering of the samples, thus reducing the effective irradiation absorbed by the reaction suspension [75,76].

Zhou et al. studied the ternary heterojunction photocatalyst CdS/M/TiO$_2$ (M = Ag, Au, Pd, Pt). Some photocatalysts enriched with the noble metals are better than pristine TiO$_2$ and binary heterojunction photocatalysts. For example, the hydrogen evolution rate of photocatalyst loaded with Pd (CdS/Pd/TiO$_2$) is 6.7 times higher than that of CdS/TiO$_2$ [77].

The photocatalytic H$_2$ generation activity also increases as greater amounts of Au are loaded onto ZnS flowers. When the Au load is less than 4%, the photocatalytic activity is able to reach 3306 μmol h^{-1}g^{-1}. Further increasing the Au load to 6% reduces the activity. Such improvement of photoactivity can be ascribed to the following reasons. First, in comparison with the conduction band minimum of ZnS, Au has a lower Fermi level. The photoexcited electrons can be transported to Au. Second, the Au(I) loaded on the ZnS lattice will extend the light absorption wavelength and enhance the light harvesting efficiency [78]. Moreover, when Au nanoparticles are incorporated on S,N-modified TiO$_2$ (SNT), the Au particles promote the visible light-driven hydrogen production activity. Because of surface plasmon resonance, 3.5 nm Au particles deposited on TiO$_2$ can increase light absorption. The amount of hydrogen generated by the 3Au-SNT is 9 times that of pure SNT [79].

In addition to Au, Pt is frequently incorporated in semiconductor photocatalysts. In a previous study by Yu et al., Pt was loaded onto a Cu$_2$ZnSnS$_4$ (CZTS) semiconductor with a maximum content of 1%, and the hydrogen production rate increased. However, the performance decreased when the Pt load was further increased, due to the optical shielding effect. The production of hydrogen by 1% Cu$_2$ZnSnS$_4$-Pt was 8 times higher than that of bare Cu$_2$ZnSnS$_4$. Intimate contact between CZTS and Pt increases the production of hydrogen [80]. Similarly, in CdS photocatalyst, the hydrogen evolution reaches a quantum efficiency of 51% and a maximum of 4800 $\mu mol\ h^{-1}$ when 0.65 wt % of Pt is loaded on CdS. However, the activity can be further increased by loadings of 0.3 wt % Pt and 0.13 wt % PdS on CdS. The co-loading of noble and non-noble metals onto pristine CdS promotes the splitting of H_2S into H_2 and S [81]. Likewise, the addition of Pt to CuS-TiO_2 enhances the evolution of hydrogen because excited electrons from the CB of CuS or through CB of TiO_2 can be transferred directly to Pt sites, resulting in the reduction of protons to hydrogen [82]. Moreover, loading Pt on novel $Cd_xCu_yZn_{1-x-y}S$ also enhances the photocatalytic performance. The presence of 0.5 wt % Pt increases the H_2 production rate to 557 $\mu mol\ h^{-1}$, as compared to 350 $\mu mol\ h^{-1}$ produced by $Cd_{0.1}Cu_{0.01}\ Zn_{0.89}S$ alone [83].

The presence of Cu in a photocatalyst facilitates carrier separation and also increases light absorption. As reported in a study of In and Cu co-doped ZnS photocatalysts by Kimi et al. [84], photocatalytic performance is strongly related to the amount of doped Cu. With the suitable amount of doped Cu (0.03), hydrogen evolution reaches a maximum that is 8 times higher than that of hydrogen produced by In(0.1)-ZnS photocatalyst. However, with the incorporation of more Cu, the photocatalytic activity becomes lower than that of single doped In(0.1)-ZnS. An excessive amount of Cu causes light scattering, and the excess Cu also acts as recombination sites to halt the photocatalytic reaction.

Although the incorporation of noble metals in photocatalysts will increase the rate of hydrogen production, different noble metals have various effects on the enhancement. When four different noble metals, Pt, Rh, Pd, and Ru, are decorated on CdS/TiO_2 photocatalyst, the amount of photogenerated H_2 by Pt loaded (640 $\mu mol\ h^{-1}$) photocatalyst is the highest [85]. Table 7 shows the photocatalytic H_2 generation for noble metal-loaded photocatalysts.

Table 7. Photocatalytic H_2 generation for noble metal-loaded photocatalyst.

Photocatalyst	Noble Metal	Synthetic Method	Activity ($\mu mol\ h^{-1}\ g^{-1}$)	Ref. (Year)
$CdS/M/TiO_2$	Au, Ag, Pt	Two-step photodeposition	-	[77] (2014)
ZnS flower	Au	Deposition-precipitation	3306	[78] (2013)
S,N-TiO_2	Au	Deposition-precipitation	267.6	[79] (2014)
Cu_2ZnSnS_4	Pt	-	1020	[80] (2014)
CdS	Pt	Photodeposition	8770 $\mu mol\ h^{-1}$	[81] (2009)
CuS-TiO_2	Pt	Hydrothermal	746	[82] (2016)
$Cd_xCu_yZn_{1-x-y}S$	Pt	Co-precipitation	557 $\mu mol\ h^{-1}$	[83] (2008)
In(0.1),Cu(x)-ZnS	Cu	Hydrothermal	16.6 $\mu mol\ h^{-1}$	[84] (2016)
CdS/TiO_2	Pt	Precipitation	Pt: 640 $\mu mol\ h^{-1}$	[85] (2007)
ZnO-CdS	Pt	Modified hydrothermal	6180	[86] (2010)

3.1.2. Transition Metal Doping

As observed from a few recent studies, transition metals (TM) have begun attracting attention because doping with TM can significantly enhance the photocatalytic performance by efficiently promoting the separation process of photoexcited holes and electrons. For instance, Chen et al. developed an in-situ photodeposition method to load Co on CdS nanorods. That work reported a highest photocatalytic activity of 1299 $\mu mol\ h^{-1}$ with the optimum loading of 1.0 wt % [87]. In comparison with nickel and iron, cobalt has the ability to improve the rate of H_2 generation. From the work done by Zhou et al., the enhancement of the activity of MoS_x was found to be in the order of Co > Ni > Fe [88]. This improved performance results from the higher amount of doped Co and the capability of Co to activate the S-edge sites [89]. A quick screening technology has been reported to find out the optimized composition of the photocatalyst. M-ZnS based photocatalysts (M = Cr, Cu, Ni, Mo, and Ag) for photoelectrochemical water oxidation applications can be screened rapidly using scanning electrochemical microscopy (SECM) with an optical fiber by finding, the spot with the highest photocurrent among the photocatalyst arrays [90].

Ni doping can enhance the photocatalytic activity of H_2 generation by increasing the absorption of light of the doped photocatalyst. In contrast, when the amount of Ni loaded on ZnS-graphene composites is increased, it will degenerate the crystalline property of the photocatalyst [91]. In addition, the incorporation of Ni on $Cd_{1-x}Zn_xS$ microsphere photocatalyst increases the rate of hydrogen production to 191 μmol h^{-1}g^{-1} when an optimum amount of 0.1 wt % Ni is loaded. The increment results from the decreased particle size and increased surface area of the photocatalyst [92]. Incorporating a metal such as Ni onto the photocatalyst can accelerate the process of transferring electrons to the surface and decrease the band gap, leading to increased photocatalytic activity. A similar trend of enhancement by Ni doping was also observed with stainless steel wire mesh@doped ZnS photocatalyst [93]. Pristine stainless steel wire mesh C60 has a hydrophobic surface (water contact angle = 103°). In contrast, the water contact angle of ZnS decorated wire mesh photocatalyst C60S0.5 is 0°. Effective contact between the sacrificial solution and the photocatalyst surface is very important because the photocatalyst is used for photocatalytic H_2 generation in aqueous sacrificial solution. Improving the hydrophilicity of the photocatalyst will lead to increased activity. Table 8 lists the photocatalytic H_2 production performances for transition metal doped photocatalysts.

Table 8. Photocatalytic hydrogen production performances for transition metal doped photocatalysts.

Photocatalyst	Dopant	Synthetic Method	Sacrificial Agent	Activity (μmol h^{-1}g^{-1})	Ref. (Year)
CdS	Co	In-situ photodeposition	$(NH_4)_2SO_3$	1299 μmol h^{-1}	[87] (2018)
MoSG	Co	Solvothermal	TEOA-H_2O	11,450	[88] (2019)
ZnS-graphene	Ni	chemical vapor deposition	Na_2S, Na_2SO_3, NaCl	8683	[89] (2015)
$Cd_{1-x}Zn_xS$	Ni	Hydrothermal	Na_2S, Na_2SO_3	191	[92] (2008)
Staninless steel@ZnS	Ni	Solvothermal	Na_2S, Na_2SO_3, NaCl	14,600	[93] (2014)

3.2. Non-Metal Doping

Gopinath et al. [94] reported that the band gap of semiconductor oxide was reduced by doping with anions because of the broadening or the upward shifting of VB. Asahi et al. [95] studied the effects of C, N, F, P, or S doping on TiO_2-based photocatalysts. Their results showed that N doping can decrease the band gap of the photocatalyst due to the energy state overlapping between the N 2p states and O 2p states. Such band gap narrowing was also found for S doped photocatalysts. Up to the present, only a few studies have demonstrated the use of anion doping to improve the visible light-induced hydrogen generation performance. As non-metal doping is more difficult to prepare by conventional chemical methods, it has received little attention. Tsuji et al. demonstrated the effectiveness of halogen codoped Pb-ZnS photocatalyst in photocatalytic activity. Although Pb-doped ZnS already has a high hydrogen evolution rate, codoping of halogen is still useful, for it facilitates the relaxation of the distortion produced by doping with large Pb ions. The photocatalytic activity of halogen and Pb codoped ZnS is three times higher than that of the Pb-doped ZnS photocatalyst [96].

3.3. Calcination

By changing the treatment temperature and ambient gas condition, the post thermal treatment of metal sulfide photocatalysts at elevated temperature in air or oxygen usually results in effective electron–hole separation and enhanced photocatalytic activity, due to the formation of metal sulfide–metal oxide heterojunction. Hong et al. [97] reported that ZnS–ZnO composite prepared by thermal treatments from preformed ZnS particles showed improved charge separation and photocatalytic activity. Optimizing oxide content in ZnS–ZnO photocatalyst by controlling O_2 partial pressure (16.9 kPa) and temperature (500 °C) can help to achieve a H_2 production rate of 494.8 μmol g^{-1} h^{-1}.

The crystallinity and surface area of a photocatalyst can be modified by calcination. Based on previous studies, the H_2 production rate increases as the calcination temperature increases. However, there is an optimum temperature for the maximum H_2 evolution rate. Further increment of temperature will cause negative effects on the H_2 production rate. This negative effect results from the decrease in the surface area after heating at high temperature. These findings have been evidenced in Ce-doped ZnO/ZnS [98], ZnS$_{1-x-0.5y}$O$_x$(OH)$_y$(1:1) [99], and CdS/TiO$_2$ [85]. Table 9 presents the influence of calcination treatment on the activity of photocatalysts.

Table 9. Effects of calcination treatment on photocatalytic performance.

Photocatalyst	Synthesis Method	Optimum Temperature (K)	Surface Area (m^2/g)	Activity (μmol h^{-1}g^{-1})	Ref. (Year)
Ce-doped ZnO/ZnS	precipitation	673	51.25	1200	[98] (2015)
ZnS$_{1-x-05y}$O$_x$(OH)$_y$(1:1)T$_1$-673	Co-precipitation	373	72.4	~375 μmol	[99] (2009)
CdS/TiO$_2$	Precipitation and sol–gel method	773	~25	~620 μmol h^{-1}	[85] (2007)

3.4. Effects of pH Level

The photocatalytic evolution of hydrogen is affected by the pH level of the sacrificial agent solution. However, the effect of pH on photocatalytic performance depends on the mechanism of the reaction. According to previous research by Markovskaya et al., the pH level of the sacrificial agent can be manipulated by adding NaOH or acetic acid. The addition of acetic acid initially increases the rate of hydrogen evolution, up to a sharp peak at pH 7.5. Further increasing the acetic acid causes a drop in the evolution rate. The H_2 production in Na$_2$S/Na$_2$SO$_3$ is described as [29]

$$Na_2S + Na_2SO_3 + 2H_2O \rightarrow H_2 + Na_2S_2O_3 + 2NaOH \tag{1}$$

According to reaction (1) of the reaction mechanism, the addition of hydroxyl ions thermodynamically impedes the H_2 production reaction, while an increase in the concentration of acetic acid promotes the reaction. The dependence of the H_2 generation rate on the pH level has been reported for Ni-doped CdS nanorods [100]. Both samples under pH 14.7 conditions performed better than the other two. These results suggested that the concentration of OH$^-$ is an important factor that will affect the efficiency of hydrogen generation.

3.5. Sacrificial Agent

The sacrificial agent plays the important role of electron donor, efficiently consuming holes to prevent recombination of charge carriers on the surface of the photocatalyst [100]. Charge recombination is one of the factors that may hinder the performance of photocatalytic reactions. Adding suitable sacrificial agents helps to solve the problem. For different reactions, different sacrificial agents should be used. For photocatalytic hydrogen production reactions, the sacrificial agent acts as the hole scavenger to reduce the charge recombination of the photoexcited electron–hole pairs. Then electrons can react with H$^+$ and enhance the photocatalytic performance for hydrogen production. Metal sulfide-based photocatalysts exhibit excellent activities in aqueous solution containing sacrificial reagents Na$_2$S and Na$_2$SO$_3$. In addition to Na$_2$S and Na$_2$SO$_3$, some other candidates—such as methanol [31], triethanolamine [20,101], lactic acid [23,44], glycerol [46], and 2-propanol [102]—can act as the sacrificial agents for metal oxysulfide or metal oxide/metal sulfide composite photocatalysts. It will be more constructive from the viewpoints of energy production and environmental protection if the sacrificial agents are sourced from the chemical waste or byproducts of industrial processes. One example is glycerol, which is a byproduct of biodiesel manufacturing. In the chemical industry, sulfur-containing side products and waste are common. In addition to the type of sacrificial agent, its concentration also affects the photocatalytic activity. As an example, the effect of glycerol concentration on the performance of photocatalysts has been studied [46]. Glycerol can react with photogenerated holes to hinder the recombination of electron–hole pairs. Optimized graphene and glycerol contents can achieve the maximum H_2 generation activity (1070 μmol h^{-1}g^{-1}). Table 10 lists the photocatalytic activity of photocatalysts using different sacrificial agents.

Table 10. Photocatalytic activity of photocatalysts using different sacrificial agents.

Photocatalyst	Morphology	Synthetic Method	Sacrificial Agent	Activity (μmol h^{-1}g^{-1})	Ref. (Year)
Ni/CdS/g-C$_3$N$_4$	Hybrid system	NaBH$_4$ reduction method	Triethanolamine	1258.7	[20] (2016)
MoS$_2$/CdS	Heterostructure	Precipitation	Lactic acid	~540	[23] (2008)
CuS/TiO$_2$	Nanocomposite	Hydrothermal	Methanol	570 μmol h^{-1}	[31] (2013)
MoS$_2$/graphene-CdS	Nanocomposite	Solution-chemistry	Lactic acid	1800 μmol h^{-1}	[44] (2014)
Graphene/ZnO-ZnS	Particle-on-sheet	Two-step heating	Glycerol	1070	[46] (2018)
ZnIn$_2$S$_4$/g-C$_3$N$_4$	Hetereojunction nanosheets	In-situ growth	Triethanolamine	5.2 μmol h^{-1}	[101] (2016)
CdS/CdSe	Nanorods	-	2-propanol	40 mmol/h-g	[102] (2010)

3.6. Morphology

The morphology of a photocatalyst also affects the performance, for the surface area and the surface active sites are influenced by the structure. Photocatalysts have different morphologies, including 3D and porous morphologies, nanosheets, nanorods, nanoflowers, and nanowires [14,19,25,32,34,47,77,103]. Recently, Amirav et al. studied tunable nanorod heterostructures. They demonstrated that a longer CdSe seeded rod was more active than a shorter rod of the same diameter, as the surface active sites were located further apart. However, nanorods with comparable rod lengths but smaller diameters will provide higher activity [102]. Panmand et al. reported that more structural defects and surface states are created on CdS decorated Bi$_2$S$_3$ nanowires. Figure 7 presents the morphology of a composite photocatalyst. The photogenerated charge carriers of photocatalysts with such defect energy levels can be effectively separated, leading to enhanced photocatalytic activity [104].

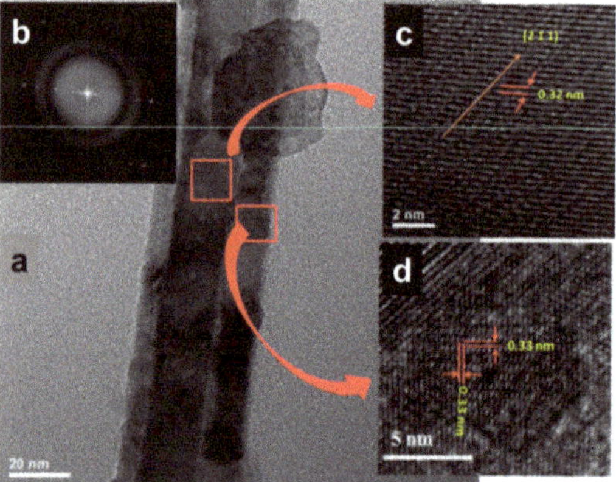

Figure 7. (**a**) HRTEM image; (**b**) SAED pattern of CdS decorated Bi$_2$S$_3$ nanowires; magnified HRTEM images of (**c**) Bi$_2$S$_3$ nanowire; (**d**) CdS nanoparticle [104].

In addition, a 2D morphology (such as a nanosheet) can help to improve the photocatalytic activity. Zhang et al. reported that ZnIn$_2$S$_4$/g-C$_3$N$_4$ heterojunction nanosheets demonstrated higher H_2

production rates compared to single heterojunction components [101]. The contact of the components to form heterojunctions is very important [105]. A nanorod array structure with a small inter-rod distance will not easily form heterojunctions with close contact between different components. In contrast, a 2D heterostructure has enhanced photocatalytic performance because the van der Waals interaction of the 2D heterostructure junction between 2D metals (1T-MoS_2) and the 2D semiconductor (O-g-C_3N_4) minimizes the Schottky barrier, thus improving the efficiency of charge transfer [106].

Moreover, the notable high hydrogen production rate and good stability achieved by mesoporous monoclinic $CaIn_2S_4$ with surface nanostructures implies the importance of structure in enhancing the H_2 evolution rate. Ding et al. reported that monoclinic $CaIn_2S_4$ (m-$CaIn_2S_4$) exhibits a lower ability than cubic $CaIn_2S_4$ (c-$CaIn_2S_4$) to absorb visible light, but it also has better photocatalytic performance. The better performance results from the larger surface area, higher pore volume, more negative conduction band potential, and efficient separation of photoexcited carriers of m-$CaIn_2S_4$ [107].

Furthermore, a core–shell structure will also benefit the photocatalyst, as it may have an increased surface area and change the surface properties. For example, Chang et al. reported that the growth of Ag_2S-ZnO@ZnS core–shell nanorods on metal wire mesh had modified the surface from hydrophobic to superhydrophilic. In addition, the H_2 production activity also increased with the increasing thickness of the ZnS shell [32]. A similar trend was also evidenced in studies of $NiCo_2O_4$@ZnS and Fe_3O_4@ZnS core shell photocatalysts [72]. Table 11 presents the performances of photocatalysts with different structures.

Table 11. Photocatalytic activity of photocatalysts with different structures.

Photocatalyst	Morphology	Synthetic Method	Sacrificial Agent	Activity (μmol h^{-1}g^{-1})	Ref. (Year)
$ZnIn_2S_4$/g-C_3N_4	Heterojunction nanosheets	In-situ growth	Triethanolamine	5.2 μmol h^{-1}	[104] (2016)
CdS/CdSe	Nanorods	-	2-propanol	40 mmol h^{-1}g^{-1}	[102] (2010)
CdS/Bi_2S_3	Nanowires	In-situ growth	H_2S, KOH	4560	[103] (2016)
$CaIn_2S_4$	Mesoporous monoclinic with surface nanostructure	High temperature sulfurization	Na_2S, Na_2SO_3	3.02 mmol h^{-1}g^{-1}	[106] (2018)
$CaIn_2S_4$/g-C_3N_4	Heterojunction nanocomposite	Two-step method	Na_2S, Na_2SO_3	102	[107] (2014)
$ZnIn_2S_4$	3D hierarchical persimmon-like shape	Oleylamine (OA)-assisted solvothermal	Na_2S, Na_2SO_3	220.45 μmol h^{-1}	[108] (2012)
Ag_2S-coupled ZnO@ZnS	Core–shell	Sulfidation	Na_2S, Na_2SO_3, NaCl	5310	[32] (2016)
$NiCo_2O_4$@ZnS	Core–shell	Solvothermal	Na_2S, Na_2SO_3, NaCl	3900	[72] (2015)
Fe_3O_4@ZnS	Core–shell	Solvothermal	Na_2S, Na_2SO_3, NaCl	880	[72] (2015)

3.6.1. Facet Effect

It has been reported that the photocatalysis reaction mainly occurs at the surface of the photocatalyst. The exposure of certain facets leads to greatly improved activity of the photocatalysts, known as the facet effect. Therefore, the preparation of photocatalysts with specific morphologies and structures is an important topic in the photocatalysis field. The facet effect has been observed for some oxide-based photocatalysts. Li et al. [109] reported that efficient separation of photoexcited electron–hole pairs can occur between different facets of photocatalytic nanomaterials. In comparison with their analogs with randomly distributed cocatalysts, selective deposition of a reduction cocatalyst and an oxidation cocatalyst on the {010} and {110} facets of $BiVO_4$ leads to higher photocatalytic activity. Ohno et al. [110] found the effect of facets on the photocatalytic activity of TiO_2 photocatalysts. For the rutile TiO_2 nanomaterials, the {110} facet can act as an effective reduction site, and the {011} facet can offer a site for effective oxidation. TiO_2 photocatalysts show high activity because of the synergistic effect between the {110} and {011} facets.

Similar results were also found for the sulfide-based photocatalysts. Song et al. found that, in comparison with 2-D Cu_2MoS_4 nanosheet with the exposed {001} facet, Cu_2MoS_4 nanotube with the exposed {010} facet exhibited effectively improved the performance for photocatalytic degradation and water splitting [111]. Shen et al. reported that the crystal facets of $ZnIn_2S_4$ with a 3D-hierarchical persimmon-like structure will influence the photocatalytic activity of $ZnIn_2S_4$. Extending the reaction time did not reveal any significant influences on the band gap or surface area of $ZnIn_2S_4$. Hence, the increase in the percentage of the {006} facet enhances hydrogen production [108]. The atomic structure of the {006} facet mainly consists of unsaturated metal cations. During the H_2 generation reaction, the exposed unsaturated Zn and In cations of $ZnIn_2S_4$ will attract S^{2-} and SO_3^- anions, which may help the oxidation process and speed up the consumption of photogenerated holes (Figure 8). This impedes the electron–hole recombination process, leading to improved photocatalytic activity.

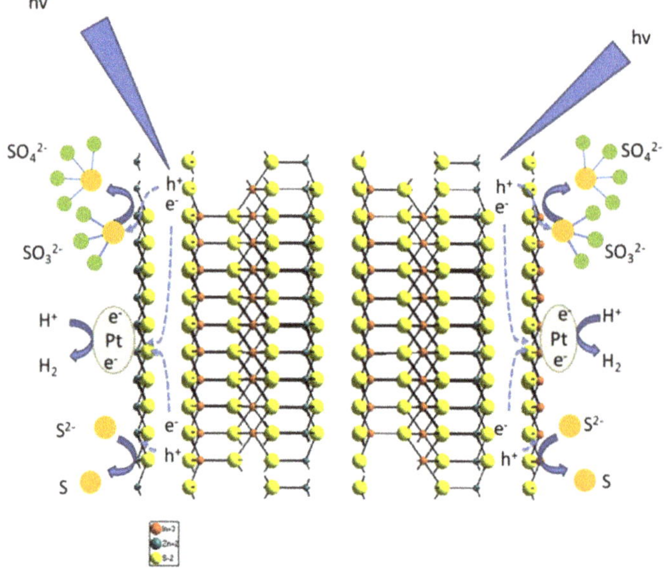

Figure 8. Schematic illustration of the hydrogen generation reaction on the {006} facets of the Pt loaded $ZnIn_2S_4$ photocatalysts. Figure adapted from reference [108].

3.6.2. Light Trapping (Light Harvesting)

It is reported that properly patterned surface textures can lead to dramatically enhanced light absorption by the photocatalyst because of the light trapping effect [32]. Some textured structures—including nanowire arrays [112], ordered mesoporous structures [113], micro-hole arrays [114], and hemisphere-array films [115,116]—are able to increase the light harvesting and photocatalytic performance of photocatalysts. Zhang et al. [112] reported that a 3D ZnO nanowire array–CdS sample exhibited substantial light-trapping enhancement in the visible light region. A schematic illustration of interface scattering when the light was irradiated on the surface of porous photocatalyst is provided in Figure 9 [116]. To enhance the light absorption efficiency, these surface textures allow multiple reflections and light scattering within the nanostructures. The incident light can travel through the cavities and decrease the optical loss. Efficient light trapping can be achieved by tuning the shape and roughness of the textured surface [117].

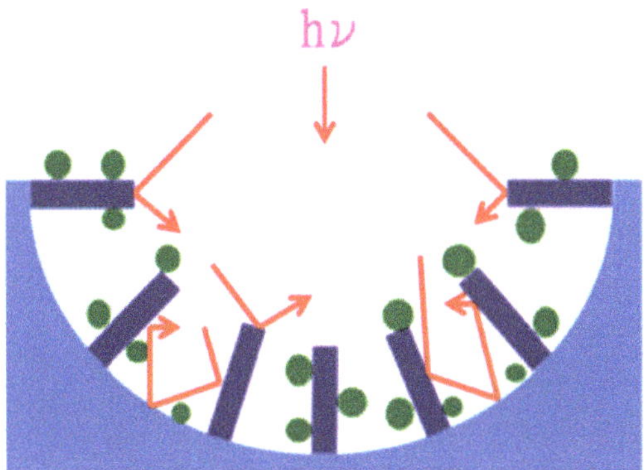

Figure 9. Schematic illustration of interface scattering due to nanograss decorated pore-array photocatalysts. Figure adapted from [116].

3.7. Fabrication Method

Photocatalysts can be prepared by numerous methods, including the coprecipitation, cation exchange, chemical bath deposition, hydrothermal, and solvothermal methods. The co-precipitation method may lead to differences between the obtained final element composition in solid solution and the stoichiometric ratio [118]. The hydrothermal method requires a great amount of time to prepare a well-crystallized solid solution [119]. In the thermolysis method, the products are treated under high temperature to achieve fast fabrication, high crystallinity, and photocatalytic performance of the solid solution [120].

Li et al. found that $Zn_{1-x}Cd_xS$ photocatalyst fabricated using the simple Zn-Cd-Tu complex thermolysis method showed better performance than did those synthesized by the coprecipitation and hydrothermal methods [118,121]. Such a thermolysis method is preferred for the following reasons: (i) the precursors can be well mixed and reacted to prepare Zn-Cd-Tu complex by the ultrasonication process; (ii) the loss of precursors can be prevented during the fabrication process; and (iii) $Zn_{1-x}Cd_xS$ with a tiny crystallite size can be prepared because thiourea releases S^{2-} ions slowly and offers some N and C atoms as the pinning points in the Zn-Cd-Tu complex [122].

Zhang et al. reported that a particular fabrication method will enable a photocatalyst to perform better in photocatalytic activity. The photocatalysts $Cd_{1-x}Zn_xS$ are prepared by three different methods: thermal sulfuration, co-precipitation without thermal treatment, and co-precipitation with thermal treatment. The photocatalyst which is synthesized by the thermal sulfuration method has better performance because the fabrication method allows uneven distribution of S^{2-} ions, thus leading to a charge gradient [123]. A similar trend is also evidenced in the work done by Park et al. on a ternary $CdS/TiO_2/Pt$ hybrid. $CdS/(Pt-sgTiO_2)$ has the highest hydrogen production because of the electron transfer from CdS to Pt through TiO_2 [124].

3.8. Crystal Size

Hydrogen production activity is also influenced by the crystal size of the photocatalyst. Li et al. [125] studied the photocatalytic hydrogen production performances of size-selected CdS nanoparticles decorated with co-catalyst Pt nanoparticles. When the size of CdS nanoparticles decreases from 4.6 to 2.8 nm, the H_2 generation quantum yield can increase from 11% to 17%. Such a dependence was observed because the driving force of photoinduced carrier transfer from CdS to vacant states of Pt nanoparticles is size-dependent. Baldovi et al. [126] prepared MoS_2 quantum dots by laser ablation of MoS_2 particles in suspension and investigated their photocatalytic hydrogen production performance. Two types of MoS_2 nanoparticles exhibited higher activity than that of bulk MoS_2. When the size of MoS_2 nanoparticles was decreased from 15–25 nm to 5 nm, the photocatalytic hydrogen generation performance was almost doubled. Holmes et al. [127] and Grigioni et al. [128] reported the dependence of the photocatalytic water splitting activity on the size of CdSe nanoparticles. The photocatalytic activity increases as the size of CdSe nanoparticles decreases. Figure 10 presents the UV–vis absorption spectra and quantum yields of CdSe quantum dots photocatalysts with various size. They also reported that the light harvesting capability and the conduction band energy should compromise to achieve maximal photocatalytic H_2 generation activity. There is an optimal size of 2.8 nm to achieve maximal photocatalytic H_2 generation activity considering the compromise among light harvesting, band structure, and charge separation.

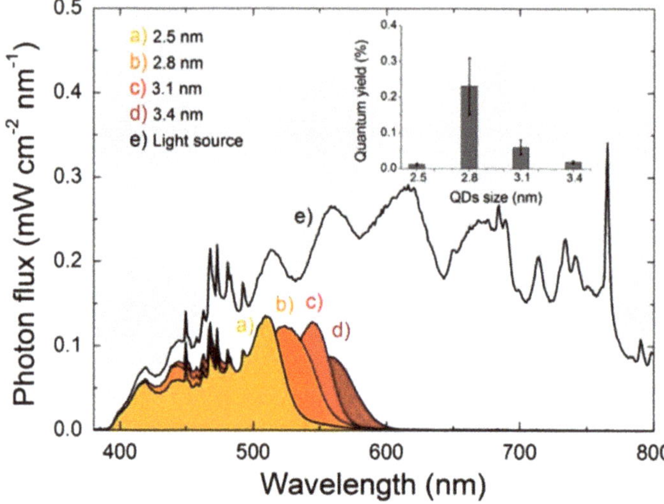

Figure 10. UV–vis absorption spectra and quantum yields of CdSe quantum dot photocatalysts with various size. Figure adapted from [128].

4. Conclusions and Perspective

In this review, we have attempted to summarize the efforts performed in the field of metal sulfide-based photocatalysts. Metal sulfide-based heterogeneous photocatalysts are promising candidates for photocatalytic hydrogen generation. Recent developments in the material design, process parameters, and performance of metal sulfide-based photocatalysts have been systematically discussed. The major problem limiting the photocatalytic H_2 generation rates of photocatalysts is the fast recombination of photoexcited electron–hole pairs. This problem can be solved by decorating with cocatalysts or incorporating noble metal nanoparticles, conductive polymers, or porous conductive substrate. The formation of heterogeneous junctions helps to promote the transport of photogenerated carriers. In-situ C K-edge NEXAFS spectra provide a new method to investigate the electronic density of the photocatalyst. It can pave the way for the rational design of photocatalysts for efficient H_2 generation. In addition, tuning the surface texture can increase the contact surface area with reactants, and the light absorption can be increased by the light trapping effect. Changing the crystal structure can also enhance the activity of a photocatalyst because of the facet effect. The fabrication of immobilized photocatalysts and magnetically separable photocatalysts makes the recycling and repeated use of photocatalysts easier to handle than photocatalyst dispersion does. The influences of doping and pH have also been discussed. The photocatalytic activity increases as the particle size of the photocatalyst decreases. This review provides a systematic overview of recent progress on the performances of various metal sulfide photocatalysts, together with some important concepts or methods to improve and characterize their performances. Key experimental parameters and an in-situ characterization method can be applied to the research of other photocatalytic materials.

In our opinion, future research efforts should be focused on the following issues. First, we should develop in-situ spectroscopy and microscopy techniques for investigating the surface active sites, electronic states, and chemical/physical changes of the photocatalysts, together with intermediates and the mechanism of the photocatalytic reactions. Second, efforts must be focused on developing outstanding materials that can achieve both high activity and excellent reusability. Possible directions for future research are developing new heterojunction structures, increasing charge transfer, enhancing light harvesting efficiency, and achieving high activity and excellent stability of recycled photocatalysts after repeated photocatalytic H_2 production processes. Finally, to implement the use of these photocatalysts for photocatalytic hydrogen generation in industry, future works should also focus on the optimized design of reactor systems for scaled-up photocatalytic processes.

Funding: This research was supported and funded by the Ministry of Science and Technology, under the contract of MOST 105-2221-E-035-087-MY3.

Acknowledgments: The authors thank the financial support from the Ministry of Science and Technology under the contract of MOST 105-2221-E-035-087-MY3.

Conflicts of Interest: The authors declare no conflict of interest.

References

1. Norskov, J.K.; Christensen, C.H. Toward efficient hydrogen production at surfaces. *Science* **2006**, *312*, 1322–1323. [CrossRef] [PubMed]
2. Das, D.; Veziroğlu, T.N. Hydrogen production by biological processes: A survey of literature. *Int. J. Hydrogen Energy* **2001**, *26*, 13–28. [CrossRef]
3. Politano, A.; Cattelan, M.; Boukhvalov, D.W.; Campi, D.; Cupolillo, A.; Agnoli, S.; Apostol, N.G.; Lacovig, P.; Lizzit, S.; Farías, D.; et al. Unveiling the mechanisms leading to H_2 production promoted by water decomposition on epitaxial graphene at room temperature. *ACS Nano* **2016**, *10*, 4543–4549. [CrossRef]
4. Christensen, C.H.; Johannessen, T.; Sørensen, R.Z.; Nørskov, J.K. Towards an ammonia-mediated hydrogen economy? *Catal. Today* **2006**, *111*, 140–144. [CrossRef]
5. Mendoza-Damián, G.; Hernández-Gordillo, A.; Fernández-García, M.E.; Acevedo-Peña, P.; Tzompantzi-Morales, F.J.; Pérez-Hernández, R. Influence of ZnS wurtzite–sphalerite junctions on ZnO Core-ZnS Shell-1D photocatalysts for H_2 production. *Int. J. Hydrogen Energy* **2019**, *44*, 10528–10540. [CrossRef]
6. Kılıç, B.; Kılıç, Ş. Hydrogen economy model for nearly net-zero cities with exergy rationale and energy-water nexus. *Energies* **2018**, *11*, 1226. [CrossRef]
7. Weidner, J.W. Solar Energy: An Enabler of Hydrogen Economy? *Electrochem. Soc. Interface* **2018**, *27*, 45. [CrossRef]
8. Viktorsson, L.; Heinonen, J.; Skulason, J.; Unnthorsson, R. A step towards the hydrogen Economy—A life cycle cost analysis of A hydrogen refueling station. *Energies* **2017**, *10*, 763. [CrossRef]
9. Ke, X.; Dai, K.; Zhu, G.; Zhang, J.; Liang, C. In situ photochemical synthesis noble-metal-free NiS on CdS-diethylenetriamine nanosheets for boosting photocatalytic H_2 production activity. *Appl. Surf. Sci.* **2019**, *481*, 669–677. [CrossRef]
10. An, Z.; Gao, J.; Wang, L.; Zhao, X.; Yao, H.; Zhang, M.; Tian, Q.; Liu, Y. Novel microreactors of polyacrylamide (PAM) CdS microgels for admirable photocatalytic H2 production under visible light. *Int. J. Hydrogen Energy* **2019**, *44*, 1514–1524. [CrossRef]
11. Yuan, Y.J.; Chen, D.; Yu, Z.T.; Zou, Z.G. Cadmium sulfide-based nanomaterials for photocatalytic hydrogen production. *J. Mater. Chem. A* **2018**, *6*, 11606–11630. [CrossRef]
12. Ahmad, H.; Kamarudin, S.K.; Minggu, L.J.; Kassim, M. Hydrogen from photo-catalytic water splitting process: A review. *Renew. Sustain. Energy Rev.* **2015**, *43*, 599–610. [CrossRef]
13. Majeed, I.; Nadeem, M.A.; Al-Oufi, M.; Nadeem, M.A.; Waterhouse, G.I.N.; Badshah, A.; Metson, J.B.; Idriss, H. On the role of metal particle size and surface coverage for photo-catalytic hydrogen production: A case study of the Au/CdS system. *Appl. Catal. B-Environ.* **2016**, *182*, 266–276. [CrossRef]

14. Cao, J.; Sun, J.Z.; Hong, J.; Li, H.Y.; Chen, H.Z.; Wang, M. Carbon Nanotube/CdS Core–Shell Nanowires Prepared by a Simple Room-Temperature Chemical Reduction Method. *Adv. Mater.* **2004**, *16*, 84–87. [CrossRef]
15. Chang, C.J.; Yang, T.L.; Weng, Y.C. Synthesis and characterization of Cr-doped ZnO nanorod-array photocatalysts with improved activity. *J. Solid State Chem.* **2014**, *214*, 101–107. [CrossRef]
16. Hsu, M.H.; Chang, C.J. Ag-doped ZnO nanorods coated metal wire meshes as hierarchical photocatalysts with high visible-light driven photoactivity and photostability. *J. Hazard. Mater.* **2014**, *278*, 444–453. [CrossRef]
17. Wang, X.; Liu, G.; Chen, Z.G.; Li, F.; Wang, L.; Lu, G.Q.; Cheng, H.M. Enhanced photocatalytic hydrogen evolution by prolonging the lifetime of carriers in ZnO/CdS heterostructures. *Chem. Commun.* **2009**, *23*, 3452–3454. [CrossRef]
18. Zhang, J.; Wang, Y.; Jin, J.; Zhang, J.; Lin, Z.; Huang, F.; Yu, J. Efficient visible-light photocatalytic hydrogen evolution and enhanced photostability of core–shell CdS/g-C$_3$N$_4$ nanowires. *ACS Appl. Mater. Interface* **2013**, *5*, 10317–10324. [CrossRef]
19. Yan, Z.; Sun, Z.; Liu, X.; Jia, H.; Du, P. Cadmium sulfide/graphitic carbon nitride heterostructure nanowire loading with a nickel hydroxide cocatalyst for highly efficient photocatalytic hydrogen production in water under visible light. *Nanoscale* **2016**, *8*, 4748–4756. [CrossRef]
20. Yue, X.; Yi, S.; Wang, R.; Zhang, Z.; Qiu, S. Cadmium sulfide and nickel synergetic co-catalysts supported on graphitic carbon nitride for visible-light-driven photocatalytic hydrogen evolution. *Sci. Rep.-UK* **2016**, *6*, 22268. [CrossRef]
21. Liu, M.; Li, F.; Sun, Z.; Ma, L.; Xu, L.; Wang, Y. Noble-metal-free photocatalysts MoS$_2$–graphene/CdS mixed nanoparticles/nanorods morphology with high visible light efficiency for H$_2$ evolution. *Chem. Commun.* **2014**, *50*, 11004–11007. [CrossRef] [PubMed]
22. Sun, Z.; Chen, H.; Zhang, L.; Lu, D.; Du, P. Enhanced photocatalytic H$_2$ production on cadmium sulfide photocatalysts using nickel nitride as a novel cocatalyst. *J. Mater. Chem. A* **2016**, *4*, 13289–13295. [CrossRef]
23. Zong, X.; Yan, H.; Wu, G.; Ma, G.; Wen, F.; Wang, L.; Li, C. Enhancement of photocatalytic H$_2$ evolution on CdS by loading MoS$_2$ as cocatalyst under visible light irradiation. *J. Am. Chem. Soc.* **2008**, *130*, 7176–7177. [CrossRef]
24. Iwashina, K.; Iwase, A.; Ng, Y.H.; Amal, R.; Kudo, A. Z-schematic water splitting into H$_2$ and O$_2$ using metal sulfide as a hydrogen-evolving photocatalyst and reduced graphene oxide as a solid-state electron mediator. *J. Am. Chem. Soc.* **2015**, *137*, 604–607. [CrossRef] [PubMed]
25. Wang, L.; Chen, H.; Xiao, L.; Huang, J. CuS/ZnS hexagonal plates with enhanced hydrogen evolution activity under visible light irradiation. *Powder Technol.* **2016**, *288*, 103–108. [CrossRef]
26. Chen, T.; Song, C.; Fan, M.; Hong, Y.; Hu, B.; Yu, L.; Shi, W. In-situ fabrication of CuS/g-C$_3$N$_4$ nanocomposites with enhanced photocatalytic H$_2$-production activity via photoinduced interfacial charge transfer. *Int. J. Hydrogen Energy* **2017**, *42*, 12210–12219. [CrossRef]
27. Chang, C.J.; Weng, H.T.; Chang, C.C. CuSZnS$_{1-x}$O$_x$/g-C$_3$N$_4$ heterostructured photocatalysts for efficient photocatalytic hydrogen production. *Int. J. Hydrogen Energy* **2017**, *42*, 23568–23577. [CrossRef]
28. Chang, C.J.; Wei, Y.H.; Kuo, W.S. Free-standing CuS–ZnS decorated carbon nanotube films as immobilized photocatalysts for hydrogen production. *Int. J. Hydrogen Energy* **2018**. [CrossRef]
29. Markovskaya, D.V.; Cherepanova, S.V.; Saraev, A.A.; Gerasimov, E.Y.; Kozlova, E.A. Photocatalytic hydrogen evolution from aqueous solutions of Na$_2$S/Na$_2$SO$_3$ under visible light irradiation on CuS/Cd$_{0.3}$Zn$_{0.7}$S and Ni$_z$Cd$_{0.3}$Zn$_{0.7}$S$_{1+z}$. *Chem. Eng. J.* **2015**, *262*, 146–155. [CrossRef]
30. Chandra, M.; Bhunia, K.; Pradhan, D. Controlled Synthesis of CuS/TiO$_2$ Heterostructured Nanocomposites for Enhanced Photocatalytic Hydrogen Generation through Water Splitting. *Inorg. Chem.* **2018**, *57*, 4524–4533. [CrossRef]
31. Wang, Q.; An, N.; Bai, Y.; Hang, H.; Li, J.; Lu, X.; Liu, Y.; Wang, F.; Li, Z.; Lei, Z. High photocatalytic hydrogen production from methanol aqueous solution using the photocatalysts CuS/TiO$_2$. *Int. J. Hydrogen Energy* **2013**, *38*, 10739–10745. [CrossRef]
32. Hsu, M.H.; Chang, C.J.; Weng, H.T. Efficient H$_2$ production using Ag$_2$S-coupled ZnO@ZnS core–shell nanorods decorated metal wire mesh as an immobilized hierarchical photocatalyst. *ACS Sustain. Chem. Eng.* **2016**, *4*, 1381–1391. [CrossRef]
33. Yue, S.; Wei, B.; Guo, X.; Yang, S.; Wang, L.; He, J. Novel Ag$_2$S/ZnS/carbon nanofiber ternary nanocomposite for highly efficient photocatalytic hydrogen production. *Catal. Commun.* **2016**, *76*, 37–41. [CrossRef]

34. Yang, X.; Xue, H.; Xu, J.; Huang, X.; Zhang, J.; Tang, Y.B.; Ng, W.T.; Kwong, H.L.; Meng, X.M.; Lee, C.S. Synthesis of porous ZnS:Ag$_2$S nanosheets by ion exchange for photocatalytic H$_2$ generation. *ACS Appl. Mater. Interface* **2014**, *6*, 9078–9084. [CrossRef] [PubMed]

35. Xin, Z.; Li, L.; Zhang, W.; Sui, T.; Li, Y.; Zhang, X. Synthesis of ZnS@CdS–Te composites with p–n heterostructures for enhanced photocatalytic hydrogen production by microwave-assisted hydrothermal method. *Mol. Catal.* **2018**, *447*, 1–12. [CrossRef]

36. Hao, X.; Zhou, J.; Cui, Z.; Wang, Y.; Wang, Y.; Zou, Z. Zn-vacancy mediated electron–hole separation in ZnS/g-C$_3$N$_4$ heterojunction for efficient visible-light photocatalytic hydrogen production. *Appl. Catal. B-Environ.* **2018**, *229*, 41–51. [CrossRef]

37. Nawaz, M. Morphology-controlled preparation of Bi$_2$S$_3$-ZnS chloroplast-like structures, formation mechanism and photocatalytic activity for hydrogen production. *J. Photochem. Photobiol. A Chem.* **2017**, *332*, 326–330. [CrossRef]

38. Jiang, D.; Sun, Z.; Jia, H.; Lu, D.; Du, P. A cocatalyst-free CdS nanorod/ZnS nanoparticle composite for high-performance visible-light-driven hydrogen production from water. *J. Mater. Chem. A* **2016**, *4*, 675–683. [CrossRef]

39. Li, X.; Yu, J.; Wageh, S.; Al-Ghamdi, A.A.; Xie, J. Graphene in photocatalysis: A review. *Small* **2016**, *12*, 6640–6696. [CrossRef]

40. Li, Q.; Li, X.; Wageh, S.; Al-Ghamdi, A.A.; Yu, J. CdS/graphene nanocomposite photocatalysts. *Adv. Energy Mater.* **2015**, *5*, 1500010. [CrossRef]

41. Gugliuzza, A.; Politano, A.; Drioli, E. The advent of graphene and other two-dimensional materials in membrane science and technology. *Curr. Opin. Chem. Eng.* **2017**, *16*, 78–85. [CrossRef]

42. Quiroz-Cardoso, O.; Oros-Ruiz, S.; Solis-Gomez, A.; López, R.; Gómez, R. Enhanced photocatalytic hydrogen production by CdS nanofibers modified with graphene oxide and nickel nanoparticles under visible light. *Fuel* **2019**, *237*, 227–235. [CrossRef]

43. Chu, K.W.; Lee, S.L.; Chang, C.J.; Liu, L. Recent Progress of Carbon Dot Precursors and Photocatalysis Applications. *Polymers* **2019**, *11*, 689. [CrossRef] [PubMed]

44. Chang, C.J.; Wang, C.W.; Wei, Y.H.; Chen, C.Y. Enhanced Photocatalytic H$_2$ Production activity of Ag-doped Bi$_2$WO$_6$-graphene Based Photocatalysts. *Int. J. Hydrogen Energy* **2018**, *43*, 11345–11354. [CrossRef]

45. Azarang, M.; Sookhakian, M.; Aliahmad, M.; Dorraj, M.; Basirun, W.J.; Goh, B.T.; Alias, Y. Nitrogen-doped graphene-supported zinc sulfide nanorods as efficient Pt-free for visible-light photocatalytic hydrogen production. *Int. J. Hydrogen Energy* **2018**, *43*, 14905–14914. [CrossRef]

46. Chang, C.J.; Lin, Y.G.; Weng, H.T.; Wei, Y.H. Photocatalytic hydrogen production from glycerol solution at room temperature by ZnO-ZnS/graphene photocatalysts. *Appl. Surf. Sci.* **2018**, *451*, 198–206. [CrossRef]

47. Chang, C.J.; Lin, Y.G.; Chao, P.Y.; Chen, J.K. AgI-BiOI-graphene composite photocatalysts with enhanced interfacial charge transfer and photocatalytic H$_2$ production activity. *Appl. Surf. Sci.* **2019**, *469*, 703–712. [CrossRef]

48. Zhang, J.; Wang, P.; Sun, J.; Jin, Y. High-Efficiency Plasmon-Enhanced and Graphene-Supported Semiconductor/Metal Core–Satellite Hetero-Nanocrystal Photocatalysts for Visible-Light Dye Photodegradation and H$_2$ Production from Water. *ACS Appl. Mater. Interface* **2014**, *6*, 19905–19913. [CrossRef]

49. Chang, C.J.; Wei, Y.H.; Huang, K.P. Photocatalytic hydrogen production by flower-like graphene supported ZnS composite photocatalysts. *Int. J. Hydrogen Energy* **2017**, *42*, 23578–23586. [CrossRef]

50. Novoselov, K.S.; Geim, A.K.; Morozov, S.V.; Jiang, D.A.; Zhang, Y.; Dubonos, S.V.; Grigorieva, I.V.; Firsov, A.A. Electric field effect in atomically thin carbon films. *Science* **2004**, *306*, 666–669. [CrossRef]

51. Nethravathi, C.; Nisha, T.; Ravishankar, N.; Shivakumara, C.; Rajamathi, M. Graphene–nanocrystalline metal sulphide composites produced by a one-pot reaction starting from graphite oxide. *Carbon* **2009**, *47*, 2054–2059. [CrossRef]

52. Tsai, M.H.; Chang, C.J.; Lu, H.H.; Liao, Y.F.; Tseng, I.H. Properties of magnetron-sputtered moisture barrier layer on transparent polyimide/graphene nanocomposite film. *Thin Solid Films* **2013**, *544*, 324–330. [CrossRef]

53. Reddy, P.M.; Chang, C.J.; Lai, C.F.; Su, M.J.; Tsai, M.H. Improved organic-inorganic/graphene hybrid composite as encapsulant for white LEDs: Role of graphene, titanium (IV) isopropoxide and diphenylsilanediol. *Compos. Sci. Technol.* **2018**, *165*, 95–105. [CrossRef]

54. Mathkar, A.; Tozier, D.; Cox, P.; Ong, P.; Galande, C.; Balakrishnan, K.; Reddy, A.L.M.; Ajayan, P.M. Controlled, stepwise reduction and band gap manipulation of graphene oxide. *J. Phys. Chem. Lett.* **2012**, *3*, 986–991. [CrossRef] [PubMed]

55. Zhu, Y.; Murali, S.; Cai, W.; Li, X.; Suk, J.W.; Potts, J.R.; Ruoff, R.S. Graphene and graphene oxide: Synthesis, properties, and applications. *Adv. Mater.* **2010**, *22*, 3906–3924. [CrossRef]

56. Dreyer, D.R.; Park, S.; Bielawski, C.W.; Ruoff, R.S. The chemistry of graphene oxide. *Chem. Soc. Rev.* **2010**, *39*, 228–240. [CrossRef]

57. Yan, J.A.; Xian, L.; Chou, M.Y. Structural and electronic properties of oxidized graphene. *Phys. Rev. Lett.* **2009**, *103*, 086802. [CrossRef]

58. Zhuo, S.; Shao, M.; Lee, S.T. Upconversion and downconversion fluorescent graphene quantum dots: Ultrasonic preparation and photocatalysis. *ACS Nano* **2012**, *6*, 1059–1064. [CrossRef]

59. Yeh, T.F.; Syu, J.M.; Cheng, C.; Chang, T.H.; Teng, H. Graphite oxide as a photocatalyst for hydrogen production from water. *Adv. Funct. Mater.* **2010**, *20*, 2255–2262. [CrossRef]

60. Peng, T.; Li, K.; Zeng, P.; Zhang, Q.; Zhang, X. Enhanced photocatalytic hydrogen production over graphene oxide–cadmium sulfide nanocomposite under visible light irradiation. *J. Phys. Chem. C* **2012**, *116*, 22720–22726. [CrossRef]

61. Hou, J.; Wang, Z.; Kan, W.; Jiao, S.; Zhu, H.; Kumar, R.V. Efficient visible-light-driven photocatalytic hydrogen production using CdS@TaON core–shell composites coupled with graphene oxide nanosheets. *J. Mater. Chem.* **2012**, *22*, 7291–7299. [CrossRef]

62. Zhang, J.; Yu, J.; Jaroniec, M.; Gong, J.R. Noble metal-free reduced graphene oxide-$Zn_xCd_{1-x}S$ nanocomposite with enhanced solar photocatalytic H_2-production performance. *Nano Lett.* **2012**, *12*, 4584–4589. [CrossRef]

63. Wang, X.; Yin, L.; Liu, G. Light irradiation-assisted synthesis of ZnO–CdS/reduced graphene oxide heterostructured sheets for efficient photocatalytic H_2 evolution. *Chem. Commun.* **2014**, *50*, 3460–3463. [CrossRef]

64. Zhang, J.; Qi, L.; Ran, J.; Yu, J.; Qiao, S.Z. Ternary NiS/$Zn_xCd_{1-x}S$/Reduced Graphene Oxide Nanocomposites for Enhanced Solar Photocatalytic H_2-Production Activity. *Adv. Energy Mater* **2014**, *4*, 1301925. [CrossRef]

65. Lee, S.L.; Chang, C.J. Recent Developments about Conductive Polymer Based Composite Photocatalysts. *Polymers* **2019**, *11*, 206. [CrossRef]

66. Chang, C.J.; Chu, K.W. ZnS/polyaniline composites with improved dispersing stability and high photocatalytic hydrogen production activity. *Int. J. Hydrogen Energy* **2016**, *41*, 21764–21773. [CrossRef]

67. Wang, C.; Hu, Z.Y.; Zhao, H.; Yu, W.; Wu, S.; Liu, J.; Chen, L.; Li, Y.; Su, B.L. Probing conducting polymers@cadmium sulfide core–shell nanorods for highly improved photocatalytic hydrogen production. *J. Colloid Interface Sci.* **2018**, *521*, 1–10. [CrossRef] [PubMed]

68. Zielińska, B.; Schmidt, B.; Mijowska, E.; Kalenczuk, R. PANI/NaTaO$_3$ composite photocatalyst for enhanced hydrogen generation under UV light irradiation. *Pol. J. Chem. Technol.* **2017**, *19*, 115–119. [CrossRef]

69. Sasikala, R.; Gaikwad, A.P.; Jayakumar, O.D.; Girija, K.G.; Rao, R.; Tyagi, A.K.; Bharadwaj, S.R. Nanohybrid MoS_2-PANI-CdS photocatalyst for hydrogen evolution from water. *Colloids Surf. A* **2015**, *481*, 485–492. [CrossRef]

70. Chang, C.J.; Chao, P.Y. Efficient photocatalytic hydrogen production by doped ZnS grown on Ni foam as porous immobilized photocatalysts. *Int. J. Hydrogen Energy* **2019**. [CrossRef]

71. Hsu, M.H.; Chang, C.J. S-doped ZnO nanorods on stainless-steel wire mesh as immobilized hierarchical photocatalysts for photocatalytic H_2 production. *Int. J. Hydrogen Energy* **2014**, *39*, 16524–16533. [CrossRef]

72. Chang, C.J.; Lee, Z.; Wei, M.; Chang, C.C.; Chu, K.W. Photocatalytic hydrogen production by magnetically separable Fe_3O_4@ZnS and $NiCo_2O_4$@ZnS core–shell nanoparticles. *Int. J. Hydrogen Energy* **2015**, *40*, 11436–11443. [CrossRef]

73. Chang, C.J.; Lee, Z.; Chu, K.W.; Wei, Y.H. $CoFe_2O_4$@ZnS core–shell spheres as magnetically recyclable photocatalysts for hydrogen production. *J. Taiwan Inst. Chem. Eng.* **2016**, *66*, 386–393. [CrossRef]

74. Murdoch, M.G.I.N.; Waterhouse, G.I.N.; Nadeem, M.A.; Metson, J.B.; Keane, M.A.; Howe, R.F.; Llorca, J.; Idriss, H. The effect of gold loading and particle size on photocatalytic hydrogen production from ethanol over Au/TiO$_2$ nanoparticles. *Nat. Chem.* **2011**, *3*, 489. [CrossRef]

75. Sun, F.; Qiao, X.; Tan, F.; Wang, W.; Qiu, X. One-step microwave synthesis of Ag/ZnO nanocomposites with enhanced photocatalytic performance. *J. Mater. Sci.* **2012**, *47*, 7262–7268. [CrossRef]

76. Sreethawong, T.; Yoshikawa, S. Comparative investigation on photocatalytic hydrogen evolution over Cu-, Pd-, and Au-loaded mesoporous TiO$_2$ photocatalysts. *Catal. Commun.* **2005**, *6*, 661–668. [CrossRef]

77. Zhou, H.; Pan, J.; Ding, L.; Tang, Y.; Ding, J.; Guo, Q.; Fan, T.; Zhang, D. Biomass-derived hierarchical porous CdS/M/TiO$_2$ (M = Au, Ag, Pt, Pd) ternary heterojunctions for photocatalytic hydrogen evolution. *Int. J. Hydrogen Energy* **2014**, *39*, 16293–16301. [CrossRef]

78. Zhang, J.; Wang, Y.; Zhang, J.; Lin, Z.; Huang, F.; Yu, J. Enhanced photocatalytic hydrogen production activities of Au-loaded ZnS flowers. *ACS Appl. Mater. Interface* **2013**, *5*, 1031–1037. [CrossRef]

79. Pany, S.; Naik, B.; Martha, S.; Parida, K. Plasmon induced nano Au particle decorated over S, N-modified TiO$_2$ for exceptional photocatalytic hydrogen evolution under visible light. *ACS Appl. Mater. Interface* **2014**, *6*, 839–846. [CrossRef]

80. Yu, X.; Shavel, A.; An, X.; Luo, Z.; Ibáñez, M.; Cabot, A. Cu$_2$ZnSnS$_4$-Pt and Cu$_2$ZnSnS$_4$-Au heterostructured nanoparticles for photocatalytic water splitting and pollutant degradation. *J. Am. Chem. Soc.* **2014**, *136*, 9236–9239. [CrossRef]

81. Yan, H.; Yang, J.; Ma, G.; Wu, G.; Zong, X.; Lei, Z.; Shi, J.; Li, C. Visible-light-driven hydrogen production with extremely high quantum efficiency on Pt–PdS/CdS photocatalyst. *J. Catal.* **2009**, *266*, 165–168. [CrossRef]

82. Manjunath, K.; Souza, V.S.; Nagaraju, G.; Santos, J.M.L.; Dupont, J.; Ramakrishnappa, T. Superior activity of the CuS–TiO$_2$/Pt hybrid nanostructure towards visible light induced hydrogen production. *New J. Chem.* **2016**, *40*, 10172–10180. [CrossRef]

83. Liu, G.; Zhao, L.; Ma, L.; Guo, L. Photocatalytic H$_2$ evolution under visible light irradiation on a novel Cd$_x$Cu$_y$Zn$_{1-x-y}$S catalyst. *Catal. Commun.* **2008**, *9*, 126–130. [CrossRef]

84. Kimi, M.; Yuliati, L.; Shamsuddin, M. Preparation and characterization of In and Cu co-doped ZnS photocatalysts for hydrogen production under visible light irradiation. *J. Energy Chem.* **2016**, *25*, 512–516. [CrossRef]

85. Jang, J.S.; Ji, S.M.; Bae, S.W.; Son, H.C.; Lee, J.S. Optimization of CdS/TiO$_2$ nano-bulk composite photocatalysts for hydrogen production from Na$_2$S/Na$_2$SO$_3$ aqueous electrolyte solution under visible light ($\lambda \geq$ 420 nm). *J. Photochem. Photobiol. A Chem.* **2007**, *188*, 112–119. [CrossRef]

86. Wang, X.; Liu, G.; Lu, G.Q.; Cheng, H.M. Stable photocatalytic hydrogen evolution from water over ZnO–CdS core–shell nanorods. *Int. J. Hydrogen Energy* **2010**, *35*, 8199–8205. [CrossRef]

87. Chen, W.; Wang, Y.; Liu, M.; Gao, L.; Mao, L.; Fan, Z.; Shangguan, W. In situ photodeposition of cobalt on CdS nanorod for promoting photocatalytic hydrogen production under visible light irradiation. *Appl. Surf. Sci.* **2018**, *444*, 485–490. [CrossRef]

88. Zhou, H.; Liu, Y.; Zhang, L.; Li, H.; Liu, H.; Li, W. Transition metal-doped amorphous molybdenum sulfide/graphene ternary cocatalysts for excellent photocatalytic hydrogen evolution: Synergistic effect of transition metal and graphene. *J. Colloid Interface Sci.* **2019**, *533*, 287–296. [CrossRef]

89. Wang, H.; Tsai, C.; Kong, D.; Chan, K.; Abild-Pedersen, F.; Nørskov, J.K.; Cui, Y. Transition-metal doped edge sites in vertically aligned MoS$_2$ catalysts for enhanced hydrogen evolution. *Nano Res.* **2015**, *8*, 566–575. [CrossRef]

90. Weng, Y.C.; Chou, Y.D.; Chang, C.J.; Chan, C.C.; Chen, K.Y.; Su, Y.F. Screening of ZnS-based photocatalysts by scanning electrochemical microscopy and characterization of potential photocatalysts. *Electrochim. Acta* **2014**, *125*, 354–361. [CrossRef]

91. Chang, C.J.; Chu, K.W.; Hsu, M.H.; Chen, C.Y. Ni-doped ZnS decorated graphene composites with enhanced photocatalytic hydrogen-production performance. *Int. J. Hydrogen Energy* **2015**, *40*, 14498–14506. [CrossRef]

92. Zhang, X.; Jing, D.; Liu, M.; Guo, L. Efficient photocatalytic H$_2$ production under visible light irradiation over Ni doped Cd$_{1-x}$Zn$_x$S microsphere photocatalysts. *Catal. Commun.* **2008**, *9*, 1720–1724. [CrossRef]

93. Chang, C.J.; Lee, Z.; Wang, C.F. Photocatalytic hydrogen production by stainless steel@ZnS core–shell wire mesh photocatalyst from saltwater. *Int. J. Hydrogen Energy* **2014**, *39*, 20754–20763. [CrossRef]

94. Rajaambal, S.; Sivaranjani, K.; Gopinath, C.S. Recent developments in solar H$_2$ generation from water splitting. *J. Chem. Sci.* **2015**, *127*, 33–47. [CrossRef]

95. Asahi, R.; Morikawa, T.; Ohwaki, T.; Aoki, K.; Taga, Y. Visible-light photocatalysis in nitrogen-doped titanium oxides. *Science* **2001**, *293*, 269–271. [CrossRef]

96. Tsuji, I.; Kudo, A. H$_2$ evolution from aqueous sulfite solutions under visible-light irradiation over Pb and halogen-codoped ZnS photocatalysts. *J. Photochem. Photobiol. A Chem.* **2003**, *156*, 249–252. [CrossRef]

97. Hong, E.; Kim, J.H. Oxide content optimized ZnS–ZnO heterostructures via facile thermal treatment process for enhanced photocatalytic hydrogen production. *Int. J. Hydrogen Energy* **2014**, *39*, 9985–9993. [CrossRef]

98. Chang, C.J.; Huang, K.L.; Chen, J.K.; Chu, K.W.; Hsu, M.H. Improved photocatalytic hydrogen production of ZnO/ZnS based photocatalysts by Ce doping. *J. Taiwan Inst. Chem. Eng.* **2015**, *55*, 82–89. [CrossRef]

99. Li, Y.; Ma, G.; Peng, S.; Lu, G.; Li, S. Photocatalytic H_2 evolution over basic zincoxysulfide $(ZnS_{1-x-0.5y}O_x(OH)_y)$ under visible light irradiation. *Appl. Catal. A-Gen.* **2009**, *363*, 180–187. [CrossRef]

100. Simon, T.; Bouchonville, N.; Berr, M.J.; Vaneski, A.; Adrović, A.; Volbers, D.; Wyrwich, R.; Döblinger, M.; Susha, A.S.; Rogach, A.L.; et al. Redox shuttle mechanism enhances photocatalytic H_2 generation on Ni-decorated CdS nanorods. *Nat. Mater.* **2014**, *13*, 1013. [CrossRef]

101. Zhang, Z.; Liu, K.; Feng, Z.; Bao, Y.; Dong, B. Hierarchical sheet-on-sheet $ZnIn_2S_4/gC_3N_4$ heterostructure with highly efficient photocatalytic H_2 production based on photoinduced interfacial charge transfer. *Sci. Rep.* **2016**, *6*, 19221. [CrossRef] [PubMed]

102. Amirav, L.; Alivisatos, A.P. Photocatalytic hydrogen production with tunable nanorod heterostructures. *J. Phys. Chem. Lett.* **2010**, *1*, 1051–1054. [CrossRef]

103. Chang, C.J.; Hsu, M.H.; Weng, Y.C.; Tsay, C.Y.; Lin, C.K. Hierarchical ZnO nanorod-array films with enhanced photocatalytic performance. *Thin Solid Films* **2013**, *528*, 167–174. [CrossRef]

104. Panmand, R.P.; Sethi, Y.A.; Deokar, R.S.; Late, D.J.; Gholap, H.M.; Baeg, J.O.; Kale, B.B. In situ fabrication of highly crystalline CdS decorated Bi_2S_3 nanowires (nano-heterostructure) for visible light photocatalyst application. *RSC Adv.* **2016**, *6*, 23508–23517. [CrossRef]

105. Chang, C.J.; Tsai, M.H.; Hsu, Y.H.; Tuan, C.S. Morphology and optoelectronic properties of ZnO rod array/conjugated polymer hybrid films. *Thin Solid Films* **2008**, *516*, 5523–5526. [CrossRef]

106. Xu, H.; Yi, J.; She, X.; Liu, Q.; Song, L.; Chen, S.; Yang, Y.; Song, Y.; Lou, J.; Li, H.; et al. 2D heterostructure comprised of metallic $1T-MoS_2/$Monolayer $O-g-C_3N_4$ towards efficient photocatalytic hydrogen evolution. *Appl. Catal. B-Environ.* **2018**, *220*, 379–385. [CrossRef]

107. Ding, J.; Hong, B.; Luo, Z.; Sun, S.; Bao, J.; Gao, C. Mesoporous monoclinic $CaIn_2S_4$ with surface nanostructure: An efficient photocatalyst for hydrogen production under visible light. *J. Phys. Chem. C* **2014**, *118*, 27690–27697. [CrossRef]

108. Shen, J.; Zai, J.; Yuan, Y.; Qian, X. 3D hierarchical $ZnIn_2S_4$: The preparation and photocatalytic properties on water splitting. *Int. J. Hydrogen Energy* **2012**, *37*, 16986–16993. [CrossRef]

109. Li, R.; Zhang, F.; Wang, D.; Yang, J.; Li, M.; Zhu, J.; Zhou, X.; Han, H.; Li, C. Spatial separation of photogenerated electrons and holes among {010} and {110} crystal facets of $BiVO_4$. *Nat. Commun.* **2013**, *4*, 1432. [CrossRef]

110. Ohno, T.; Sarukawa, K.; Matsumura, M. Crystal faces of rutile and anatase TiO_2 particles and their roles in photocatalytic reactions. *New J. Chem.* **2002**, *26*, 1167–1170. [CrossRef]

111. Zhang, K.; Lin, Y.; Muhammad, Z.; Wu, C.; Yang, S.; He, Q.; Zheng, X.; Chen, S.; Ge, B.; Song, L. Active {010} facet-exposed Cu_2MoS_4 nanotube as high-efficiency photocatalyst. *Nano Res.* **2017**, *10*, 3817–3825. [CrossRef]

112. Bai, Z.; Yan, X.; Li, Y.; Kang, Z.; Cao, S.; Zhang, Y. 3D-Branched ZnO/CdS Nanowire Arrays for Solar Water Splitting and the Service Safety Research. *Adv. Energy Mater.* **2016**, *6*, 1501459. [CrossRef]

113. Li, W.; Wu, Z.; Wang, J.; Elzatahry, A.A.; Zhao, D. A perspective on mesoporous TiO_2 materials. *Chem. Mater.* **2013**, *26*, 287–298. [CrossRef]

114. Liu, H.; Zhou, H.; Ding, J.; Zhang, D.; Zhu, H.; Fan, T. Hydrogen evolution via sunlight water splitting on an artificial butterfly wing architecture. *Phys. Chem. Chem. Phys.* **2011**, *13*, 10872–10876. [CrossRef]

115. Chang, C.J.; Kuo, E.H. Light-trapping effects and dye adsorption of ZnO hemisphere-array surface containing growth-hindered nanorods. *Colloids Surf. A Physicochem. Eng. Asp.* **2010**, *363*, 22–29. [CrossRef]

116. Hung, S.T.; Chang, C.J.; Hsu, M.H. Improved photocatalytic performance of ZnO nanograss decorated pore-array films by surface texture modification and silver nanoparticle deposition. *J. Hazard. Mater.* **2011**, *198*, 307–316. [CrossRef]

117. Kluth, O.; Rech, B.; Houben, L.; Wieder, S.; Schöpe, G.; Beneking, C.; Wagner, H.; Löffl, A.; Schock, H.W. Texture etched ZnO: Al coated glass substrates for silicon based thin film solar cells. *Thin Solid Films* **1999**, *351*, 247–253. [CrossRef]

118. Xing, C.; Zhang, Y.; Yan, W.; Guo, L. Band structure-controlled solid solution of $Cd_{1-x}Zn_xS$ photocatalyst for hydrogen production by water splitting. *Int. J. Hydrogen Energy* **2006**, *31*, 2018–2024. [CrossRef]

119. Wang, X.; Liu, G.; Chen, Z.G.; Li, F.; Lu, G.Q.M.; Cheng, H.M. Efficient and stable photocatalytic H_2 evolution from water splitting by $(Cd_{0.8}Zn_{0.2})$ S nanorods. *Electrochem. Commun.* **2009**, *11*, 1174–1178. [CrossRef]

120. Sung, Y.M.; Lee, Y.J.; Park, K.S. Kinetic Analysis for Formation of $Cd_{1-x}Zn_xSe$ Solid-Solution Nanocrystals. *J. Am. Chem. Soc.* **2006**, *128*, 9002–9003. [CrossRef]
121. Wang, L.; Wang, W.; Shang, M.; Yin, W.; Sun, S.; Zhang, L. Enhanced photocatalytic hydrogen evolution under visible light over $Cd_{1-x}Zn_xS$ solid solution with cubic zinc blend phase. *Int. J. Hydrogen Energy* **2010**, *35*, 19–25. [CrossRef]
122. Li, Q.; Meng, H.; Zhou, P.; Zheng, Y.; Wang, J.; Yu, J.; Gong, J. $Zn_{1-x}Cd_xS$ Solid Solutions with Controlled Bandgap and Enhanced Visible-Light Photocatalytic H_2-Production Activity. *ACS Catal.* **2013**, *3*, 882–889. [CrossRef]
123. Zhang, K.; Jing, D.; Xing, C.; Guo, L. Significantly improved photocatalytic hydrogen production activity over $Cd_{1-x}Zn_xS$ photocatalysts prepared by a novel thermal sulfuration method. *Int. J. Hydrogen Energy* **2007**, *32*, 4685–4691. [CrossRef]
124. Park, H.; Choi, W.; Hoffmann, M.R. Effects of the preparation method of the ternary $CdS/TiO_2/Pt$ hybrid photocatalysts on visible light-induced hydrogen production. *J. Mater. Chem.* **2008**, *18*, 2379–2385. [CrossRef]
125. Li, W.; O'Dowd, G.; Whittles, T.J.; Hesp, D.; Gründer, Y.; Dhanak, V.R.; Jäckel, F. Colloidal dual-band gap cell for photocatalytic hydrogen generation. *Nanoscale* **2015**, *7*, 16606–16610. [CrossRef]
126. Baldoví, H.G.; Latorre-Sánchez, M.; Esteve-Adell, I.; Khan, A.; Asiri, A.M.; Kosa, S.A.; Garcia, H. Generation of MoS_2 quantum dots by laser ablation of MoS_2 particles in suspension and their photocatalytic activity for H_2 generation. *J. Nanopart. Res.* **2016**, *18*, 240–248. [CrossRef]
127. Holmes, M.A.; Townsend, T.K.; Osterloh, F.E. Quantum confinement controlled photocatalytic water splitting by suspended CdSe nanocrystals. *Chem. Commun.* **2012**, *48*, 371–373. [CrossRef]
128. Grigioni, I.; Bernareggi, M.; Sinibaldi, G.; Dozzi, M.V.; Selli, E. Size-dependent performance of CdSe quantum dots in the photocatalytic evolution of hydrogen under visible light irradiation. *Appl. Catal. A-Gen.* **2016**, *518*, 176–180. [CrossRef]

 *catalysts*

Article

Wastewater Treatment and Electricity Production in a Microbial Fuel Cell with Cu–B Alloy as the Cathode Catalyst

Paweł P. Włodarczyk * and Barbara Włodarczyk

Institute of Technical Sciences, Faculty of Natural Sciences and Technology, University of Opole, Dmowskiego Str. 7-9, 45-365 Opole, Poland
* Correspondence: pawel.wlodarczyk@uni.opole.pl; Tel.: +48-077-401-6717

Received: 5 May 2019; Accepted: 25 June 2019; Published: 28 June 2019

Abstract: The possibility of wastewater treatment and electricity production using a microbial fuel cell with Cu–B alloy as the cathode catalyst is presented in this paper. Our research covered the catalyst preparation; measurements of the electroless potential of electrodes with the Cu–B catalyst, measurements of the influence of anodic charge on the catalytic activity of the Cu–B alloy, electricity production in a microbial fuel cell (with a Cu–B cathode), and a comparison of changes in the concentration of chemical oxygen demand (COD), NH_4^+, and NO_3^- in three reactors: one excluding aeration, one with aeration, and during microbial fuel cell operation (with a Cu–B cathode). During the experiments, electricity production equal to 0.21–0.35 mA·cm^{-2} was obtained. The use of a microbial fuel cell (MFC) with Cu–B offers a similar reduction time for COD to that resulting from the application of aeration. The measured reduction of NH_4^+ was unchanged when compared with cases employing MFCs, and it was found that effectiveness of about 90% can be achieved for NO_3^- reduction. From the results of this study, we conclude that Cu–B can be employed to play the role of a cathode catalyst in applications of microbial fuel cells employed for wastewater treatment and the production of electricity.

Keywords: non-precious metal catalysts; Cu–B alloy; microbial fuel cell; cathode; environmental engineering; oxygen electrode; renewable energy sources

---

1. Introduction

At present, the power industry faces difficulties ensuring the production of greater volumes of energy to meet the increased demand. Simultaneously, the production of waste and wastewater increases considerably. This means that large amounts of industrial and municipal wastewater may be generated. The traditional design of a wastewater treatment plant consumes a lot of energy to perform efficiently, and this generates considerable costs. Approximately \$23 billion is spent annually by the United States on domestic wastewater treatment and improving the quality of publicly owned treatment infrastructure costs another \$200 billion [1]. In this context, it is clear that it is important to decrease the costs of wastewater treatment. Nowadays, there are different ideas for the use of wastewater as a raw material for other technologies, and there has been fast development in renewable sources of energy using wastewater. A technical device that can combine electricity production with wastewater treatment is a microbial fuel cell (MFC) [2]. MFCs are ecological sources of electric energy which produce electricity from wastewater [2–4]. While the first observation of an electrical current generated by bacteria is generally credited to Potter [5], very few practical advances were achieved in this field prior to the 1960s [6–8]. In the 1990s, there was increased interest in MFC research [9–11], but significant development of MFCs only occurred in recent years [2–4,12–15].

MFCs are bio-electrochemical systems in the form of devices that use bacteria as catalysts to oxidize organic and inorganic matter and generate a current [2,7]. Activated sludge is capable of producing electrons e^- and H^+ ions. In an MFC, organic material is oxidized on the anode, and the product of oxidation is CO_2 and electrons. For a glucose reaction, we obtain [2,16,17].

$$\text{ANODE } C_6H_{12}O_6 + 6H_2O \rightarrow 6CO_2 + 24H^+ + 24e^- \tag{1}$$

$$\text{CATHODE } 24H^+ + 24e^- + 6O_2 \rightarrow 12H_2O \tag{2}$$

$$\text{Summary reaction}: C_6H_{12}O_6 + 6O_2 \rightarrow 6CO_2 + 6H_2O + \text{electricity} \tag{3}$$

Electron-producing bacteria that are capable of wastewater treatment play a key role in the effective performance of MFCs [2,4]. Such bacteria include *Geobacter*, *Shewanella*, or *Pseudomonas*, among many other genera [18–25]. An analysis of reports in this field demonstrates that the highest values of capacity are generated by MFCs comprising multispecies aggregates, where microorganisms grow in the form of biofilms. Mixed cultures seem to provide a solid and more efficient solution compared to cultures based on a single strain, and their isolation from natural sources is a much less complex task. In contrast, the use of single-strain cultures is associated with technical limitations, mainly resulting from the need for ensuring sterile growth conditions, and the process usually involves high costs [26]. Figure 1 shows a diagram of the microbial fuel cell.

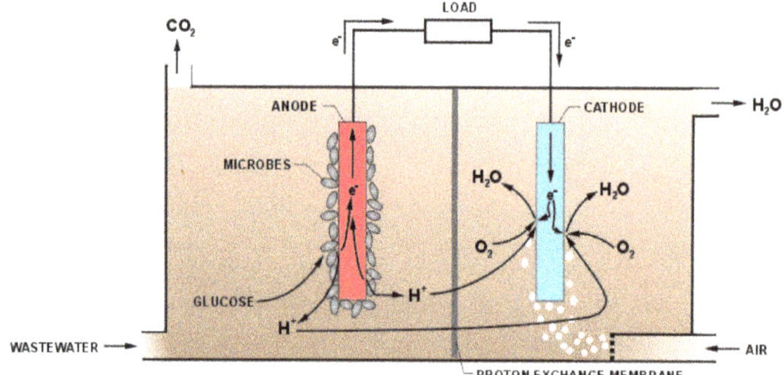

Figure 1. Operating principles of a microbial fuel cell (MFC). Figure is not to scale.

Currently, several theoretical and practical works connected to increasing the MFCs' power have been presented, not only in the field of the microorganism selection. The upper limit of the power level that is achievable in MFCs is not yet known because there are many reasons for power limitations. The reasons limiting the maximum power density may be different, e.g., high internal resistance or low speed of reactions on electrodes [2,4,27,28]. The speed of the process depends on the catalyst used. In an MFC, the catalyst at the anode is microbes (on a carbon electrode). Thus, it is important to find a catalyst for the cathode. Due to its excellent catalytic properties, platinum is most commonly used as the catalyst. However, due to the high price of platinum, we should look for other catalysts, such as the non-precious metals. In previous works [29–34] were compared the performances of microbial fuel cells (MFCs) equipped with different cheap electrode materials (graphite, carbon felt, foam, and cloth and carbon nanotube sponges, and Polypyrrole/carbon black composite) during two-month-long tests in which they were operated under the same operating conditions. Despite using sp2 carbon materials (carbon felt, foam, and cloth) as the anode in the different MFCs, the results demonstrated that there were important differences in the performance, pointing out the relevance of the surface area and other physical characteristics to the efficiency of MFCs. Differences were found not only in the production of electricity but also in the consumption of fuel. Carbon felt was found to be the most efficient anode material, whereas the worst results were obtained with carbon cloth. Performance seems to have a direct relationship with the specific area of the anode materials. In comparing the performances of the MFCs equipped with carbon felt and stainless steel as the cathodes, the latter showed the worse performance, which clearly indicates how the cathodic process may become the bottleneck of the MFC performance. Besides platinum, graphite, carbon felt, foam, cloth, etc., Ni or metal borides are frequently used as the catalyst of electrodes. Due to costs, in MFCs, carbon or carbon cloth with platinum is most often used as the cathode catalyst. It is also possible to use metal catalysts for the cathodes of MFCs [35,36]. The theoretical current density is described by the Butler–Volmer exponential function [37]. Unfortunately, in real conditions, the choice of catalyst is mainly come to by experimental methods [37,38]. For this reason, experimental research on the selection of new catalysts for MFCs is still conducted [14,29,30,33–40]. Herein, we demonstrate the possibility of using Cu–B alloy as a cathode catalyst for MFCs for municipal wastewater treatment and electricity production.

2. Results and Discussion

Figures 2–5 show the trend in time of electroless potential measured at Cu–B alloys in alkaline electrolyte (KOH). Cu–B alloys that contained 3%, 6%, 9%, and 12% of B, oxidized for 1, 3, 6, and 8 h, were selected and the experimental set up in 3.2 chapter was adopted for these measurements.

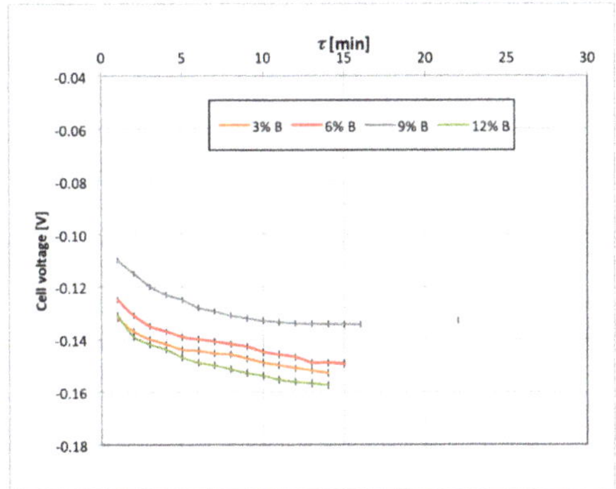

Figure 2. The electroless potential of electrodes with Cu–B catalyst which were oxidized for 1 h.

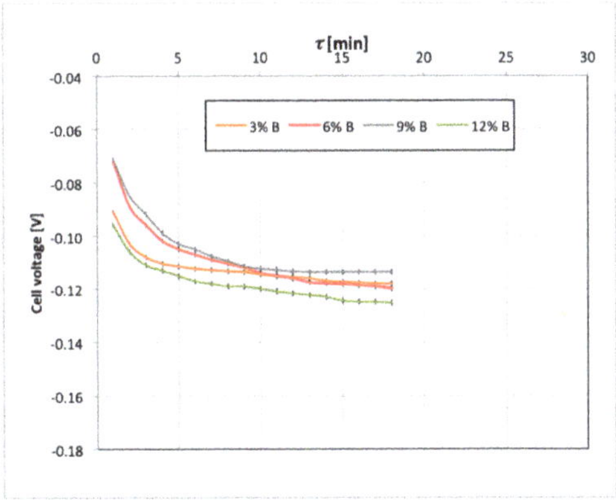

Figure 3. The electroless potential of electrodes with Cu–B catalyst which were oxidized for 3 h.

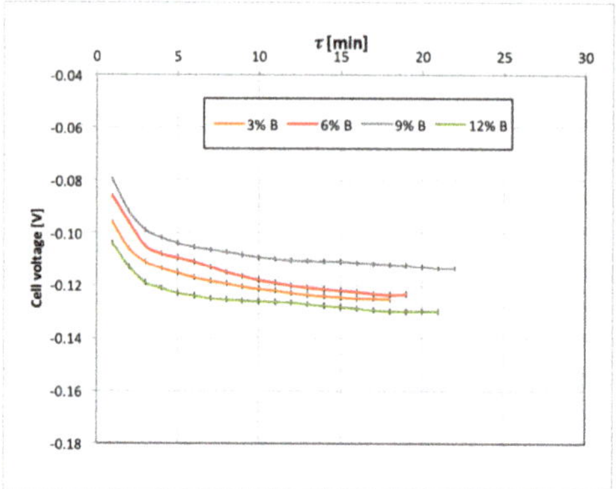

Figure 4. The electroless potential of electrodes with Cu–B catalyst which were oxidized for 6 h.

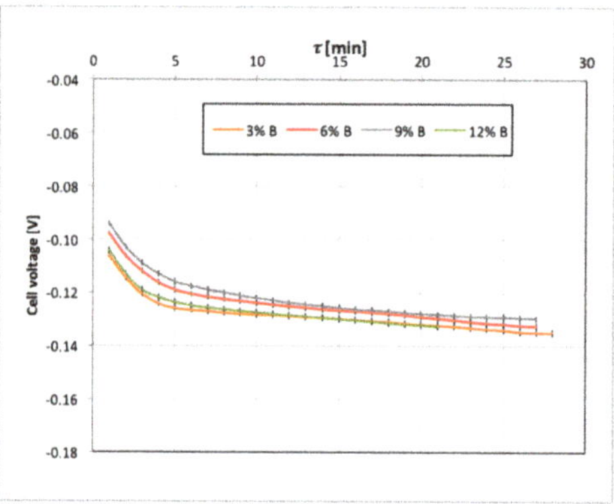

Figure 5. The electroless potential of electrodes with Cu–B catalyst which were oxidized for 8 h.

For all concentrations of boride in the Cu–B catalyst, higher current density was obtained when it was oxidized for 6 h. Thus, measurements of the effect of anodic charge on the catalytic activity of the Cu–B catalyst were performed for the samples oxidized for 6 h, shown in Figures 6–10. By colored lines (1-4) was marked the subsequent anodic charge.

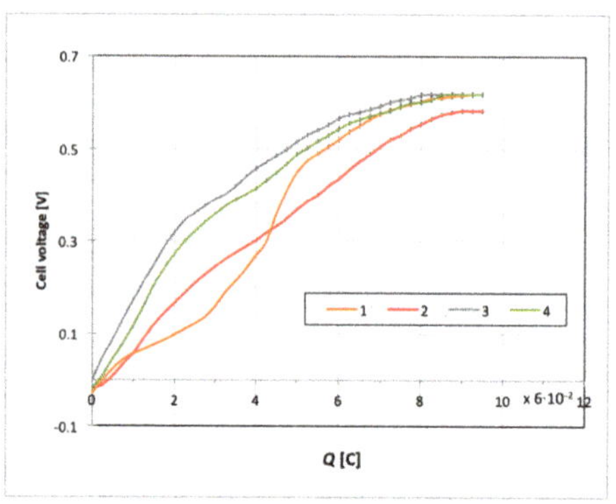

Figure 6. Influence of anodic charge on the catalytic activity of Cu–B alloy containing 3% of B.

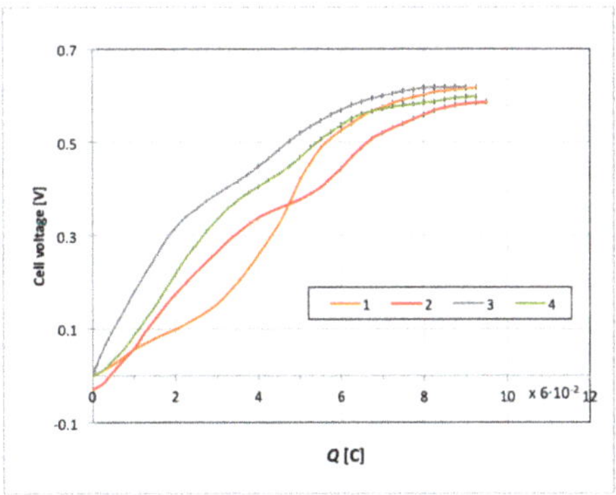

Figure 7. Influence of anodic charge on the catalytic activity of Cu–B alloy containing 6% of B.

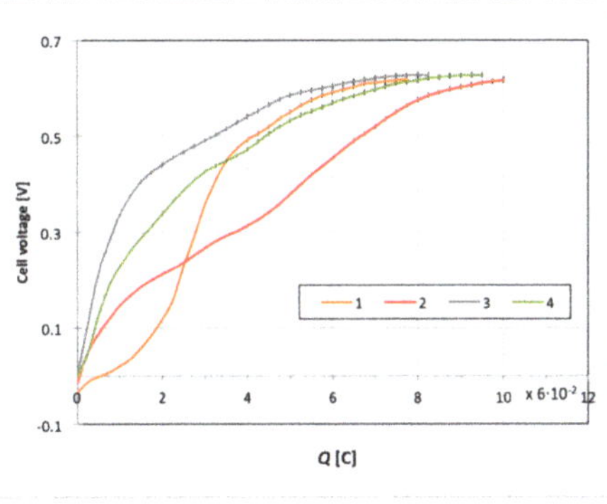

Figure 8. Influence of anodic charge on the catalytic activity of Cu–B alloy containing 9% of B.

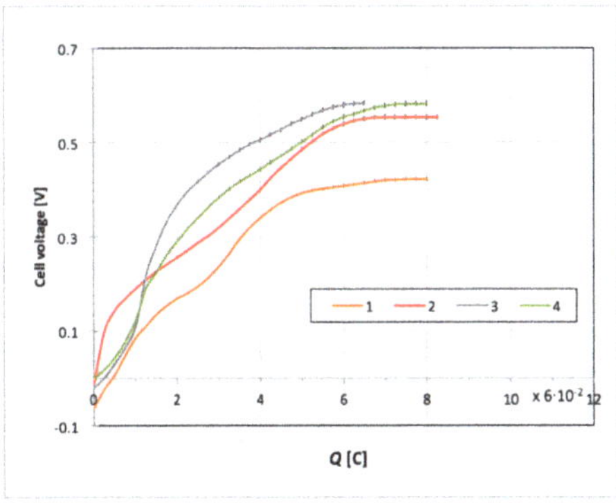

Figure 9. Influence of anodic charge on the catalytic activity of Cu–B alloy containing 12% of B.

Based on the data (Figures 2–5), it should be noted that in any case, the electroless potential is the highest for the alloy with 9% B concentration. Moreover, the cell voltage is also the highest for the alloy with 9% B concentration after the third anodic charging of the electrode (Figures 6–9). Such parameters ensure high efficiency of the electrode's functioning. Therefore, based on the data analysis (Figures 2–9), the electrode with 9% B concentration after the third anodic charge was chosen for further measurements of the MFC.

Figures 10–12 show the trends of chemical oxygen demand (COD), NH_4^+, and NO_3^- concentration during wastewater treatment in the three reactors (R1-R3, 3.3 chapter), in which the MFC was equipped with a Cu–B cathode and with a carbon cloth cathode.

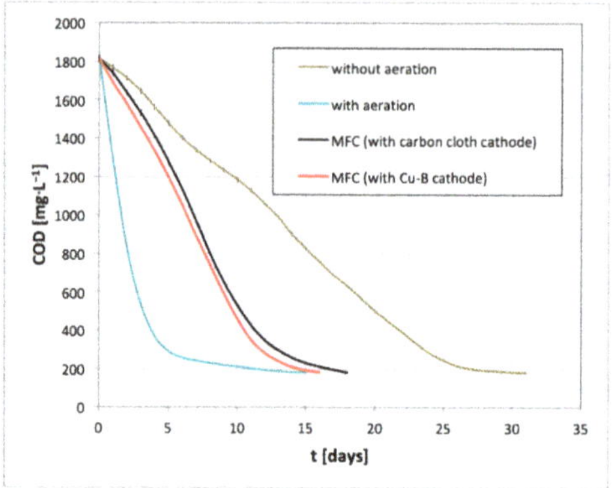

Figure 10. Trend of chemical oxygen demand (COD) reduction during wastewater treatment performed at the different reactors.

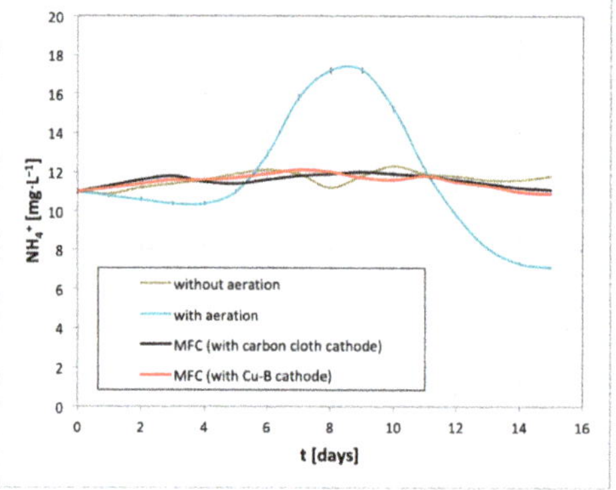

Figure 11. Trend of NH_4^+ reduction during wastewater treatment performed at the different reactors.

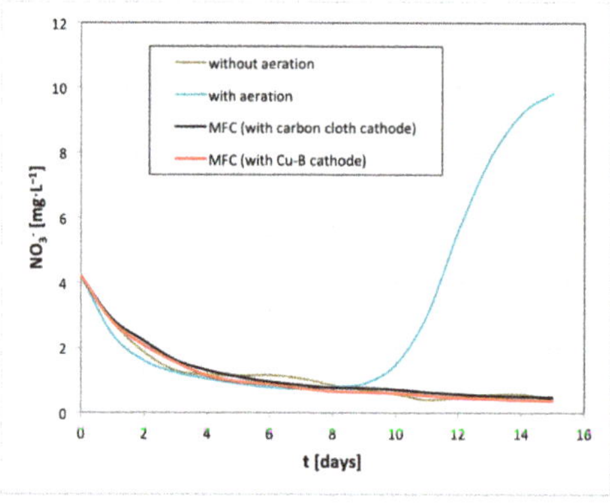

Figure 12. Trend of NO_3^- reduction during wastewater treatment performed at the different reactors.

Figure 13 shows power curves of the MFC (with a Cu–B cathode and with a carbon cloth cathode). The data shown in Figure 13 were obtained during the operation of the MFC (R3 in 3.3 chapter).

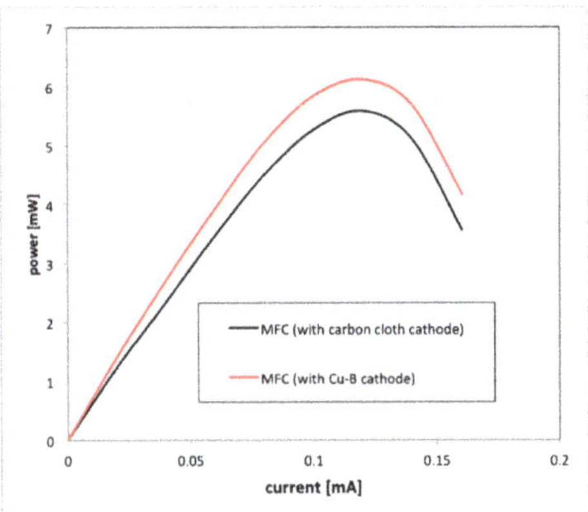

Figure 13. Power curves of the MFC: effect of the cathode material.

An analysis of the data indicated that, for all the examined concentrations of B, Cu–B alloy oxidized for 6 h showed high electroless potential. The best result was obtained after triple anodic oxidation, at a temperature of 673 K and when the concentration of boride is equal to 9%. Therefore, the electrode with a 9% concentration of B after the third anodic charge was chosen for further measurements with the MFC. Wastewater from a wastewater treatment plant was fed into the MFC. The R3 reactor was analyzed in two cases: using the Cu–B cathode and using a carbon cloth cathode. A removal level of COD of 90% was recorded in all reactors (R1, R2, and R3) (Figure 10). The characteristics of the curves were found to be different. Better performance was found in the characteristic curve of COD removal during aeration (Figure 10, blue line) than in the curve of COD removal during MFC operation (Figure 10, red and black line) since around 81% effectiveness of COD reduction after about 5 days was obtained in this period.

However, over time, a COD reduction level of 90% was achieved by the application of the MFC (over a period of 16 days resulting from the use of the Cu–B cathode, compared to 18 days for the case of the carbon cloth cathode). These results are similar to the case of the reduction time during the aeration process (15 days). It should be noted that the performance of the Cu–B cathode is better than that of carbon cloth. A faster decrease in the COD concentration is always measured at Cu-B sample. The measurement of NH_4^+ reduction shows no changes during the MFC operation (R3) (Figure 11) for either the MFC with a Cu–B cathode or that with a carbon cloth cathode. A similar situation occurred for the R1 without aeration.

However, the measurements (Figure 12) also show the effectiveness of NO_3^- reduction (during 15 days) in the MFC with a Cu–B cathode (effectiveness of 90.71%), the MFC with a carbon cloth cathode (effectiveness of 88.33%), and the reactor without aeration (effectiveness of 89.52%). The increased NH_4^+ concentration in R2 results from the attachment of hydrogen molecules to ammonia ions during wastewater putrefaction (Figure 11) [41,42]. The increased NO_3^- concentration (Figure 12) is the result of nitrification during the growth of bacteria [43]. During the MFC experiments, current densities of 0.21 mA·cm^{-2} for the carbon cloth cathode and 0.35 mA·cm^{-2} for the Cu–B cathode were obtained. Power levels of 5.58 mW in the MFC with the carbon cloth cathode and 6.11 mW in the MFC with the Cu–B cathode were obtained. The power obtained in the MFC with Cu–B alloy as the cathode catalyst is similar to the power obtained in an MFC with a Ni–Co cathode [44,45]. However, in the case of using the Ni–Co cathode, the MFC was powered with process wastewater from a yeast factory, while the MFC (with Cu–B alloy as the cathode catalyst) analyzed in this work was powered with municipal wastewater from a wastewater treatment plant.

3. Materials and Methods

3.1. Preparation of a Cathode with Cu–B Catalyst

The Cu–B alloys were obtained by the method of electrochemical deposition and were deposited on copper mesh electrodes. The alloys were deposited from a mixture of mainly $NaBH_4$ and $CuSO_4$ [46,47]. The alloys were obtained at temperatures of 355–365 K and at a current density 1–3 A·dm^{-2} [42,46,47]. The composition of the mixture used for electrochemical catalyst deposition is summarized in Table 1.

Table 1. Composition of the mixture applied for catalyst deposition (Cu–B alloy).

Component	Volume
$NaBH_4$	0.02 mol·L^{-1}
$CuSO_4 \cdot 7H_2O$	0.05 mol·L^{-1}
NaOH	1.00 mol·L^{-1}
Trilon B	0.12 mol·L^{-1}

Before the deposition of the alloy, the copper electrode was prepared in several steps [42,44,48,49]: the surface was mechanically purified (to a shine) and then degreased in 25% aqueous solution of KOH (after degreasing, the surface should be completely wettable with water); then, the electrode was digested in acetic acid and subsequently washed with alcohol.

To obtain different contents of B in the alloys, the temperature and current density were selected experimentally. Electrodes with Cu–B alloy as the catalyst (a selection of alloys with different contents of B) for further measurements were selected by the XRD method using a single-crystal X-ray diffractometer (Xcalibur, Oxford Diffraction, UK). During the electrochemical deposition, 12 alloys with different concentrations of boride were obtained. Figure 14 shows the concentrations of components in the samples obtained during electrochemical deposition.

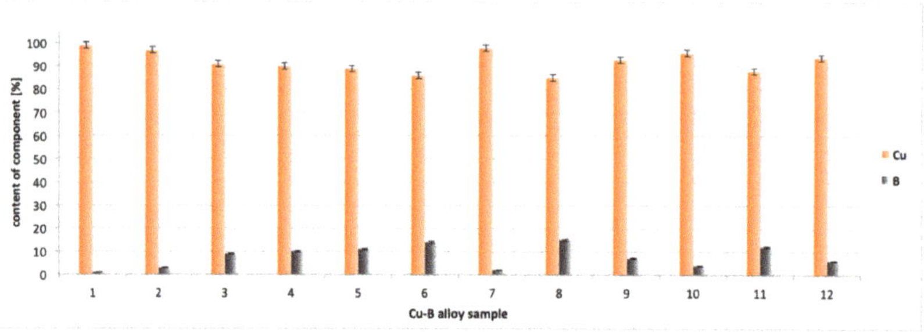

Figure 14. Concentration of components in the samples obtained during electrochemical deposition.

For further research, Samples 2 (3% of B), 3 (9% of B), 11 (12% of B), and 12 (6% of B) were selected. A further increase in B concentration (over 12%) did not cause an increase in efficiency of the MFC (i.e., an increase in cell power and current density). These samples were selected based on previous studies [42,46,47] and to ensure an even increase in B concentration to 12% (in this case, every 3%). Thus, the Cu–B alloys with 3%, 6%, 9%, and 12% of B were used in measurements.

3.2. Selection of the Electrodes (with Cu–B Catalyst) for Measurements

To assess the Cu–B alloy oxygen activity, first, the oxidation of the alloy was carried out with measurements of the stationary potential of the oxidized electrode. Due to the fact that the cathode is constantly oxygenated during MFC operation, it is necessary to pre-oxidize it. Without pre-oxygenation, the electrode would oxidize during MFC operation and there would be an efficiency decrease (and, thus, also a decrease in the current density and the cell's power). The Cu–B alloy was oxidized at a temperature of 673 K. The oxidation times were 1, 3, 6, and 8 h. The KS 520/14 silt furnace (ELIOG Industrieofenbau GmbH, Römhild, Germany) was used for electrode oxidation. Next, we measured the influence of anodic charge on the catalytic activity of the Cu–B alloy. Initial anode charging avoids a drop in the cell (MFC) efficiency during operation. Figure 15 shows a schematic view of the measurement of the catalytic activity of the Cu–B alloy.

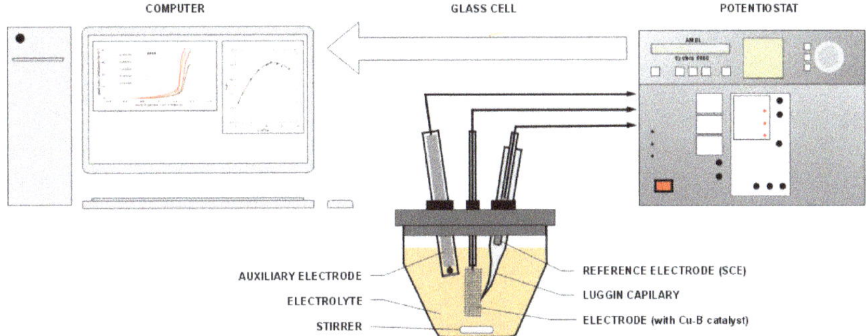

Figure 15. Schematic view of the reactor for the measurement of the electroless potential and the influence of anodic charge of electrodes with a Cu–B catalyst.

These measurements were carried out in a glass cell with the use of a potentiostat. An aqueous solution of KOH (2 M) was used as the electrolyte. A saturated calomel electrode (SCE) was used as the reference electrode. The experiments were conducted using an AMEL System 500 potentiostat (Amel S.l.r., Milano, Italy) with CorrWare software (Scribner Associates Inc., Southern Pines, NC, USA).

3.3. Measurements of Wastewater Treatment and Electricity Production in the MFC with a Cu–B Cathode

Wastewater samples from a municipal wastewater treatment plant were used in the measurements applied for the purposes of this study. Table 2 contains a summary of the parameters of the wastewater applied in the measurements.

Table 2. Parameters of wastewater applied for measurements.

Parameter	Value
COD [mg·L^{-1}]	1811.0
NH$_4^+$ [mg·L^{-1}]	11.1
NO$_3^-$ [mg·L^{-1}]	4.2
pH	6.5

The initial phase of the analysis involved measuring the reduction in the chemical oxygen demand (COD) in the investigated samples. Subsequently, variations in the NH$_4^+$ and NO$_3^-$ concentrations were measured. These measurements were conducted with regard to three types of reactors: one excluding aeration (Reactor 1—R1), one with aeration (Reactor 2—R2), and in the form of a continuous measurement performed in an MFC (Reactor 3—R3). Figure 16 shows these three types of reactors used in the measurement of wastewater parameters during wastewater treatment.

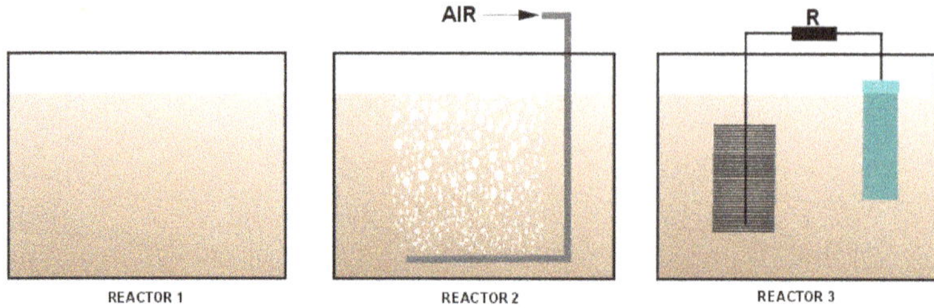

Figure 16. The three types of reactors used in the measurement of wastewater parameters during wastewater treatment: one excluding aeration (Reactor 1—R1), one with aeration (Reactor 2—R2), and as an MFC (Reactor 3—R3).

All reactors had the same dimensions (length/width/height: 40 cm × 20 cm × 20 cm). Thus, each reactor contained an equal volume of wastewater of 15 L. Each reactor worked separately but at the same time. The measurements of COD reduction were carried out (in each reactor, Figure 16) to a point where a 90% decrease in the concentration was obtained (Figure 10) [44,50], while the measurements of NH_4^+ and NO_3^- concentrations were carried out over the same time as the shortest COD reduction time (15 days; reduction time for R1 reactor with aeration, Figure 10). Determining the measurement time for NH_4^+ and NO_3^- parameters allowed a comparison of the results for all three reactors (Figures 11 and 12). In the reactor excluding aeration (R1), an interface between the wastewater and air occurred only at the wastewater surface. In the reactor with aeration (R2), aeration of wastewater was achieved as a result of using a pump with a capacity of 270 L·h^{-1}. In the last reactor design (R3), the treatment of wastewater occurred as a result of using an MFC. The wastewater parameters and electrical parameters were measured during the MFC operation. Figure 17 shows a schematic view of the MFC (Reactor 3).

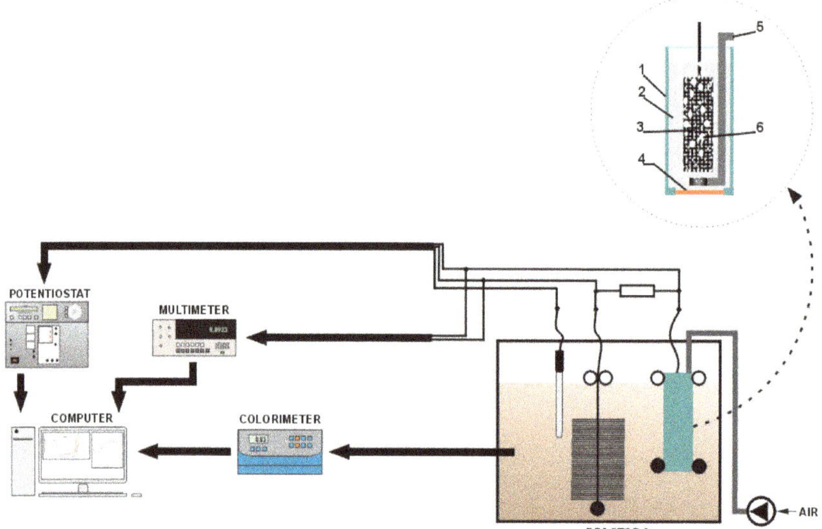

Figure 17. Schematic view of the MFC (Reactor 3, Figure 16): 1, casing; 2, electrolyte; 3, Cu–B cathode or the carbon cloth electrode; 4, proton exchange membrane (PEM); 5, air supply; 6, air bubbles.

Carbon cloth was used in the MFC in the anode, and the metal mesh with Cu–B catalyst formed the material applied in the design of the cathode of the system. For comparison, measurements were also carried out for a carbon cloth cathode. The surface area of the anode was 20 cm^2, while it was 15 cm^2

for the cathode. The electrical circuit of the MFC was constantly connected with a 10 Ω resistor [44,51]. The acclimation time of the microorganisms was 5 days [2,14,50].

The cathode was placed in a casing that was printed using 3D technology. The thickness of a single print layer was 0.09 mm. ABS (acrylonitrile butadiene styrene) filament was used as the material for 3D printing. The bottom wall was printed as a perforated wall in order to install a proton exchange membrane (PEM). After the PEM was installed, a catholyte with aqueous KOH solution (0.1 N) was filled into the casing. Subsequently, the cathode with Cu–B catalyst (or a carbon cloth cathode) was applied as the catholyte. Throughout the course of the experiment employing MFC (R3), the cathode was aerated at a capacity of 10 L·h^{-1}.

Nafion PF 117 (The Chemours Company, Wilmington, DE, USA), 183 µm thick, was used as the PEM. A Zortrax M200 printer (Zortrax S.A, Olsztyn, Poland) with Z-Suite software (Zortrax S.A, Olsztyn, Poland) was used to print the casing. A Hanna HI 83224 colorimeter (HANNA Instruments, Woonsocket, RI, USA) was applied for the measurement of wastewater parameters. An AMEL System 500 potentiostat (Amel S.l.r., Milano, Italy) with CorrWare software (Scribner Associates Inc, Southern Pines, NC, USA) and a Fluke 8840A multimeter (Fluke Corporation, Everett, WA, USA) were applied for the electrical measurements.

4. Conclusions

As demonstrated by the measurements, copper boride alloy is most suitable when oxidized for 6 h at a temperature of 673 K and when the concentration of boride is equal to 9%. In the MFC with a Cu–B cathode (9% of B), a current density of 0.35 mA·cm^{-2} and power level of 6.11 mW were obtained. A shorter (by two days) COD reduction time (assumed reduction level: 90%) was shown in the MFC with a Cu–B cathode than in the MFC with a carbon cloth cathode. Moreover, the effectiveness of NO_3^- reduction (during a 15 day period) in the MFC with a Cu–B cathode was higher than that in the MFC with a carbon cloth cathode. Thus, the higher catalytic activity of the Cu–B catalyst compared to carbon cloth and the possibility of using Cu–B alloy as a cathode catalyst in MFCs powered with municipal wastewater were shown in this paper.

Author Contributions: Data curation, P.P.W. and B.W.; investigation, P.P.W. and B.W.; methodology, P.P.W.; writing—review and editing, P.P.W. and B.W.; supervision, P.P.W. and B.W.

Funding: This research received no external funding.

References

1. US EPA Report. *Clean Watersheds Needs Survey Overview*; United States Environmental Protection Agency: Galloway, NJ, USA, 2008.
2. Logan, B. *Microbial Fuel Cells*; Wiley: Hoboken, NJ, USA, 2008.
3. Rabaey, K.; Verstraete, W. Microbial fuel cells: Novel biotechnology for energy generation. *Trends Biotechnol.* **2005**, *23*, 291–298. [CrossRef] [PubMed]
4. Sivasankar, V.; Mylsamy, P.; Omine, K. *Microbial Fuel Cell Technology for Bioelectricity*, 1st ed.; Springer: Berlin, Germany, 2018.
5. Potter, M.C. Electrical effects accompanying the decomposition organic compounds. *Proc. Roy. Soc. Lond. Ser.* **1911**, *B84*, 260–276. [CrossRef]
6. Davis, J.B.; Yarbrough, H.F., Jr. Preliminary experiments on a microbial fuel cell. *Science* **1962**, *137*, 615–616. [CrossRef] [PubMed]
7. Berk, R.S.; Canfield, J.H. Bioelectrochemical energy conversion. *Appl. Microbiol.* **1964**, *12*, 10–12. [PubMed]
8. Lewis, K. Symposium on bioelectrochemistry of microorganisms: IV. Biochemical fuel cells. *Bacteriol. Rev.* **1966**, *30*, 101–113. [PubMed]
9. Allen, R.M.; Benetto, H.P. Microbial fuel cells: Electricity production from carbohydrates. *Appl. Biochem. Biotechnol.* **1993**, *39*, 27–40. [CrossRef]

10. Kim, B.H.; Park, H.S.; Shin, P.K.; Chang, I.S.; Kim, H.J. Mediator-Less Biofuel Cell. U.S. Patent 5,976,719, 2 November 1999.

11. Kim, H.J.; Hyun, M.S.; Chang, I.S.; Kim, B.H. A microbial fuel cell type lactate biosensor using a metal-reducing bacterium. *Shewanella Putrefaciens J. Microbiol. Biotechnol.* **1999**, *9*, 365–367.

12. Min, B.; Logan, B.E. Continuous electricity generation from domestic wastewater and organic substrates in a at plate microbial fuel cell. *Environ. Sci. Technol.* **2004**, *38*, 5809–5814. [CrossRef]

13. Min, B.; Cheng, S.; Logan, B.E. Electricity generation using membrane and salt bridge microbial fuel cells. *Water Res.* **2005**, *39*, 1675–1686. [CrossRef]

14. Logan, B.E.; Hamelers, B.; Rozendal, R.; Schroder, U.; Keller, J.; Verstraete, W.; Rabaey, K. Microbial Fuel Cells: Methodology and Technology. *Environ. Sci. Technol.* **2006**, *40*, 5181–5192. [CrossRef]

15. Franks, A.E.; Nevin, K.P. Microbial fuel cells, a current review. *Energies* **2010**, *3*, 899–919. [CrossRef]

16. Liu, H.; Ramnarayanan, R.; Logan, B.E. Production of electricity during wastewater treatment using a single chamber microbial fuel cell. *Environ. Sci. Technol.* **2004**, *38*, 2281–2285. [CrossRef] [PubMed]

17. Rabaey, K.; Alterman, P.; Clauwaert, P.; De Schamphelaire, L.; Boon, N.; Verstraete, W. Microbial fuel cells in relation to conventional anaerobic digestion technology. *Eng. Life Sci.* **2006**, *6*, 285–292.

18. Bond, D.R.; Lovley, D.R. Electricity production by *Geobacter sulfurreducens* attached to electrodes. *Appl. Environ. Microbiol.* **2003**, *69*, 1548–1555. [CrossRef]

19. Chaudhuri, S.K.; Lovley, D.R. Electricity generation by direct oxidation of glucose in mediatorless microbial fuel cells. *Nat. Biotechnol.* **2003**, *21*, 1229–1232. [CrossRef] [PubMed]

20. Kim, H.J.; Park, H.S.; Hyun, M.S.; Chang, I.S.; Kim, M.; Kim, B.H. A Mediator-Less Microbial Fuel Cell Using a Metal Reducing Bacterium. *Shewanella Putrefaciens. Enzym. Microb. Technol.* **2002**, *30*, 145–152. [CrossRef]

21. Park, H.S.; Kim, B.H.; Kim, H.S.; Kim, H.J.; Kim, G.T.; Kim, M.; Chang, I.S.; Park, Y.K.; Chang, H.I. A novel electrochemically active and Fe(III)-reducing bacterium phylogenetically related to *Clostridium butyricum* isolated from a microbial fuel cell. *Anaerobe* **2001**, *7*, 297–306. [CrossRef]

22. Pham, C.A.; Jung, S.J.; Phung, N.T.; Lee, J.; Chang, I.S.; Kim, B.H.; Yi, H.; Chun, J. A novel electrochemically active and Fe(III)-reducing bacterium phylogenetically related to *Aeromonas hydrophila*, isolated from a microbial fuel cell. *FEMS Microbiol. Lett.* **2003**, *223*, 129–134. [CrossRef]

23. Bond, D.R.; Lovley, D.R. Evidence for involvement of an electron shuttle in electricity generation by *Geothrix fermentans. Appl. Environ. Microbiol.* **2005**, *71*, 2186–2189. [CrossRef]

24. Reguera, G.; McCarthy, K.D.; Mehta, T.; Nicoll, J.S.; Tuominen, M.T.; Lovley, D.R. Extracellular electron transfer via microbial nanowires. *Nature* **2005**, *435*, 1098–1101. [CrossRef]

25. Reguera, G.; Nevin, K.P.; Nicoll, J.S.; Covalla, S.F.; Woodard, T.L.; Lovley, D.R. Biofilm and nanowire production leads to incre- ased current in *Geobacter sulfurreducens* fuel cells. *Appl. Environ. Microbiol.* **2006**, *72*, 7345–7348. [CrossRef] [PubMed]

26. Patil, S.A.; Surakasi, V.P.; Koul, S.; Ijmulwar, S.; Vivek, A.; Shouche, Y.S.; Kapadnis, B.P. Electricity generation using chocolate industry wastewater and its treatment in activated sludge based microbial fuel cell and analysis of developed microbial community in the anode chamber. *Bioresour. Technol.* **2009**, *100*, 5132–5139. [CrossRef] [PubMed]

27. Logan, B.E.; Regan, J.M. Electricity-producing bacterial communities in microbial fuel cells. *Trends Microbiol.* **2006**, *14*, 512–518. [CrossRef] [PubMed]

28. Logan, B.E.; Regan, J.M. Microbial fuel cells—Challenges and applications. *Environ. Sci. Technol.* **2006**, *40*, 5172–5180. [CrossRef] [PubMed]

29. Asensio, Y.; Montes, I.B.; Fernández-Marchante, C.M.; Lobato, J.; Cañizares, P.; Rodrigo, M.A. Selection of cheap electrodes for two-compartment microbial fuel cells. *J. Electroanal. Chem.* **2017**, *785*, 235–240. [CrossRef]

30. Erbay, C.; Yang, G.; de Figueiredo, P.; Sadr, R.; Yu, C.; Han, A. Three-dimensional porous carbon nanotube sponges for high-performance anodes of microbial fuel cells. *J. Power Sources* **2015**, *298*, 177–183. [CrossRef]

31. Liu, Y.; Harnisch, F.; Fricke, K.; Schroeder, U.; Climent, V.; Feliu, J.M. The study of electrochemically active microbial biofilms on different carbon-based anode materials in microbial fuel cells. *Biosens. Bioelectron.* **2010**, *25*, 2167–2171. [CrossRef] [PubMed]

32. Penteado, E.D.; Fernández-Marchante, C.M.; Zaiat, M.; Cañizares, P.; Gonzalez, E.R.; Rodrigo, M.A. Influence of carbon electrode material on energy recovery from winery wastewater using a dual-chamber microbial fuel cell. *Environ. Technol.* **2017**, *38*, 1333–1341. [CrossRef]

33. Yuan, Y.; Zhou, S.; Zhuang, L. Polypyrrole/carbon black composite as a novel oxygen reduction catalyst for microbial fuel cells. *J. Power Sources* **2010**, *195*, 3490–3493. [CrossRef]

34. Zuo, K.; Liang, S.; Liang, P.; Zhou, X.; Sun, D.; Zhang, X.; Huang, X. Carbon filtration cathode in microbial fuel cell to enhance wastewater treatment. *Bioresour. Technol.* **2015**, *185*, 426–430. [CrossRef]

35. Dumas, C.; Mollica, A.; Féron, D.; Basséguy, R.; Etcheverry, L.; Bergel, A. Marine microbial fuel cell: Use of stainless steel electrodes as anode and cathode materials. *Electrochim. Acta* **2006**, *53*, 468–473. [CrossRef]

36. Martin, E.; Tartakovsky, B.; Savadogo, O. Cathode materials evaluation in microbial fuel cells: A comparison of carbon, Mn_2O_3, Fe_2O_3 and platinum materials. *Electrochim. Acta* **2011**, *58*, 58–66. [CrossRef]

37. Bockris, J.O.M.; Reddy, A.K.N. *Modern Electrochemistry*; Kulwer Academic/Plenum Publishers: New York, NY, USA, 2000.

38. Twigg, M.V. *Catalyst Handbook*; Wolfe Publishing Ltd.: London, UK, 1989.

39. Cheng, S.; Liu, H.; Logan, B.E. Power densities using different cathode catalysts (Pt and CoTMPP) and polymer binders (Nafion and PTFE) in single chamber microbial fuel cells. *Environ. Sci. Technol.* **2006**, *40*, 364–369. [CrossRef] [PubMed]

40. Wu, W.; Niu, H.; Yang, D.; Wang, S.-B.; Wang, J.; Lin, J.; Hu, C. Controlled layer-by-layer deposition of carbon nanotubes on electrodes for microbial fuel cells. *Energies* **2019**, *12*, 363. [CrossRef]

41. McMurry, J.E.; Ballantine, D.S.; Hoeger, C.A.; Peterson, V.E.; Castellion, M.E. *Fundamentals of Inorganic, Organic and Biological Chemistry*; Pearson: London, UK, 2009.

42. Włodarczyk, B.; Włodarczyk, P.P. Microbial fuel cell with Cu–B cathode powering with wastewater from yeast production. *J. Ecol. Eng.* **2017**, *18*, 224–230. [CrossRef]

43. Łomotowski, J.; Szpindor, A. *Modern Wastewater Treatment Systems (in Polish)*; Arkady: Warsaw, Poland, 2002.

44. Włodarczyk, B.; Włodarczyk, P.P. Microbial fuel cell with Ni-Co cathode powered with yeast wastewater. *Energies* **2018**, *11*, 3194. [CrossRef]

45. Włodarczyk, B.; Włodarczyk, P.P. Use of Ni-Co alloy as cathode catalyst in single chamber microbial fuel cell. *Ecol. Eng.* **2017**, *18*, 210–216. [CrossRef]

46. Włodarczyk, B.; Włodarczyk, P.P. Methanol electrooxidation with Cu–B catalyst. *Infrastruct. Ecol. Rural Areas (Polish Acad. Sci. Crac.)* **2016**, *4*, 1483–1492. [CrossRef]

47. Włodarczyk, B.; Włodarczyk, P.P. Electricity production in microbial fuel cell with Cu–B alloy as catalyst of anode. *Civ. Eng. QUAESTI* **2015**, *3*, 305–308. [CrossRef]

48. Włodarczyk, P.P.; Włodarczyk, B. Possibility of Using Ni-Co Alloy as Catalyst for Oxygen Electrode of Fuel Cell. *Chin. Bus. Rev.* **2015**, *14*, 159–167. [CrossRef]

49. Włodarczyk, P.P.; Włodarczyk, B. Ni-Co alloy as catalyst for fuel electrode of hydrazine fuel cell. *China-USA Bus. Rev.* **2015**, *14*, 269–279. [CrossRef]

50. Huggins, T.; Fallgren, P.H.; Jin, S.; Ren, Z.J. Energy and performance comparison of microbial fuel cell and conventional aeration treating of wastewater. *J. Microb. Biochem. Technol.* **2013**, *S6:002*, 1–5. [CrossRef]

51. Ren, Z.; Yan, H.; Wang, W.; Mench, M.M.; Regan, J.M. Characterization of microbial fuel cells at microbially and electrochemically meaningful time scales. *Environ. Sci. Technol.* **2011**, *45*, 2435–2441. [CrossRef] [PubMed]

 *catalysts*

Article

Preparation and Analysis of Ni–Co Catalyst Use for Electricity Production and COD Reduction in Microbial Fuel Cells

Paweł P. Włodarczyk * and Barbara Włodarczyk *

Institute of Environmental Engineering and Biotechnology, Faculty of Natural Sciences and Technology, University of Opole, Kominka Str. 6, 45-032 Opole, Poland
* Correspondence: pawel.wlodarczyk@uni.opole.pl (P.P.W.); barbara.wlodarczyk@uni.opole.pl (B.W.);
  Tel.: +48-077-401-6717 (P.P.W.)

Received: 31 October 2019; Accepted: 5 December 2019; Published: 8 December 2019

Abstract: Microbial fuel cells (MFCs) are devices than can contribute to the development of new technologies using renewable energy sources or waste products for energy production. Moreover, MFCs can realize wastewater pre-treatment, e.g., reduction of the chemical oxygen demand (COD). This research covered preparation and analysis of a catalyst and measurements of changes in the concentration of COD in the MFC with a Ni–Co cathode. Analysis of the catalyst included measurements of the electroless potential of Ni–Co electrodes oxidized for 1–10 h, and the influence of anodic charge on the catalytic activity of the Ni–Co alloy (for four alloys: 15, 25, 50, and 75% concentration of Co). For the Ni–Co alloy containing 15% of Co oxidized for 8 h, after the third anodic charge the best catalytic parameters was obtained. During the MFC operation, it was noted that the COD reduction time (to 90% efficiency) was similar to the reduction time during wastewater aeration. However, the characteristic of the aeration curve was preferred to the curve obtained during the MFC operation. The electricity measurements during the MFC operation showed that power equal to 7.19 mW was obtained (at a current density of 0.47 mA·cm^{-2}).

Keywords: Ni–Co catalyst; oxygen electrode; microbial fuel cell; environmental engineering; electricity production

---

1. Introduction

Microbial fuel cells (MFCs) are a technology of electric energy generation using organic matter contained in wastewater [1,2]. Thus, this technology allows the production of energy from waste products (e.g., wastewater, also industry wastewater) [1,3–9]. Moreover, MFCs allow for wastewater pretreatment, e.g., reduction of the chemical oxygen demand (COD) concentration. The first observations of electricity production by bacteria were conducted by Potter [10]. However, greater progress in the development of this technology was only achieved in the 1960s [11–13]. Because of the increasing pollution of the environment, research on MFC technology resumed in the 1990s [14–16], but real development of these technology has only happened in recent years [1–7,17–25].

MFC forms a bio-electrochemical system in which bacteria oxidize organic matter by acting as a biocatalyst. Therefore, organic matter constitutes the fuel applied for the production of electricity [1–4,12,22]. This system includes electrodes in the electrode chambers, usually separated by a proton exchange membrane (PEM) [26]. On the anode, bacteria oxidize organic matter to produce e$^-$ electrons and H$^+$ ions (as well as an additional volume of CO$_2$), whereas H$_2$O is generated on the cathode through the reaction of e$^-$ ions combined with H$^+$ and O$_2$ [1–4,22]. This is a result of cathode aeration in the cathode chamber. Among many other genera, the *Shewanella* spp., *Pseudomonas* spp., or *Geobacter* spp. bacteria are capable of generating electrons [21,27–34]. Studies have demonstrated that the highest current values in MFC are generated by using multicultural microorganisms. They have higher efficiency in relation to microorganisms accumulated in monocultures, as a result of the competition between the bacterial cultures [35]. Research into the peak limits of the current density and power level of MFC is still at an experimental stage. The limitation of the maximum power density or current density is the result of the low rate of reactions on electrodes [1–4,36,37]. The rate of the processes taking place on the electrodes is primarily influenced by the type of applied catalyst. Since the role of the anode catalyst is taken over by microorganisms, it is important to look for an adequate cathode catalyst. Because of the excellent catalytic properties, platinum is most commonly used as the catalyst [38,39]. Unfortunately, it is also characterized by high price. Therefore, it is necessary to look for other catalysts that do not contain precious metals [40–47]. As electrodes (also as catalysts) do not contain precious metals, different electrode materials (carbon fiber brush, carbon felt, carbon foam, carbon cloth, graphite paper, and others) were analyzed in various works [48–56]. Of these materials, the most efficient anode material is carbon felt. The lowest parameters during MFC operation were obtained using carbon cloth as electrodes. However, it was found that it was also possible to use metal electrodes (also as metal electrodes with metal catalysts) as the cathodes of microbial fuel cells [1,7,8,22,41,57,58]. Nickel is also characterized by good catalytic properties. However, pure nickel (Raney Ni, the most commonly used form of nickel catalyst) is difficult to use, e.g., it should never be exposed to air. Even after preparation, Raney Ni still contains small amounts of hydrogen gas and may spontaneously ignite when exposed to air. Therefore, Ni (mainly as Raney Ni) is supplied as an aqueous suspension [59]. Therefore, nickel alloys should be easier (also safer) to use while maintaining good catalytic properties. Additionally, other metals and metal alloys are used as catalysts. One of them is cobalt and its alloys. These materials have also been used or analyzed during research as a catalytic material [60–65].

The description of the current density applies the Butler–Volmer exponential function. However, this function only leads to an output in the form of a theoretical value, which usually deviates significantly from the values that are obtained experimentally in comparable circumstances [38,66]. Therefore, it is necessary to conduct experimental research concerned with the selection of a catalyst suitable for a specific substance (in this case a substance employed as the waste material) that further constitutes the fuel applied in MFC [1,2,38,48–55,66]. The selection of a catalyst has an effect on the final cost of electricity production and the expenses associated with pre-treatment of wastewater for the needs of MFC. Because of the large amount of wastewater, it is necessary to provide its treatment, which incurs huge expenses [67]. However, the energy potential of wastewater allows it to be considered in the role of a potential source of energy in future applications, primarily as fuel for MFCs. MFCs are understood as an element that can support the traditional wastewater treatment techniques, as its role is primarily concerned with reducing COD concentration.

In this work the possibility of using a nickel–cobalt alloy (Ni–Co) in MFC was analyzed. The analysis concerned the use of Ni–Co alloy as a cathode catalyst for electricity production and COD reduction.

2. Results and Discussion

Figures 1–4 show the concentrations of Ni and Co in the alloy samples obtained by the method of electrochemical deposition for planned 15, 25, 50, and 75% concentrations of Co.

Figures 1–4 allow visualization and understanding that it is necessary not only to preserve the deposition parameters, but also to sort and select the resulting alloys. This necessity arises from the fact that despite maintaining constant deposition parameters, different alloy compositions were obtained in corresponding cases.

For further research, the following Ni–Co alloy samples were selected: 7 (15% of Co; Figure 1), 5 (25% of Co; Figure 2), 2 (50% of Co; Figure 3), and 1 (75% of Co; Figure 4).

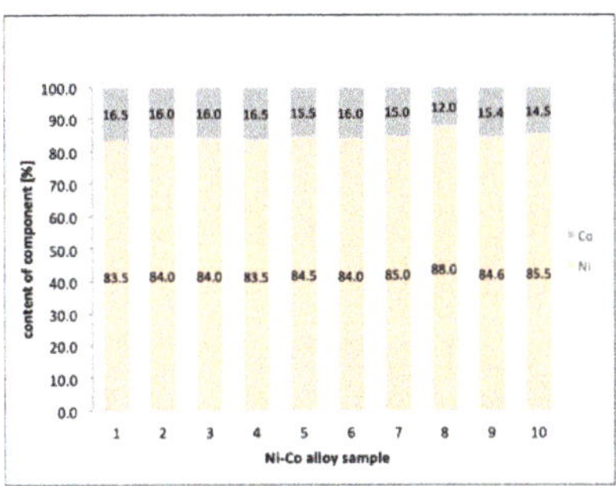

Figure 1. Concentration of Ni and Co in the alloy samples obtained by electrochemical deposition for planned 15% concentration of Co.

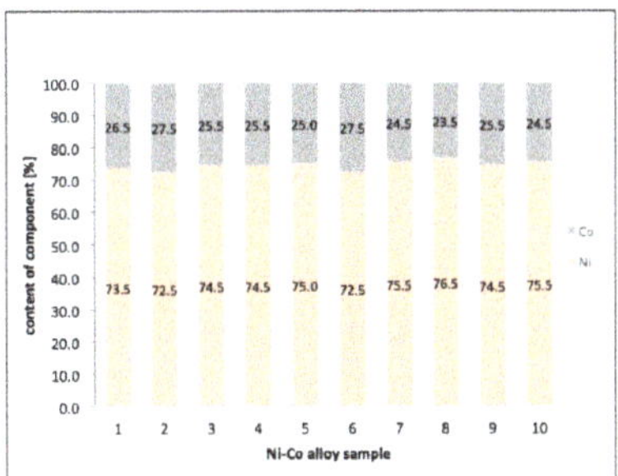

Figure 2. Concentration of Ni and Co in the alloy samples obtained by electrochemical deposition for planned 25% concentration of Co.

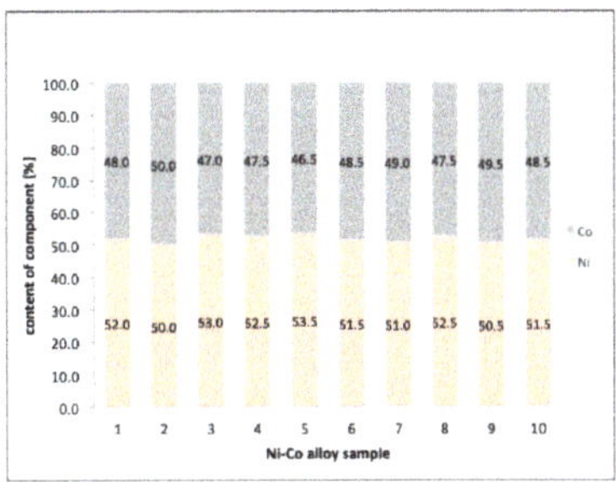

Figure 3. Concentration of Ni and Co in the alloy samples obtained by electrochemical deposition for planned 50% concentration of Co.

Ni–Co alloys that contained exactly 15, 25, 50, and 75% of Co were selected. These alloys were oxidized for 1, 2, 4, 6, 8, and 10 h. Next, measurements were carried out according to the methodology presented in Section 3.2. Before oxidation, all samples were gray, but after oxidation over 4, 6, and 8 h, the surfaces of the samples were multi-colored, while after oxidation over 10 h, the color of the sample surfaces changed to black.

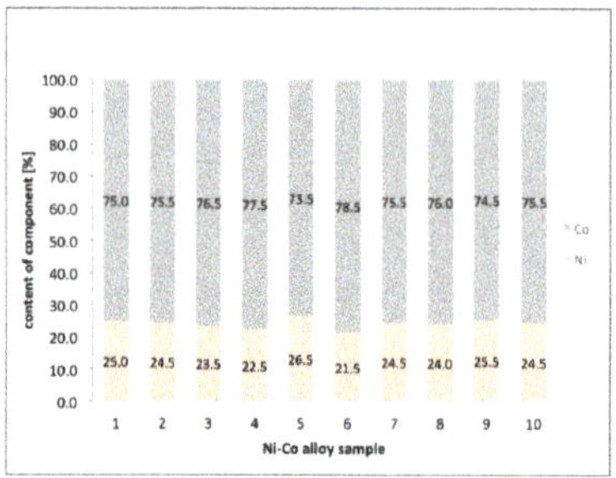

Figure 4. Concentration of Ni and Co in the alloy samples obtained by electrochemical deposition for planned 75% concentration of Co.

Figures 5 and 6 show stabilization of the electroless potential in the glass half-cell, with the Ni–Co work electrode (oxidized for 1 and 2 h) in alkaline electrolyte (aqueous solution of KOH). 7,8 show alloy oxidized for 4 and 6 h. 9,10 show alloys oxidized for 8 and 10 h.

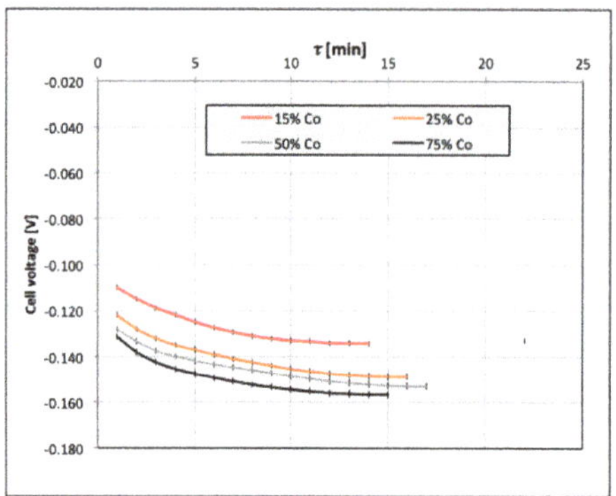

Figure 5. The electroless potential of Ni–Co electrodes (oxidized for 1 h).

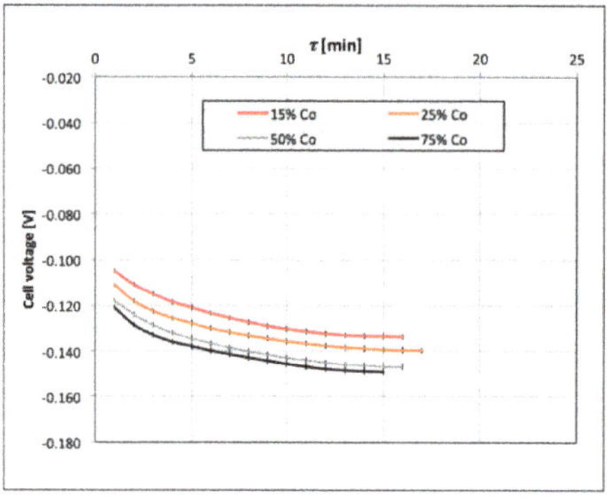

Figure 6. The electroless potential of Ni–Co electrodes (oxidized for 2 h).

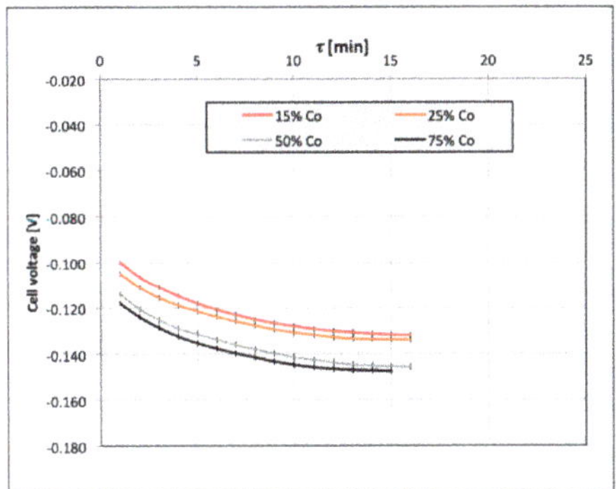

Figure 7. The electroless potential of Ni–Co electrodes (oxidized for 4 h).

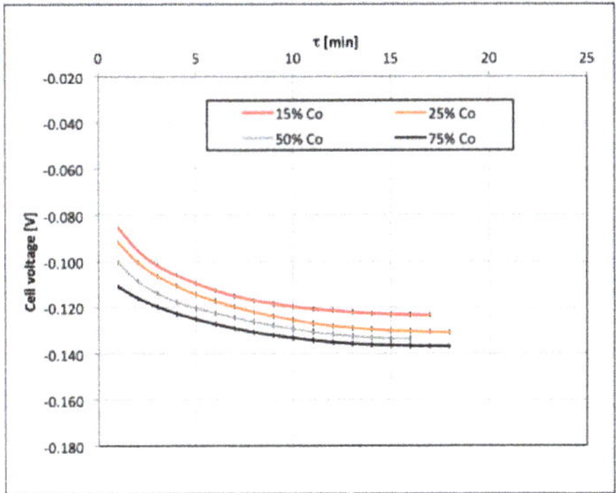

Figure 8. The electroless potential of Ni–Co electrodes (oxidized for 6 h).

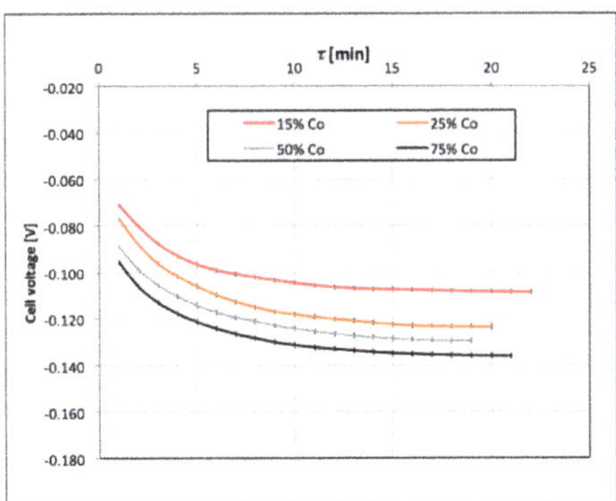

Figure 9. The electroless potential of Ni–Co electrodes (oxidized for 8 h).

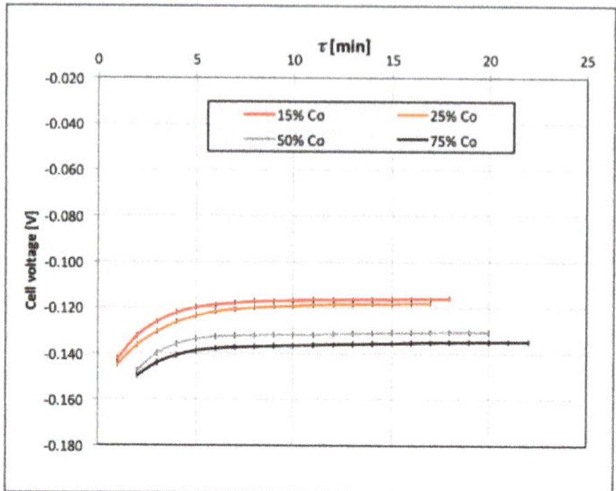

Figure 10. The electroless potential of Ni–Co electrodes (oxidized for 10 h).

The data (Figures 5–10) show that during each measurement, the electroless potential was the highest for the Ni–Co catalyst with 15% of Co. Moreover, the highest electroless potential was obtained for Ni–Co electrodes oxidized for 8 h (Figure 9).

The analysis of the samples also demonstrated that oxides of lower order, NiO and CoO, were generated during oxidation lasting from 1 to 8 h. However, during oxidation taking over 10 h, $NiCo_2O_4$ and Co_3O_4 oxides were formed, which decreased the electroless potential (Figure 10) [68–74]. Therefore, the next measurements, including the influence of anodic charge on the catalytic activity of the Ni–Co catalyst (described in Section 3.2), were performed for the alloy samples oxidized for 8 h (Figures 11–14). Colored lines (1–4) in Figures 11–14 refer to the subsequent anodic charges.

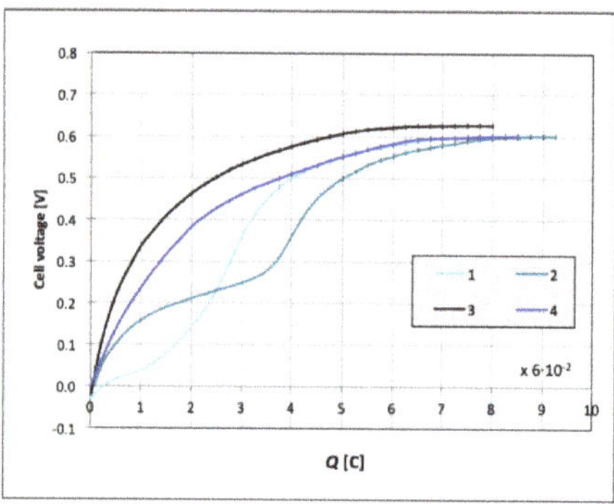

Figure 11. Voltage of a half-cell with Ni–Co electrode (alloy with 15% of Co); influence of anodic charge on the catalytic activity of Ni–Co catalyst (lines 1–4 show subsequent anodic charges).

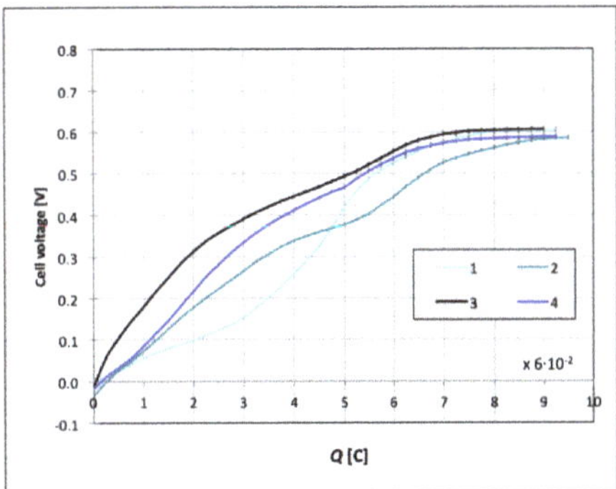

Figure 12. Voltage of a half-cell with Ni–Co electrode (alloy with 25% of Co); influence of anodic charge on the catalytic activity of Ni–Co catalyst (lines 1–4 show subsequent anodic charges).

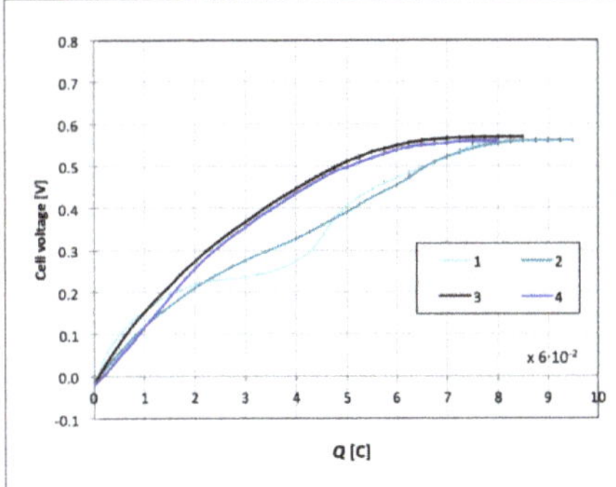

Figure 13. Voltage of a half-cell with Ni–Co electrode (alloy with 50% of Co); influence of anodic charge on the catalytic activity of Ni–Co catalyst (lines 1–4 show subsequent anodic charges).

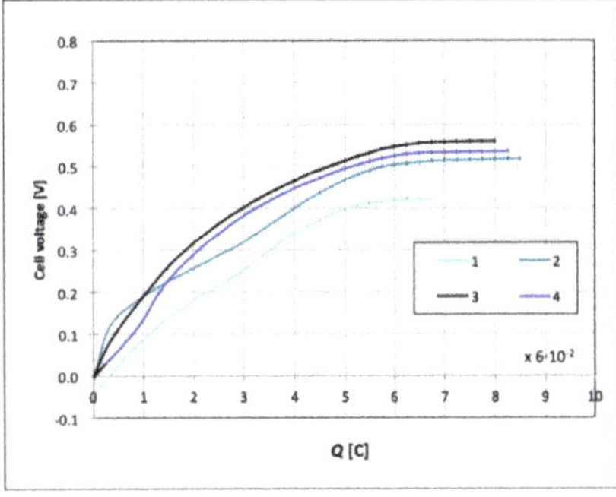

Figure 14. Voltage of a half-cell with Ni–Co electrode (alloy with 75% of Co); influence of anodic charge on the catalytic activity of Ni–Co catalyst (lines 1–4 show subsequent anodic charges).

In the next step, the MFCs (described in Section 3.3) with different cathodes (with different contents of Co, and after oxidation of electrodes for 1–10 h) were built. Table 1 shows the maximum power and average cell voltage obtained in MFCs, with Ni–Co cathodes oxidized for 1–8 h, without anodic charge before using the electrodes (cathodes).

Table 1. Maximum power and average cell voltage obtained in microbial fuel cells (MFCs), with Ni–Co cathodes oxidized for 1–8 h, and without anodic charge before using the electrodes (cathodes).

Oxidizing Time of the Ni–Co Electrode (h)	Max Power Obtained in MFC (mW)				Average Voltage of the MFC (V)			
	Content of Co in the Alloy (%)				Content of Co in the Alloy (%)			
	15	25	50	75	15	25	50	75
1	5.38	4.67	4.60	4.15	0.88	0.81	0.80	0.76
2	5.41	4.62	4.61	4.34	0.89	0.80	0.78	0.75
4	5.55	4.75	4.66	4.66	0.91	0.83	0.79	0.77
6	5.76	5.01	4.67	4.68	0.91	0.83	0.81	0.78
8	**6.02**	**5.08**	**4.76**	**4.70**	**0.93**	**0.87**	**0.86**	**0.82**
10	4.62	4.79	4.60	4.33	0.88	0.79	0.80	0.69

The highest parameters (power and cell voltage) for the MFC with electrodes oxidized for 8 h were obtained. Thus, these electrodes were chosen for further measurements of the MFCs (with electrodes after the anodic charge). Table 1 shows the maximum power and average cell voltage obtained in MFCs, with Ni–Co cathodes (15% of Co), and oxidized for 8 h, with anodic charge before using the electrodes (cathodes).

The cell voltage of the third anodic charging of the electrode was also highest for the same catalyst (15% Co) (Figures 11–14, Table 1). Such parameters (Figures 5–14, Tables 1 and 2) show that the Ni–Co electrode with 15% of Co showed high performance. Figure 15 shows the cell voltage and Figure 16 shows the power curves during the MFC with the highest parameter electrode (Ni–Co cathode with 15% of Co, and after triple the anodic charge) operation.

Table 2. Maximum power and average cell voltage obtained in MFCs, with Ni–Co cathodes oxidized for 8 h, and with anodic charge before using the electrodes (cathodes).

Content of Co in the Ni–Co Electrode (%)	Max Power Obtained in MFC (mW)				Average Voltage of the MFC (V)			
	Number of Anodic Charge				Number of Anodic Charge			
	1	2	3	4	1	2	3	4
15	6.34	6.66	**7.19**	7.09	0.99	0.99	**1.03**	1.01
25	5.45	5.52	6.72	6.63	0.96	0.97	0.99	0.96
50	5.44	5.63	6.69	6.59	0.93	0.94	0.96	0.91
75	5.03	5.59	6.68	6.65	0.88	0.91	0.92	0.88

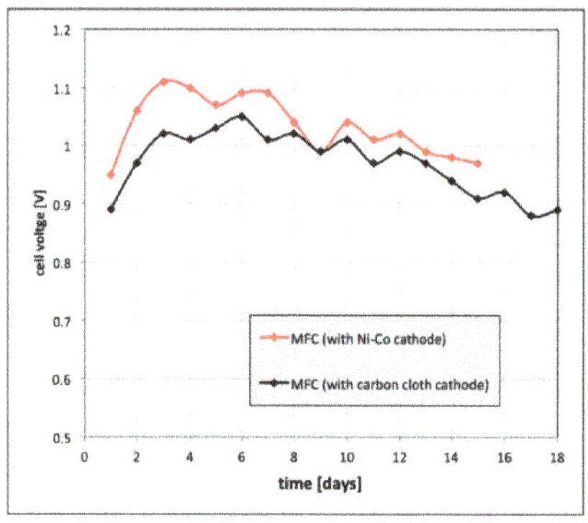

Figure 15. Cell voltage of the MFC in time: red line—with Ni–Co cathode (15% Co), black line—carbon cloth cathode.

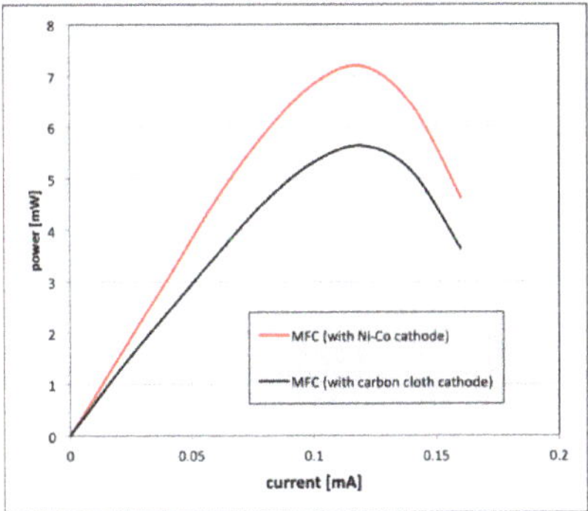

Figure 16. Power curves during the MFC operation: red line—with Ni–Co cathode (15% Co), black line—carbon cloth cathode.

Figure 17 shows the COD reduction in time in the three reactors without aeration, with aeration, and during the operation of MFC with the highest parameter electrode (Ni–Co cathode with 15% of Co, and after triple the anodic charge) and with the carbon cloth electrode.

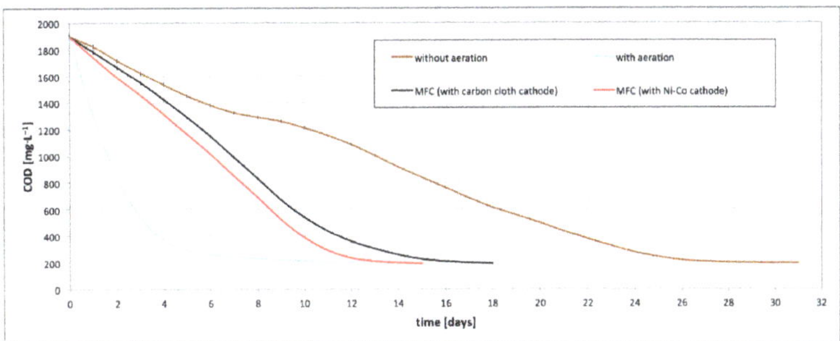

Figure 17. Chemical oxygen demand (COD) reduction in time in the three reactors: without aeration, with aeration, and during the MFC operation (with Ni–Co and carbon cloth electrodes).

On the basis of the results of the measurements of the electrode potential, we can state that among all analyzed Ni and Co concentrations, the highest value was obtained for oxidized alloys over 8 h at 673 K. In addition, for all oxidation times, the highest values of electrode potential were obtained by an alloy containing 15% Co. A similar outcome was obtained when the electrode potential was performed after anodic charging. In each case, the alloy with the composition comprising 15% Co had the higher catalytic activity of the analyzed alloys. In the case of triple anode charging using this alloy (15% Co), the most favorable potential curve was derived. Therefore, the alloy containing 15% Co, after 8 h of oxidation and three times anode loading, was selected for further measurements by application of MFC. MFC (third reactor, see Section 3.3, Figure 18) was investigated in a setup combined with a Ni–Co cathode and for comparison with a cathode made of carbon cloth. During the MFC operation, the levels of COD reduction to the effectiveness of 90% were recorded: 15 days for a Ni–Co catalyst, and 18 days for the carbon cloth catalyst. In comparison to the data regarding the reactor with aeration (Figure 15, blue line), this time was longer by 3 days for the carbon cloth catalyst (Figure 17, black line), and the same for a Ni–Co catalyst (Figure 17, red line). However, the characteristics of the COD decline curve were comparably less favorable when MFC was applied, as using the aeration taking 5 days resulted in an 84.8% decrease in the initial COD level (Figure 17, blue line). We should note that similar results were obtained for the Cu–B catalyst in the conditions when similar wastewater parameters were employed; however, the use of Ni–Co alloy as a catalyst provided slightly better parameters (shorter COD reduction time) compared to that based on the Cu–B alloy [8,22,75].

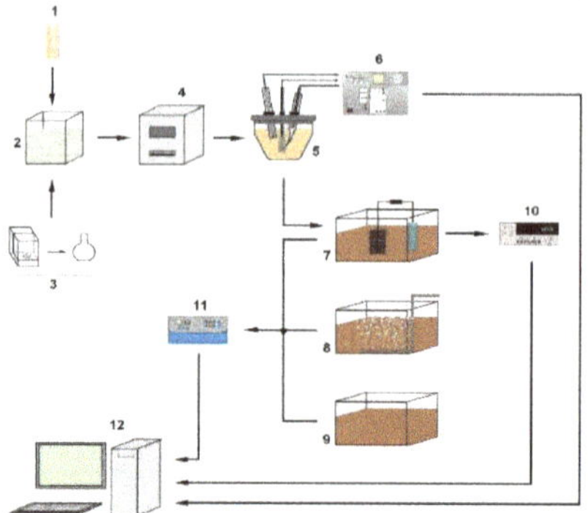

Figure 18. Scheme of the measurements: 1—copper mesh (electrode) to deposit the catalyst, 2—electrolysis bath to deposit the catalyst, 3—electrolyte preparation, 4—furnace for electrode (catalyst) oxidation, 5—glass cell for electroless potential analysis and for the anodic charge analysis, 6—potentiostat, 7—third reactor (MFC), 6—second reactor (with aeration), 9—first reactor (without aeration), 10—multimeter, 11—colorimeter, 12—computer.

The measurement of the cell potential demonstrated higher potential during MFC operation of the Ni–Co catalyst than for the case when the carbon cloth was employed (Figure 15). During MFC operation, 5.63 mW of power (with the carbon cloth cathode) and 7.19 mW of power (with the Ni–Co cathode) were obtained (Figure 16). The analysis of the current density (based on data from power measurements and the area of electrodes) showed that the current density for the Ni–Co catalyst reached a maximum of 0.47 mA·cm^{-2}, and 0.23 mA·cm^{-2} for carbon cloth. The maximum power obtained during the operation of MFC with Ni–Co alloy, was slightly higher (by 1.08 mW) compared to the maximum power obtained during the operation of MFC with a Cu–B cathode (6.11 mW) [22,75].

3. Materials and Methods

3.1. Preparation of a Ni–Co Cathode

The electrochemical deposition technique was applied to obtain Ni–Co alloy. A copper mesh was used as the electrode to apply the catalyst (i.e., Ni–Co alloy). The electrolyte for alloying mainly consisted of a mixture of $NiSO_4$ and $CoSO_4$ [7,76]. The electrochemical deposition was carried out at temperatures of 293–323 K, at a current density of 1–3 A·dm^{-2}, and at pH 2.0–5.5 [7]. Prior to the application process of electrolytic alloy, the copper mesh was first washed with 25% aqueous KOH to ensure its adequate wettability. The next step involved its washing in acetic acid and then in alcohol [22].

On the basis of previous studies [7], alloys containing 15, 25, 50, and 75% content of Co were selected for measurement. The temperature, pH, and current density of electrochemical deposition (to obtain different contents of Ni and Co in the alloys) were selected experimentally, but according to the previously proposed methodology for obtaining the alloy [7]. For further measurements, the Ni–Co electrodes with different contents of Co were selected by the XRD method. During electrochemical deposition, for all of the planned alloys (15, 25, 50, and 75% of Co) the ten different alloys with similar concentrations of Co were obtained according to a previously developed methodology [7,76]. Based on the research results (Figures 1–4), for further research, the following alloy samples were selected: 7 (15% of Co; Figure 1), 5 (25% of Co; Figure 2), 2 (50% of Co; Figure 3), and 1 (75% of Co; Figure 4).

The selection was carried out using a single-crystal X-ray diffractometer (Xcalibur, Oxford Diffraction, UK).

3.2. Selection of the Ni–Co Electrodes for Measurement

First, measurements of the stationary potential of the electrodes were performed in order to assess the oxidation activity of the Ni–Co electrodes (with different contents of Ni and Co). Since the electrode is constantly oxidized during the operation of MFC, it was necessary to pre-oxidize the electrodes, otherwise, the electrodes would change their catalytic properties (along with the level of oxidation of the electrode surface) throughout their operation of MFC. The oxidation temperature of all of the Ni–Co alloy samples was 673 K. The oxidation times were 1, 2, 4, 6, 8, and 10 h. Next, we measured the influence of anodic charge on the catalytic activity of the Ni–Co alloy. These measurements were carried out in a glass half-cell (250 mL^3) with the use of a potentiostat (Figure 18; 5). An aqueous solution of KOH (2 M) was used as the electrolyte. A saturated calomel electrode (SCE) was used as the reference electrode [7,22,77].

The KS 520/14 silt furnace (ELIOG Industrieofenbau GmbH, Römhild, Germany) was used for electrode oxidation. The experiments of the influence of anodic charge on the catalytic activity of the Ni–Co alloy were conducted using an AMEL System 500 potentiostat (Figure 18; 6) (Amel S.l.r., Milano, Italy). The potentiostat was controlled by a computer using CorrWare software (Scribner Associates Inc., Southern Pines, NC, USA).

3.3. Measurements of Electricity Production and COD Reduction during the Operation of MFC (with Ni–Co Cathode)

The following phase involved the analysis of the Ni–Co alloy in the function of the cathode catalyst in MFC. The research included measurements of COD decline and electric energy production during the operation of the microbial fuel cell.

Waste material was applied to serve as fuel for MFC with Ni–Co cathodes. For this purpose, wastewater (WW) derived from the municipal wastewater treatment plant (WWTP) was used. Table 3 contains the COD concentration and pH of the WW applied in the measurements.

Table 3. Parameters of wastewater (WW) (from the wastewater treatment plant (WWTP)) used in measurements of electricity production and COD reduction.

Parameter	Value
COD (mg·L^{-1})	1899.0
pH	6.5

To assess the effectiveness of COD reduction in the microbial fuel cell, the following were compared: COD reduction time in MFC, COD reduction time in WW without interference, and COD reduction time accompanying WW aeration. To this end, three reactors were applied: one without aeration (Figure 18; 9), one with aeration (Figure 18; 8), and one containing MFC (Figure 18; 7). Figure 18 shows the scheme of the measurements.

Electrical parameters of the MFC were measured in parallel (at the same time) with the COD reduction [7,22,78]. All reactors had the same dimensions, and therefore the same volume of WW (15 L) [7,22]. The reactors worked at the same time, and were filled with the same WW. Measurements of COD were performed in all reactors until obtaining a 90% reduction of COD [20,78]. In the reactor with aeration, an interface between the WW and air occurred only at the WW surface [5,20,75]. For the aeration of the WW, the air pump was used (air pump capacity: 270 L·h^{-1}) [7,22]. The MFC included a carbon cloth anode, a Ni–Co cathode, and a Nafion PEM. After the measurements of the MFC with the Ni–Co cathode, measurements of the MFC with the carbon cloth cathode were also made. In both cases (for measurements of the MFC), the surface area of the anode was 20 cm^2, and the surface area of the cathode was 15 cm^2. The cathode was immersed in 0.1 N aqueous solution KOH and aerated (air pump capacity: 10 L·h^{-1}) [22]. The electrodes of the MFC were constantly connected with a 10 Ω resistor [5,20]. Microorganisms were acclimated for 5 days [1,2,22,78].

Nafion PF 117 (183 µm thick) (The Chemours Company, Wilmington, DE, USA) was used as the PEM. A Zortrax M200 printer (Zortrax S.A, Olsztyn, Poland) was used to print the housing of the cathode (Figure 18; 7). To prepare the 3D object (cathode housing) for printing, Z-Suite software (Zortrax S.A, Olsztyn, Poland) was used. A Hanna HI 83224 colorimeter (HANNA Instruments, Woonsocket, RI, USA) was applied for the COD measurement in WW. A Fluke 8840A multimeter (Fluke Corporation, Everett, WA, USA) was used for the electrical measurements.

4. Conclusions

This paper reports the results of a study concerned with methodology applicable for the production and selection of Ni–Co alloy as a cathode catalyst for MFC fed by municipal WW. Following that, measurements were performed with the application of the resulting alloys. As was demonstrated by the research, the most favorable catalytic parameters were obtained when a Ni–Co alloy containing 15% Co was applied, which was oxidized over 8 h at 673 K. This alloy demonstrated the highest value of the electroless potential (Figures 5–10) and the cell potential after charging as an anode (Figures 11–14). For this reason, this alloy was applied as a cathode catalyst (Figure 18; 7) in further measurements using MFC.

The measurements concerned with COD reduction also demonstrated that in each of the analyzed cases (without aeration, with aeration, and using MFC), the assumed level of reduction was 90%. The time of COD reduction using MFC with Ni–Co cathode (15% Co) to the assumed level of 90% was 15 days (Figure 15). The time obtained was shorter (by 3 days) than the time of COD reduction using MFC with a carbon cloth electrode, and also shorter than when an electrode with a Cu–B catalyst was applied [22,75].

During MFC (with Ni–Co cathode, 15% Co) operation, a current density of 0.47 mA·cm^{-2} and a power of 7.19 mW were recorded (Figure 17). For comparison, in MFC with the carbon cloth cathode, a current density of 0.23 mA·cm^{-2} and a power of 5.63 mW were recorded.

The maximum power obtained during the operation of MFC with Ni–Co alloy was slightly higher (by 1.56 mW) than the maximum power obtained during the operation of MFC with the carbon cloth cathode (Figure 16), and slightly higher (by 1.08 mW) than the maximum power obtained during the operation of MFC with the Cu–B cathode [22,75].

For these reasons, this article demonstrated the applicability of a Ni–Co alloy (containing 15% of Co) as a cathode catalyst in a microbiological fuel cell, which is based on the supply of municipal WW. At the same time, the higher catalytic activity of a Ni–Co catalyst (containing 15% of Co) was demonstrated compared to a carbon cloth, as well as compared to a Cu–B catalyst.

Author Contributions: Data curation, P.P.W. and B.W.; investigation, P.P.W. and B.W.; methodology, P.P.W.; writing—Review and editing, P.P.W. and B.W.; supervision, P.P.W. and B.W.

Funding: This research received no external funding.

Conflicts of Interest: The authors declare no conflict of interest.

References

1. Logan, B. *Microbial Fuel Cells*; Wiley: Hoboken, NJ, USA, 2008.
2. Logan, B.E.; Hamelers, B.; Rozendal, R.; Schroder, U.; Keller, J.; Verstraete, W.; Rabaey, K. Microbial Fuel Cells: Methodology and Technology. *Environ. Sci. Technol.* **2006**, *40*, 5181–5192. [CrossRef] [PubMed]
3. Rabaey, K.; Verstraete, W. Microbial fuel cells: Novel biotechnology for energy generation. *Trends Biotechnol.* **2005**, *23*, 291–298. [CrossRef] [PubMed]
4. Sivasankar, V.; Mylsamy, P.; Omine, K. *Microbial Fuel Cell Technology for Bioelectricity*, 1st ed.; Springer: Berlin, Germany, 2018.

5. Permana, D. Performance of Single Chamber Microbial Fuel Cell (SCMFC) for biological treatment of tofu wastewater. In Proceedings of the IOP Conference Series: Earth and Environmental Science, Saint Petersburg, Russia, 17–18 April 2019; Volume 277, p. 012008. [CrossRef]

6. Włodarczyk, B.; Włodarczyk, P.P. Comparison of powering the microbial fuel cell with various kinds of wastewater. *Infrastruct. Ecol. Rural Areas* **2019**, *2*, 131–140. [CrossRef]

7. Włodarczyk, P.P.; Włodarczyk, B. Microbial fuel cell with Ni-Co cathode powered with yeast wastewater. *Energies* **2018**, *11*, 3194. [CrossRef]

8. Włodarczyk, B.; Włodarczyk, P.P. Microbial fuel cell with Cu–B cathode powering with wastewater from yeast production. *J. Ecol. Eng.* **2017**, *18*, 224–230. [CrossRef]

9. Włodarczyk, B.; Włodarczyk, P.P. Analysis of the Potential of an Increase in Yeast Output Resulting from the Application of Additional Process Wastewater in the Evaporator Station. *Appl. Sci.* **2019**, *9*, 2282. [CrossRef]

10. Potter, M.C. Electrical effects accompanying the decomposition organic compounds. *Proc. R. Soc. Lond. Ser. B* **1911**, *84*, 260–276. [CrossRef]

11. Davis, J.B.; Yarbrough, H.F., Jr. Preliminary experiments on a microbial fuel cell. *Science* **1962**, *137*, 615–616. [CrossRef]

12. Berk, R.S.; Canfield, J.H. Bioelectrochemical energy conversion. *Appl. Microbiol.* **1964**, *12*, 10–12.

13. Lewis, K. Symposium on bioelectrochemistry of microorganisms: IV. Biochemical fuel cells. *Bacteriol. Rev.* **1966**, *30*, 101–113.

14. Kim, H.J.; Hyun, M.S.; Chang, I.S.; Kim, B.H. A microbial fuel cell type lactate biosensor using a metal-reducing bacterium. *Shewanella Putrefaciens J. Microbiol. Biotechnol.* **1999**, *9*, 365–367.

15. Kim, B.H.; Park, H.S.; Shin, P.K.; Chang, I.S.; Kim, H.J. Mediator-Less Biofuel Cell. U.S. Patent 5,976,719, 2 November 1999.

16. Allen, R.M.; Benetto, H.P. Microbial fuel cells: Electricity production from carbohydrates. *Appl. Biochem. Biotechnol.* **1993**, *39*, 27–40. [CrossRef]

17. Min, B.; Logan, B.E. Continuous electricity generation from domestic wastewater and organic substrates in a at plate microbial fuel cell. *Environ. Sci. Technol.* **2004**, *38*, 5809–5814. [CrossRef] [PubMed]

18. Liu, H.; Ramnarayanan, R.; Logan, B.E. Production of electricity during wastewater treatment using a single chamber microbial fuel cell. *Environ. Sci. Technol.* **2004**, *38*, 2281–2285. [CrossRef]

19. Min, B.; Cheng, S.; Logan, B.E. Electricity generation using membrane and salt bridge microbial fuel cells. *Water Res.* **2005**, *39*, 1675–1686. [CrossRef]

20. Rabaey, K.; Alterman, P.; Clauwaert, P.; De Schamphelaire, L.; Boon, N.; Verstraete, W. Microbial fuel cells in relation to conventional anaerobic digestion technology. *Eng. Life Sci.* **2006**, *6*, 285–292.

21. Franks, A.E.; Nevin, K.P. Microbial fuel cells, a current review. *Energies* **2010**, *3*, 899–919. [CrossRef]

22. Włodarczyk, P.P.; Włodarczyk, B. Wastewater Treatment and Electricity Production in a Microbial Fuel Cell with Cu–B Alloy as the Cathode Catalyst. *Catalysts* **2019**, *9*, 572. [CrossRef]

23. Cheng, S.; Liu, H.; Logan, B.E. Increased Power Generation in a Continuous Flow MFC with Advective Flow through the Porous Anode and Reduced Electrode Spacing. *Environ. Sci. Technol.* **2006**, *40*, 2426–2432. [CrossRef]

24. Kimberlynn, C. *Microbial Fuel Cell Possibilities on American Indian Tribal Lands*; Sandia National Lab.(SNL-NM): Albuquerque, NM, USA, 2016. [CrossRef]

25. Saha, T.C.; Protity, A.T.; Zohora, F.T.; Shaha, M.; Ahmed, I.; Barua, E.; Sarker, P.K.; Mukharjee, S.K.; Barua, A.; Salimullah, M.; et al. Microbial Fuel Cell (MFC) Application for Generation of Electricity from Dumping Rubbish and Identification of Potential Electrogenic Bacteria. *Adv. Ind. Biotechnol.* **2019**, *2*, 10. [CrossRef]

26. Flimban, S.G.A.; Ismail, I.M.I.; Kim, T.; Oh, S.-E. Overview of Recent Advancements in the Microbial Fuel Cell from Fundamentals to Applications: Design, Major Elements, and Scalability. *Energies* **2019**, *12*, 3390. [CrossRef]

27. Bond, D.R.; Lovley, D.R. Evidence for involvement of an electron shuttle in electricity generation by *Geothrix fermentans*. *Appl. Environ. Microbiol.* **2005**, *71*, 2186–2189. [CrossRef] [PubMed]

28. Bond, D.R.; Lovley, D.R. Electricity production by *Geobacter sulfurreducens* attached to electrodes. *Appl. Environ. Microbiol.* **2003**, *69*, 1548–1555. [CrossRef] [PubMed]

29. Kim, H.J.; Park, H.S.; Hyun, M.S.; Chang, I.S.; Kim, M.; Kim, B.H. A Mediator-Less Microbial Fuel Cell Using a Metal Reducing Bacterium. *Shewanella Putrefaciens. Enzym. Microb. Technol.* **2002**, *30*, 145–152. [CrossRef]

30. Park, H.S.; Kim, B.H.; Kim, H.S.; Kim, H.J.; Kim, G.T.; Kim, M.; Chang, I.S.; Park, Y.K.; Chang, H.I. A novel electrochemically active and Fe(III)-reducing bacterium phylogenetically related to *Clostridium butyricum* isolated from a microbial fuel cell. *Anaerobe* **2001**, *7*, 297–306. [CrossRef]

31. Chaudhuri, S.K.; Lovley, D.R. Electricity generation by direct oxidation of glucose in mediatorless microbial fuel cells. *Nat. Biotechnol.* **2003**, *21*, 1229–1232. [CrossRef]

32. Pham, C.A.; Jung, S.J.; Phung, N.T.; Lee, J.; Chang, I.S.; Kim, B.H.; Yi, H.; Chun, J. A novel electrochemically active and Fe(III)-reducing bacterium phylogenetically related to *Aeromonas hydrophila*, isolated from a microbial fuel cell. *FEMS Microbiol. Lett.* **2003**, *223*, 129–134. [CrossRef]

33. Reguera, G.; McCarthy, K.D.; Mehta, T.; Nicoll, J.S.; Tuominen, M.T.; Lovley, D.R. Extracellular electron transfer via microbial nanowires. *Nature* **2005**, *435*, 1098–1101. [CrossRef]

34. Jenol, M.A.; Ibrahim, M.F.; Kamal Bahrin, E.; Kim, S.W.; Abd-Aziz, S. Direct Bioelectricity Generation from Sago Hampas by Clostridium beijerinckii SR1 Using Microbial Fuel Cell. *Molecules* **2019**, *24*, 2397. [CrossRef]

35. Patil, S.A.; Surakasi, V.P.; Koul, S.; Ijmulwar, S.; Vivek, A.; Shouche, Y.S.; Kapadnis, B.P. Electricity generation using chocolate industry wastewater and its treatment in activated sludge based microbial fuel cell and analysis of developed microbial community in the anode chamber. *Bioresour. Technol.* **2009**, *100*, 5132–5139. [CrossRef]

36. Logan, B.E.; Regan, J.M. Microbial fuel cells—Challenges and applications. *Environ. Sci. Technol.* **2006**, *40*, 5172–5180. [CrossRef] [PubMed]

37. Logan, B.E.; Regan, J.M. Electricity-producing bacterial communities in microbial fuel cells. *Trends Microbiol.* **2006**, *14*, 512–518. [CrossRef] [PubMed]

38. Bockris, J.O.M.; Reddy, A.K.N. *Modern Electrochemistry*; Kulwer Academic/Plenum Publishers: New York, NY, USA, 2000.

39. Cheng, S.; Hong Liu, H.; Logan, B.E. Power Densities Using Different Cathode Catalysts (Pt and CoTMPP) and Polymer Binders (Nafion and PTFE) in Single Chamber Microbial Fuel Cells. *Environ. Sci. Technol.* **2006**, *40*, 364–369. [CrossRef] [PubMed]

40. Sanchez, D.V.P.; Huynh, P.; Kozlov, M.E.; Baughman, R.H.; Vidic, R.D.; Yun, M. Carbon Nanotube/Platinum (Pt) Sheet as an Improved Cathode for Microbial Fuel Cells. *Energy Fuels* **2010**, *24*, 5897–5902. [CrossRef]

41. Martin, E.; Tartakovsky, B.; Savadogo, O. Cathode materials evaluation in microbial fuel cells: A comparison of carbon, Mn_2O_3, Fe_2O_3 and platinum materials. *Electrochim. Acta* **2011**, *58*, 58–66. [CrossRef]

42. Morris, J.M.; Jin, S.; Wang, J.; Zhu, C.; Urynowiczcz, M.A. Lead dioxide as an alternative catalyst to platinum in microbial fuel cells. *Electrochem. Commun.* **2007**, *9*, 1730–1734. [CrossRef]

43. Liew, K.B.; Daud, W.R.W.; Ghasemia, M.; Leong, J.X.; Lim, S.S.; Ismail, M. Non-Pt catalyst as oxygen reduction reaction in microbial fuel cells: A review. *Int. J. Hydrog. Energy* **2014**, *39*, 4870–4883. [CrossRef]

44. Zhang, L.; Liu, C.; Zhuang, L.; Li, W.; Zhou, S.; Zhang, J. Manganese dioxide as an alternative cathodic catalyst to platinum in microbial fuel cells. *Biosens. Bioelectron.* **2009**, *24*, 2825–2829. [CrossRef]

45. Santoro, C.; Lei, Y.; Li, B.; Cristianid, P. Power generation from wastewater using single chamber microbial fuel cells (MFCs) with platinum-free cathodes and pre-colonized anodes. *Biochem. Eng. J.* **2012**, *62*, 8–16. [CrossRef]

46. Wang, Z.; Huang, H.; Liu, H.; Zhou, X. Self-sustained electrochemical promotion catalysts for partial oxidation reforming of heavy hydrocarbons. *Int. J. Hydrog. Energy* **2012**, *37*, 17928–17935. [CrossRef]

47. Huang, H.; Wang, Z.; Zhou, X.; Liu, H.; Wei, Y.; Pramuanjaroenkij, A.; Bordas, A.; Page, M.; Cai, S.; Zhang, X. Development and study of self-sustained electrochemical promotion catalysts for hydrocarbon reforming. In *ECS Transactions*, 2nd ed.; Electrochemical Society Inc.: Pennington, NJ, USA, 2013; Volume 58, pp. 243–254. [CrossRef]

48. Liu, Y.; Harnisch, F.; Fricke, K.; Schroeder, U.; Climent, V.; Feliu, J.M. The study of electrochemically active microbial biofilms on different carbon-based anode materials in microbial fuel cells. *Biosens. Bioelectron.* **2010**, *25*, 2167–2171. [CrossRef] [PubMed]

49. Yuan, Y.; Zhou, S.; Zhuang, L. Polypyrrole/carbon black composite as a novel oxygen reduction catalyst for microbial fuel cells. *J. Power Sources* **2010**, *195*, 3490–3493. [CrossRef]

50. Erbay, C.; Yang, G.; de Figueiredo, P.; Sadr, R.; Yu, C.; Han, A. Three-dimensional porous carbon nanotube sponges for high-performance anodes of microbial fuel cells. *J. Power Sources* **2015**, *298*, 177–183. [CrossRef]

51. Zuo, K.; Liang, S.; Liang, P.; Zhou, X.; Sun, D.; Zhang, X.; Huang, X. Carbon filtration cathode in microbial fuel cell to enhance wastewater treatment. *Bioresour. Technol.* **2015**, *185*, 426–430. [CrossRef]

52. Penteado, E.D.; Fernández-Marchante, C.M.; Zaiat, M.; Cañizares, P.; Gonzalez, E.R.; Rodrigo, M.A. Influence of carbon electrode material on energy recovery from winery wastewater using a dual-chamber microbial fuel cell. *Environ. Technol.* **2017**, *38*, 1333–1341. [CrossRef]

53. Asensio, Y.; Montes, I.B.; Fernández-Marchante, C.M.; Lobato, J.; Cañizares, P.; Rodrigo, M.A. Selection of cheap electrodes for two-compartment microbial fuel cells. *J. Electroanal. Chem.* **2017**, *785*, 235–240. [CrossRef]

54. Wu, W.; Niu, H.; Yang, D.; Wang, S.-B.; Wang, J.; Lin, J.; Hu, C. Controlled layer-by-layer deposition of carbon nanotubes on electrodes for microbial fuel cells. *Energies* **2019**, *12*, 363. [CrossRef]

55. Sudirjo, E.; Buisman, C.J.; Strik, D.P. Activated Carbon Mixed with Marine Sediment is Suitable as Bioanode Material for *Spartina anglica* Sediment/Plant Microbial Fuel Cell: Plant Growth, Electricity Generation, and Spatial Microbial Community Diversity. *Water* **2019**, *11*, 1810. [CrossRef]

56. Ling, J.; Xu, Y.; Lu, C.; Lai, W.; Xie, G.; Zheng, L.; Talawar, M.P.; Du, Q.; Li, G. Enhancing Stability of Microalgae Biocathode by a Partially Submerged Carbon Cloth Electrode for Bioenergy Production from Wastewater. *Energies* **2019**, *12*, 3229. [CrossRef]

57. Dumas, C.; Mollica, A.; Féron, D.; Basséguy, R.; Etcheverry, L.; Bergel, A. Marine microbial fuel cell: Use of stainless steel electrodes as anode and cathode materials. *Electrochim. Acta* **2006**, *53*, 468–473. [CrossRef]

58. Osorio de la Rosa, E.; Vázquez Castillo, J.; Carmona Campos, M.; Barbosa Pool, G.R.; Becerra Nuñez, G.; Castillo Atoche, A.; Ortegón Aguilar, J. Plant Microbial Fuel Cells–Based Energy Harvester System for Self-powered IoT Applications. *Sensors* **2019**, *19*, 1378. [CrossRef] [PubMed]

59. Armour, M.A. *Hazarodous Laboratory Chemicals Disposal Guide*; CRC Press: Boca Raton, FL, USA, 2003.

60. HaoYu, E.; Cheng, S.; Scott, K.; Logan, B.E. Microbial fuel cell performance with non-Pt cathode catalysts. Journal of Power Sources. *J. Power Sources* **2007**, *171*, 275–281. [CrossRef]

61. Li, B.; Wang, M.; Zhou, X.; Wang, X.; Liu, B.; Li, B. Pyrolyzed binuclear-cobalt-phthalocyanine as electrocatalyst for oxygen reduction reaction in microbial fuel cells. *Bioresour. Technol.* **2015**, *193*, 545–548. [CrossRef] [PubMed]

62. Tang, X.; Ng, H.Y. Cobalt and nitrogen-doped carbon catalysts for enhanced oxygen reduction and power production in microbial fuel cells. *Electrochim. Acta* **2017**, *247*, 193–199. [CrossRef]

63. Ahmed, J.; Yuan, Y.; Zhou, L.; Kim, S. Carbon supported cobalt oxide nanoparticles–iron phthalocyanine as alternative cathode catalyst for oxygen reduction in microbial fuel cells. *J. Power Sources* **2012**, *208*, 170–175. [CrossRef]

64. Williams, H.; Gnanamani, M.K.; Jacobs, G.; Shafer, W.D.; Coulliette, D. Fischer–Tropsch Synthesis: Computational Sensitivity Modeling for Series of Cobalt Catalysts. *Catalysts* **2019**, *9*, 857. [CrossRef]

65. Jo, S.B.; Kim, T.Y.; Lee, C.H.; Woo, J.H.; Chae, H.J.; Kang, S.-H.; Kim, J.W.; Lee, S.C.; Kim, J.C. Selective CO Hydrogenation Over Bimetallic Co-Fe Catalysts for the Production of Light Paraffin Hydrocarbons (C_2–C_4): Effect of Space Velocity, Reaction Pressure and Temperature. *Catalysts* **2019**, *9*, 779. [CrossRef]

66. Twigg, M.V. *Catalyst Handbook*; Wolfe Publishing Ltd.: London, UK, 1989.

67. US EPA Report. *Clean Watersheds Needs Survey Overview*; United States Environmental Protection Agency: Washington, DC, USA, 2008.

68. Zhou, J.; Zhang, Y.; Li, S.; Chen, J. Ni/NiO Nanocomposites with Rich Oxygen Vacancies as High-Performance Catalysts for Nitrophenol Hydrogenation. *Catalysts* **2019**, *9*, 944. [CrossRef]

69. Wang, Y.; Guo, J.; Wang, T.; Shao, J.; Wang, D.; Yang, Y.-W. Mesoporous Transition Metal Oxides for Supercapacitors. *Nanomaterials* **2015**, *5*, 1667–1689. [CrossRef]

70. Huang, L.; Chen, D.C.; Ding, Y.; Feng, S.; Wang, Z.L.; Liu, M.L. Nickel-cobalt hydroxide nanosheets coated on $NiCo_2O_4$ nanowires grown on carbon fiber paper for high-performance pseudocapacitors. *Nano Lett.* **2013**, *13*, 3135–3139. [CrossRef]

71. Rakhi, R.B.; Chen, W.; Hedhili, M.N.; Cha, D.; Alshareef, H.N. Enhanced rate performance of mesoporous Co_3O_4 nanosheet supercapacitor electrodes by hydrous RuO_2 nanoparticle decoration. *ACS Appl. Mater. Interfaces* **2014**, *6*, 4196–4206. [CrossRef] [PubMed]

72. Yuan, C.Z.; Zhang, X.G.; Su, L.H.; Gao, B.; Shen, L.F. Facile synthesis and self-assembly of hierarchical porous NiO nano/micro spherical superstructures for high performance supercapacitors. *J. Mater. Chem.* **2009**, *19*, 5772–5777. [CrossRef]

73. Lee, J.W.; Ahn, T.; Kim, J.H.; Ko, J.M.; Kim, J. Nanosheets based mesoporous NiO microspherical structures via facile and template-free method for high performance supercapacitors. *Electrochim. Acta* **2011**, *56*, 4849–4857. [CrossRef]

74. Yang, M.; Li, J.X.; Li, H.H.; Su, L.W.; Wei, J.P.; Zhou, Z. Mesoporous slit-structured NiO for high-performance pseudocapacitors. *Phys. Chem. Chem. Phys.* **2012**, *14*, 11048–11052. [CrossRef]

75. Włodarczyk, B.; Włodarczyk, P.P. Comparsion of Cu-B alloy and stainless steel as electrode material for microbial fuel cell. In *Renewable Energy Sources: Engineering, Technology, Innovation ICORES 2018*; Wróbel, M., Jewiarz, M., Szlek, A., Eds.; Springer Nature Switzerland AG: Basel, Switzerland, 2020. [CrossRef]

76. Włodarczyk, P.P.; Włodarczyk, B. Ni-Co alloy as catalyst for fuel electrode of hydrazine fuel cell. *China-USA Bus. Rev.* **2015**, *14*, 269–279. [CrossRef]

77. Holtzer, M.; Staronka, A. *Physical chemistry (in Polish)*; AGH University of Science and Technology Press: Cracow, Poland, 2000.

78. Huggins, T.; Fallgren, P.H.; Jin, S.; Ren, Z.J. Energy and performance comparison of microbial fuel cell and conventional aeration treating of wastewater. *J. Microb. Biochem. Technol.* **2013**, *S6:002*, 1–5. [CrossRef]

Review

Synthesis and Surface Modification of TiO_2-Based Photocatalysts for the Conversion of CO_2

Samar Al Jitan [1,2], Giovanni Palmisano [1,2,*] and Corrado Garlisi [1,2]

[1] Department of Chemical Engineering, Khalifa University, Abu Dhabi P.O. Box 127788, UAE; samar.aljitan@ku.ac.ae (S.A.J.); corrado.garlisi@ku.ac.ae (C.G.)

[2] Research and Innovation Center on CO_2 and H_2 (RICH Center), Khalifa University, Abu Dhabi P.O. Box 127788, UAE

* Correspondence: giovanni.palmisano@ku.ac.ae; Tel.: +971-02-810-9246

Received: 12 December 2019; Accepted: 23 January 2020; Published: 14 February 2020

Abstract: Among all greenhouse gases, CO_2 is considered the most potent and the largest contributor to global warming. In this review, photocatalysis is presented as a promising technology to address the current global concern of industrial CO_2 emissions. Photocatalysis utilizes a semiconductor material under renewable solar energy to reduce CO_2 into an array of high-value fuels including methane, methanol, formaldehyde and formic acid. Herein, the kinetic and thermodynamic principles of CO_2 photoreduction are thoroughly discussed and the CO_2 reduction mechanism and pathways are described. Methods to enhance the adsorption of CO_2 on the surface of semiconductors are also presented. Due to its efficient photoactivity, high stability, low cost, and safety, the semiconductor TiO_2 is currently being widely investigated for its photocatalytic ability in reducing CO_2 when suitably modified. The recent TiO_2 synthesis and modification strategies that may be employed to enhance the efficiency of the CO_2 photoreduction process are described. These modification techniques, including metal deposition, metal/non-metal doping, carbon-based material loading, semiconductor heterostructures, and dispersion on high surface area supports, aim to improve the light absorption, charge separation, and active surface of TiO_2 in addition to increasing product yield and selectivity.

Keywords: photocatalytic reduction; CO_2; TiO_2 photocatalysts; surface modification; solar fuel

1. Introduction

The global emission of greenhouse gases, such as CO_2, continues to rise by ~3% each year [1]. Higher atmospheric concentrations of greenhouse gases lead to surface warming of the land and oceans [2]. According to computational results based on climate models, the average global value of the temperature rise of +2 °C will be surpassed when the concentration of CO_2 reaches 550 ppm [1]. If current trends are kept, the CO_2 concentration will reach this threshold value by 2050. In addition to warming the Earth, CO_2 emissions have increased the ocean's acidity. Thirty million of the 90 million tons of CO_2 discharged each day end up in the oceans as carbonic acid, lowering the ocean's pH level [3]. Both the increased temperatures and the higher acidity of the ocean have initiated a set of other impacts including the melting of glaciers, rising sea levels, deeper and longer droughts, more and larger forest fires, migration of tropical diseases, accelerated extinction rates, increased destructive power of tropical storms, and increasingly large downpours of rain and snow [3,4]. Natural processes can potentially remove most of the CO_2 that human activities are adding to the atmosphere; however, these processes operate very slowly and will take too long to prevent rapid climate change and its impacts [2].

Three theoretical approaches have been widely suggested for solving the climate problem: (1) direct reduction of CO_2 emissions by changing industrial and urban processes and habits, (2) CO_2 capture and storage (CCS), and (3) CO_2 capture and utilization (CCU). Due to the increasing population rate and the increasing demand for high-quality life, the direct reduction of CO_2 emissions seems infeasible [5]. To keep up with energy needs of a growing population, governments are compelled to continue with their current industrial activities despite the large amounts of resulting greenhouse gas emissions. Carbon capture and sequestration (CCS) is considered to be a very promising solution to removing excess CO_2 from the atmosphere. The idea behind CCS lies in storing large quantities of captured CO_2 in underground geological formations. The stored CO_2 may be used for recovering oil and gas from partly exploited fields, thus giving an economic value to CO_2 [6]. The two major issues associated with CCS are cost and storage. The high cost of CCS is due to the large amounts of energy required to separate CO_2 from the emission stream. It is estimated that this separation process could account for 70–90% of the total operating cost of CCS [7]. Also, the storage of CO_2 is considered challenging due to the large amounts of CO_2 that needs to be stored and the risk of leakage [2]. Although CCS technologies may appear to be very promising in removing excess CO_2 from the atmosphere, they are expensive, energy-intensive, and require large capital investment for industrial application [8].

To make CCS technologies more economically feasible, carbon capture and utilization (CCU) technologies have emerged as a feasible and promising technique that can complement the storage of huge quantities of CO_2 in geological and ocean formations. Industrial applications of CO_2 are present in numerous sectors including chemical, oil and gas, energy, pulp and paper, steel, food, and pharmaceuticals [9]. Currently, the commercial CCU technologies use CO_2 for enhanced oil recovery (EOR) applications. The use of CO_2 as a raw input material in the chemical industry is limited to a few processes such as the production of salicylic acid, urea and polycarbonates [8,10]. Recently, new research is looking into converting the captured CO_2 into valuable products including chemicals, polymers and fuel. It is estimated that 5 to 10% of the total CO_2 emissions may be utilized for the synthesis of value-added products [11]. The specific application of converting CO_2 into fuel is being actively studied by researches and is showing great promise for future industrial applications. A wide variety of fuels, including methanol, ethanol, methane, dimethyl ether, formic acid, petroleum-equivalent fuels, and others may be produced through different CO_2 conversion processes. However, these conversion processes produce large quantities of CO_2 and will therefore increase the concentration of CO_2 in the atmosphere instead of reducing it [11,12]. Thus, to stay within the targeted goal of decreasing CO_2 emissions, low-carbon energy sources, such as renewable resources, must be used as the primary energy input in the CO_2-to-fuels conversion process. Other drawbacks of CO_2-to-fuels conversion processes include intensive energy requirements and low energy conversion efficiency [12]. The main methods of CO_2 conversion are thermochemical [13–15], electrochemical [12,13], biological [16–18], and photocatalytic [19–23].

Photocatalysis is considered to be a promising method for the conversion of CO_2 into valuable products, such as methane, hydrogen, methanol, formaldehyde, ethanol, and higher hydrocarbons [6,19]. One major advantage of photocatalysis over other conversion methods is that it can take place at room temperature and under atmospheric pressure conditions [19]. In addition to that, it utilizes a renewable and sustainable form of energy, namely solar energy, for the conversion of CO_2. Note that, unlike conventional processes, the photoreduction of CO_2 does not increase net CO_2 emissions or consume additional energy [23]. Despite the intensive research, the photocatalytic reduction of CO_2 is still considered inefficient. This is mainly due to the absence of scalable reactor designs able to simultaneously introduce reactants, photons, and visible light-responsive catalysts to produce specific fuels in significant quantities [23]. The main challenge in developing an economically feasible process for the photocatalytic conversion of CO_2 is finding a suitable catalyst [24]. The most common type of catalysts used in the photocatalytic conversion of CO_2 are the inexpensive and naturally abundant transition-metal oxides, such as titanium dioxide (TiO_2). TiO_2 is one of the most widely used and commonly investigated semiconductors for photocatalytic applications [25]. This is attributed to its low toxicity, low cost, high efficiency, and high stability. Other several types of metal oxide semiconductors, including zirconium oxide, gallium oxide and thallium oxide, have also been investigated for their use in the photocatalytic reduction of CO_2 [25]. The poor light absorption and the low product selectivity of these photocatalysts pose a challenge for researchers aiming to efficiently reduce CO_2 [26]. Currently, the development of new, stable, inexpensive, abundant, nontoxic, selective, and visible light-responsive photocatalysts is being actively investigated [19,27]. In this regard, some state-of-the-art photocatalysts, such are perovskite oxides and III-V semiconductors, have shown some great promise in driving the photocatalytic reduction of CO_2 specifically under direct sunlight. Perovskite oxides exhibit interesting compositional flexibility that allows for precise band gap tuning and defect engineering [28]. On the other hand, III-V semiconductors can easily meet the thermodynamic requirements of CO_2 reduction to CO due to their higher (less positive) conduction band when compared to that of other metal oxide semiconductors [29]. Nonetheless, the current photocatalytic efficiency of perovskite oxides and III-V semiconductors is too low for practical CO_2 reduction applications. Furthermore, their low selectivity and high instability have urged researches to continue with their investigations on TiO_2-based photocatalysts for the reduction of CO_2 [28].

Besides enhanced photocatalysts, researchers have been also coupling photocatalysis with other CO_2 conversion methods aiming at improving the process efficiency. For instance, in a study proposed by Zhang et al. [30], a photo-thermochemical process was used for the reduction of CO_2. Combining both processes together allowed for the utilization of both solar and thermal energy. The major limitation of thermochemical conversion techniques is the high temperature (>1273 K) step required to reduce the metal oxide catalyst and create an oxide vacancy. In the photo-thermochemical process, this step is replaced by photocatalysis where UV light is used instead of high temperatures to form vacant sites. Next, the temperature needs to be only slightly raised (573–873 K) to accelerate the dissociation of CO_2 to CO. Electrocatalysis is another CO_2 conversion method that has been coupled with photocatalysis. As previously mentioned in this work, photocatalysts with outstanding structural and optical properties need to be developed for an efficient CO_2 photoreduction process. From a thermodynamic perspective, these photocatalysts must possess favorable band gap energies for CO_2 reduction. In photo-electrocatalysis, this limitation may be overcome if an appropriate external bias is applied to the system [31]. Therefore, a broader range of photocatalysts with lower thermodynamic restrictions can be used for the reduction of CO_2. Halmann [32] was the first to report the successful photo-electrocatalytic conversion of CO_2 into formic acid, formaldehyde, and methanol.

In this review, the current advances on CO_2 photoreduction over TiO_2-based catalysts are critically discussed. The photocatalytic mechanism of CO_2 reduction has been explained particularly in the case when water is used as a reductant. Furthermore, the various modification techniques of TiO_2, including metal deposition, metal/non-metal doping, carbon-based material loading, formation of semiconductor heterostructures, and dispersion on high surface area supports, have been summarized. Although a number of review articles [33–35] have highlighted the possible surface modifications of TiO_2 photocatalysts, this paper provides emphasis on modification techniques that will specifically enhance the CO_2 photoreduction efficiency of TiO_2 photocatalysts. Future directions toward efficient photocatalytic systems for the reduction of CO_2 have been also presented.

2. Photocatalytic Conversion of CO_2: Thermodynamics and Kinetics

In the photocatalytic approach, semiconductors irradiated by UV or visible light are used for the reduction of CO_2. As can be seen in Figure 1, the light energy (*hv*) absorbed by the photocatalyst is used to produce electron–hole pairs, which are generated when that energy is equal or greater than the band gap energy (E_g) of the photocatalyst. Only then will the electrons (e^-) be promoted from the valence band (VB) to the conduction band (CB) of the semiconducting material, simultaneously creating a hole (h^+) at the VB [20]. The e^- and h^+ are then transferred to active redox species present across the photocatalytic interface in order to participate in the conversion process [26].

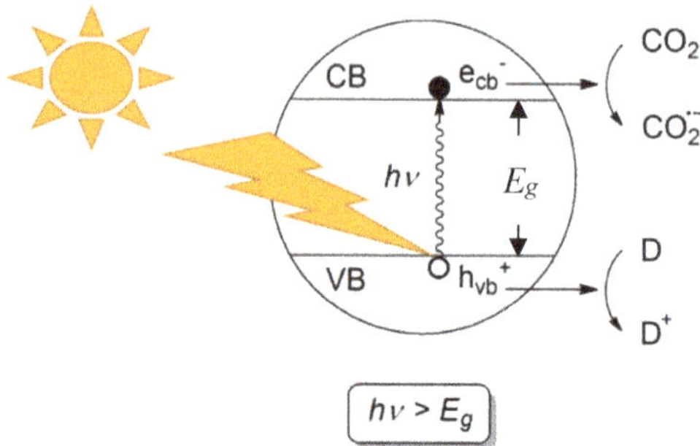

Figure 1. Generation of an electron–hole pair in a photocatalyst. Reproduced from work in [20]. Copyright 2011 Royal Society of Chemistry.

The photocatalytic reduction of CO_2 into value-added products involves radical-chain reactions that form cation, anion, and electrically neutral or charged radicals. These radicals are produced as a result of the reaction with photogenerated electrons and holes between the metal oxide photocatalyst and the reactants [26,36]. As CO_2 will be reduced, the presence of a co-reagent, or an electron donor, that will be simultaneously oxidized, is necessary [37]. A sacrificial donor (D), usually water, is oxidized by the photogenerated holes in the valence band, while the electrons in the conduction band reduce CO_2 [38]. Hydroxyl groups present on the surface of the photocatalyst might also be oxidized by the holes [39]. The end products of the photocatalytic reduction of CO_2 depend on both the redox potential (thermodynamics), and surface electron density and photoadsorption/photodesorption thermodynamics and kinetics [40].

The overall efficiency of the photocatalytic reduction of CO_2 is governed by the effectiveness of three processes: the photogeneration of the electron/hole pair, the interfacial transfer of electrons between the photocatalyst surface and the adsorbed CO_2, and the conversion of redox species to valuable products [20]. The features of the photocatalyst, the energy of light, and the concentration of reactants are all factors that greatly influence the reaction rate of the photocatalytic process. Other important parameters include pH (when the reaction takes place in water), which affects the charge of the photocatalytic surface, and temperature, which impacts the collision frequency between the reactants and the photocatalyst [21]. Moreover, the surface of the photocatalyst must be thoroughly cleaned prior to any photocatalytic test. Otherwise, carbon-containing species, resulting from the synthesis procedure, will be present on the photocatalyst's surface and will contribute to product formation. Dilla et al. [41] developed an efficient photocatalytic cleaning step that may be performed right before the start of the CO_2 photoreduction test. Herein, humid helium is flushed through the reactor under light irradiation. The cleaning progress is then monitored and, as soon as the concentration of products from the cleaning reaction is low, CO_2 is fed into the reactor. This method not only ensures a clean photocatalytic surface but also rids the reactor from any trace hydrocarbon contaminants that might be present inside.

If the participating species are adsorbed on the semiconductor surface, then the photocatalytic process will certainly be more efficient due to the decrease in activation energy and increase in substrate concentration near reactive sites. When CO_2 adsorbs onto the photocatalyst surface, its structure transforms from linear to the more reactive bent form (Figure 2). More specifically, the carbon atom of CO_2 binds to a surface oxygen site, while the oxygen atom binds to the surface metal center of the metal oxide [42]. As a result of the structural transformation, the lowest unoccupied molecular orbital (LUMO) level of the metal oxide decreases lowering its activation energy [39]. However, density functional theory (DFT) calculations performed by Sorescu et al. [43] demonstrated that CO_2 preferentially binds to the metal oxide surface in a linear geometry. This is most probably due to the low binding energy (−46 kJ/mol) associated with the linear configuration, the considerable energy required to bend CO_2 (38.6 kcal/mol) and the significant deformation the metal oxide surface undergoes when binding to bent CO_2 [42,43].

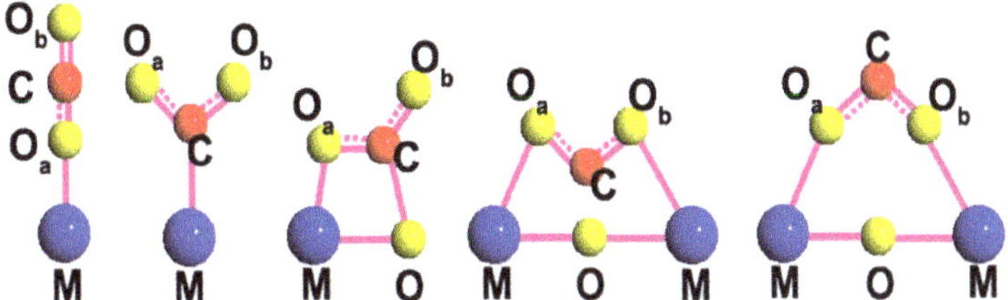

Figure 2. Five possible CO_2 adsorption configuration models on metal oxide semiconductors. M: Metal C: Carbon O: Oxygen. Reproduced from work in [44]. Copyright 2014 Wiley-VCH.

To enhance the adsorption of CO_2, basic sites and/or micropores must be introduced on the surface of the photocatalyst. The introduction of sites that can store electrons and acid sites that can stabilize the species derived from CO_2 must also be considered [37]. Therefore, researchers have to focus on developing such complex photocatalysts with a combination of acidic and basic sites. Additionally, the adsorption of CO_2 could also be improved by synthesizing metal-oxide photocatalysts with oxygen vacancy sites. These vacancies may then be filled by the oxygen atoms of CO_2 promoting the adsorption of the latter [45]. The generation of an unexpected attraction between the oxygen vacancy and CO_2 subsequently lowers the reactive barrier [44].

A large and negative reduction potential ($E = -1.90$ V) is required for the single electron reaction that coverts CO_2 to $CO_2^{\bullet-}$ [46]:

$$CO_2 + e^- \rightarrow CO_2^{\bullet-}, \tag{1}$$

This reaction is thermodynamically unfavorable due to the high stability of CO_2 and the large amount of energy needed to change the geometrical configuration of CO_2 from linear to bent. $CO_2^{\bullet-}$ is a highly unstable radical anion that almost immediately converts to CO_3^{2-}, CO, and $C_2O_4^{2-}$ [46]. If the transfer of the single electron occurs simultaneously with a proton transfer, the CO_2 reduction potential can be significantly lowered (i.e., more positive) due to the stabilization of $CO_2^{\bullet-}$ by the proton [37]. The reduction potential can be further reduced to a more positive value when multiple electrons and holes are simultaneously transferred to CO_2. Therefore, the multiple electron reduction of CO_2 is considered more thermodynamically favorable. Here, CO_2 typically converts to CO, CH_2O_2, CH_2O, CH_3OH, and CH_4. Although more thermodynamically favorable, these multiple electron reactions are kinetically more challenging and lead to difficulties in process efficiency and selectivity.

As depicted in Figure 3, any photocatalytic reaction mechanism generally involves the following steps [47]; (1) reactant adsorption onto active sites, (2) light absorption from the catalyst with subsequent photogeneration of e^- and h^+, (3) interaction between charges and adsorbed reactants, (4) redox reactions, and (5) product desorption. Nonetheless, the particular reaction mechanism of CO_2 photocatalysis is fairly complex. The reduction process, which is significantly influenced by the type of photocatalyst used, can give rise to a range of various possible products. The mechanism is also highly dependent on the energy of the photoexcited charges. If the excited charge carriers do not have the sufficient energy to react with the intermediate species, the reaction will not proceed and the final expected product from the photocatalytic reaction pathway will not be formed. Other factors that may inhibit the formation of products include product accumulation on the photocatalytic surface and photocatalyst deactivation. Moreover, the formed products might re-oxidize back to CO_2 in oxygen-rich environments. As a result of the aforementioned challenges, the yield of CO_2 photoreduction is usually very low and, as such, the formed products are generally difficult to detect [39].

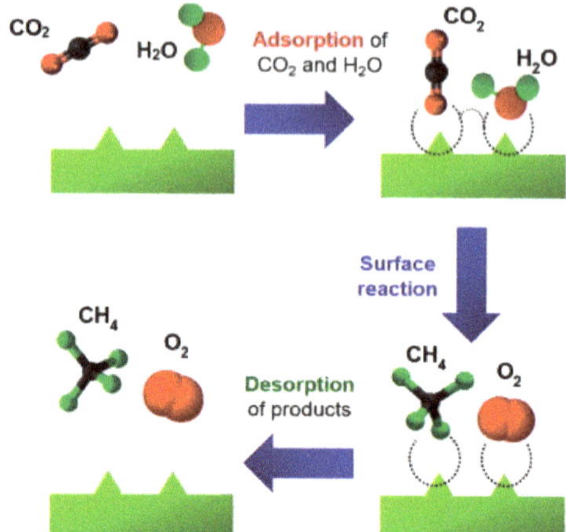

Figure 3. Basic illustration describing the reaction mechanism of CO$_2$ photoreduction. Reproduced from work in [47]. Copyright 2016 Elsevier B.V.

The kinetic modeling of CO$_2$ photoreduction has only been investigated by few. In these studies, two insights have been suggested on the rate limiting step. The first proposition claims that the activation of CO$_2$ through charge transfer is rate limiting [48]. In their mechanistic study, Uner et al. [49] suggested that the rate of the overall process of CO$_2$ photoreduction was limited by the production of electrons and protons. The second proposition claims that the reactant adsorption/product desorption is rate limiting [48]. A kinetic model proposed by Saladin et al. [50] demonstrated that product desorption was rate limiting at low temperatures while reactant adsorption was rate limiting at high temperatures. Further studies need to be conducted to help advance the understanding of reaction kinetics in CO$_2$ photoreduction.

Due to its availability and low cost, water is a desirable reducing agent to provide hydrogen atoms in the photocatalytic conversion of CO$_2$. It reacts with holes (h$^+$) to form O$_2$ and H$^+$. The H$^+$ ions later interact with the excited electrons (e$^-$) producing H$_2$ [51]:

$$2H_2O + 4h^+ \rightarrow O_2 + 4H^+ \qquad E = 0.81 \text{ V,} \tag{2}$$

$$2H^+ + 2e^- \rightarrow H_2 \qquad E = -0.42 \text{ V,} \tag{3}$$

Together, the CO$_2$ and the H$^+$ react and lead to the formation of stable molecules as shown in the equations below [51].

$$CO_2 + 2H^+ + 2e^- \rightarrow CO + H_2O \qquad E = -0.53 \text{ V,} \tag{4}$$

$$CO_2 + 2H^+ + 2e^- \rightarrow HCOOH \qquad E = -0.61 \text{ V,} \tag{5}$$

$$CO_2 + 4H^+ + 4e^- \rightarrow HCHO + H_2O \qquad E = -0.48 \text{ V,} \tag{6}$$

$$CO_2 + 6H^+ + 6e^- \rightarrow CH_3OH + H_2O \qquad E = -0.38 \text{ V,} \tag{7}$$

$$CO_2 + 8H^+ + 8e^- \rightarrow CH_4 + 2H_2O \qquad E = -0.24 \text{ V,} \tag{8}$$

Studies have shown that the presence of water in the reaction system favors the formation of methane as the major CO$_2$ reduction product [37]. Two fundamental mechanisms have been proposed for this specific reaction. One mechanism suggests that methane is produced in series [52]:

$$CO_2 \rightarrow HCOOH \rightarrow HCHO \rightarrow CH_3OH \rightarrow CH_4, \tag{9}$$

The second mechanism assumes that methane forms in parallel with methanol [53]:

$$CO_2 \rightarrow CO \rightarrow CH_3OH/CH_4, \tag{10}$$

Evidently, the reaction pathway of CO_2 photoconversion also greatly depends on the phase (liquid/vapor) of water, which has to be considered as an important reaction parameter [39]. The major product of CO_2 photoreduction in aqueous media is usually formic acid (Figure 4), whereas the main product in gaseous phase is typically CO (Figure 5). It is important to note that although methanol and methane are the desired end products, the reaction usually does not proceed to that point. Instead, it typically stops after the formation of formic acid or CO.

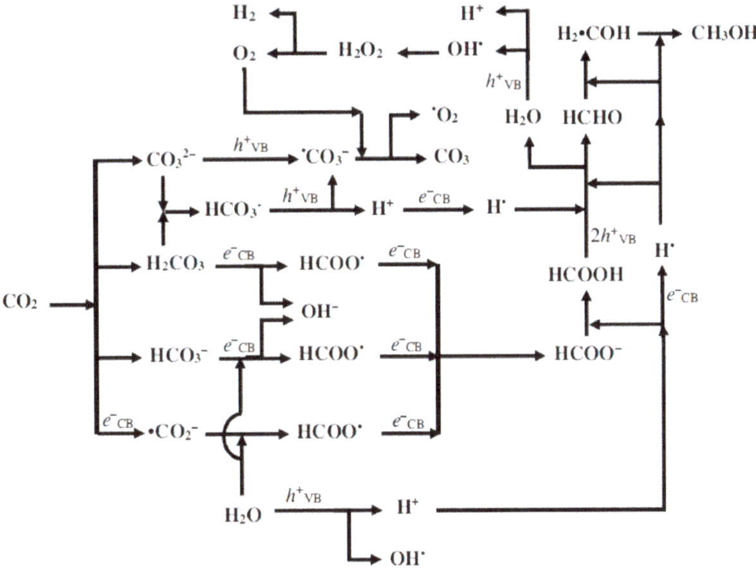

Figure 4. Aqueous phase photocatalysis of CO_2 and water: proposed reaction pathway. Reproduced from work in [39]. Copyright 2016 Elsevier Ltd.

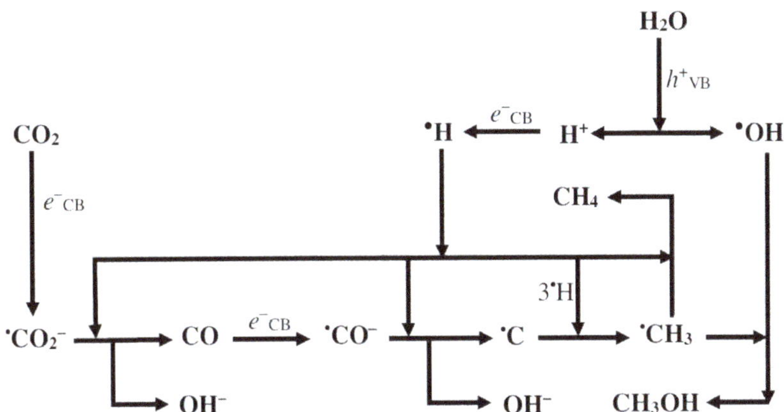

Figure 5. Gaseous phase photocatalysis of CO_2 and water: proposed reaction pathway. Reproduced from work in [39]. Copyright 2016 Elsevier Ltd.

Despite the aforementioned, the use of water has some limitations, including low CO_2 solubility (relevant when H_2O is used as the solvent), competing reduction reaction leading to hydrogen, and weak reducibility, eventually lowering the CO_2 conversion efficiency [38]. The photoreduction of water to hydrogen requires lower energy ($E = 0$ V) than the photoreduction of CO_2 ($E = -1.90$ V) and is, therefore, considered a strongly competing reduction process [39]. Furthermore, the photoreduction of CO_2 involves a multi-step charge transfer mechanistic approach in contrast to the single electron transfer step sufficient to initiate the photoreduction of water to hydrogen [37]. Due to these thermodynamic and kinetic limitations, and depending on the reductant used to obtain hydrogen, the generation of hydrogen from water photoreduction might occur with efficiency much higher than that of CO_2 photoreduction. Thus, when water is used as a reductant in the photocatalytic conversion of CO_2, hydrogen can appear as the predominant product. In addition, the higher dipole moment of water (D = 1.85), when compared to that of CO_2 (D = 0), favors its adsorption on the surface of the photocatalyst leading to another competing process [54]. The adsorption equilibrium constants of CO_2 and water vapor were determined in a study performed by Tan et al. [47]. Results showed that water vapor had a considerably higher adsorption equilibrium constant (8.07 bar^{-1}) than that of CO_2 (0.019 bar^{-1}). This implies that CO_2 adsorbs weakly on the surface of the photocatalyst. Ideally, and as discussed earlier in this section, CO_2 adsorbs to a photocatalyst surface in a bent configuration, whereby the carbon atom of CO_2 binds to a surface oxygen site while the oxygen atom binds to the surface metal center. This, however, is only applicable in the absence of water. Molecular dynamics studies were employed by Klyukin et al. [55] to help investigate the surface adsorption of CO_2 in the presence of water. Simulations showed that CO_2 adsorbed at oxygen sites while water saturated the metal sites. To manage the competitive adsorption between CO_2 and water, an optimum concentration of both reactants must be carefully considered. If CO_2 was present at extremely high concentrations, then it would effectively compete with water for the reactive sites. On the other hand, at CO_2 concentrations lower than that of water, only a limited number of CO_2 molecules would adsorb on the photocatalyst surface [47]. The formation of water-soluble products that are difficult to recover is another challenge faced in presence of water [37].

To help overcome the drawbacks of water as a reducing agent, many researches are currently investigating the use of alternative reductants for the photocatalytic conversion of CO_2. In this respect, the dielectric constant (κ) of the reducing agent significantly impacts the conversions and yields of the photocatalytic process. Solvents with varying dielectric constants (κ), including water (78.5), acetonitrile (37.5), 2-propanol (18.3), and dichloromethane (9.1), were tested as reducing agents in the photocatalytic reduction of CO_2 [56]. As κ increased, the conversion to formate increased while that to carbon monoxide decreased. This may be explained by the stabilization of the $CO_2^{\bullet-}$ radical by solvents with higher κ. Consequently, the more stable anion radical will interact more weakly with the photocatalytic surface.

As methanol is a strong reducing agent and has high CO_2 solubility, Qin et al. [38] investigated its use as a reductant for the photoreduction of CO_2 on a CuO-TiO_2 composite catalyst. Results show that CO_2 was reduced to produce methyl formate with a yield of 1602 µmol/g-cat/h. In addition, other studies on the photocatalytic reduction of CO_2 in a solution of NaOH showed that the use of NaOH as a reductant enhances the photocatalytic efficiency [57,58]. In this respect, CO_2 absorbs chemically in NaOH solutions due to its acidic properties leading to a concentration much higher than in DI water, and moreover the OH^- present in solution may serve as strong hole scavenger, enhancing charge separation. The optimal concentration of NaOH was reported to be 0.2 M [57,58].

Studies have also shown that the conversion of CO_2 using hydrogen as a reducing agent is higher than that when water is used [59]. This is because the photoreduction of CO_2 with hydrogen is more thermodynamically favorable than that with water [39]. Lo et al. [60] studied the photocatalytic conversion of CO_2 using bare TiO_2 (Evonik P-25) under UV irradiation. Three different reductants were tested: water, hydrogen, and water + hydrogen. The results, which are presented in Figure 6, show that the highest product yields of CO_2 photoreduction were obtained when hydrogen and saturated water vapor were used together as reductants, producing methane, carbon monoxide and ethane with a yield of 4.11, 0.14, and 0.10 μmol/g-cat/h, respectively. The authors propose that water accelerated the photoreduction of CO_2 with hydrogen by donating electrons and subsequently inhibiting charge recombination. They also suggest that water supplied more hydrogen atoms for the photoreduction of CO_2.

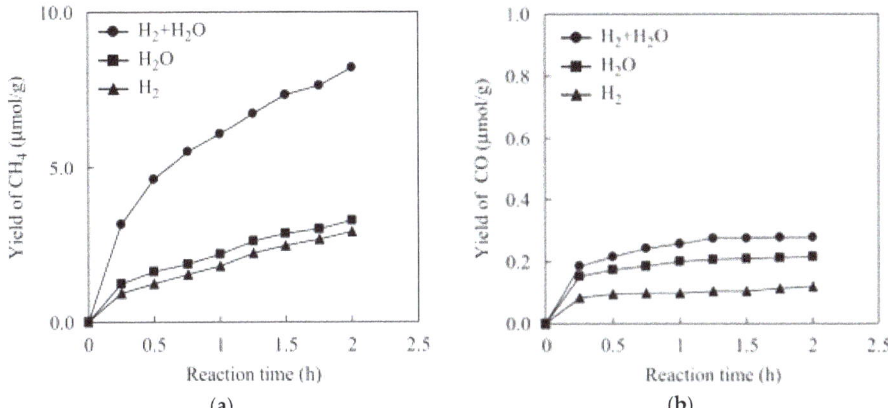

Figure 6. Effect of reductant on the yield of CO_2 photoreduction products, i.e., CH_4 (**a**) and CO (**b**). Reproduced from work in [60]. Copyright 2007 Elsevier B.V.

The photocatalytic reduction of compressed CO_2 (either liquid or supercritical) might be considered as an alternative solution to the use of organic solvents, but it has been rarely investigated [61]. Kaneco et al. [62] performed a study on the photoreduction of liquefied CO_2 using TiO_2 as the catalyst and water as the reducing agent. Results show the highly selective formation of formic acid with no other reduction products detected. Kaneco et al. [63] also investigated the photocatalytic reduction of CO_2 into formic acid in supercritical CO_2. The photoreduction reaction was conducted at 9.0 MPa and 35 °C with TiO_2 as the catalyst. Under these conditions, the production rate of formic acid reached a maximum yield of 1.76 μmol/g-cat/h. Kometani et al. [64] examined the photocatalytic reduction of CO_2 in a supercritical mixture of water and carbon dioxide (400 °C and 30 MPa) using Pt–TiO_2 as the catalyst. Carbon monoxide, methane, formic acid, and formaldehyde were all detected as reaction products with yields much higher than those obtained from the same reaction but at room temperature. Despite the enhanced product yields, the major drawback in using compressed CO_2 is the large amount of energy required to change its state from gas to liquid or to supercritical. One possible suggestion to help mitigate this limitation could be to target industrial processes that already utilize CO_2 in its compressed form. Thus, the overall energy efficiency of the photoreduction process could be greatly enhanced.

3. Titanium Dioxide: Synthesis and Surface Modification Strategies for Enhanced CO_2 Photoreduction

As shown in Figure 7, TiO_2 exists as one of three mineral phases: anatase, brookite, and rutile. Anatase-phase TiO_2 is the most common phase of TiO_2, but has a band gap of 3.2 eV, and is therefore weakly active under visible light. Rutile-phase, with a band gap of 3.0 eV, has the strongest visible light absorption, whereas brookite (band gap = 3.3 eV) has the weakest [65]. The use of mixed-phase TiO_2 enhances both the visible light harvesting ability and the electron–hole separation of TiO_2 [66]. The crystalline phase of TiO_2 is highly dependent on its preparation technique.

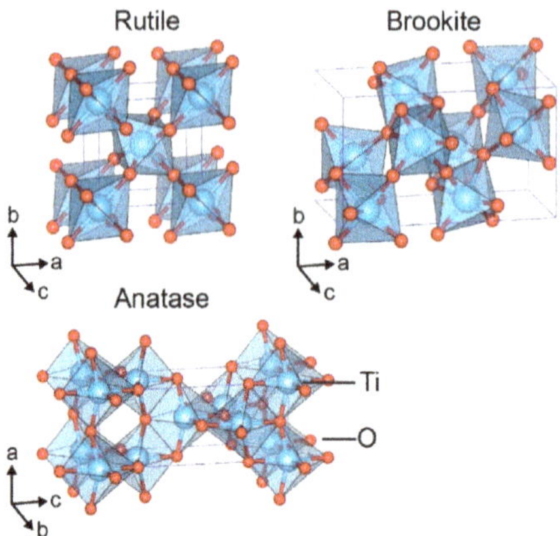

Figure 7. Crystal structures of TiO_2. Reproduced from work in [67].

The sol–gel method is widely used among researches for the preparation of bare and doped TiO_2-based photocatalysts both as thin films or powders [68]. This is attributed to the many benefits of this simple and cost effective method, which include the synthesis of nano-sized crystallized powder of high purity and homogeneity at relatively low temperature, possibility of tuning stoichiometry and morphology and easy preparation of composite materials [69,70]. Other photocatalysts such as ZrO_2 [71], ZnO [72], and WO_3 [73] have also been fabricated via the sol–gel process.

In general, the one-pot sol–gel procedure, shown in Figure 8, consists of two processes: hydrolysis and condensation. During these processes, the metallic atoms of the precursor molecules bind to form metal oxides and metal hydroxides. In the case of TiO_2, titanium alkoxides, such as titanium isopropoxide or titanium n-butoxide, are used as the precursor. Alcohol and acid must also be introduced into the reaction system as reaction modifiers. A densely cross-linked 3D TiO_2 gel with large specific surface area is formed as an end product after allowing the mixture of precursor, alcohol and acid to stir for several hours [70]. The choice of reaction parameters including reactants molar ratio, solution pH, reaction temperature, and reaction time, significantly affect the morphology of the final catalyst [74,75]. For instance, different reactant molar ratios would result in different hydrolysis rates which would in turn return structurally different TiO_2 catalysts [76,77]. Another example is the choice of acid where using acetic acid in the sol–gel synthesis of TiO_2 was shown to favor the formation of anatase-phase TiO_2, whereas using hydrochloric acid favored the formation of brookite and rutile mixed-phase TiO_2. Post-synthesis treatment techniques such as aging, drying and annealing are sometimes utilized to enhance the activity of the synthesized TiO_2 photocatalyst [78]. To develop TiO_2 films, the viscous sol may be deposited on a substrate via film coating techniques [79,80].

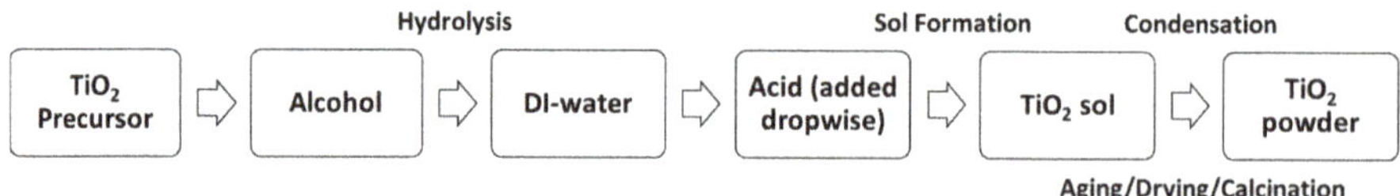

Figure 8. Sol–gel method for the preparation of bare and doped TiO_2-based photocatalysts.

The hydro/solvo thermal method is another one-pot TiO_2 synthesis technique that takes place in an aqueous solution above room temperature and atmospheric pressure [81]. More specifically, the TiO_2 precursor (titanium alkoxides) is mixed with an aqueous solution of an acid or a base for 16 to 72 h at high reaction temperature (110–180 °C). Post-synthesis treatment, which includes washing the obtained precipitate with DI water and then dispersing it in HCl solution before calcination, is usually performed to enhance the nanostructure of TiO_2. Quenching, or rapid cooling, is also commonly applied post-synthesis to enhance the photocatalytic activity of TiO_2 [82]. The hydro/solvo thermal process yields highly pure and well-defined TiO_2 nanocrystals with narrow particle size distribution [70]. Depending on the synthesis process parameters, which include the choice of precursor, the concentration of acidic/alkaline solution and the reaction temperature and time, different TiO_2 morphologies may be obtained. This ability to control the synthesis process to obtain TiO_2 as nanoparticles, nanotubes, nanoribbons, or nanowires is a great advantage of the hydro/solvo thermal treatment technique [83–85].

A number of studies have reported the significant CO_2 photocatalytic conversion improvement of TiO_2-based catalysts when synthesized under supercritical conditions. Camarillo et al. [86] prepared TiO_2 catalysts in supercritical CO_2 via a hydrothermal method. The synthesized TiO_2 catalyst was used for the photoreduction of CO_2 to methane, where the catalyst exhibited better photoactivity than the standard reference catalyst Evonik P-25. When prepared in supercritical CO_2, TiO_2 displayed improved CO_2 adsorption, enhanced charge separation and stronger visible light absorption. The highest methane production rate of 1.13 μmol/g-cat/h was obtained when diisopropoxititanium bis(acetylacetonate) precursor and isopropyl alcohol were used in the catalyst preparation.

The choice of precursor significantly affects the photocatalytic property of the synthesized catalyst. Bellardita et al. [87] prepared TiO_2 photocatalysts via the hydrothermal method using two different precursors, titanium tetrachloride and titanium butoxide. The TiO_2 photocatalysts were tested for the reduction of CO_2 and results showed that the photocatalysts synthesized from titanium tetrachloride favored the formation of formaldehyde, whereas those synthesized from titanium butoxide favored the formation of methane. The same study also examined the use of copper-doped TiO_2 for the photoreduction of CO_2. Results showed that doping TiO_2, synthesized with titanium tetrachloride precursor, with 1 wt.% copper enhanced the formaldehyde production rate. However, doping TiO_2, synthesized with titanium butoxide precursor, with 1 wt.% copper decreased the methane production rate.

The catalyst morphology plays a crucial role in determining its photocatalytic efficiency. TiO_2 nanostructures (nanosheets, nanorods, nanowires, nanotubes, nanobelts, etc.) have been shown to exhibit superior performance for photocatalytic reduction of CO_2 [26]. These nanostructures provide large surface area, reduced grain boundaries, and facile charge transport paths. Specifically, reduced grain boundaries and defects usually lead to a direct pathway for electron transport to catalytic sites, inhibiting the electron–hole recombination rate.

The use of unmodified TiO_2 photocatalysts for the reduction of CO_2 under UV light has been reported by many to be inefficient [37]. This is mainly attributed to the reduction potential (-0.5 V) of electrons in the TiO_2 CB which is much lower (i.e., more positive) than the theoretical thermodynamic requirement for the single electron reduction of CO_2 (-1.90 V). To drive any reduction process, the potential of the CB must be more negative than that of the reduction reaction [39]. This is true in the case of the multiple electron reduction of CO_2 to methane (-0.24 V) or to methanol (-0.38 V). On the other hand, the co-presence of strong reducing electrons and free protons is detrimental as they may interact to produce molecular hydrogen. This may be avoided by introducing spatially separated centers within the TiO_2 lattice in order to trap the charges and reduce their recombination [37].

Several modification strategies have been suggested to help overcome the drawbacks of TiO_2. Many of these strategies are shown in Figure 9 and they include metal deposition, metal/non-metal doping, loading of carbon-based material, formation of semiconductor heterostructures, and dispersion on high surface area supports. All of the aforementioned techniques serve to enhance the separation of electrons and holes and to extend the absorption of light into the visible range. Loading TiO_2 with carbon-based materials and dispersing TiO_2 on inert supports offers the added advantages of increasing the concentration of surface electrons, improving the surface adsorption of reactants (specifically CO_2), and reducing the agglomeration of TiO_2 nanoparticles. Nonetheless, the utilization of carbon-based materials hinders the absorption of light while high surface area supports lower the light utilization efficiency since they absorb and scatter part of the radiation resulting in a waste of photons. Although TiO_2-based heterostructures are complex to synthesize and are considered to be relatively unstable, they effectively separate oxidation and reduction sites improving the photocatalytic performance of TiO_2 semiconductors. To overcome limitations and drawbacks, many researches have suggested to employ hybrid systems which combine two or more of these different modification techniques [40]. The advantages and disadvantages of the main modification strategies along with the performance results from the most important CO_2 photoreduction studies are summarized in Tables 1 and 2 below [19,26,37,88–90]. Further details on the pros and cons of the mentioned TiO_2 surface modification techniques are discussed in the following individual Sections 3.1–3.5

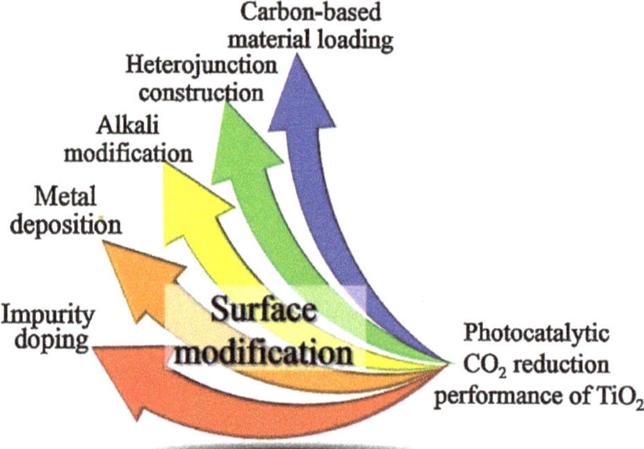

Figure 9. Photocatalyst modification strategies. Reproduced from work in [40]. Copyright 2016 Elsevier B.V.

Table 1. Advantages and disadvantages of TiO_2 modification techniques.

Modification Strategy	Advantages	Disadvantages
Metal Deposition	-Enhances electron–hole separation	-Expensive -Rare
Metal Doping	-Extends light absorption into visible range -Enhances electron–hole separation	-Act as recombination centers -Leads to structural defects -Possibility of metal leaching
Non-Metal Doping	-Extends light absorption into visible range -Enhances electron–hole separation	-Act as recombination centers
Carbon-based Material Loading	-Extends light absorption into visible range -Enhances electron–hole separation -Increases electron concentration on TiO_2 surface -Improves CO_2 adsorption on catalytic surface -Reduces agglomeration of TiO_2 nanoparticles	-Forbids light absorption by TiO_2
Heterostructures	-Extends light absorption into visible range -Enhances electron–hole separation -Separates reduction and oxidation sites -Enhances the catalyst's product selectivity, pore structure and electronic properties	-Complex synthesis method -Instability
Dispersion on Supports	-Eliminates need for post treatment separation -Provides high surface area -Enhances dispersion of TiO_2 nanoparticles -Improves adsorption of reactants on catalytic surface	-Low light utilization efficiency
Alkali Modification	-Enhances electron–hole separation -Enhances the chemisorption of CO_2	-Encapsulation of TiO_2 nanoparticles

Table 2. Photocatalytic performance of various TiO_2-based catalysts in the reduction of CO_2*.

Modification Strategy	Photocatalyst	Synthesis Method	Reductant	Light Source	Main Product Yield (μmol/g-cat/h)	Ref.
Metal Deposition *Platinum*	One-dimensional TiO_2 single crystals coated with ultrafine Pt NPs	CVD	Water	UV	Methane: 1361	[91]
	0.2 wt.% Pt on mesoporous TiO_2	Soft Template	Water	UV	Methane: 2.85	[36]
Metal Deposition *Gold*	0.5 wt.% Au on TiO_2 NWs	HT	Hydrogen	Visible	Carbon monoxide: 1237 Methanol: 12.65	[92]
Metal Deposition *Silver*	Ag-electrodeposited on TiO_2 NRs	HT	Water	UV	Methane: 2.64	[93]
Metal Deposition *Copper*	1.7 wt.% Cu and 0.9 wt.% Pt NPs on TiO_2 (Evonik P-25)	-	Water	UV + Visible	Methane: 33	[94]
	Cu/C–TiO_2	Sol–gel	Water	UV	Methane: 2.526	[95]
Metal Doping *Copper*	1.2 wt.% Cu-TiO_2	HT	Water	UV	Methanol: 0.45	[96]
	2 wt.% Cu-TiO_2	Sol–gel	NaOH	UV	Methanol: 12.5	[58]
	3 wt.% Cu-TiO_2 (Evonik P-25)	-	$KHCO_3$	UV	Methanol: 194	[97]
Metal Doping *Nickel*	0.1 mol % Ni-TiO_2	ST	Water	UV	Methane: 14	[98]
Metal Doping *Cerium*	0.28 mol % Ce-TiO_2	Sol–gel	NaOH	UV	Methane: 0.889	[99]
Non-Metal Doping *Nitrogen*	N-TiO_2 NTs	HT	NaOH	Visible	Formic acid: 1039 Methanol: 94.4 Formaldehyde: 76.8	[100]
Non-Metal Doping *Carbon*	C–TiO_2	IMM	Na_2SO_3	Solar	Formic acid: 439	[101]
Carbon-based Material Loading *CNTs*	MWCNT/TiO_2 nanocomposite	-	Water	Visible	Methane: 0.17	[102]
	Anatase-TiO_2 NPs/MWCNT	Sol–gel	Water	UV	Ethanol: 29.872	[103]
	Rutile-TiO_2 NRs/MWCNT	HT	Water	UV	Formic acid: 25.02	[103]
	CNT-Ni/TiO_2 nanocomposites	CVD	Water	Visible	Methane: 0.145	[104]
Carbon-based Material Loading *Graphene*	Graphene/ N-TiO_2	ST	Water	Visible	Methane: 0.37	[105]
Hetero-structures	1 wt.% CuO-TiO_2 composite	-	Methanol	UV	Methyl formate: 1602	[38]
Dispersion on Supports	0.5 wt.% Cu/TiO_2-silica nanocomposite	Sol–gel	Water	UV	Carbon monoxide: 60 Methane: 10	[106]
Alkali Modification	3 wt.% NaOH-TiO_2	IMP	Water	UV	Methane: 8.667	[90]

* Abbreviations: NP: nanoparticle; NW: nanowire; NR: nanorod; NT: nanotube; CVD: chemical vapor deposition; HT: hydrothermal; ST: solvothermal; IMM: immersion; IMP: impregnation; MWCNT: multiwall carbon nanotubes; UV: ultraviolet.

3.1. Metal Deposition

Although considered an expensive modification technique, metal deposition enhances the electron–hole separation in a semiconductor, improving the photoreduction efficiency of CO_2 [40,107]. As the work function of the metal increases, the metal's ability to accept the photogenerated electrons also increases [36,40]. This in turn enhances the electron–hole separation and the overall photocatalytic activity of TiO_2. Some of the most commonly deposited metals arranged in order of highest work function include platinum (5.93 eV), palladium (5.60 eV), gold (5.47 eV), and silver (4.74 eV) [25,40].

Wang et al. [91] demonstrated high CO_2 photoreduction efficiency of one-dimensional (1D) TiO_2 single crystals coated with ultrafine Pt nanoparticles. This efficient Pt–TiO_2 nanofilm, shown in Figure 10, was synthesized via chemical vapor deposition and exhibited selective formation of methane with a maximum yield of 1361 μmol/g-cat/h. Deposition of Pt nanoparticles enhanced the electron–hole separation leading to a more efficient photocatalytic performance. Li et al. [36] also proposed the use of Pt-deposited TiO_2 for the photoreduction of CO_2 to methane. Results showed that depositing 0.2 wt.% Pt on mesoporous TiO_2 yields methane with a production rate of 2.85 μmol/g-cat/h. Tostón et al. [108]

examined the photocatalytic activity of 1 wt.% Pt–TiO$_2$ photocatalysts prepared in supercritical CO$_2$. The synthesized photocatalyst was used to reduce CO$_2$ into methane. The study indicated that the use of supercritical CO$_2$ results in a photocatalyst with higher surface area, crystallization degree, pore volume, visible light absorbance, and methane production rate (0.245 µmol/g-cat/h).

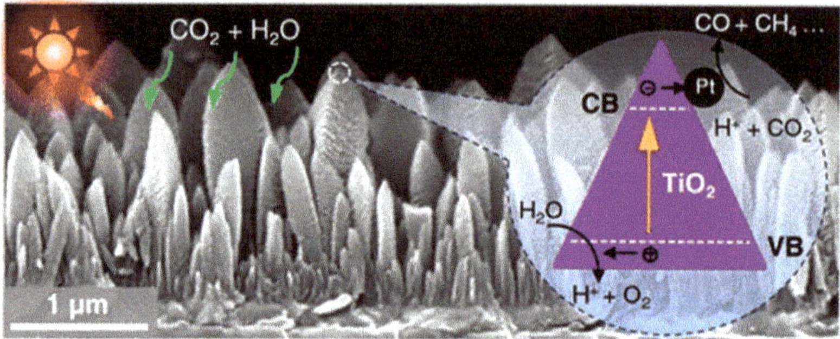

Figure 10. Enhanced CO$_2$ photoreduction efficiency by ultrafine Pt nanoparticles deposited on TiO$_2$ crystals. Reproduced from work in [91]. Copyright 2012 American Chemical Society.

Owing to the effect of localized surface plasmon resonance (LSPR) and to that of Schottky barrier formation, Au and Ag nanoparticles have been shown to enhance the visible-light activity of several semiconducting materials. In LSPR, an electromagnetic field is created and the photoreaction is improved via photon scattering, plasmon resonance energy transfer and hot electron excitation. On the other hand, the formation of a Schottky barrier enhances the photoactivity by trapping and prolonging the electron life [109]. Tahir et al. [92] described the photoreduction of CO$_2$ using Au-decorated TiO$_2$ nanowires, prepared by hydrothermal method. Deposition of 0.5 wt.% Au on the TiO$_2$ nanowires yielded 1237 µmol carbon monoxide/g-cat/h and 12.65 µmol methanol/g-cat/h. Surface deposition of Au nanoparticles enhanced charge separation and improved photocatalytic activity under visible light through plasmon excitation. More specifically, the LSPR effect of Au nanoparticles promotes electrons to the CB of TiO$_2$. The positively charged plasmas of Au nanoparticles then trap photogenerated CB electrons enhancing the separation of charges. Next, the LSPR effect of Au improves the energy of these trapped electrons and, as a result, the efficiency of CO$_2$ photoreduction increases [109]. Kong et al. [93] proposed the use of Ag-electrodeposited TiO$_2$ nanorods, prepared by hydrothermal method, for the photocatalytic reduction of CO$_2$ to methane. The photocatalyst exhibited enhanced photoactivity with a methane yield of 2.64 µmol/g-cat/h. This improved performance was attributed to the plasmonic characteristics of the Ag nanoparticles.

Murakami et al. [110] photocatalytically reduced CO$_2$ using Ag/Au-TiO$_2$. The deposited metal nanoparticles act as reductive sites increasing the production of methanol compared to that of bare TiO$_2$. Zhai et al. [94] photodeposited Cu and Pt nanoparticles on TiO$_2$ (Evonik P-25) using copper sulfate and chloroplatinic acid, respectively, as the metal precursors. Results showed the photo-depositing TiO$_2$ with 1.7 wt.% Cu and 0.9 wt.% Pt allowed to produce methane from CO$_2$ with a yield of 33 µmol/g-cat/h. Yan et al. [95] used a sol–gel method to synthesize a novel Cu/C–TiO$_2$ catalyst for the photoreduction of CO$_2$ in water under UV irradiation. Co-deposition of Cu and C onto the TiO$_2$ surface extended the light absorption into the visible range, reduced the electron–hole recombination rate and provided an increased number of reaction sites on the photocatalytic surface. The enhanced Cu/C–TiO$_2$ photocatalyst reduced CO$_2$ to methane with a production rate of 2.53 µmol/g-cat/h.

3.2. Doping

One of the many strategies used to enhance the activity of TiO_2 photocatalysts is doping with metals such as copper, nickel, silver and cerium. This technique is the most widely applied method used to extend the light absorption range and suppress the recombination rate of TiO_2. Although introducing metal nanoparticles into the TiO_2 matrix leads to structural defects, their presence reduces the TiO_2 bandgap, shifting the absorption threshold to visible [40]. The reduction in bandgap occurs as a consequence of the new energy level produced by the dispersion of dopants in the TiO_2 lattice [111]. Not only that, but the doped metal may also act as an electron trap, enhancing the separation of the photogenerated electrons and holes during irradiation as illustrated in Figure 11. One major drawback of metal doping is the possibility of metal leaching and, consequently, catalyst deactivation. This might occur as a result of the photocorrosion of doped TiO_2, especially when water is used as a reductant [37].

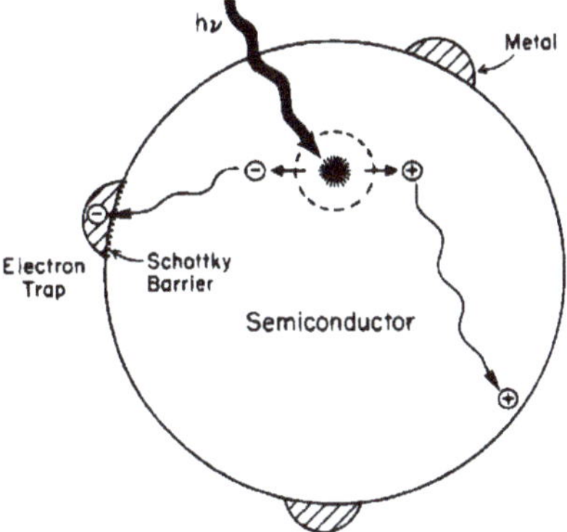

Figure 11. Electron–hole separation of metal-doped semiconductors. Reproduced from work in [112]. Copyright 1995 American Chemical Society.

To prepare metal-doped TiO_2, the sol–gel method is commonly followed due to its versatility and simplicity. In a typical process, a TiO_2 precursor is mixed with a metal-dopant precursor which has been dissolved in alcohol. After hydrolysis for several hours and at elevated temperatures, the reaction mixture is dried and a metal-doped catalyst in powder form is obtained [111]. Chemical vapor deposition and hydro/solvo-thermal techniques may also be used for the synthesis of metal-doped TiO_2.

One of the most commonly used metal dopants incorporated into TiO_2 semiconductors is copper. Different metal loading levels of Cu-doped TiO_2 were tested for the photocatalytic reduction of CO_2 under UV light. For instance, Wu et al. [96] reduced CO_2 to methanol using a Cu-doped TiO_2 catalyst under UV irradiation. The catalyst was synthesized via a hydrothermal method and exhibited a band gap of 3.3 eV. Doping TiO_2 with 1.2 wt.% copper produced methanol with a yield of 0.45 μmol/g-cat/h. Tseng et al. [58] reported the photocatalytic reduction of CO_2 into methanol under UV irradiation using copper-doped TiO_2, prepared using the sol–gel method. The results of this study showed that doping the TiO_2 catalyst with 2 wt.% copper gave the highest methanol yield of 12.5 μmol/g-cat/h. The copper doping lowered the electron–hole recombination probability, which consequently boosted the photocatalytic efficiency. Nasution et al. [97] studied the use of Cu-doped TiO_2 (Evonik P-25), prepared by an impregnation method, for the photocatalytic reduction of CO_2 under UV irradiation. Results showed that doping TiO_2 with 3 wt.% copper yields a maximum methanol production rate of 194 μmol/g-cat/h. All of the aforementioned studies suggest that the photocatalytic reduction of CO_2 using a Cu-doped TiO_2 catalyst under UV illumination yields methanol as the main reaction product.

Due to the LSPR effect of copper nanoparticles, Cu-doped TiO_2 has been shown to exhibit enhanced photocatalytic activity under visible light. The strong local electron field from the LSPR effect improves the energy of the trapped electrons resulting in an enhanced photocatalytic CO_2 reduction process. The concentration of copper dopant, however, must be carefully considered. Although excess copper nanoparticles improve the separation of charges, they also lead to the formation of larger copper particles. These larger particles have a weaker LSPR effect and as such considerably lower the photocatalytic efficiency [113].

The metal dopant concentration significantly affects the photocatalytic activity of doped TiO_2. Depending on its concentration, the metal dopant can either hinder or promote the anatase-to-rutile transformation during calcination. The crystal structure transformation to rutile reduces the catalyst's surface area and removes any active species present on the anatase surface, thereby lowering the catalytic performance of doped TiO_2 [114]. To maintain high anatase levels and therefore better photocatalytic activity, low concentrations of metals (0.1–0.5 mol %) are commonly doped into TiO_2. In addition to the structural transformation, the dopant loading level also influences the recombination of photogenerated charges. By acting as recombination centers, metal dopants, can significantly increase the electron–hole recombination especially when present at high concentrations (>3 mol %), therefore lowering the photocatalytic efficiency [115]. Kwak et al. [98] investigated the photocatalytic reduction of CO_2 to methane using a nickel-doped TiO_2 catalyst, prepared via a solvothermal method. Doping with 0.1 mol % nickel gave the highest methane yield of 14 µmol/g-cat/h. Matějová et al. [99] examined the photoactivity of cerium-doped TiO_2, synthesized by the sol–gel method, for the reduction of CO_2. The doped cerium shifted the light absorption of the catalyst into the visible range and enhanced charge separation. Results showed that doping TiO_2 with 0.28 mol % cerium gave the highest methane yield of 0.889 µmol/g-cat/h.

Nonmetal dopants, such as nitrogen, carbon, sulfur, and fluoride, are commonly used to help reduce the TiO_2 band gap, consequently increasing the absorption of visible light [26]. Substitutional nonmetal doping typically introduces defect states localized at the impurity site which reduce the band gap and induce absorption in the visible region. Furthermore, these defect states, which form below the conduction band, have been shown to be good acceptors of electrons. Nonetheless, nonmetal doping also leads to the formation of oxygen vacancies. These oxygen vacancy defects generally act as charge recombination centers which are detrimental in photocatalytic reactions and must therefore be avoided [116]. As the quantity of nonmetal dopant increases, the amount of defects increases and the photocatalytic activity consequently decreases. Therefore, in nonmetal doping, extra care must be taken in optimizing the dopant concentration for enhanced visible light absorption and improved photocatalytic activity with an acceptable extent of defects [117]. One strategy commonly used to reduce the recombination of charges in nonmetal doped TiO_2 is co-doping with an electron donor–acceptor pair [118]. Similar to metal-doping, non-metal doping is usually performed via the sol–gel process, although other techniques including hydro/solvo-thermal methods may also be applied.

Studies have shown that the main reaction product resulting from the photocatalytic conversion of CO_2 using non-metal dopants is formic acid. For example, Zhao et al. [100] studied the photocatalytic activity of nitrogen-doped TiO_2 nanotubes prepared by a hydrothermal method, where titanium (III) chloride and hexamethylene tetramine were used as precursors. As a doping agent, nitrogen extended the light absorption of TiO_2 into the visible range. The formic acid yield was 1039 µmol/g-cat/h, the methanol yield was 94.4 µmol/g-cat/h, and the formaldehyde yield was 76.8 µmol/g-cat/h. Xue et al. [101] investigated the non-metal, substitutional doping of TiO_2 with carbon using citric acid as a precursor. The doped carbon lowered the TiO_2 band gap, shifting the light absorption into the visible range, and improved the charge separation efficiency. CO_2 was photocatalytically reduced to yield 439 µmol formic acid/g-cat/h.

3.3. Carbon-Based Material Loading

The incorporation of carbon-based materials into TiO_2 catalysts is gaining wide interest in the field of photocatalysis. This is mainly due to the carbon-based materials' properties such as their abundancy, low cost, good corrosion resistance, high electron conductivity, large specific surface area, and tunable surface properties. Although in most cases they reduce the light absorbed by TiO_2, loading the catalyst with carbon based-materials, such as carbon nanotubes or graphene, enhances the electron–hole separation, the electron concentration on the TiO_2 surface, and the CO_2 adsorption through π–π conjugations between CO_2 molecules and the carbon-based material [40].

Carbon nanotubes are regarded to be among the most remarkable emerging materials mainly due to their unique electronic, adsorption, mechanical, chemical, and thermal characteristics [119]. Advances in the synthesis techniques of multiwalled carbon nanotubes (MWCNTs) have resulted in a significant reduction of the material's cost, consequently allowing for their use on a large industrial scale [103]. In addition to that, the mesoporosity of carbon nanotubes, which favors the diffusion of reacting species, is an added value of this widely investigated material [120]. Many recent studies have focused on developing synthesis methods to combine carbon nanotubes with semiconducting catalysts, such as TiO_2. This new hybrid composite is believed to enhance the efficiency of many photocatalytic applications, especially with its improved charge transfer properties and the formation of new active sites [103,120]. The main disadvantage in synthesizing MWCNT/TiO_2 hybrids is the need for treating the carbon nanotubes with strong acids in order to introduce the functional groups to the surface [120]. A range of different techniques has been employed for the fabrication of MWCNT/TiO_2 composites: mechanical mixing [121], sol–gel [122], electrospinning [123,124], and chemical vapor deposition [125]. Depending on the choice of the synthetic strategy, composite materials with different coating uniformity and varying physical characteristics are prepared. More specifically, uniform TiO_2 coatings on carbon nanotubes may be achieved through electrospinning and chemical vapor deposition, whereas other methods such as sol–gel typically yield nonuniform coatings with TiO_2 aggregates being randomly dispersed on the carbon nanotube surface [126]. On the other hand, chemical vapor deposition and electrospinning are complex techniques which require the use of specialized equipment.

Using the process shown in Figure 12, Gui et al. [102] prepared a MWCNT/TiO_2 nanocomposite for the photocatalytic reduction of CO_2. The MWCNTs enhanced the TiO_2 photoactivity under visible light, yielding 0.17 µmol methane/g-catalyst/h. Xia et al. [103] studied the effect of using two different methods of synthesis—sol–gel and hydrothermal—for the preparation of MWCNT-supported TiO_2 composite. The sol–gel method formed anatase-phase TiO_2 nanoparticles on the MWCNT, whereas rutile-phase TiO_2 nanorods were formed when the hydrothermal method was used. In addition to that, the MWCNT/TiO_2 composite prepared via sol–gel mainly reduced CO_2 to ethanol (29.87 µmol/g-cat/h), whereas the composite prepared hydrothermally mainly reduced CO_2 to formic acid (25.02 µmol/g-cat/h). The efficiency of the CO_2 photoreduction reaction was significantly enhanced with the incorporation of MWCNTs. More specifically, the MWCNTs helped reduce the agglomeration of TiO_2 nanoparticles and helped increase the separation of the photogenerated electron–hole pairs. Ong et al. [104] demonstrated the photocatalytic activity of CNT-Ni/TiO_2 nanocomposites in reducing CO_2 to methane under visible light. The nanocomposite, prepared by chemical vapor deposition, exhibited a reduced band gap of 2.22 eV and a methane yield of 0.145 µmol/g-cat/h.

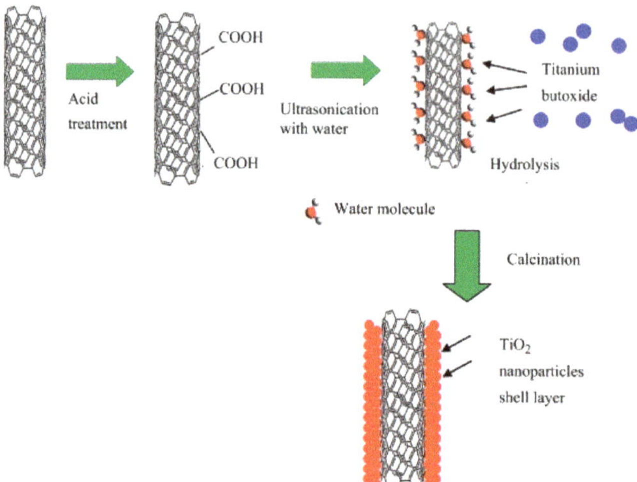

Figure 12. Enhanced visible light responsive MWCNT/TiO$_2$ core–shell nanocomposites for photocatalytic reduction of CO$_2$. Reproduced from work in [102]. Copyright 2013 Elsevier B.V.

Graphene is a 2-D, single layer of graphite with outstanding physiochemical properties including low production costs, ease of scalability, exceptional catalytic performance, high mechanical strength, high porosity, high thermal and electrical conductivity, remarkably high CO$_2$ adsorption capacity, high transparency, high flexibility, and high specific surface area. Furthermore, the exceptional photocatalytic properties of graphene, which include zero band gap, large BET area and high electron mobility, have encouraged its use in many photocatalytic applications [127]. Many different types of semiconducting photocatalysts have been coupled with graphene with the aim of enhancing the overall efficiency of photocatalysts [128,129]. The preparation process greatly affects the morphology, structure, size, properties, and activity of the composite photocatalyst [130]. The methods commonly used for the preparation of graphene-based TiO$_2$ are generally divided into two main categories: in situ crystallization (including hydrothermal/solvothermal, sol–gel, microwave assisted and others) and ex situ hybridization (including solution mixing, self-assembly, electrospinning and others). It is important to note that molecular linkers between TiO$_2$ and the graphene sheets are not required during the synthesis of the composite [131]. This is a great benefit since molecular linkers might act as electron traps decreasing the overall photocatalytic efficiency of the graphene-based TiO$_2$ composite.

In general, both the hydrothermal and the solvothermal processes are characterized by their high reactivity, low energy requirement, mild reaction conditions, relatively environmental set-up, and simple solvent control. Under optimal conditions, both techniques can yield large quantities of graphene-based photocatalysts at low cost [129]. Furthermore, TiO$_2$ nanostructures with high crystallinity are typically produced through the one-pot hydrothermal/solvothermal approach without the need for post-synthesis calcination [132]. Experimental conditions, including precursor concentration, pH, temperature, and time, significantly affect the reaction pathway and the nanomaterial's crystallinity and, therefore, must be carefully considered [129,132].

In sol–gel techniques and through a series of hydrolysis and condensation steps, a liquid colloidal solution "sol" containing graphene/graphene oxide and TiO$_2$ precursor is transitioned into a xerogel [129,132]. The hydroxyl groups present on the surface of the graphene oxide sheets provide nucleation sites for hydrolysis [129]. Consequently, the TiO$_2$ nanoparticles will be strongly bonded to graphene, offering a great advantage over other synthesis methods [133].

Solution mixing is fairly simple and involves only the mixing of a TiO_2 precursor in a suspension of graphene oxide under vigorous/ultrasonic agitation. Simple treatments, including drying and calcination, may be utilized after synthesis to improve the photocatalytic efficiency of the final composite material [130]. Furthermore, graphene oxide may be reduced to graphene sheets by the addition of agents such as amines, sodium borohydride, and ascorbic acid [129]. It is important to note that detrimental defects in the graphene sheets may arise as a result of long exposure time and high power ultrasonication. In addition to that, formation of chemical bonds between TiO_2 and graphene may be difficult due to the mild operating conditions [130]. Nonetheless, this technique provides the benefit of uniformly distributing the TiO_2 nanoparticles over the graphene sheets, ultimately improving the photocatalyst's activity [129].

Self-assembly is also another highly efficient, time-saving and cost effective synthesis technique [134,135]. This method offers structure and size control via component design and is applicable in various fields. The major disadvantages include the need for surfactants, sensitivity to environmental factors and production of graphene with low mechanical strength [136]. Surfactants are required in order to help disperse graphene oxide sheets and in order to improve TiO_2 nanoparticle loading [132]. Self-assembly based on the electrostatic attraction between negatively charged graphene oxide sheets and positively charged TiO_2 nanoparticles has been used to synthesize layered graphene-based TiO_2 semiconductors in a simple and cost-effective manner [132,137].

Tu et al. [138] loaded TiO_2 nanoparticles onto graphene nanosheets via an in situ method. The addition of graphene significantly increased the specific surface area of the catalyst, consequently generating more adsorption and reaction sites on the catalytic surface. The 2 wt.% graphene/TiO_2 catalyst was used to reduce CO_2 to ethane with a yield of 16.8 μmol/g-cat/h. Ong et al. [105] deposited nitrogen-doped TiO_2 nanoparticles onto graphene sheets using a solvothermal method. The nitrogen-doped TiO_2/graphene catalyst reduced CO_2 to methane with a yield of 0.37 μmol/g-cat/h. The efficient charge separation of graphene and the enhanced absorption of visible light were attributed to the improved photocatalytic performance of TiO_2.

3.4. Heterostructures

Sensitizing wide bandgap semiconductors with narrow bandgap semiconductors or dye molecules to form heterostructures is another method used to improve the photoactivity of many photocatalysts. This strategy provides means of increasing absorption toward the visible light region, at the same time enhancing the electron–hole pair separation, setting apart the reduction and oxidation sites for an improved performance [40,139]. Examples of narrow bandgap semiconductors that have been coupled with TiO_2 for photocatalytic reduction of CO_2 under visible light irradiation include CdS [140], Bi_2S_3 [141], CdSe quantum dots (QDs) [142], PbS QDs [143], and AgBr [144]. To enhance the adsorption of CO_2 molecules, materials with high intrinsic basicity, such as cobalt aluminum hydroxide, can be applied as sensitizers of TiO_2 [145]. Four different types of heterostructures exist depending on the charge carrier separation mechanism: conventional type-II, p-n, direct Z-scheme and surface heterojunction. However, these mechanisms will not be further discussed as they are considered to be outside the scope of this review. More details may be found elsewhere [40,146,147].

To synthetize heterostructured photocatalysts, several fabrication methods, including chemical vapor deposition, atomic layer deposition, hydrothermal/solvothermal, and ion exchange reactions, have been developed [147]. Chemical vapor deposition allows for the sequential deposition of multiple materials on a substrate surface. This synthesis route follows a two-step growth procedure: (1) growth of inner core semiconductor on a suitable substrate, and (2) deposition of outer layer shell semiconductor on the core surface. Compared to chemical vapor deposition, atomic layer deposition is usually more favored due to its enhanced benefits which include precise control of film thickness at the atomic level and conformal growth of complex nanostructures [148,149]. Studies have also shown that atomic layer deposition enhances the light trapping and carrier separation properties of the photocatalytic heterostructure [150,151]. A more convenient preparation technique is the hydrothermal/solvothermal method. Both the precursor dissolution and the reaction rate are enhanced by this process. Ion exchange is another approach used to synthesize photocatalytic heterostructures. In this novel process, cations at the interface of the two semiconducting materials are exchanged [152,153]. Without changing the shape of the ionic semiconductor, ion exchange reactions selectively yield a dimer structure with an epitaxial hetero-interface [147].

Li et al. [141] prepared a Bi_2S_3/TiO_2 nanotube heterostructure for the photocatalytic reduction of CO_2 into methanol. The incorporation of Bi_2S_3 enhanced the visible light absorption and the photocatalytic performance of the TiO_2 nanotubes. The methanol yield of the Bi_2S_3/TiO_2 nanotube heterostructure was reported to be 44.92 μmol/g-cat/h. Qin et al. [38] investigated the conversion of CO_2 to methyl formate on a $CuO-TiO_2$ heterostructured composite in methanol. The photocatalytic activity was greatly enhanced due to the heterojunction between CuO and TiO_2. More specifically, the surface-junction improved the transfer of charges by facilitating the migration of electrons from the TiO_2 CB to the CuO VB as depicted in Figure 13a. Consequently, the probability of charge recombination was lowered and the photocatalytic efficiency was enhanced. As can be seen in Figure 13b, the TiO_2 catalyst loaded with 1 wt.% CuO exhibited the highest methyl formate yield of 1602 μmol/g-cat/h. One of the most significant approaches was carried out by Nguyen et al. [154], who employed a metal doped TiO_2 catalyst sensitized with ruthenium dye (N3 dye) for the photoreduction of CO_2 into methane. The methane yield of the N3-Dye-Cu (0.5 wt.%)-Fe (0.5 wt.%)/TiO_2 was 0.617 μmol/g-cat/h. The improved photoreduction of the dye-sensitized TiO_2 is attributed to the full absorption of visible light by the N3-dye along with the efficient charge transfer in the N3 dye-TiO_2 system.

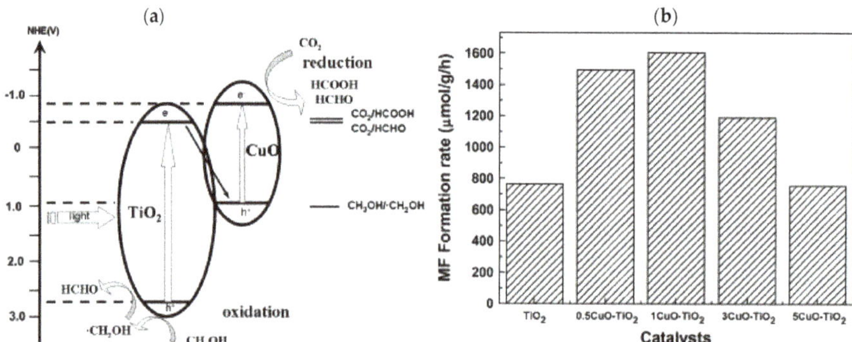

Figure 13. Photocatalytic conversion of CO_2 to methyl formate (MF) over $CuO–TiO_2$. (**a**) Band positions of CuO and TiO_2. (**b**) Formation rate of MF for the different catalysts. Reproduced from work in [38]. Copyright 2010 Elsevier Inc.

3.5. Dispersion on Supports

The dispersion of TiO_2 as a nanoparticle on various types of supports enhances the catalyst's product selectivity, pore structure, and electronic properties. In addition to that, this modification technique eliminates the need for post treatment separation and provides high surface area and mass transfer rate. An ideal supported-TiO_2 photocatalyst must have effective light absorption properties, be resistant to degradation induced by the immobilization technique and provide firm adhesion between the support and the catalyst. Key challenges of this strategy include mass transfer limitations and low light utilization efficiency. TiO_2 photocatalysts may be immobilized by dip or spin coating onto substrates such as fibers, membranes, glass, monolithic ceramics, silica, and clays. In dip coating, the thoroughly cleaned supports are immersed in a coating precursor solution as is demonstrated in Figure 14a. After being slowly pulled out of the solution, the coated support is then dried out to remove excess solution and moisture [155]. The withdrawal speed of the substrate, number of coating cycles and the TiO_2 solution viscosity determine the catalyst film thickness [23]. In spin coating, the precursor is deposited on the substrate by a dispenser while the substrate is rotated as demonstrated in Figure 14b. The rotation continues until uniform distribution of the precursor layer is achieved. Although the spinning process will result in the partial evaporation of the precursor solvent, the substrate will still require additional thermal treatment to stabilize the layer. The thickness of the films obtained by spin coating depends on the concentration of the precursor solution and the rotation speed [156]. Due to its scale-up applicability and higher controllability, dip coating is usually preferred over the spin coating technique [157]. In addition to that, superior structural and optical properties, such as enhanced crystallinity and higher average particle size, have been observed for dip-coated films in comparison to spin-coated ones [158,159].

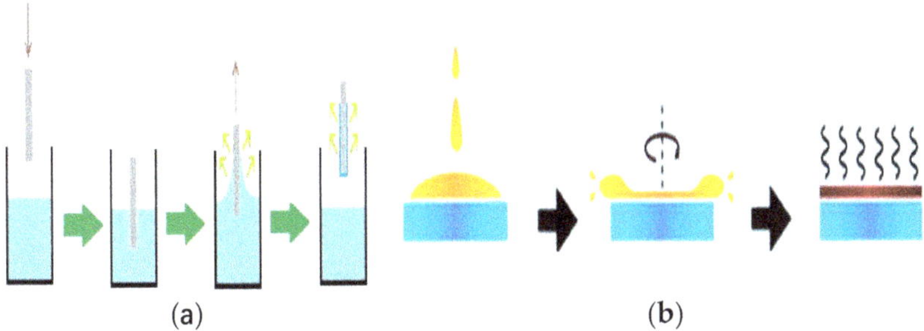

(a)   (b)

Figure 14. Dip-coating (**a**) and spin-coating (**b**) of TiO_2 photocatalyst on an inert support. Reproduced from work in [155,160]. Copyright 2017 Elsevier Ltd., 2016 Elsevier B.V.

Bellardita et al. [87] investigated the photocatalytic activity of silica-supported TiO_2 for the reduction of CO_2. The primary product obtained from the photocatalytic reaction was acetaldehyde. Li et al. [106] prepared an ordered mesoporous silica-supported Cu/TiO_2 nanocomposites via a sol–gel method for the photocatalytic reduction of CO_2 in water under UV irradiation (Figure 15a). As previously discussed, one of the many benefits of sol–gel synthesis concerns the control of pore size and the pore size distribution of the final catalyst. The sol–gel method used to prepare the silica-supported Cu/TiO_2 was carefully formulated to produce ordered mesoporous silica with a pore size of ~15 nm [161]. The ordered mesoporous structure of silica was clearly observed in transmission electron microscopy (TEM) images as shown in Figure 15c,d. Mesoporous silica, with its ordered pore networks, has a much higher surface area than agglomerates of silica nanoparticles with irregular mesopores and is therefore considered a better support for catalysts. The high surface area of the ordered mesoporous silica support (>300 m^2/g) enhanced the dispersion of TiO_2 nanoparticles and improved the adsorption of reactants on the catalytic surface. The main product of the CO_2 photoreduction reaction was carbon monoxide when bare TiO_2 was supported on silica. However, when Cu/TiO_2 was supported on silica, the production rate of methane significantly increased. The deposition of Cu enhanced the separation of charges and improved the kinetics of the multi-electron reactions, consequently increasing methane selectivity. As can be deduced from Figure 15b, the 0.5 wt.% Cu/TiO_2-Silica composite produced carbon monoxide and methane with a yield of 60 and 10 µmol/g-cat/h, respectively.

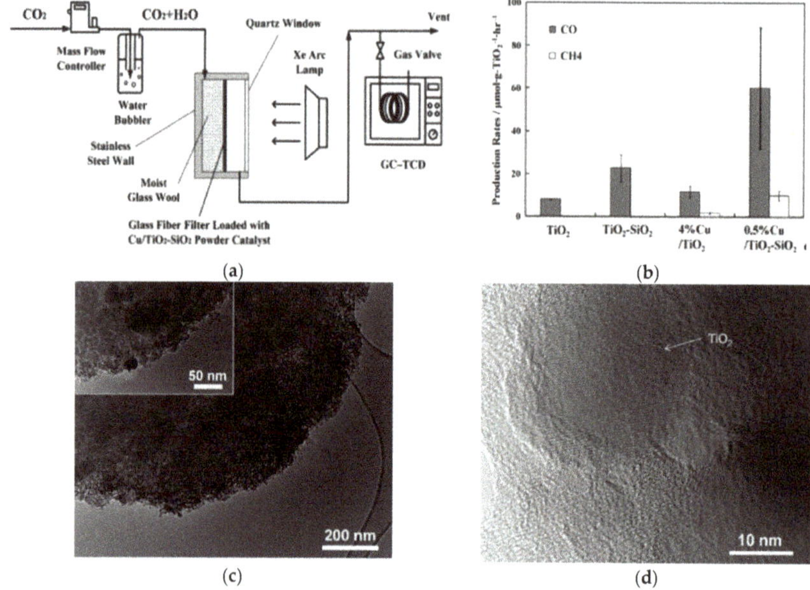

Figure 15. Photocatalytic reduction of CO_2 on ordered mesoporous silica supported Cu/TiO_2. (**a**) Schematic representation of experimental set-up; (**b**) production rates of CO and methane; (**c**) transmission electron microscopy (TEM) image of the 0.5 wt.% Cu/TiO_2-Silica composite; (**d**) high resolution TEM image of a single TiO_2 nanoparticle. Reproduced from work in [106]. Copyright 2010 Elsevier B.V.

4. Conclusions and Future Perspective

In this review, recent advances in the area of CO_2 photoreduction have been discussed and the basic mechanism of CO_2 photoreduction, particularly with water, has been described. Limited studies exist on the reaction mechanism and kinetics of CO_2 photoreduction as a result of the complexity of the reaction which is due mainly to the possibility of forming various products and the presence of many reaction pathways. In situ techniques, such as FTIR and NMR, would help investigate crucial reaction kinetic parameters such as active sites, electronic states and reaction intermediates. Additionally, various modification strategies of TiO_2 that may help overcome the limitations of current CO_2 photoreduction technologies have been described. A comparative assessment between the different TiO_2-based catalysts that have been reported in literature so far for CO_2 photoreduction was made. This overview of the latest research trends in CO_2 photoreduction may be used as a tool to help develop low-cost and highly-efficient TiO_2-based semiconductors for an improved photocatalytic reduction of CO_2.

Regardless of the significant theoretical advancements, the practical application of TiO_2 photocatalysts for the photoreduction of CO_2 is still far from being attained. This is mainly due to the lack of stable and visible light active photocatalysts. Researchers must focus on engineering a photocatalytic nanocomposite with an optimum band gap that simultaneously enhances charge separation and visible light absorption. A few suggested strategies include coupling wide band gap semiconductors with narrow ones, introduction of oxygen vacancies and defect sites and functionalization of the TiO_2 surface. To enhance the adsorption and reduction of CO_2 especially in the presence of water, a basic and hydrophobic TiO_2 surface must be developed. The latest studies clearly demonstrate the considerable improvement in the yield and efficiency of the CO_2 photoreduction process with modified TiO_2 catalysts. More specifically, the major limitations of TiO_2, including weak visible light absorption, increased charge recombination, and low CO_2 adsorption capacity, have been greatly overcome by synthesizing nanocomposites with exposed reactive sites and improved morphology.

Surface modifications to TiO_2-based photocatalysts cannot solely suffice the technical and economic feasibility requirements of CO_2 photoreduction applications, especially if they are to be applied on a large industrial scale. One promising solution could be to combine photocatalysis with other CO_2 conversion methods such as thermochemical and/or electrochemical catalysis. Another suggested solution could be to couple modified TiO_2 with other visible-light active photocatalysts with the aim of improving the photocatalytic process under solar irradiation. Further enhancement might be attained if compressed CO_2 (liquid or supercritical) is used for the photocatalytic reduction process. Although this increases the CO_2 conversion efficiency, the overall cost will be considerably high due to intensive compression requirements. In this case, industries already utilizing compressed CO_2 must be targeted.

Funding: We acknowledge financial support from Khalifa University of Science and Technology through the Research and Innovation Center on CO_2 and H_2 (RICH) (grant RC2-2019-007). We also acknowledge financial support from Abu Dhabi Education and Knowledge through the Award for Research Excellence (grant AARE17-8434000095).

Conflicts of Interest: The authors declare no conflict of interest.

References

1. Ryaboshapko, A.G.; Revokatova, A.P. A potential role of the negative emission of carbon dioxide in solving the climate problem. *Russ. Meteorol. Hydrol.* **2015**, *40*, 443–455. [CrossRef]
2. Nordhaus, W.D. *The Climate Casino: Risk, Uncertainty, and Economics for a Warming World*; Yale University Press: New Haven, CT, USA, 2013.
3. Gore, A. The Crisis Comes Ashore. *New Repub.* **2010**, *241*, 10–12.
4. Shin, D.Y.; Jo, J.H.; Lee, J.-Y.; Lim, D.-H. Understanding mechanisms of carbon dioxide conversion into methane for designing enhanced catalysts from first-principles. *Comput. Chem.* **2016**, *1083*, 31–37. [CrossRef]
5. Holdren, J.P. Population and the energy problem. *Popul. Environ.* **1991**, *12*, 231–255. [CrossRef]
6. Sanchez, J.A. *Carbon Dioxide Capture: Processes, Technology and Environmental Implications*; Nova Science Publishers, Inc.: Hauppauge, NY, USA, 2016.
7. Spigarelli, B.P.; Kawatra, S.K. Opportunities and challenges in carbon dioxide capture. *J. Co2 Util.* **2013**, *1*, 69–87. [CrossRef]
8. Cheah, W.Y.; Ling, T.C.; Juan, J.C.; Lee, D.-J.; Chang, J.-S.; Show, P.L. Biorefineries of carbon dioxide: From carbon capture and storage (CCS) to bioenergies production. *Bioresour. Technol.* **2016**, *215*, 346–356. [CrossRef]
9. Koytsoumpa, E.I.; Bergins, C.; Kakaras, E. The CO_2 economy: Review of CO_2 capture and reuse technologies. *J. Supercrit. Fluids* **2018**, *132*, 3–16. [CrossRef]
10. Sahebdelfar, S.; Takht Ravanchi, M. Carbon dioxide utilization for methane production: A thermodynamic analysis. *J. Pet. Sci. Eng.* **2015**, *134*, 14–22. [CrossRef]
11. Centi, G.; Perathoner, S. Opportunities and prospects in the chemical recycling of carbon dioxide to fuels. *Catal. Today* **2009**, *148*, 191–205. [CrossRef]
12. Muradov, N. Industrial Utilization of CO_2: A Win–Win Solution. In *Liberating Energy from Carbon: Introduction to Decarbonization*; Springer: New York, NY, USA, 2014; pp. 325–383. [CrossRef]
13. Ganesh, I. Conversion of carbon dioxide into methanol—A potential liquid fuel: Fundamental challenges and opportunities (a review). *Renew. Sustain. Energy Rev.* **2014**, *31*, 221–257. [CrossRef]
14. Arakawa, H.; Aresta, M.; Armor, J.N.; Barteau, M.A.; Beckman, E.J.; Bell, A.T.; Bercaw, J.E.; Creutz, C.; Dinjus, E.; Dixon, D.A.; et al. Catalysis Research of Relevance to Carbon Management: Progress, Challenges, and Opportunities. *Chem. Rev.* **2001**, *101*, 953–996. [CrossRef]
15. Liu, X.-M.; Lu, G.Q.; Yan, Z.-F.; Beltramini, J. Recent Advances in Catalysts for Methanol Synthesis via Hydrogenation of CO and CO_2. *Ind. Eng. Chem. Res.* **2003**, *42*, 6518–6530. [CrossRef]
16. Farrelly, D.J.; Everard, C.D.; Fagan, C.C.; McDonnell, K.P. Carbon sequestration and the role of biological carbon mitigation: A review. *Renew. Sustain. Energy Rev.* **2013**, *21*, 712–727. [CrossRef]
17. Alaswad, A.; Dassisti, M.; Prescott, T.; Olabi, A.G. Technologies and developments of third generation biofuel production. *Renew. Sustain. Energy Rev.* **2015**, *51*, 1446–1460. [CrossRef]

18. Ramachandriya, K.D.; Kundiyana, D.K.; Wilkins, M.R.; Terrill, J.B.; Atiyeh, H.K.; Huhnke, R.L. Carbon dioxide conversion to fuels and chemicals using a hybrid green process. *Appl. Energy* **2013**, *112*, 289–299. [CrossRef]

19. de_Richter, R.K.; Ming, T.; Caillol, S. Fighting global warming by photocatalytic reduction of CO_2 using giant photocatalytic reactors. *Renew. Sustain. Energy Rev.* **2013**, *19*, 82–106. [CrossRef]

20. Hoffmann, M.R.; Moss, J.A.; Baum, M.M. Artificial photosynthesis: Semiconductor photocatalytic fixation of CO_2 to afford higher organic compounds. *Dalton Trans.* **2011**, *40*, 5151–5158. [CrossRef]

21. Koči, K.; Obalová, L.; Lacný, Z. Photocatalytic reduction of CO_2 over TiO_2 based catalysts. *Chem. Pap.* **2008**, *62*, 1–9. [CrossRef]

22. Habisreutinger, S.N.; Schmidt-Mende, L.; Stolarczyk, J.K. Photocatalytic reduction of CO_2 on TiO_2 and other semiconductors. *Angew. Chem. Int. Ed.* **2013**, *52*, 7372–7408. [CrossRef]

23. Ola, O.; Maroto-Valer, M.M. Review of material design and reactor engineering on TiO_2 photocatalysis for CO_2 reduction. *J. Photochem. Photobiol. C* **2015**, *24*, 16–42. [CrossRef]

24. Kumar, A.; Ashok, A.; Shurair, M.S.; Yussuf, K.; van den Broeke, L.J. Carbon Management: Regional Solutions Based on Carbon Dioxide Utilization and Process Integration. In Proceedings of the 4th International Gas Processing Symposium, Doha, Qatar, 26–27 October 2014; Elsevier: Oxford, UK, 2015.

25. Li, K.; An, X.; Park, K.H.; Khraisheh, M.; Tang, J. A critical review of CO_2 photoconversion: Catalysts and reactors. *Catal. Today* **2014**, *224*, 3–12. [CrossRef]

26. Meryem, S.S.; Nasreen, S.; Siddique, M.; Khan, R. An overview of the reaction conditions for an efficient photoconversion of CO_2. *Rev. Chem. Eng.* **2018**, *34*, 409–425. [CrossRef]

27. Mallouk, T.E. The Emerging Technology of Solar Fuels. *J. Phys. Chem. Lett.* **2010**, *1*, 2738–2739. [CrossRef]

28. Shi, R.; Waterhouse, G.I.N.; Zhang, T. Recent Progress in Photocatalytic CO_2 Reduction Over Perovskite Oxides. *Sol. Rrl* **2017**, *1*, 1700126. [CrossRef]

29. AlOtaibi, B.; Fan, S.; Wang, D.; Ye, J.; Mi, Z. Wafer-level artificial photosynthesis for CO_2 reduction into CH_4 and CO using GaN nanowires. *ACS Catal.* **2015**, *5*, 5342–5348. [CrossRef]

30. Zhang, Y.; Xu, C.; Chen, J.; Zhang, X.; Wang, Z.; Zhou, J.; Cen, K. A novel photo-thermochemical cycle for the dissociation of CO_2 using solar energy. *Appl. Energy* **2015**, *156*, 223–229. [CrossRef]

31. Wang, P.; Wang, S.; Wang, H.; Wu, Z.; Wang, L. Recent Progress on Photo-Electrocatalytic Reduction of Carbon Dioxide. *Part. Part. Syst. Char.* **2018**, *35*, 1700371. [CrossRef]

32. Halmann, M. Photoelectrochemical reduction of aqueous carbon dioxide on p-type gallium phosphide in liquid junction solar cells. *Nature* **1978**, *275*, 115–116. [CrossRef]

33. Kumar, S.G.; Devi, L.G. Review on Modified TiO_2 Photocatalysis under UV/Visible Light: Selected Results and Related Mechanisms on Interfacial Charge Carrier Transfer Dynamics. *J. Phys. Chem. A* **2011**, *115*, 13211–13241. [CrossRef]

34. Daghrir, R.; Drogui, P.; Robert, D. Modified TiO_2 For Environmental Photocatalytic Applications: A Review. *Ind. Eng. Chem. Res.* **2013**, *52*, 3581–3599. [CrossRef]

35. Wen, J.; Li, X.; Liu, W.; Fang, Y.; Xie, J.; Xu, Y. Photocatalysis fundamentals and surface modification of TiO_2 nanomaterials. *Chin. J. Catal.* **2015**, *36*, 2049–2070. [CrossRef]

36. Li, X.; Zhuang, Z.; Li, W.; Pan, H. Photocatalytic reduction of CO_2 over noble metal-loaded and nitrogen-doped mesoporous TiO_2. *Appl. Catal. A* **2012**, *429–430*, 31–38. [CrossRef]

37. Corma, A.; Garcia, H. Photocatalytic reduction of CO_2 for fuel production: Possibilities and challenges. *J. Catal.* **2013**, *308*, 168–175. [CrossRef]

38. Qin, S.; Xin, F.; Liu, Y.; Yin, X.; Ma, W. Photocatalytic reduction of CO_2 in methanol to methyl formate over $CuO–TiO_2$ composite catalysts. *J. Colloid Interface Sci.* **2011**, *356*, 257–261. [CrossRef]

39. Karamian, E.; Sharifnia, S. On the general mechanism of photocatalytic reduction of CO_2. *J. Co2 Util.* **2016**, *16*, 194–203. [CrossRef]

40. Low, J.; Cheng, B.; Yu, J. Surface modification and enhanced photocatalytic CO_2 reduction performance of TiO_2: A review. *Appl. Surf. Sci.* **2017**, *392*, 658–686. [CrossRef]

41. Dilla, M.; Schlögl, R.; Strunk, J. Photocatalytic CO_2 Reduction Under Continuous Flow High-Purity Conditions: Quantitative Evaluation of CH4 Formation in the Steady-State. *ChemCatChem* **2017**, *9*, 696–704. [CrossRef]

42. Mino, L.; Spoto, G.; Ferrari, A.M. CO_2 capture by TiO_2 anatase surfaces: A combined DFT and FTIR study. *J. Phys. Chem. C* **2014**, *118*, 25016–25026. [CrossRef]

43. Sorescu, D.C.; Al-Saidi, W.A.; Jordan, K.D. CO_2 adsorption on TiO_2 (101) anatase: A dispersion-corrected density functional theory study. *J. Chem. Phys.* **2011**, *135*, 124701. [CrossRef]

44. Tu, W.; Zhou, Y.; Zou, Z. Photocatalytic conversion of CO_2 into renewable hydrocarbon fuels: State-of-the-art accomplishment, challenges, and prospects. *Adv. Mater.* **2014**, *26*, 4607–4626. [CrossRef]

45. Yuan, L.; Xu, Y.-J. Photocatalytic conversion of CO_2 into value-added and renewable fuels. *Appl. Surf. Sci.* **2015**, *342*, 154–167. [CrossRef]

46. Halmann, M. 15—Photochemical Fixation of Carbon Dioxide. In *Energy Resources Through Photochemistry and Catalysis*; Grätzel, M., Ed.; Academic Press: Cambridge, MA, USA, 1983; pp. 507–534. Available online: https://linkinghub.elsevier.com/retrieve/pii/B9780122957208500193 (accessed on 1 December 2019). [CrossRef]

47. Tan, L.-L.; Ong, W.-J.; Chai, S.-P.; Mohamed, A.R. Photocatalytic reduction of CO_2 with H_2O over graphene oxide-supported oxygen-rich TiO_2 hybrid photocatalyst under visible light irradiation: Process and kinetic studies. *Chem. Eng. J.* **2017**, *308*, 248–255. [CrossRef]

48. Liu, L.; Li, Y. Understanding the reaction mechanism of photocatalytic reduction of CO_2 with H_2O on TiO_2-based photocatalysts: A review. *Aerosol Air Qual. Res.* **2014**, *14*, 453–469. [CrossRef]

49. Uner, D.; Oymak, M.M. On the mechanism of photocatalytic CO_2 reduction with water in the gas phase. *Catal. Today* **2012**, *181*, 82–88. [CrossRef]

50. Saladin, F.; Alxneit, I. Temperature dependence of the photochemical reduction of CO_2 in the presence of H_2O at the solid/gas interface of TiO_2. *J. Chem. Soc. Faraday Trans.* **1997**, *93*, 4159–4163. [CrossRef]

51. Handoko, A.D.; Li, K.; Tang, J. Recent progress in artificial photosynthesis: CO_2 photoreduction to valuable chemicals in a heterogeneous system. *Curr. Opin. Chem. Eng.* **2013**, *2*, 200–206. [CrossRef]

52. Subrahmanyam, M.; Kaneco, S.; Alonso-Vante, N. A screening for the photo reduction of carbon dioxide supported on metal oxide catalysts for C1–C3 selectivity. *Appl. Catal. B* **1999**, *23*, 169–174. [CrossRef]

53. Sasirekha, N.; Basha, S.J.S.; Shanthi, K. Photocatalytic performance of Ru doped anatase mounted on silica for reduction of carbon dioxide. *Appl. Catal. B* **2006**, *62*, 169–180. [CrossRef]

54. Akhter, P.; Hussain, M.; Saracco, G.; Russo, N. Novel nanostructured-TiO_2 materials for the photocatalytic reduction of CO_2 greenhouse gas to hydrocarbons and syngas. *Fuel* **2015**, *149*, 55–65. [CrossRef]

55. Klyukin, K.; Alexandrov, V. CO_2 Adsorption and Reactivity on Rutile TiO_2(110) in Water: An Ab Initio Molecular Dynamics Study. *J. Phys. Chem. C* **2017**, *121*, 10476–10483. [CrossRef]

56. Liu, B.-J.; Torimoto, T.; Matsumoto, H.; Yoneyama, H. Effect of solvents on photocatalytic reduction of carbon dioxide using TiO_2 nanocrystal photocatalyst embedded in SiO_2 matrices. *J. Photochem. Photobiol. A* **1997**, *108*, 187–192. [CrossRef]

57. Kočí, K.; Obalová, L.; Matějová, L.; Plachá, D.; Lacný, Z.; Jirkovský, J.; Šolcová, O. Effect of TiO_2 particle size on the photocatalytic reduction of CO_2. *Appl. Catal. B* **2009**, *89*, 494–502. [CrossRef]

58. Tseng, I.-H.; Chang, W.-C.; Wu, J.C. Photoreduction of CO_2 using sol–gel derived titania and titania-supported copper catalysts. *Appl. Catal. B* **2002**, *37*, 37–48. [CrossRef]

59. Sastre, F.; Corma, A.; García, H. 185 nm Photoreduction of CO_2 to Methane by Water. Influence of the Presence of a Basic Catalyst. *J. Am. Chem. Soc.* **2012**, *134*, 14137–14141. [CrossRef] [PubMed]

60. Lo, C.-C.; Hung, C.-H.; Yuan, C.-S.; Wu, J.-F. Photoreduction of carbon dioxide with H_2 and H_2O over TiO_2 and ZrO_2 in a circulated photocatalytic reactor. *Sol. Energy Mater. Sol. Cells* **2007**, *91*, 1765–1774. [CrossRef]

61. Grills, D.C.; Fujita, E. New directions for the photocatalytic reduction of CO_2: Supramolecular, scCO$_2$ or biphasic ionic liquid—scCO$_2$ systems. *J. Phys. Chem. Lett.* **2010**, *1*, 2709–2718. [CrossRef]

62. Kaneco, S.; Kurimoto, H.; Ohta, K.; Mizuno, T.; Saji, A. Photocatalytic reduction of CO_2 using TiO_2 powders in liquid CO_2 medium. *J. Photochem. Photobiol. A* **1997**, *109*, 59–63. [CrossRef]

63. Kaneco, S.; Kurimoto, H.; Shimizu, Y.; Ohta, K.; Mizuno, T. Photocatalytic reduction of CO_2 using TiO_2 powders in supercritical fluid CO_2. *Energy* **1999**, *24*, 21–30. [CrossRef]

64. Kometani, N.; Hirata, S.; Chikada, M. Photocatalytic reduction of CO_2 by Pt-loaded TiO_2 in the mixture of sub-and supercritical water and CO_2. *J. Supercrit. Fluids* **2017**, *120*, 443–447. [CrossRef]

65. Liu, L.; Zhao, H.; Andino, J.M.; Li, Y. Photocatalytic CO_2 reduction with H_2O on TiO_2 nanocrystals: Comparison of anatase, rutile, and brookite polymorphs and exploration of surface chemistry. *ACS Catal.* **2012**, *2*, 1817–1828. [CrossRef]

66. Chen, L.; Graham, M.E.; Li, G.; Gentner, D.R.; Dimitrijevic, N.M.; Gray, K.A. Photoreduction of CO_2 by TiO_2 nanocomposites synthesized through reactive direct current magnetron sputter deposition. *Thin Solid Film.* **2009**, *517*, 5641–5645. [CrossRef]

67. Haggerty, J.E.S.; Schelhas, L.T.; Kitchaev, D.A.; Mangum, J.S.; Garten, L.M.; Sun, W.; Stone, K.H.; Perkins, J.D.; Toney, M.F.; Ceder, G.; et al. High-fraction brookite films from amorphous precursors. *Sci. Rep.* **2017**, *7*, 15232. [CrossRef] [PubMed]

68. Zhang, P.; Tang, B.; Xia, W.; Wang, F.; Wei, X.; Xiang, J.; Li, J. Preparation and Characterizations of TiO_2 Nanoparticles by Sol-Gel Process using DMAC Solvent. *MMECEB* **2015**, *2016*, 892–895.

69. Akpan, U.; Hameed, B. The advancements in sol–gel method of doped-TiO_2 photocatalysts. *Appl. Catal. A* **2010**, *375*, 1–11. [CrossRef]

70. Wang, Y.; He, Y.; Lai, Q.; Fan, M. Review of the progress in preparing nano TiO_2: An important environmental engineering material. *J. Environ. Sci.* **2014**, *26*, 2139–2177. [CrossRef] [PubMed]

71. Chen, Y.; Lunsford, S.K.; Song, Y.; Ju, H.; Falaras, P.; likodimos, V.; Kontos, A.G.; Dionysiou, D.D. Synthesis, characterization and electrochemical properties of mesoporous zirconia nanomaterials prepared by self-assembling sol–gel method with Tween 20 as a template. *Chem. Eng. J.* **2011**, *170*, 518–524. [CrossRef]

72. Sun, Y.; Seo, J.H.; Takacs, C.J.; Seifter, J.; Heeger, A.J. Inverted polymer solar cells integrated with a low-temperature-annealed sol-gel-derived ZnO film as an electron transport layer. *Adv. Mater.* **2011**, *23*, 1679–1683. [CrossRef]

73. Djaoued, Y.; Balaji, S.; Beaudoin, N. Sol–gel synthesis of mesoporous WO3–TiO_2 composite thin films for photochromic devices. *J. Sol-Gel Sci. Technol.* **2013**, *65*, 374–383. [CrossRef]

74. Chen, C.; Cai, W.; Long, M.; Zhang, J.; Zhou, B.; Wu, Y.; Wu, D. Template-free sol–gel preparation and characterization of free-standing visible light responsive C, N-modified porous monolithic TiO_2. *J. Hazard. Mater.* **2010**, *178*, 560–565. [CrossRef]

75. Tseng, T.K.; Lin, Y.S.; Chen, Y.J.; Chu, H. A review of photocatalysts prepared by sol-gel method for VOCs removal. *Int. J. Mol. Sci.* **2010**, *11*, 2336–2361. [CrossRef]

76. Lu, J.; Li, L.; Wang, Z.; Wen, B.; Cao, J. Synthesis of visible-light-active TiO_2-based photo-catalysts by a modified sol–gel method. *Mater. Lett.* **2013**, *94*, 147–149. [CrossRef]

77. You, Y.; Zhang, S.; Wan, L.; Xu, D. Preparation of continuous TiO_2 fibers by sol–gel method and its photocatalytic degradation on formaldehyde. *Appl. Surf. Sci.* **2012**, *258*, 3469–3474. [CrossRef]

78. Xie, Y.; Zhao, Q.; Zhao, X.J.; Li, Y. Low temperature preparation and characterization of N-doped and NS-codoped TiO_2 by sol–gel route. *Catal. Lett.* **2007**, *118*, 231–237. [CrossRef]

79. Šegota, S.; Ćurković, L.; Ljubas, D.; Svetličić, V.; Houra, I.F.; Tomašić, N. Synthesis, characterization and photocatalytic properties of sol–gel TiO_2 films. *Ceram. Int.* **2011**, *37*, 1153–1160. [CrossRef]

80. Kim, M.-S.; Liu, G.; Nam, W.K.; Kim, B.-W. Preparation of porous carbon-doped TiO_2 film by sol–gel method and its application for the removal of gaseous toluene in the optical fiber reactor. *J. Ind. Eng. Chem.* **2011**, *17*, 223–228. [CrossRef]

81. Macwan, D.; Dave, P.N.; Chaturvedi, S. A review on nano-TiO_2 sol–gel type syntheses and its applications. *Asian J. Mater. Sci.* **2011**, *46*, 3669–3686. [CrossRef]

82. Payakgul, W.; Mekasuwandumrong, O.; Pavarajarn, V.; Praserthdam, P. Effects of reaction medium on the synthesis of TiO_2 nanocrystals by thermal decomposition of titanium (IV) n-butoxide. *Ceram. Int.* **2005**, *31*, 391–397. [CrossRef]

83. Liu, Z.; Sun, D.D.; Guo, P.; Leckie, J.O. One-step fabrication and high photocatalytic activity of porous TiO_2 hollow aggregates by using a low-temperature hydrothermal method without templates. *Chem. Eur. J.* **2007**, *13*, 1851–1855. [CrossRef]

84. Ou, H.-H.; Lo, S.-L. Review of titania nanotubes synthesized via the hydrothermal treatment: Fabrication, modification, and application. *Sep. Purif. Technol.* **2007**, *58*, 179–191. [CrossRef]

85. Fumin, W.; Zhansheng, S.; Feng, G.; Jinting, J.; Adachi, M. Morphology control of anatase TiO_2 by surfactant-assisted hydrothermal method. *Chin. J. Chem. Eng.* **2007**, *15*, 754–759.

86. Camarillo, R.; Tostón, S.; Martínez, F.; Jiménez, C.; Rincón, J. Preparation of TiO_2-based catalysts with supercritical fluid technology: Characterization and photocatalytic activity in CO_2 reduction. *J. Chem. Technol. Biotechnol.* **2017**, *92*, 1710–1720. [CrossRef]

87. Bellardita, M.; Di Paola, A.; García-López, E.; Loddo, V.; Marcì, G.; Palmisano, L. Photocatalytic CO_2 reduction in gas-solid regime in the presence of bare, SiO_2 supported or Cu-loaded TiO_2 samples. *Curr. Org. Chem.* **2013**, *17*, 2440–2448. [CrossRef]

88. Maeda, K. Photocatalytic water splitting using semiconductor particles: History and recent developments. *J. Photochem. Photobiol. C* **2011**, *12*, 237–268. [CrossRef]

89. Humayun, M.; Raziq, F.; Khan, A.; Luo, W. Modification strategies of TiO_2 for potential applications in photocatalysis: A critical review. *Green Chem. Lett. Rev.* **2018**, *11*, 86–102. [CrossRef]

90. Meng, X.; Ouyang, S.; Kako, T.; Li, P.; Yu, Q.; Wang, T.; Ye, J. Photocatalytic CO_2 conversion over alkali modified TiO_2 without loading noble metal cocatalyst. *Chem. Commun.* **2014**, *50*, 11517–11519. [CrossRef]

91. Wang, W.-N.; An, W.-J.; Ramalingam, B.; Mukherjee, S.; Niedzwiedzki, D.M.; Gangopadhyay, S.; Biswas, P. Size and structure matter: Enhanced CO_2 photoreduction efficiency by size-resolved ultrafine Pt nanoparticles on TiO_2 single crystals. *J. Am. Chem. Soc.* **2012**, *134*, 11276–11281. [CrossRef]

92. Tahir, M.; Tahir, B.; Amin, N.A.S. Gold-nanoparticle-modified TiO_2 nanowires for plasmon-enhanced photocatalytic CO_2 reduction with H_2 under visible light irradiation. *Appl. Surf. Sci.* **2015**, *356*, 1289–1299. [CrossRef]

93. Kong, D.; Tan, J.Z.Y.; Yang, F.; Zeng, J.; Zhang, X. Electrodeposited Ag nanoparticles on TiO_2 nanorods for enhanced UV visible light photoreduction CO_2 to CH4. *Appl. Surf. Sci.* **2013**, *277*, 105–110. [CrossRef]

94. Zhai, Q.; Xie, S.; Fan, W.; Zhang, Q.; Wang, Y.; Deng, W.; Wang, Y. Photocatalytic conversion of carbon dioxide with water into methane: Platinum and copper (I) oxide co-catalysts with a core–shell structure. *Angew. Chem.* **2013**, *125*, 5888–5891. [CrossRef]

95. Yan, Y.; Yu, Y.; Cao, C.; Huang, S.; Yang, Y.; Yang, X.; Cao, Y. Enhanced photocatalytic activity of TiO_2–Cu/C with regulation and matching of energy levels by carbon and copper for photoreduction of CO_2 into CH_4. *CrystEngComm* **2016**, *18*, 2956–2964. [CrossRef]

96. Wu, J.C.S.; Lin, H.-M.; Lai, C.-L. Photo reduction of CO_2 to methanol using optical-fiber photoreactor. *Appl. Catal. A* **2005**, *296*, 194–200. [CrossRef]

97. Nasution, H.W.; Purnama, E.; Kosela, S.; Gunlazuardi, J. Photocatalytic reduction of CO_2 on copper-doped Titania catalysts prepared by improved-impregnation method. *Catal. Commun.* **2005**, *6*, 313–319. [CrossRef]

98. Kwak, B.S.; Vignesh, K.; Park, N.-K.; Ryu, H.-J.; Baek, J.-I.; Kang, M. Methane formation from photoreduction of CO_2 with water using TiO_2 including Ni ingredient. *Fuel* **2015**, *143*, 570–576. [CrossRef]

99. Matějová, L.; Kočí, K.; Reli, M.; Čapek, L.; Hospodková, A.; Peikertová, P.; Matěj, Z.; Obalová, L.; Wach, A.; Kuśtrowski, P.; et al. Preparation, characterization and photocatalytic properties of cerium doped TiO_2: On the effect of Ce loading on the photocatalytic reduction of carbon dioxide. *Appl. Catal. B* **2014**, *152–153*, 172–183. [CrossRef]

100. Zhao, Z.; Fan, J.; Wang, J.; Li, R. Effect of heating temperature on photocatalytic reduction of CO_2 by N–TiO_2 nanotube catalyst. *Catal. Commun.* **2012**, *21*, 32–37. [CrossRef]

101. Xue, L.M.; Zhang, F.H.; Fan, H.J.; Bai, X.F. Preparation of C Doped TiO_2 Photocatalysts and their Photocatalytic Reduction of Carbon Dioxide. *Adv. Mater. Res.* **2011**, *183–185*, 1842–1846. [CrossRef]

102. Gui, M.M.; Chai, S.-P.; Xu, B.-Q.; Mohamed, A.R. Enhanced visible light responsive MWCNT/TiO_2 core–shell nanocomposites as the potential photocatalyst for reduction of CO_2 into methane. *Sol. Energy Mater. Sol. Cells* **2014**, *122*, 183–189. [CrossRef]

103. Xia, X.-H.; Jia, Z.-J.; Yu, Y.; Liang, Y.; Wang, Z.; Ma, L.-L. Preparation of multi-walled carbon nanotube supported TiO_2 and its photocatalytic activity in the reduction of CO_2 with H_2O. *Carbon* **2007**, *45*, 717–721. [CrossRef]

104. Ong, W.-J.; Gui, M.M.; Chai, S.-P.; Mohamed, A.R. Direct growth of carbon nanotubes on Ni/TiO_2 as next generation catalysts for photoreduction of CO_2 to methane by water under visible light irradiation. *RSC Adv.* **2013**, *3*, 4505–4509. [CrossRef]

105. Ong, W.-J.; Tan, L.-L.; Chai, S.-P.; Yong, S.-T.; Mohamed, A.R. Self-assembly of nitrogen-doped TiO_2 with exposed {001} facets on a graphene scaffold as photo-active hybrid nanostructures for reduction of carbon dioxide to methane. *Nano Res.* **2014**, *7*, 1528–1547. [CrossRef]

106. Li, Y.; Wang, W.-N.; Zhan, Z.; Woo, M.-H.; Wu, C.-Y.; Biswas, P. Photocatalytic reduction of CO_2 with H_2O on mesoporous silica supported Cu/TiO_2 catalysts. *Appl. Catal. B* **2010**, *100*, 386–392. [CrossRef]

107. Sakthivel, S.; Shankar, M.V.; Palanichamy, M.; Arabindoo, B.; Bahnemann, D.W.; Murugesan, V. Enhancement of photocatalytic activity by metal deposition: Characterisation and photonic efficiency of Pt, Au and Pd deposited on TiO_2 catalyst. *Wat. Res.* **2004**, *38*, 3001–3008. [CrossRef] [PubMed]

108. Tostón, S.; Camarillo, R.; Martínez, F.; Jiménez, C.; Rincón, J. Supercritical synthesis of platinum-modified titanium dioxide for solar fuel production from carbon dioxide. *Chin. J. Catal.* **2017**, *38*, 636–650. [CrossRef]

109. Khan, M.R.; Chuan, T.W.; Yousuf, A.; Chowdhury, M.; Cheng, C.K. Schottky barrier and surface plasmonic resonance phenomena towards the photocatalytic reaction: Study of their mechanisms to enhance photocatalytic activity. *Catal. Sci. Technol.* **2015**, *5*, 2522–2531. [CrossRef]

110. Murakami, N.; Saruwatari, D.; Tsubota, T.; Ohno, T. Photocatalytic reduction of carbon dioxide over shape-controlled titanium (IV) oxide nanoparticles with co-catalyst loading. *Curr. Org. Chem.* **2013**, *17*, 2449–2453. [CrossRef]

111. Zaleska, A. Doped-TiO_2: A review. *Recent Pat. Eng.* **2008**, *2*, 157–164. [CrossRef]

112. Linsebigler, A.L.; Lu, G.; Yates, J.T. Photocatalysis on TiO_2 Surfaces: Principles, Mechanisms, and Selected Results. *Chem. Rev.* **1995**, *95*, 735–758. [CrossRef]

113. Liu, E.; Qi, L.; Bian, J.; Chen, Y.; Hu, X.; Fan, J.; Liu, H.; Zhu, C.; Wang, Q. A facile strategy to fabricate plasmonic Cu modified TiO_2 nano-flower films for photocatalytic reduction of CO_2 to methanol. *Mater. Res. Bull.* **2015**, *68*, 203–209. [CrossRef]

114. Kemp, T.J.; McIntyre, R.A. Transition metal-doped titanium (IV) dioxide: Characterisation and influence on photodegradation of poly (vinyl chloride). *Polym. Degrad. Stab.* **2006**, *91*, 165–194. [CrossRef]

115. Feng, H.; Zhang, M.-H.; Yu, L.E. Hydrothermal synthesis and photocatalytic performance of metal-ions doped TiO_2. *Appl. Catal. A* **2012**, *413–414*, 238–244. [CrossRef]

116. Di Valentin, C.; Pacchioni, G. Trends in non-metal doping of anatase TiO_2: B, C, N and F. *Catal. Today* **2013**, *206*, 12–18. [CrossRef]

117. Marschall, R.; Wang, L. Non-metal doping of transition metal oxides for visible-light photocatalysis. *Catal. Today* **2014**, *225*, 111–135. [CrossRef]

118. Sun, L.; Zhao, X.; Cheng, X.; Sun, H.; Li, Y.; Li, P.; Fan, W. Evaluating the C, N, and F pairwise codoping effect on the enhanced photoactivity of ZnWO4: The charge compensation mechanism in donor–acceptor pairs. *J. Phys. Chem. C* **2011**, *115*, 15516–15524. [CrossRef]

119. Wang, W.; Serp, P.; Kalck, P.; Silva, C.G.; Faria, J.L. Preparation and characterization of nanostructured MWCNT-TiO_2 composite materials for photocatalytic water treatment applications. *Mater. Res. Bull.* **2008**, *43*, 958–967. [CrossRef]

120. Oh, W.-C.; Zhang, F.-J.; Chen, M.-L. Preparation of MWCNT/TiO_2 composites by using MWCNTs and Titanium (IV) alkoxide precursors in Benzene and their photocatalytic effect and bactericidal activity. *Bull. Korean Chem. Soc.* **2009**, *30*, 2637–2642.

121. Kuo, C.-Y. Prevenient dye-degradation mechanisms using UV/TiO_2/carbon nanotubes process. *J. Hazard. Mater.* **2009**, *163*, 239–244. [CrossRef] [PubMed]

122. Wang, W.; Serp, P.; Kalck, P.; Faria, J.L. Photocatalytic degradation of phenol on MWNT and titania composite catalysts prepared by a modified sol–gel method. *Appl. Catal. B* **2005**, *56*, 305–312. [CrossRef]

123. Hu, G.; Meng, X.; Feng, X.; Ding, Y.; Zhang, S.; Yang, M. Anatase TiO_2 nanoparticles/carbon nanotubes nanofibers: Preparation, characterization and photocatalytic properties. *Asian J. Mater. Sci.* **2007**, *42*, 7162–7170. [CrossRef]

124. Aryal, S.; Kim, C.K.; Kim, K.-W.; Khil, M.S.; Kim, H.Y. Multi-walled carbon nanotubes/TiO_2 composite nanofiber by electrospinning. *Mater. Sci. Eng. C* **2008**, *28*, 75–79. [CrossRef]

125. Yu, H.; Quan, X.; Chen, S.; Zhao, H. TiO_2- multiwalled carbon nanotube heterojunction arrays and their charge separation capability. *J. Phys. Chem. C* **2007**, *111*, 12987–12991. [CrossRef]

126. Chen, M.-L.; Zhang, F.-J.; Oh, W.-C. Synthesis, characterization, and photocatalytic analysis of CNT/TiO_2 composites derived from MWCNTs and titanium sources. *New Carbon Mater.* **2009**, *24*, 159–166. [CrossRef]

127. Tang, B.; Chen, H.; Peng, H.; Wang, Z.; Huang, W. Graphene modified TiO_2 composite photocatalysts: Mechanism, progress and perspective. *Nanomaterials* **2018**, *8*, 105. [CrossRef]

128. Xiang, Q.; Yu, J.; Jaroniec, M. Graphene-based semiconductor photocatalysts. *Chem. Soc. Rev.* **2012**, *41*, 782–796. [CrossRef] [PubMed]

129. Chowdhury, S.; Balasubramanian, R. Graphene/semiconductor nanocomposites (GSNs) for heterogeneous photocatalytic decolorization of wastewaters contaminated with synthetic dyes: A review. *Appl. Catal. B* **2014**, *160*, 307–324. [CrossRef]

130. Jianwei, C.; Jianwen, S.; Xu, W.; Haojie, C.; Minglai, F. Recent progress in the preparation and application of semiconductor/graphene composite photocatalysts. *Chin. J. Catal.* **2013**, *34*, 621–640.

131. Singh, V.; Joung, D.; Zhai, L.; Das, S.; Khondaker, S.I.; Seal, S. Graphene based materials: Past, present and future. *Prog. Mater Sci.* **2011**, *56*, 1178–1271. [CrossRef]

132. Hu, C.; Lu, T.; Chen, F.; Zhang, R. A brief review of graphene–metal oxide composites synthesis and applications in photocatalysis. *J. Chin. Adv. Mater. Soc.* **2013**, *1*, 21–39. [CrossRef]

133. Ozer, L.Y.; Garlisi, C.; Oladipo, H.; Pagliaro, M.; Sharief, S.A.; Yusuf, A.; Almheiri, S.; Palmisano, G. Inorganic semiconductors-graphene composites in photo(electro)catalysis: Synthetic strategies, interaction mechanisms and applications. *J. Photochem. Photobiol. C* **2017**, *33*, 132–164. [CrossRef]

134. Wang, L.-Q.; Wang, D.; Liu, J.; Exarhos, G.J. Probing Porosity and Pore Interconnectivity in Self-Assembled TiO_2–Graphene Hybrid Nanostructures Using Hyperpolarized 129Xe NMR. *J. Phys. Chem. C* **2012**, *116*, 22–29. [CrossRef]

135. Ying, J.; Yang, X.-Y.; Tian, G.; Janiak, C.; Su, B.-L. Self-assembly: An option to nanoporous metal nanocrystals. *Nanoscale* **2014**, *6*, 13370–13382. [CrossRef]

136. Yu, X.; Lin, D.; Li, P.; Su, Z. Recent advances in the synthesis and energy applications of TiO_2-graphene nanohybrids. *Sol. Energy Mater. Sol. Cells* **2017**, *172*, 252–269. [CrossRef]

137. Kim, I.Y.; Lee, J.M.; Kim, T.W.; Kim, H.N.; Kim, H.I.; Choi, W.; Hwang, S.J. A Strong Electronic Coupling between Graphene Nanosheets and Layered Titanate Nanoplates: A Soft-Chemical Route to Highly Porous Nanocomposites with Improved Photocatalytic Activity. *Small* **2012**, *8*, 1038–1048. [CrossRef] [PubMed]

138. Tu, W.; Zhou, Y.; Liu, Q.; Yan, S.; Bao, S.; Wang, X.; Xiao, M.; Zou, Z. An In Situ Simultaneous Reduction-Hydrolysis Technique for Fabrication of TiO_2-Graphene 2D Sandwich-Like Hybrid Nanosheets: Graphene-Promoted Selectivity of Photocatalytic-Driven Hydrogenation and Coupling of CO_2 into Methane and Ethane. *Adv. Funct. Mater.* **2013**, *23*, 1743–1749. [CrossRef]

139. Tian, J.; Zhao, Z.; Kumar, A.; Boughton, R.I.; Liu, H. Recent progress in design, synthesis, and applications of one-dimensional TiO_2 nanostructured surface heterostructures: A review. *Chem. Soc. Rev.* **2014**, *43*, 6920–6937. [CrossRef] [PubMed]

140. Beigi, A.A.; Fatemi, S.; Salehi, Z. Synthesis of nanocomposite CdS/TiO_2 and investigation of its photocatalytic activity for CO_2 reduction to CO and CH_4 under visible light irradiation. *J. Co2 Util.* **2014**, *7*, 23–29. [CrossRef]

141. Li, X.; Liu, H.; Luo, D.; Li, J.; Huang, Y.; Li, H.; Fang, Y.; Xu, Y.; Zhu, L. Adsorption of CO_2 on heterostructure $CdS(Bi_2S_3)/TiO_2$ nanotube photocatalysts and their photocatalytic activities in the reduction of CO_2 to methanol under visible light irradiation. *Chem. Eng. J.* **2012**, *180*, 151–158. [CrossRef]

142. Wang, C.; Thompson, R.L.; Baltrus, J.; Matranga, C. Visible light photoreduction of CO_2 using $CdSe/Pt/TiO_2$ heterostructured catalysts. *J. Phys. Chem. Lett.* **2009**, *1*, 48–53. [CrossRef]

143. Wang, C.; Thompson, R.L.; Ohodnicki, P.; Baltrus, J.; Matranga, C. Size-dependent photocatalytic reduction of CO_2 with PbS quantum dot sensitized TiO_2 heterostructured photocatalysts. *J. Mater. Chem.* **2011**, *21*, 13452–13457. [CrossRef]

144. Asi, M.A.; He, C.; Su, M.; Xia, D.; Lin, L.; Deng, H.; Xiong, Y.; Qiu, R.; Li, X.-Z. Photocatalytic reduction of CO_2 to hydrocarbons using $AgBr/TiO_2$ nanocomposites under visible light. *Catal. Today* **2011**, *175*, 256–263. [CrossRef]

145. Kumar, S.; Isaacs, M.A.; Trofimovaite, R.; Durndell, L.; Parlett, C.M.; Douthwaite, R.E.; Coulson, B.; Cockett, M.C.; Wilson, K.; Lee, A.F. P25@ CoAl layered double hydroxide heterojunction nanocomposites for CO_2 photocatalytic reduction. *Appl. Catal. B* **2017**, *209*, 394–404. [CrossRef]

146. Low, J.; Yu, J.; Jaroniec, M.; Wageh, S.; Al-Ghamdi, A.A. Heterojunction photocatalysts. *Adv. Mater.* **2017**, *29*, 1601694. [CrossRef] [PubMed]

147. Li, H.; Zhou, Y.; Tu, W.; Ye, J.; Zou, Z. State-of-the-art progress in diverse heterostructured photocatalysts toward promoting photocatalytic performance. *Adv. Funct. Mater.* **2015**, *25*, 998–1013. [CrossRef]

148. George, S.M. Atomic layer deposition: An overview. *Chem. Rev.* **2010**, *110*, 111–131. [CrossRef] [PubMed]

149. Parsons, G.N.; George, S.M.; Knez, M. Progress and future directions for atomic layer deposition and ALD-based chemistry. *MRS Bull.* **2011**, *36*, 865–871. [CrossRef]

150. Lin, Y.; Xu, Y.; Mayer, M.T.; Simpson, Z.I.; McMahon, G.; Zhou, S.; Wang, D. Growth of p-type hematite by atomic layer deposition and its utilization for improved solar water splitting. *J. Am. Chem. Soc.* **2012**, *134*, 5508–5511. [CrossRef] [PubMed]

151. Wang, T.; Luo, Z.; Li, C.; Gong, J. Controllable fabrication of nanostructured materials for photoelectrochemical water splitting via atomic layer deposition. *Chem. Soc. Rev.* **2014**, *43*, 7469–7484. [CrossRef]

152. Son, D.H.; Hughes, S.M.; Yin, Y.; Paul Alivisatos, A. Cation exchange reactions in ionic nanocrystals. *Sci. (New York N.Y.)* **2004**, *306*, 1009–1012. [CrossRef]

153. Rivest, J.B.; Jain, P.K. Cation exchange on the nanoscale: An emerging technique for new material synthesis, device fabrication, and chemical sensing. *Chem. Soc. Rev.* **2013**, *42*, 89–96. [CrossRef]

154. Nguyen, T.-V.; Wu, J.C.S.; Chiou, C.-H. Photoreduction of CO_2 over Ruthenium dye-sensitized TiO_2-based catalysts under concentrated natural sunlight. *Catal. Commun.* **2008**, *9*, 2073–2076. [CrossRef]

155. Srikanth, B.; Goutham, R.; Badri Narayan, R.; Ramprasath, A.; Gopinath, K.P.; Sankaranarayanan, A.R. Recent advancements in supporting materials for immobilised photocatalytic applications in waste water treatment. *J. Environ. Manag.* **2017**, *200*, 60–78. [CrossRef]

156. Suárez, S. Immobilised Photocatalysts. In *Design of Advanced Photocatalytic Materials for Energy and Environmental Applications*; Coronado, J.M., Fresno, F., Hernández-Alonso, M.D., Portela, R., Eds.; Springer: London, UK, 2013; pp. 245–267. [CrossRef]

157. Wu, L.; Yang, D.; Fei, L.; Huang, Y.; Wu, F.; Sun, Y.; Shi, J.; Xiang, Y. Dip-Coating Process Engineering and Performance Optimization for Three-State Electrochromic Devices. *Nanoscale Res. Lett.* **2017**, *12*, 390. [CrossRef] [PubMed]

158. Kyesmen, P.I.; Nombona, N.; Diale, M. Influence of coating techniques on the optical and structural properties of hematite thin films. *Surf. Interfaces* **2019**, *17*, 100384. [CrossRef]

159. Deepa, M.; Saxena, T.K.; Singh, D.P.; Sood, K.N.; Agnihotry, S.A. Spin coated versus dip coated electrochromic tungsten oxide films: Structure, morphology, optical and electrochemical properties. *Electrochim. Acta* **2006**, *51*, 1974–1989. [CrossRef]

160. Matsushima, T.; Inoue, M.; Fujihara, T.; Terakawa, S.; Qin, C.; Sandanayaka, A.S.D.; Adachi, C. High-coverage organic-inorganic perovskite film fabricated by double spin coating for improved solar power conversion and amplified spontaneous emission. *Chem. Phys. Lett.* **2016**, *661*, 131–135. [CrossRef]

161. Pitoniak, E.; Wu, C.-Y.; Londeree, D.; Mazyck, D.; Bonzongo, J.-C.; Powers, K.; Sigmund, W. Nanostructured silica-gel doped with TiO_2 for mercury vapor control. *J. Nanopart. Res.* **2003**, *5*, 281–292. [CrossRef]

 *catalysts* 

Article

Enhanced Photocatalytic Activity of Semiconductor Nanocomposites Doped with Ag Nanoclusters Under UV and Visible Light

Jorge González-Rodríguez [1,*], Lucía Fernández [1], Yanina B. Bava [2], David Buceta [2], Carlos Vázquez-Vázquez [2], Manuel Arturo López-Quintela [2], Gumersindo Feijoo [1] and Maria Teresa Moreira [1]

[1] CRETUS Institute, Department of Chemical Engineering, Universidade de Santiago de Compostela, 15782 Santiago de Compostela, Spain; luciaf.fernandez@usc.es (L.F.); gumersindo.feijoo@usc.es (G.F.); maite.moreira@usc.es (M.T.M.)

[2] Laboratory of Magnetism and Nanotechnology, Department of Physical Chemistry, Faculty of Chemistry, Universidade de Santiago de Compostela, 15782 Santiago de Compostela, Spain; ybava@quimica.unlp.edu.ar (Y.B.B.); david.buceta@usc.es (D.B.); carlos.vazquez.vazquez@usc.es (C.V.-V.); malopez.quintela@usc.es (M.A.L.-Q.)

* Correspondence: jorgegonzalez.rodriguez@usc.es; Tel.: +34-8818-16771

Received: 18 November 2019; Accepted: 23 December 2019; Published: 26 December 2019

Abstract: Emerging contaminants (ECs) represent a wide range of compounds, whose complete elimination from wastewaters by conventional methods is not always guaranteed, posing human and environmental risks. Advanced oxidation processes (AOPs), based on the generation of highly oxidizing species, lead to the degradation of these ECs. In this context, TiO_2 and ZnO are the most widely used inorganic photocatalysts, mainly due to their low cost and wide availability. The addition of small amounts of nanoclusters may imply enhanced light absorption and an attenuation effect on the recombination rate of electron/hole pairs, resulting in improved photocatalytic activity. In this work, we propose the use of silver nanoclusters deposited on ZnO nanoparticles (ZnO–Ag), with a view to evaluating their catalytic activity under both ultraviolet A (UVA) and visible light, in order to reduce energetic requirements in prospective applications on a larger scale. The catalysts were produced and then characterized by scanning electron microscopy (SEM), X-ray diffractometry (XRD) and inductively coupled plasma-optical emission spectrometry (ICP-OES). As proof of concept of the capacity of photocatalysts doped with nanoclusters, experiments were carried out to remove the azo dye Orange II (OII). The results demonstrated the high photocatalytic efficiency achieved thanks to the incorporation of nanoclusters, especially evident in the experiments performed under white light.

Keywords: AOPs; zinc oxide; nanoclusters; photocatalysis; UVA; visible light

---

1. Introduction

Environmental awareness has identified water scarcity as a problem of increasing magnitude in many areas due to its decisive and essential role in life. The increase in world population, changes in consumption patterns, the high demand for water in intensive irrigation agriculture or the frequent events of floods and droughts leads to the depletion of many water resources and the unequal distribution of water in different regions of the planet [1]. In this context, the development of novel analytical methods for the analysis of water and the improvement of existing ones reveal the presence of ECs in both drinking water and wastewater effluents. Some of these compounds may be toxic to terrestrial and aquatic organisms at low concentrations [2]. These compounds represent a scientific-technological challenge, since the existing plants have not been designed for their elimination.

On the other hand, the regulation foreseen for the elimination of this type of contaminants implies that the new facilities must face their efficient removal [3,4]. For example, the release of wastewater with pharmaceutical products and bacteria caused the increased resistance to amoxicillin/clavulanic acid in *Salmonella enterica* strains from 1% to 7–16% within 15 years (2003–2018). Furthermore, other compounds such as endocrine disrupting chemicals (EDCs) can be released into the environment, stored in the organisms due to their recalcitrant properties and stability, and accumulated in certain organs producing long-term effects [5]. In view of the above, it is necessary to ensure treatment systems that are capable of eliminating ECs in the different water matrices due to the adverse effects on human health and ecosystem [6].

Thus, advanced oxidation processes (AOPs) represent an alternative to conventional methods to remove these contaminants. AOPs are physicochemical processes that involve the generation of reactive oxygen species (ROS), which are effective against oxidation of organic matter because they have high oxidation potentials capable of reacting and degrading a wide range of contaminants [7]. In recent years, the use of heterogeneous photocatalysts has been intensively studied for their wide application for environmental protection, with special attention to wastewater treatment [8]. Some authors have studied the degradation of the ECs using different types of catalyst based on metallic oxides such as ZnO or TiO_2 for the degradation of pharmaceuticals and personal care products (PPCPs) [9], pesticides [10] or industrial contaminants [7]. However, photocatalysis may be limited by costly energy requirements associated with the use of ultraviolet (UV) lamps, which also have limitations related to low quantum efficiency. Thus, photoactive materials are being developed, whose catalytic activity takes place in the optical window of visible light [11].

When semiconductor materials are used in photocatalysis, the photocatalyst is irradiated with light (hν) of equal or greater energy than its characteristic band gap, so that the electrons (e^-) of the valence band (VB) are promoted to the conduction band (CB), thereby generating electron/hole (e^-/h^+) pairs. Other mechanistic aspects are based on two broad types of simultaneously occurring photochemical reactions on the surface of the catalyst. The first involves reduction, from the photo-induced negative electrons at the CB to the dissolved O_2 present in the medium, producing superoxide ions ($O_2^{-\bullet}$), which can form hydroxyl radicals ($OH^\bullet$) and hydrogen peroxide (H_2O_2) in acidic medium. H_2O_2 may also decompose to $OH^\bullet$ under irradiation. The second involves oxidation, from the photoinduced positive holes at the VB, which react with H_2O or hydroxyl ions (OH^-) to produce $OH^\bullet$. The active oxygen species $O_2^{\bullet-}$ and $OH^\bullet$, and h^+ react with organic molecules, triggering their consequent degradation (Figure 1) [12].

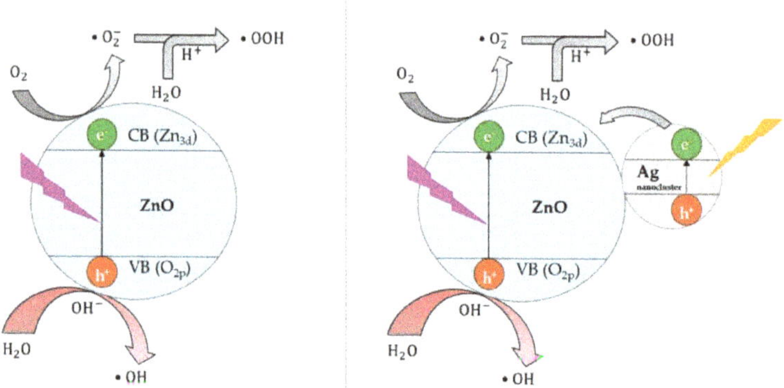

Figure 1. General mechanism of semiconductor photocatalysis (**left**) and modification of energy levels with the incorporation of silver nanoclusters (**right**).

Among the different photocatalysts, ZnO has received considerable attention due to its exceptional optoelectronic properties, strong oxidation capability, abundance and physicochemical stability [8]. In this context and with a view to their application in wastewater treatment, the photocatalytic efficiency of ZnO nanoparticles (NPs) has been evaluated in the degradation of pharmaceutical and personal care products (PPCPs) [13] or dyes [14].

However, the photocatalytic efficiency of ZnO NPs is often limited by an inefficient absorption of visible light and a rapid recombination rate of e^-/h^+ pairs. To avoid these disadvantages, the incorporation of metallic and non-metallic elements inside the crystal lattice of semiconductors reduces the recombination of electron-hole pairs. These compounds act by creating new energy levels between the CB and the VB that act as electron traps. Examples of doping with metallic silver or copper on ZnO or TiO_2 were reported to improve photocatalytic efficiencies [7,15–19].

On the other hand, these drawbacks can be circumvented also by depositing noble metals on the surface of the catalyst, which trap electrons of the CB of the semiconductor, thus reducing the possibilities of e^-/h^+ recombination and increasing the photocatalytic activity (Figure 1) [7]. In the search for novel materials capable of degrading organic pollutants in water under sunlight, ZnO NPs have been synthesized and functionalized with silver nanoclusters by a simple and green deposition method in water, conducted at ambient conditions.

Nanoclusters of metal elements are particles with low numbers of atoms, from 2 to ≤100 atoms, with sizes below ≈1.5–2 nm, and properties dramatically different from what would be expected from the scaling laws that govern the behavior of bulk and metal nanoparticles [20]. Nanoclusters of metal elements show the presence of discrete energy states and a sizable HOMO-LUMO bandgap, similar to the conduction band–valence band in semiconductors and lose the metallic behavior. This bandgap can be tuned by changing the number of atoms, the type of metal and the supporting material, and they can be used for different catalytic applications (heterogeneous catalysis, photocatalysis and electrocatalysis) [20–23].

In this article, the crystallinity, optical properties and morphology of the nanostructures obtained, ZnO–Ag, have been evaluated. Finally, the photocatalytic activity of ZnO–Ag nanocomposites with different Ag loadings has been studied in the removal of the dye Orange II (OII), used as model compound of organic pollutant, under UVA and white light.

2. Results and Discussion

2.1. Characterization of Catalysts

A sample containing silver nanoclusters of ≤10 atoms was used for the deposition onto the ZnO NPs. These small nanoclusters show planar geometries, as it can be shown by atomic force microscopy (Figure S1), confirming the presence of nanoclusters of ≤10 atoms [24].

Different Ag loads were applied on the surface of the ZnO NPs, so that four types of ZnO–Ag NCs with an Ag content of 1.3, 2.9, 3.2 and 7.4% (*w/w*) were obtained. The samples were structurally characterized by X-ray diffraction patterns. The two crystalline phases present in the samples were metallic silver (Ag, JCPDS PDF-2 card number 04-0783, peaks highlighted with red down-pointing triangles in Figure 2) and zincite (ZnO, JCPDS PDF-2 card number 36-1451, peaks highlighted with black up-pointing triangles in Figure 2) with hexagonal wurtzite structure.

No additional peaks were observed in the patterns, revealing the absence of impurity phases in the catalyst. Furthermore, there was no significant shift of the diffraction peaks, proving that silver atoms did not substitute any Zn sites in the lattice but were deposited onto the surface of ZnO. Figure 3 shows the morphology of the ZnO NPs (left) and ZnO–Ag NCs (right) observed by FE-SEM. ZnO NPs are present in the form of spherical aggregates of different sizes, between 50 and 500 nm. These aggregates are composed of smaller ZnO NPs (10–15 nm). In the case of the ZnO–Ag NCs, the presence of separate Ag nanoparticles along with the spherical aggregates is shown in Figure 3 (right).

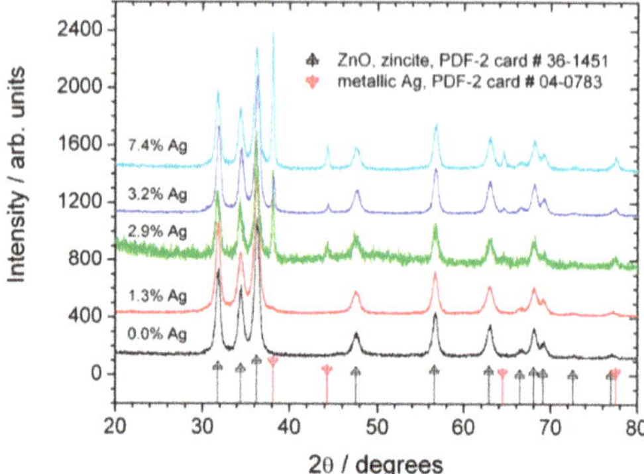

Figure 2. X-ray diffraction patterns of ZnO NPs and hybrid ZnO–Ag NCs. The main reflections from zincite (ZnO, JCPDS PDF-2 card number 36-1451) and metallic silver (Ag, JCPDS PDF-2 card number 04-0783) are included as drop lines.

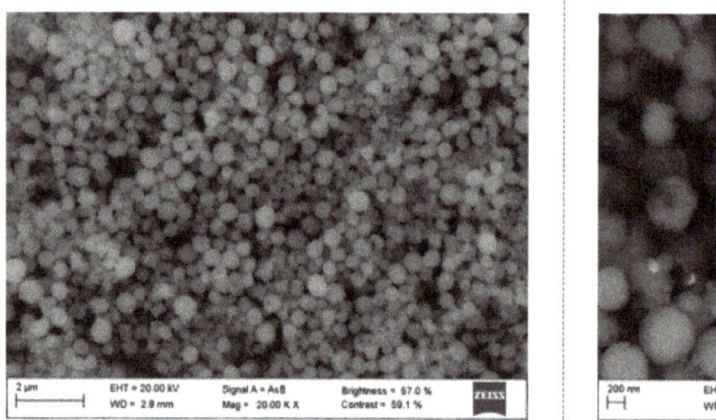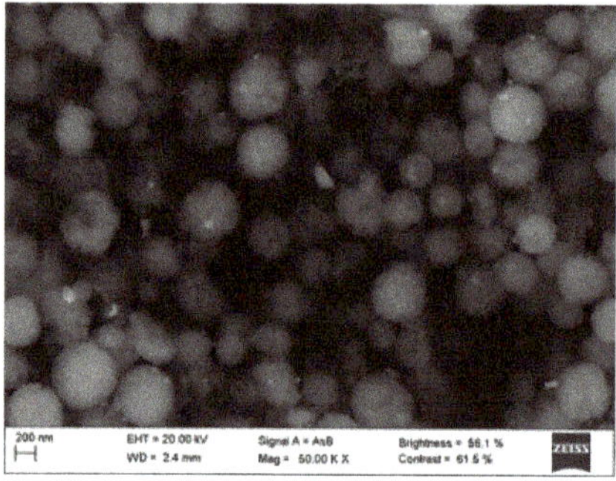

Figure 3. Scanning electron micrograph of the spherical ZnO aggregates (**left**); Detailed scanning electron micrograph of the ZnO–Ag nanocomposites, showing the presence of small Ag nanoparticles as brighter spots (**right**).

2.2. Influence of Ag Nanoclusters Loading onto Photocatalytic Activity of ZnO

Experiments were performed using fixed concentrations of a photocatalyst (200 mg L^{-1}) and OII (10 mg L^{-1}) in 10 mL of aqueous solutions, under UVA light for 60 min or white light for 180 min (Figure 4). In the absence of photocatalysts, photolysis controls resulted in OII degradation of 9% under UVA and 5% under white light, while adsorption studies of samples kept in dark conditions showed no OII removal (data not shown). When comparing the decolorization results, using ZnO nanoparticles as photocatalyst, no significant improvement in dye removal was observed, with maximum percentages of 16% and 9% under UVA and white light, respectively. As a general rule for all the experiments, OII degradation exhibited accelerated kinetic rates under UVA irradiation, which is attributed to a strong light absorption of these wavelengths by ZnO, while ZnO absorption in the visible region is weaker. It can be noted that the photocatalytic performance of the ZnO NPs improved with the addition of Ag nanoclusters, obtaining 97% and 49% of OII removal in the presence of ZnO–Ag with 1.3% (*w/w*) using UVA or white light, respectively. Photocatalytic performance gradually decreased with increasing Ag loads in the ZnO NPs. Therefore, there is evidence of the existence of an optimal silver loading to enhance the photocatalytic activity of the NC. This can be explained by the specific surface of ZnO available to interact with incident light, being lower with increasing concentrations of Ag nanoclusters in the NCs [25]. In fact, the decoration of the ZnO NPs with Ag nanoclusters leads to a color modification, from white to brownish and grey, due to the formation of Ag nanoparticles, which improves the absorption of the ZnO–Ag NCs in the visible region (Figure 4).

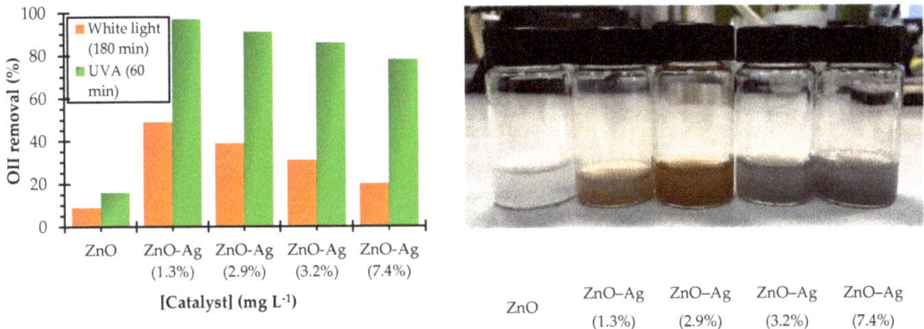

Figure 4. Silver loading effect on photocatalytic performance. The values in brackets correspond to the percentage of Ag in each NC (**left**); Aqueous suspensions of ZnO-NPs and ZnO–Ag NCs with different silver loadings (**right**).

However, according to previous findings, above the optimum Ag loading effectively deposited onto the NC, this can lead to an enhancement of the e^-/h^+ recombination rate, acting itself as a recombination center, thus contributing to a decrease in photocatalytic efficiency [26].

2.3. Influence of Photocatalyst Concentration on OII Removal

From the previous results, ZnO–Ag with 1.3% (*w/w*) of Ag was selected as the most effective photocatalyst. The effect of ZnO–Ag (1.3%) concentration was evaluated in 10 mL of aqueous samples containing an initial OII concentration ($C_{OII,i}$) of 10 mg L^{-1}, which were subjected to white light with 200–1000 mg L^{-1} of ZnO NPs or ZnO–Ag (1.3%) NCs, and to UVA light with 50–500 mg L^{-1} of photocatalyst (Figure 5). After 3 h of white light irradiation, samples containing from 200 to 750 mg L^{-1} of ZnO–Ag (1.3%) showed increasing photocatalytic rates, obtaining 49 to 78% of OII elimination, respectively. This can be explained by the enhancement of the active sites in the catalyst by increasing their concentration in the samples.

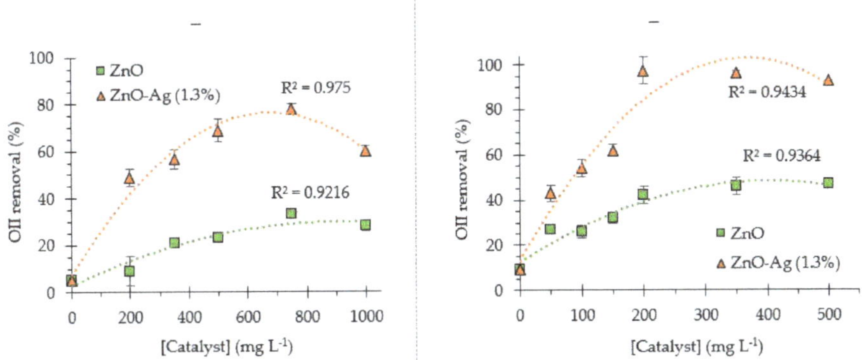

Figure 5. Influence of photocatalyst concentration under white light irradiation for 180 min (**left**) and UVA light irradiation for 60 min (**right**). Error bars were calculated considering a normal distribution, for $p < 0.01$ [obtained from kinetic data].

After 5 h of white light irradiation of the sample containing ZnO–Ag (1.3%) NCs at 750 mg L^{-1}, it was found that the OII concentration was below the detection limit of the spectrophotometric method, which was evidenced by the complete color removal of the sample. However, by increasing the concentration of the photocatalyst to a value of 1000 mg L^{-1} of ZnO–Ag (1.3%), the extent of decolorization was inferior to that of the maximum: 60% OII removal. This phenomenon was observed by other authors for different photocatalysts [27,28] and is probably due to the effects of the NP aggregation and the reduction of the available surface area for photon absorption. A similar trend, but with slower reaction kinetics, was observed in samples containing ZnO NPs. As expected, the tests carried out under UVA light showed full removal of OII after 1 h, using ZnO–Ag (1.3%) NCs at 200 mg L^{-1}. These data were adjusted following a second order equation with an adequate fitting of data, achieving R^2 up to 0.92.

Accordingly, the negative natural logarithm of the ratio between OII concentration and its initial concentration, ln (C$_{OII,0}$/C$_{OII,t}$), was plotted as a function of the irradiation time and a linear regression was obtained. Correlation coefficients (R^2), half-lives (t$_{1/2}$) and apparent pseudo-first order rate constants (k) are presented in Table 1. R^2 ranges from 0.9323 to 0.9990, confirming the suitability of the pseudo-first order model to describe the kinetics of OII removal in the presence of ZnO–Ag (1.3%) and ZnO NPs, also applied by other authors to model the photocatalytic degradation of dyes and emerging contaminants. [16,27,29–31].

Table 1. Kinetic parameters for the photodegradation of OII with ZnO–Ag (1.3%) and ZnO NPs.

Irradiation Source	[Catalyst] (mg L^{-1})	ZnO–Ag (1.3%)			ZnO		
		R^2	k (h^1)	t$_{1/2}$ (h)	R^2	k (h^{-1})	t$_{1/2}$ (h)
White light	200	0.9861	0.235 ± 0.006	2.96	0.9425	0.036 ± 0.002	19.25
	350	0.9862	0.290 ± 0.007	2.41	0.9728	0.083 ± 0.003	8.25
	500	0.9930	0.390 ± 0.007	1.78	0.9847	0.089 ± 0.002	7.70
	750	0.9624	0.455 ± 0.017	1.52	0.9748	0.141 ± 0.005	4.81
	1000	0.9984	0.308 ± 0.003	2.27	0.9635	0.121 ± 0.004	5.78
UVA light	50	0.9859	0.541 ± 0.018	1.28	0.9323	0.346 ± 0.023	1.99
	100	0.9854	0.758 ± 0.026	0.92	0.9923	0.311 ± 0.008	2.22
	150	0.9823	1.006 ± 0.037	0.69	0.9988	0.388 ± 0.004	1.78
	200	0.9722	3.436 ± 0.176	0.20	0.9990	0.552 ± 0.005	1.26
	350	0.9778	3.019 ± 0.137	0.23	0.9951	0.633 [a] ± 0.012	1.10
	500	0.9946	2.650 ± 0.055	0.26	0.9925	0.642 [a] ± 0.016	1.08

[a] Not significantly different ($p < 0.05$).

Other works reported the photocatalytic degradation of OII using modified ZnO catalyst to improve kinetics. Chen et al. [27] obtained a kinetic constant of 0.033 h^{-1} using micro-structured ZnO, which amounted to 0.065 h^{-1} for ZnO decorated with Ag, both values lower than those obtained in these experiments, and they also used a larger catalyst concentration: 1500 mg L^{-1}. The enhancement of the results using nanostructured ZnO could take place because nanostructured catalysts have more surface/volume ratio than the micro-sized material, increasing the number of active sites per mass unit. Moreover, Siuleiman et al. [30] have structured ZnO in nanowires, obtaining kinetic constants of 0.092 and 0.112 h^{-1} for UVA and visible light irradiation with a catalyst concentration of 500 mg L^{-1}, respectively.

In view of the above, the incorporation of the Ag nanoclusters causes an improvement of the kinetic constants of 3–6 times, both under UVA and white light radiation. In addition, the improvement in degradation rates by comparing the same catalyst concentrations under white and UVA light is 7–10 times greater. For all cases, the most notable differences occur for a catalyst concentration in the range of 200–500 mg L^{-1}. These ratios are similar to those obtained by Sornalingam et al. [31] using Au-TiO$_2$ NCs with UVA and cold white light. Reuse tests cannot be carried out due to the losses of catalyst at the recovery stage.

3. Materials and Methods

3.1. Materials

All chemicals used in this work were reagent-grade and were used without further purification. Diethylene glycol ((HOCH$_2$CH$_2$)$_2$O, DEG, 99%) was supplied by Alfa Aesar (Thermo Fisher, Kandel, Germany); Orange II (C$_{16}$H$_{10}$N$_2$Na$_2$O$_7$S$_2$, >85%), zinc(II) acetate dihydrate (Zn(CH$_3$CO$_2$)$_2$·2H$_2$O, >98%), silver nitrate (AgNO$_3$, >99%) and absolute ethanol (CH$_3$CH$_2$OH, >99.8%) were supplied by Sigma-Aldrich (St. Louis, MI, USA). Silver nanoclusters (Ag-AQCs DS0481) were provided by NANOGAP SUB-NM POWDER, S.A (ZIP code 15895 O Milladoiro, A Coruña, Spain). This sample contains a mixture of Ag nanoclusters with ≤10 silver atoms per nanocluster (100 mg L^{-1}) and Ag(I) ions (400 mg L^{-1}).

3.2. Synthesis of Nanostructured Photocatalysts

3.2.1. ZnO Nanoparticles

The synthesis of ZnO NPs is based on the preparation of polyol-mediated ZnO [32]. In particular, 100 mL of 90 mM Zn(II) acetate solution in DEG were placed in a round-bottom flask and heated to 180 °C for 2 h under mechanical agitation. The obtained NPs were centrifuged at 7500 rpm for 15 min. Then, ZnO NPs were washed four times with ethanol. Finally, ZnO NPs were redispersed in water at a concentration ca. 0.83% (*w/w*) (determined by thermogravimetric analysis).

3.2.2. ZnO–Ag Nanocomposites

10 mL of the Ag nanoclusters stock solution (100 mg L^{-1} in water) was placed in a 20 mL glass vial and pH was adjusted to 5 with NH$_4$OH (28–30% *w/w*). To obtain nanocomposites (NCs) with different silver loadings, a given volume (1.1–4.1 mL) of the previous prepared stock solution of ZnO NPs (8.3 g L^{-1}) was added. The reaction mixture was incubated in an orbital shaker for 15 min (220 rpm, 24 °C). Then, the NC was centrifuged (7000 rpm, 25 min), the supernatant was removed, and the solids were re-dispersed in 20 mL of water. The sample was again centrifuged and the solids re-dispersed in 20 mL of water. Duplicate samples were prepared and after the first centrifugation step, the dispersion was subjected to a photochemical treatment at 254 nm for 15 min in order to reduce the residual Ag(I) ions present in the dispersion. The samples were then centrifuged and the solids re-dispersed in 20 mL of water. Blank samples of ZnO NPs without Ag nanoclusters were prepared following the same procedure but using 10 mL of an AgNO$_3$ solution (400 mg L^{-1}) to show the different behavior of clusters and Ag ions/nanoparticles formed in the blank.

3.3. Characterization of the ZnO–Ag Nanocomposites

The study of the crystalline phases was carried out by X-ray diffraction (XRD) in powder samples with a Philips PW1710 diffractometer (Cu Kα radiation source, λ = 1.54186 Å). Measurements were collected between 20° < 2θ < 80°, with steps of 0.020° and time per step of 5 s. The concentration of aqueous stocks of ZnO NPs was obtained by thermogravimetric analysis (TGA). The thermogravimetric curves were recorded with a Perkin Elmer TGA 7 thermobalance, operating under N$_2$ atmosphere, from room temperature to 850 °C, at a scanning rate of 10 °C min^{-1}.

Field-emission scanning electron micrographs were taken with a ZEISS FE-SEM ULTRA Plus microscope using the angle selective backscatter electron detector (AsB detector). The final concentrations of Zn and Ag were determined by inductively coupled plasma-optical emission spectrometry (ICP-OES) using a Perkin Elmer Model Optima 3300 DV spectrometer, equipped with an AS91 autosampler.

Atomic force microscopy (AFM) measurements were conducted under normal ambient conditions using an XE-100 instrument (Park Systems, Suwon, Korea) in non-contact mode. The AFM tips were aluminum-coated silicon ACTA from Park Systems with a resonance frequency of 325 kHz. For AFM imaging, a drop of the Ag nanoclusters diluted sample was deposited onto a freshly cleaved mica sheet (Grade V-1 Muscovite) (Park Systems, Suwon, Korea), which was thoroughly washed with Milli-Q water and dried under nitrogen flow.

3.4. Photocatalytic Degradation of Orange II under UVA and White Light

The photocatalytic activities of ZnO and ZnO–Ag were evaluated by exposure of samples in glass beakers to UVA (365 nm wavelength UVP pen Ray model 11SC-1L) or white irradiation (fluorescent lamp PL G23 11 W 6400 K, 400–730 nm). Photocatalytic tests were performed on 10 mL of aqueous samples containing 50–1000 mg L^{-1} of photocatalyst and 10 mg L^{-1} of OII, at pH 7 and room temperature. The fluorescent lamp is located externally on one side, at approximately 3 cm from the vial, and the UVA lamp is in the center of the samples, using a submerged quartz tube. Two types of control samples were performed in parallel: direct photolysis control samples of OII under the same irradiation conditions; and adsorption control samples containing photocatalysis and OII, under the same sample preparation but kept in darkness. The solutions were stirred for 30 min in dark to achieve adsorption equilibrium. At regular intervals, spectrophotometric measurements were performed to monitor OII concentration in a BioTek PowerWave XS2 micro-plate spectrophotometer (Winooski, VT, USA). The photodegradation yield (%) was determined using the following equation:

$$\text{Yield (\%)} = (C_{OII,I} - C_{OII,t})/C_{OII,0} \times 100\% \tag{1}$$

3.5. Determination of Kinetic Parameters

The determination of kinetic parameters was performed by adjusting a pseudo-first order kinetic model (Equation (2)) to each set of photocatalyst concentration used in the UVA and white light studies. The linearization of this equation (Equation (3)) and the expression used to calculate the half-life (Equation (4)) are shown below:

$$C_{OII,t} = C_{OII,i}\, e^{-kt}, \tag{2}$$

$$\ln (C_{OII,i}/C_{OII,t}) = kt, \tag{3}$$

$$t_{1/2} = \ln(2)/k, \tag{4}$$

being k the kinetic constant; t the time of the experiment and $t_{1/2}$ the half-life of the compound under study.

4. Conclusions

ZnO nanoparticles were prepared by a simple polyol-mediated method and successfully decorated with Ag nanoclusters, obtaining a novel nanocomposite (ZnO–Ag) with different degrees of silver loadings (1.3–7.4% *w/w*). In addition, the final dispersions of nanoparticles received a photochemical treatment to remove the residual Ag, avoiding the interferences in the subsequent photodegradation step. The influence of the Ag content on ZnO regarding the removal of Orange II was studied, obtaining that the presence of this noble metal at 1.3% greatly enhanced the photocatalytic activity, which suggests the potential of this nanocomposite to be applied in prospective applications in the field of water treatment, both in drinking and wastewater treatment plants. Semiconductor photocatalysis represents a promising alternative to conventional technologies since the use of chemicals would be avoided and solar energy could be used as photon source. In addition, the unspecific oxidation mechanisms in AOPs allow degradation and mineralization of a wide range of pollutants.

In this work, photocatalytic studies were performed under UVA and white light, obtaining the optimum concentrations of catalyst and nanoclusters that achieved removal percentages up to 75% for visible light after 3 h and nearly complete removal for UVA after 1 h. Further research is needed to fully explore this photocatalysis in practical applications. One of the main drawbacks of this catalyst is its separation from the water matrix after treatment. Its immobilization on a suitable support that avoids additional steps such as centrifugation, which is the method used so far, will allow the reuse of the photocatalyst in different water treatment cycles, bringing this research closer to a real wastewater treatment plant. An alternative for this immobilization is the deposition of the NPs onto magnetic nanoparticles, which can be easily separated by applying a magnetic field. Moreover, immobilization over supports such as silica, zeolites or alumina would improve the recovery of the catalysts towards their industrial applications.

Supplementary Materials: The following are available online at http://www.mdpi.com/2073-4344/10/1/31/s1. Figure S1: AFM topography image and line profiles of small Ag nanoclusters deposited on mica.

Author Contributions: Conceptualization and supervision, M.T.M., G.F., M.A.L.-Q. and C.V.-V.; investigation, J.G.-R.; resources, Y.B.B., D.B.; writing–original draft, L.F.; writing–review & editing, J.G.-R., M.T.M., C.V.-V. All authors have read and agreed to the published version of the manuscript.

Funding: This research was supported by two projects granted by Spanish Ministry of Science, Innovation and Universities: MODENA Project CTQ2016-79461-R and CLUSTERCAT Project MAT2015-67458-P, and Fundación Ramón Areces, Spain (Project CIVP18A3940).

Acknowledgments: J.G.-R. thanks the Xunta de Galicia Counseling of Education, Universities and Vocational Training for his predoctoral fellowship. M.T.M., G.F. and J.G.-R. belong to CRETUS Institute. The authors also thank Xunta de Galicia for the CRETUS (AGRUP2015/02) and AEMAT (ED431E-2018/08) Strategic Partnerships, and the use of RIAIDT-USC analytical facilities. The authors belong to the Galician Competitive Research Groups ED431C-2017/22 and ED431C-2017/29, co-funded by FEDER.

Conflicts of Interest: The authors declare no conflict of interest.

References

1. Mekonnen, M.M.; Hoekstra, A.Y. Sustainability: Four billion people facing severe water scarcity. *Sci. Adv.* **2016**, *2*, 1–6. [CrossRef] [PubMed]

2. Mirzaei, A.; Chen, Z.; Haghighat, F.; Yerushalmi, L. Removal of pharmaceuticals from water by homo/heterogonous Fenton-type processes—A review. *Chemosphere* **2017**, *174*, 665–688. [CrossRef]

3. Ahuja, S. *Water Reclamation and Sustainability*, 1st ed.; Elsevier: San Diego, CA, USA, 2014.

4. Rodriguez-Narvaez, O.M.; Peralta-Hernandez, J.M.; Goonetilleke, A.; Bandala, E.R. Treatment technologies for emerging contaminants in water: A review. *Chem. Eng. J.* **2017**, *323*, 361–380. [CrossRef]

5. Martín-Pozo, L.; de Alarcón-Gómez, B.; Rodríguez-Gómez, R.; García-Córcoles, M.T.; Çipa, M.; Zafra-Gómez, A. Analytical methods for the determination of emerging contaminants in sewage sludge samples. A review. *Talanta* **2019**, *192*, 508–533. [CrossRef] [PubMed]

6. Hernández, F.; Bakker, J.; Bijlsma, L.; de Boer, J.; Botero-Coy, A.M.; Bruinen de Bruin, Y.; Fischer, S.; Hollender, J.; Kasprzyk-Hordern, B.; Lamoree, M.; et al. The role of analytical chemistry in exposure science: Focus on the aquatic environment. *Chemosphere* **2019**, *222*, 564–583. [CrossRef]

7. Lee, K.M.; Lai, C.W.; Ngai, K.S.; Juan, J.C. Recent developments of zinc oxide based photocatalyst in water treatment technology: A review. *Water Res.* **2016**, *88*, 428–448. [CrossRef]

8. Zhou, C.; Lai, C.; Huang, D.; Zeng, G.; Zhang, C.; Cheng, M.; Hu, L.; Wan, J.; Xiong, W.; Wen, M.; et al. Highly porous carbon nitride by supramolecular preassembly of monomers for photocatalytic removal of sulfamethazine under visible light driven. *Appl. Catal. B Environ.* **2018**, *220*, 202–210. [CrossRef]

9. Bohdziewicz, J.; Kudlek, E.; Dudziak, M. Influence of the catalyst type (TiO_2 and ZnO) on the photocatalytic oxidation of pharmaceuticals in the aquatic environment. *Desalin. Water Treat.* **2016**, *57*, 1552–1563. [CrossRef]

10. Kosera, V.S.; Cruz, T.M.; Chaves, E.S.; Tiburtius, E.R.L. Triclosan degradation by heterogeneous photocatalysis using ZnO immobilized in biopolymer as catalyst. *J. Photochem. Photobiol. A Chem.* **2017**, *344*, 184–191. [CrossRef]

11. Fernández, L.; Borzecka, W.; Lin, Z.; Schneider, R.J.; Huvaere, K.; Esteves, V.I.; Cunha, Â.; Tomé, J.P.C. Nanomagnet-photosensitizer hybrid materials for the degradation of 17β-estradiol in batch and flow modes. *Dye Pigment* **2017**, *142*, 535–543. [CrossRef]

12. Andreozzi, R.; Caprio, V.; Insola, A.; Marotta, R. Advanced oxidation processes (AOP) for water purification and recovery. *Catal. Today* **1999**, *53*, 51–59. [CrossRef]

13. El-Kemary, M.; El-Shamy, H.; El-Mehasseb, I. Photocatalytic degradation of ciprofloxacin drug in water using ZnO nanoparticles. *J. Lumin.* **2010**, *130*, 2327–2331. [CrossRef]

14. Ameen, S.; Shaheer Akhtar, M.; Shik Shin, H. Speedy photocatalytic degradation of bromophenol dye over ZnO nanoflowers. *Mater. Lett.* **2017**, *209*, 150–154. [CrossRef]

15. Papari, G.P.; Silvestri, B.; Vitiello, G.; De Stefano, L.; Rea, I.; Luciani, G.; Aronne, A.; Andreone, A. Morphological, Structural, and Charge Transfer Properties of F-Doped ZnO: A Spectroscopic Investigation. *J. Phys. Chem. C* **2017**, *121*, 16012–16020. [CrossRef]

16. Han, F.; Kambala, V.S.R.; Dharmarajan, R.; Liu, Y.; Naidu, R. Photocatalytic degradation of azo dye acid orange 7 using different light sources over Fe^{3+}-doped TiO_2 nanocatalysts. *Environ. Technol. Innov.* **2018**, *12*, 27–42. [CrossRef]

17. Ziashahabi, A.; Prato, M.; Dang, Z.; Poursalehi, R.; Naseri, N. The effect of silver oxidation on the photocatalytic activity of Ag/ZnO hybrid plasmonic/metal-oxide nanostructures under visible light and in the dark. *Sci. Rep.* **2019**, *9*, 1–12. [CrossRef]

18. Vitiello, G.; Clarizia, L.; Abdelraheem, W.; Esposito, S.; Bonelli, B.; Ditaranto, N.; Vergara, A.; Nadagouda, M.; Dionysiou, D.D.; Andreozzi, R.; et al. Near UV-Irradiation of CuOx-Impregnated TiO_2 Providing Active Species for H_2 Production Through Methanol Photoreforming. *ChemCatChem* **2019**, *11*, 4314–4326. [CrossRef]

19. Clarizia, L.; Vitiello, G.; Pallotti, D.K.; Silvestri, B.; Nadagouda, M.; Lettieri, S.; Luciani, G.; Andreozzi, R.; Maddalena, P.; Marotta, R. Effect of surface properties of copper-modified commercial titanium dioxide photocatalysts on hydrogen production through photoreforming of alcohols. *Int. J. Hydrogen Energy* **2017**, *42*, 28349–28362. [CrossRef]

20. Buceta, D.; Piñeiro, Y.; Vázquez-Vázquez, C.; Rivas, J.; López-Quintela, M.A. Metallic clusters: Theoretical background, properties and synthesis in microemulsions. *Catalysts* **2014**, *4*, 356–374. [CrossRef]

21. Attia, Y.A.; Buceta, D.; Requejo, F.G.; Giovanetti, L.J.; López-Quintela, M.A. Photostability of gold nanoparticles with different shapes: The role of Ag clusters. *Nanoscale* **2015**, *7*, 11273–11279. [CrossRef]

22. Vilar-Vidal, N.; Rey, J.R.; Quintela, M.A.L. Green emitter copper clusters as highly effi cient and reusable visible degradation photocatalysts. *Small* **2014**, *10*, 3632–3636. [CrossRef] [PubMed]

23. Attia, Y.A.; Buceta, D.; Blanco-Varela, C.; Mohamed, M.B.; Barone, G.; López-Quintela, M.A. Structure-Directing and High-Efficiency Photocatalytic Hydrogen Production by Ag Clusters. *J. Am. Chem. Soc.* **2014**, *136*, 1182–1185. [CrossRef] [PubMed]

24. Lee, H.M.; Ge, M.; Sahu, B.R.; Tarakeshwar, P.; Kim, K.S. Geometrical and Electronic Structures of Gold, Silver, and Gold–Silver Binary Clusters: Origins of Ductility of Gold and Gold–Silver Alloy Formation. *J. Phys. Chem. B* **2003**, *107*, 9994–10005. [CrossRef]

25. Wang, J.; Fan, X.M.; Zhou, Z.W.; Tian, K. Preparation of Ag nanoparticles coated tetrapod-like ZnO whisker photocatalysts using photoreduction. *Mater. Sci. Eng. B* **2011**, *176*, 978–983. [CrossRef]

26. Vaiano, V.; Matarangolo, M.; Murcia, J.J.J.; Rojas, H.; Navío, J.A.A.; Hidalgo, M.C.C. Enhanced photocatalytic removal of phenol from aqueous solutions using ZnO modified with Ag. *Appl. Catal. B Environ.* **2018**, *225*, 197–206. [CrossRef]

27. Chen, P.K.; Lee, G.J.; Anandan, S.; Wu, J.J. Synthesis of ZnO and Au tethered ZnO pyramid-like microflower for photocatalytic degradation of orange II. *Mater. Sci. Eng. B* **2012**, *177*, 190–196. [CrossRef]

28. Lin, J.C.T.; Sopajaree, K.; Jitjanesuwan, T.; Lu, M.C. Application of visible light on copper-doped titanium dioxide catalyzing degradation of chlorophenols. *Sep. Purif. Technol.* **2018**, *191*, 233–243. [CrossRef]

29. Butler, E.B.; Chen, C.C.; Hung, Y.T.; Al Ahmad, M.S.; Fu, Y.P. Effect of Fe-doped TiO_2 photocatalysts on the degradation of acid orange 7. *Integr. Ferroelectr.* **2016**, *168*, 1–9. [CrossRef]

30. Siuleiman, S.; Kaneva, N.; Bojinova, A.; Papazova, K.; Apostolov, A.; Dimitrov, D. Photodegradation of Orange II by ZnO and TiO_2 powders and nanowire ZnO and ZnO/TiO_2 thin films. *Colloids Surf. A Physicochem. Eng. Asp.* **2014**, *460*, 408–413. [CrossRef]

31. Sornalingam, K.; McDonagh, A.; Zhou, J.L.; Johir, M.A.H.; Ahmed, M.B. Photocatalysis of estrone in water and wastewater: Comparison between Au-TiO_2 nanocomposite and TiO_2, and degradation by-products. *Sci. Total Environ.* **2018**, *610–611*, 521–530. [CrossRef]

32. Feldmann, C.; Jungk, H.O. Polyol-Mediated Preparation of Nanoscale Oxide Particles. *Angew. Chem. Int. Ed.* **2001**, *40*, 359–362. [CrossRef]

 *catalysts*

Article

Niobium Oxide Catalysts as Emerging Material for Textile Wastewater Reuse: Photocatalytic Decolorization of Azo Dyes

Alexsandro Jhones dos Santos [1], Luana Márcia Bezerra Batista [2], Carlos Alberto Martínez-Huitle [1,*], Ana Paula de Melo Alves [3,*] and Sergi Garcia-Segura [4,*]

[1] Laboratório de Eletroquímica Ambiental e Aplicada (LEAA), Institute of Chemistry, Federal University of Rio Grande do Norte, Lagoa Nova, 59078-970 Natal, Brazil; alexsandrojhones@hotmail.com

[2] Instituto Federal de Pernambuco, Campus Afogados da Ingazeira, Rua Edson Barbosa de Araújo, s/n, Bairro Manoela Valadares, 56800-000 Afogados da Ingazeira-PE, Brazil; luanamarciaufrn@gmail.com

[3] Laboratório de Combustíveis e Materiais (LACOM), Department of Chemistry, Federal University of Paraíba, Campus I—Lot. Cidade Universitária, 58051-900 João Pessoa, Brazil

[4] Nanosystems Engineering Research Center for Nanotechnology-Enabled Water Treatment, School of Sustainable Engineering and the Built Environment, Arizona State University, Tempe, AZ 85287-3005, USA

* Correspondence: carlosmh@quimica.ufrn.br (C.A.M.-H.); anachemistry@pq.cnpq.br (A.P.d.M.A.); sergio.garcia.segura@asu.edu (S.G.-S.)

Received: 2 December 2019; Accepted: 12 December 2019; Published: 14 December 2019

Abstract: Niobium-based metal oxides are emerging semiconductor materials with barely explored properties for photocatalytic wastewater remediation. Brazil possesses the greatest reserves of niobium worldwide, being a natural resource that is barely exploited. Environmental applications of solar active niobium photocatalysts can provide opportunities in the developing areas of Northeast Brazil, which receives over 22 MJ m^2 of natural sunlight irradiation annually. The application of photocatalytic treatment could incentivize water reuse practices in small and mid-sized textile businesses in the region. This work reports the facile synthesis of Nb_2O_5 catalysts and explores their performance for the treatment of colored azo dye effluents. The high photoactivity of this alternative photocatalyst makes it possible to quickly obtain complete decolorization, in less than 40 min of treatment. The optimal operational conditions are defined as 1.0 g L^{-1} Nb_2O_5 loading in slurry, 0.2 M of H_2O_2, pH 5.0 to treat up to 15 mg L^{-1} of methyl orange solution. To evaluate reutilization without photocatalytic activity loss, the Nb_2O_5 was recovered after the experience and reused, showing the same decolorization rate after several cycles. Therefore, Nb_2O_5 appears to be a promising photocatalytic material with potential applicability in wastewater treatment due to its innocuous character and high stability.

Keywords: advanced oxidation processes; azo dye; sustainable resources; niobium; water reuse; water treatment

---

1. Introduction

New trends in photocatalysis aim to identify niche applications for emerging semiconductor materials within a sustainable context. Titanium dioxide is the most studied photocatalyst and photoelectrocatalyst for water treatment [1,2] and water splitting applications [3,4]. However, recent research efforts have explored alternative semiconductor materials to overcome the most stringent barrier for the implementation of TiO_2 in developing countries: its low activation under visible light irradiation [5,6]. Enabling the use of the natural sunlight irradiation would reduce operational expenditure, providing the opportunity for technology adaptation by regions with high solar radiation according to solar maps. For example, the northeastern region of Brazil receives >22 MJ m^2 of sunlight irradiation annually due to its proximity to the equatorial line, which corresponds to approximately 10 h of sunlight every day [7].

Niobium-based oxide semiconductors have promising characteristics for environmental applications due to their hypoallergenic character, low cytotoxicity, and physiological and chemical inertness, along with their high thermodynamic stability [8,9]. However, the most remarkable aspect for their application in Brazil is their wide availability in the country, which is the major producer of niobium worldwide, with 99.0% of the world's niobium being located in Brazil [10]. Sustainable exploitation of this resource for environmental applications may positively affect socio-economic aspects, since few applications of niobium have been identified.

Brazil is the fourth biggest cotton textile exporter worldwide and the fifth biggest global manufacturer. Textile fiber dyeing process requires large volumes of water, resulting in the release of large amounts of colored wastewater effluent [11,12]. Green and sustainable manufacturing approaches require the minimization of the usage of water resources. Removal of organic dyes can incentivize water reuse in dyeing baths [13]. Photocatalytic treatment could be a suitable low-cost alternative for small/mid-sized Brazilian textile facilities [14,15]. Photocatalytic treatment is classified as an Advanced Oxidation Process, since it allows the in situ generation of highly oxidant species such as hydroxyl radicals ($^\bullet$OH) [16,17]. The light irradiation of a semiconductor with photons with a higher energy than its band gap enables the photo-excitation of an electron from the filled valence band of the semiconductor to the empty conduction band (e_{cb}^-), leaving a positively charged vacancy or hole (h_{vb}^+) following Reaction (1) [18,19]. Organic pollutants (i.e., dyes) can then be oxidized by the photogenerated hole, as well as by heterogeneous $^\bullet$OH formed on the photocatalyst surface from the water oxidation by the h_{vb}^+ according to Reaction (2) [20,21].

$$\text{Semiconductor} + h\nu \rightarrow h_{vb}^+ + e_{cb}^- \tag{1}$$

$$h_{vb}^+ + H_2O \rightarrow {^\bullet}OH + H^+ \tag{2}$$

The recombination of e_{cb}^- with unreacted h_{vb}^+ through Reaction (3) is responsible for the oxidative power loss in photocatalytic systems [22,23]. Dissolved oxygen can react with e_{cb}^-, yielding superoxide radicals ($O_2^{\bullet-}$) through Reaction (4), which contributes to slowing down the recombination reaction [24,25]. However, the use of e_{cb}^- scavengers can further minimize the extent of this undesirable reaction while enhancing performance. For example, the weak oxidant H_2O_2 can react with e_{cb}^- and $O_2^{\bullet-}$, producing additional $^\bullet$OH from Reactions (5) and (6), respectively [26].

$$e_{cb}^- + h_{vb}^+ \rightarrow \text{heat} \tag{3}$$

$$e_{cb}^- + O_2 \rightarrow O_2^{\bullet-} \tag{4}$$

$$e_{cb}^- + H_2O_2 \rightarrow {^\bullet}OH + OH^- \tag{5}$$

$$O_2^{\bullet-} + H_2O_2 \rightarrow {^\bullet}OH + OH^- + O_2 \tag{6}$$

Dye bath effluent decolorization enables water reuse for additional textile dyeing processes, minimizing the environmental footprint and the capital costs associated with water usage. This work studies the applicability of novel niobium oxide photocatalysts (Nb_2O_5) in the efficient decolorization of wastewaters containing azo-dyes. The synthesized materials were characterized in accordance with calcination methodologies. Decolorization capabilities were evaluated, and operational variables were optimized. This innovative alternative for water treatment processes, which is emerging as a new trend in photocatalytic technologies, meets sustainable technology manufacturing needs through the use of regionally abundant natural resources.

2. Results and Discussions

2.1. Characterization of Nb₂O₅ Photocatalyst

The synthesized Nb_2O_5 photocatalyst powder was characterized by SEM. The micrograph in Figure 1a exhibits a uniform particle size distribution of ca. 200 nm. The magnified catalyst surface depicted in Figure 1b exhibits a high roughness and porosity, which could be induced by the calcination process of the niobium oxalate complex ammonium salt. The mapping of the chemical composition of the catalyst particles (Figure 1c) indicated that they are composed of niobium with a homogenous distribution.

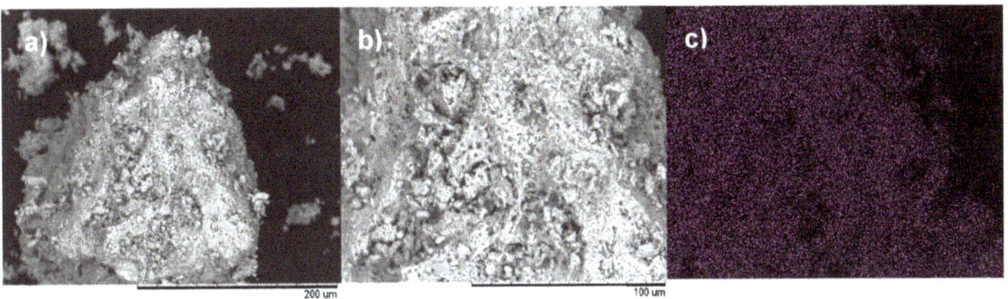

Figure 1. Scanning electron micrographs of Nb_2O_5-synthesized photocatalyst with magnifications of (**a**) 500× and (**b**) 1000×, and (**c**) niobium mapping micrograph.

The FTIR spectrum of Nb_2O_5 in Figure 2 depicts the characteristic bands of niobium oxides: one shoulder at 900 cm^{-1} is attributed to the stretch Nb-O and the bands at 661 and 600 cm^{-1} are attributed to the angular vibration Nb-O-Nb [27]. Additionally, a band can be observed at 1622 cm^{-1} that is related to the water adsorbed on the surface of Nb_2O_5 [8,28], and a small band can be observed at 1531 cm^{-1} that is usually associated with impurities from the precursor salt of niobium. These impurities in the precursor are considered to be beneficial in the literature, since they act as stabilizers of the niobium oxide catalyst [29,30].

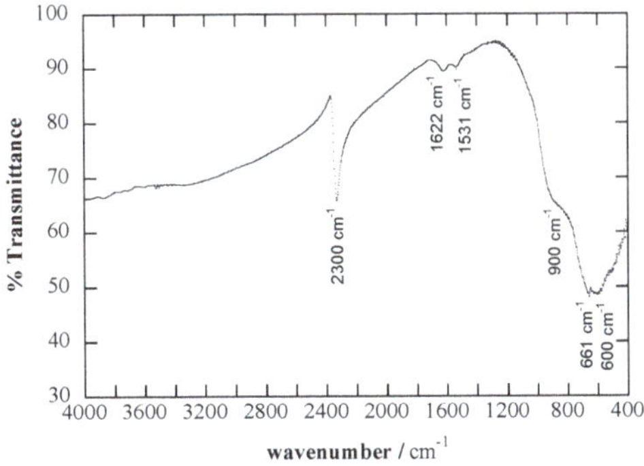

Figure 2. FTIR spectra of synthesized Nb_2O_5.

Niobium oxide has a complex crystal morphology, wherein at least 12 crystallographic structures have been identified to date [31]. The X-ray diffraction pattern obtained from the synthesized Nb_2O_5 powder catalyst is shown in Figure 3a. It can be seen that the diffractogram presents reflections $2\theta = 22.6°$, $28.5°$, $36.7°$, $46.3°$, $50.4°$, and $55.3°$, which correspond to the crystallographic planes of miller index (001), (100), (101), (002), (110), and (102), respectively. These planes are characteristic of the pseudohexagonal structure of niobium oxide [27,32], proving the obtention of a catalyst with a defined crystalline microstructure with a crystallite size of 2.2 nm estimated from the Scherrer formula. The electronic spectroscopy of diffuse reflectance made it possible to determine the band-gap energy required for the photo-promotion of one electron (see Figure 3b). The results indicated that Nb_2O_5 presents a band gap of 3.1 eV, similar to the characteristic band gap of titanium dioxide photocatalysts [9,33], and at the low end of the Nb_2O_5 bandgap values [34].

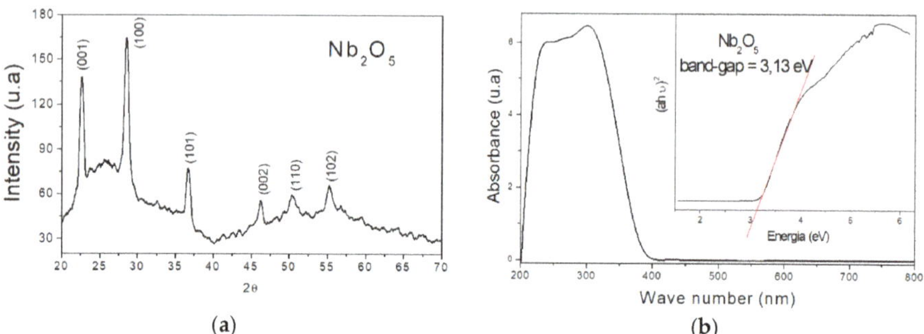

Figure 3. (**a**) X-ray diffractogram of Nb_2O_5. (**b**) UV-vis DRS absorption spectra of synthesized Nb_2O_5. The inset panel shows the Tauc plot for the band-gap energy determination of 3.13 eV.

The isotherm of nitrogen adsorption-desorption of Nb_2O_5 in Figure 4a presents a hysteresis loop characteristic of type IV isotherms, which is typical of mesoporous materials. This is in agreement with the characteristic morphologies observed by SEM in Figure 1. The specific surface area of 42.36 m^2 g^{-1} was calculated from the BET method (S_{BET}). Meanwhile, the BJH analysis (Figure 4b) revealed the presence of uniformly sized mesopores with an average diameter (d_p) of 10.1 nm and a mean pore volume (V_p) equal to 0.102 cm^3 g^{-1}. These results are in agreement with those reported by Wang et al. [32] for commercial niobium oxide calcined at a temperature of 500 °C (S_{BET} = 49.9 m^2 g^{-1}, V_p = 0.12 cm^3 g^{-1} e d_p = 9.6 nm), even when the niobium oxide precursor selected was different.

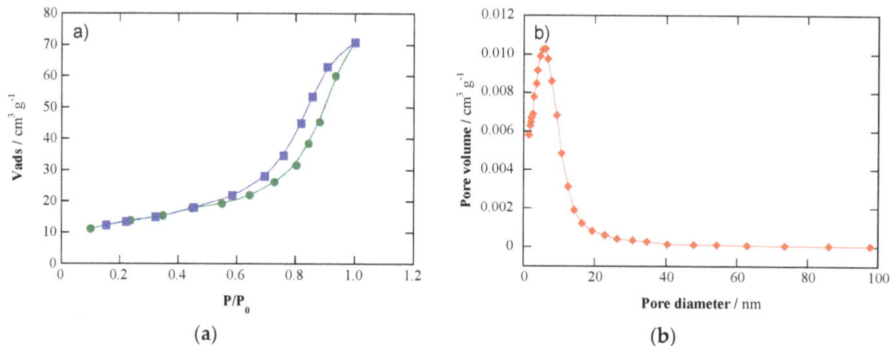

Figure 4. (**a**) Nb_2O_5 photocatalyst nitrogen (●) adsorption–(■) desorption isotherms. (**b**) Pore size distribution plot that depicts the characteristic response of mesoporous materials.

2.2. Photocatalytic Activity on Azo Dye Decolorization

The photocatalytic activity of the synthesized Nb_2O_5 was verified on the basis of the corresponding decolorization of solutions containing 5 mg L^{-1} methyl orange (MO) as model azo dye (see Table 1). MO is highly photostable, and is not degraded under direct sunlight irradiation, as depicted in Figure 5 [12]. Meanwhile, ca. 6.0% decolorization after 100 min was observed when 1.0 g L^{-1} of Nb_2O_5 catalyst suspended in solution was exposed to sunlight irradiation. Photocatalytic degradation can be assumed, since only a discrete 0.6% removal was observed under dark conditions due to adsorption on the porous Nb_2O_5. The slight photocatalytic removal under solar irradiation could be justified by the faster recombination Reaction (3), which diminishes the available oxidants (i.e., h_{vb}^+ and $^\bullet OH$) [17,22]. The strategic use of selective e_{cb}^- scavengers may synergistically enhance the photocatalytic response by inhibiting the extent of Reaction (3) [16,26]. As shown in Figure 5, the addition of H_2O_2 boosts the performance of Nb_2O_5, which attains complete decolorization after 40 min. It is important to remark that the sole addition of H_2O_2 under dark conditions had no effect on dye solution decolorization, because of the weak oxidative capacity of H_2O_2 ($E°(H_2O_2/H_2O) = 1.76$ V/SHE). Nevertheless, decolorization was observed in the presence of H_2O_2 under direct solar irradiation due to the photolytic decomposition of H_2O_2 according to Reaction (7). Please note that this reaction only occurs under UVC irradiation, which is a component of the solar light in Northeast Brazil [24,35].

$$H_2O_2 + h\nu \rightarrow 2\,^\bullet OH \tag{7}$$

Table 1. Chemical structure and characteristics of Methyl Orange azo dye.

Property	Characteristics
IUPAC name	Sodium 4-{[4-(dimethylamino) phenyl] diazenyl} benzene-1-sulfonate
Common name	Methyl Orange
CAS number	547-58-0
Color Index number	13,025
M/g mol^{-1}	327.3
λ_{max}/nm	464
pK_a	3.45
Chemical formula	$C_{14}H_{14}N_3SO_3Na$
Chemical structure	

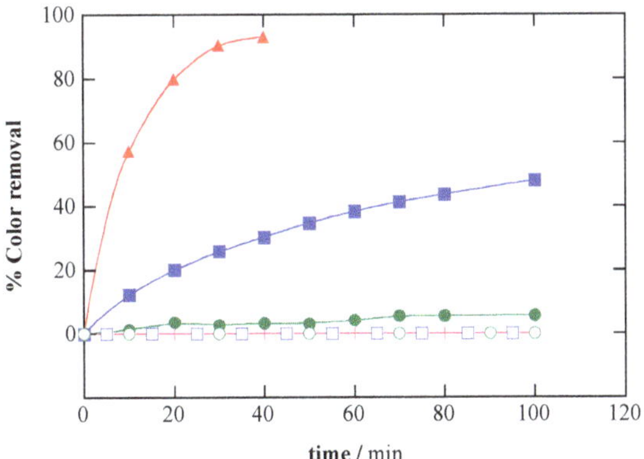

Figure 5. Evaluation of the photocatalytic degradation of 100 mL of 5 mg L^{-1} of MO at pH 5.0 with ($\bigcirc$, $\bullet$) 1.0 g L^{-1} of Nb_2O_5, ($\square$, $\blacksquare$) 0.20 M H_2O_2, ($\triangle$, $\blacktriangle$) Nb_2O_5 g L^{-1} with 0.20 M H_2O_2, ($\bigcirc$, $\square$, $\triangle$) in dark conditions or ($\bullet$, $\blacksquare$, $\blacktriangle$) under sunlight irradiation. The photostability of the dye solution was also tested (+).

The synergetic effect observed between Nb_2O_5 and e_{cb}^- scavenger H_2O_2 can be explained by the enhanced generation of $^\bullet OH$ from Reaction (2) due to the significant reduction of the recombination rate of Reaction (3) [26,36]. This effect results in a consequent increase in the availability of reactive oxygen species on the Nb_2O_5 photocatalyst surface, and then the increased oxidation capabilities of the system degrading the azo dye molecule [17,23].

The synergetic effect of H_2O_2 evidences its role in Nb_2O_5 photocatalytic efficiency performance. For this reason, the role of H_2O_2 concentration in solution was studied as a rate driving parameter for decolorization. Figure 6 shows the percentage of color removal attained for increasing doses of H_2O_2 acting as e_{cb}^- scavenger. Higher color removal percentages of 6.0%, 71.5% and 91.0% were observed for increasing concentrations of H_2O_2 of 0 M, 0.10 M and 0.20 M, respectively. It should be noted that further increase in H_2O_2 concentration resulted in a decrease in performance, achieving a lower color removal of only 77.2% after 80 min of Nb_2O_5 photocatalytic treatment. This phenomenon can be explained by the acceleration of concomitant waste reactions [17,37]. One of the main processes that decreases performance is the oxidation of excess H_2O_2 by $^\bullet OH$ following Reaction (8) [24,26]. This undesired side-reaction not only consumes $^\bullet OH$, which will subsequently not be able to oxidize the target azo dye, but it also diminishes the available amount of H_2O_2. Moreover, the excessive accumulation of radical species can promote their dimerization following Reactions (9) and (10) [12,38]. These undesired reactions do not contribute to the overall photocatalytic performance, and may slow down the decolorization kinetics, as was observed experimentally (see Figure 6). Therefore, an optimal dose of 0.2 M H_2O_2 was defined for the following experiments to ensure faster solution decolorization and higher photocatalytic efficiency.

$$H_2O_2 + {}^\bullet OH \rightarrow HO_2{}^\bullet + H_2O \tag{8}$$

$$HO_2{}^\bullet + {}^\bullet OH \rightarrow H_2O + O_2 \tag{9}$$

$$2\,{}^\bullet OH \rightarrow H_2O_2 \tag{10}$$

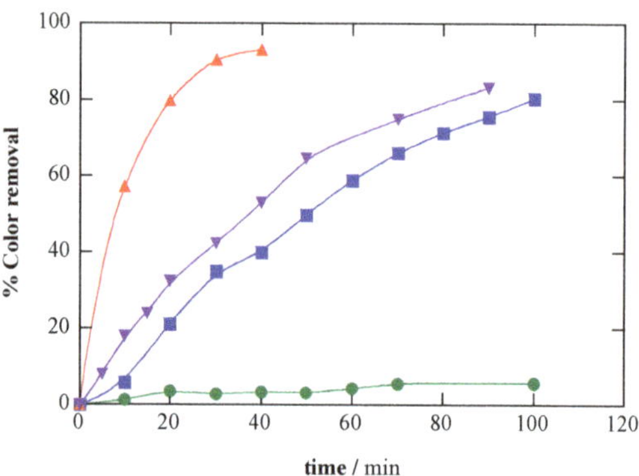

Figure 6. Impact of the e_{cb}^- scavenger dose on the solar photocatalytic decolorization of 100 mL of 5 mg L^{-1} of MO with 1.0 g L^{-1} of Nb_2O_5 at pH 5.0. Initial H_2O_2 concentration: (●) 0 M, (■) 0.10 M, (▲) 0.20 M, and (▼) 0.30 M.

2.3. Effect of pH on the Photocatalytic Decolorization of Azo Dye Methyl Orange

The pH of the treated solution is one of the variables with the greatest influence on the photocatalytic degradation of pollutants, because it affects several physical-chemical properties of the catalysts that enhance or reduce the degradation efficiency, including the catalyst surface charge and the organic adsorptivity [17,23,39]. The point of zero charge (PZC) $pH_{PZC} = 4.86$ of Nb_2O_5 was determined by the pH drift method [24,40], as described in the methodology section. The pH_{PZC} is an intrinsic characteristic of the catalyst that makes it possible to determine whether the surface of Nb_2O_5 is negatively or positively charged as a function of the pH. If the working pH conditions are above of the pH_{PZC} of the Nb_2O_5, the catalyst surface is negatively charged. Conversely, the surface will be positively charged when the pH is below the pH_{PZC}.

Figure 7 shows the effect of the initial pH on the decolorization efficiency of MO using 1.0 g L^{-1} of Nb_2O_5 photocatalyst and 0.20 M of H_2O_2. It is important to note that in the range of pH under consideration, the sulfonic group of azo dye MO is deprotonated, with the pollutant molecule being negatively charged (see Table 1). However, at pH = 3.0, one of the molecule's N is protonated, and the global charge of the molecule will be neutral (pK_a = 3.45). This consideration is an important fact that could justify the lower decolorization achieved at alkaline pH. Above the pH_{PZC}, the Nb_2O_5 surface is negatively charged, and the dye molecule will also be negatively charged, thus leading to electrostatic charge repulsion, making the adsorption processes more difficult, along with the approach of the molecule towards the photocatalyst surface. Thus, with increasing values of pH, the number of negatively charged sites on the Nb_2O_5 also increases, further reducing the photodecolorization process in the following sequence 7.0 > 9.0 > 11.0, with percentages of color removal of 39.3%, 16.0% and 7.5%, respectively. In addition, it is well known that the rate of decomposition of unstable H_2O_2 increases with increasing pH in accordance with Equations (11) or (12) in strongly alkaline media [41]. This loss of H_2O_2 by chemical decomposition affects the overall photocatalytic performance. First, it reduces the availability of e_{cb}^- scavenger in solution. Second, it consequently reduces the generation of $^\bullet OH$ radicals. In contrast, at pH 3.0, the surface is positively charged. This stimulates the molecules' approach to and absorption onto the surface through the negatively charged sulfonic group. Thus, below pH 3.0, complete color removal can be achieved at lower treatment times of 10 min.

$$2\,H_2O_2 \rightarrow 2\,H_2O + O_2 \tag{11}$$

$$HO_2^- \rightarrow OH^- + \frac{1}{2}\,O_2 \tag{12}$$

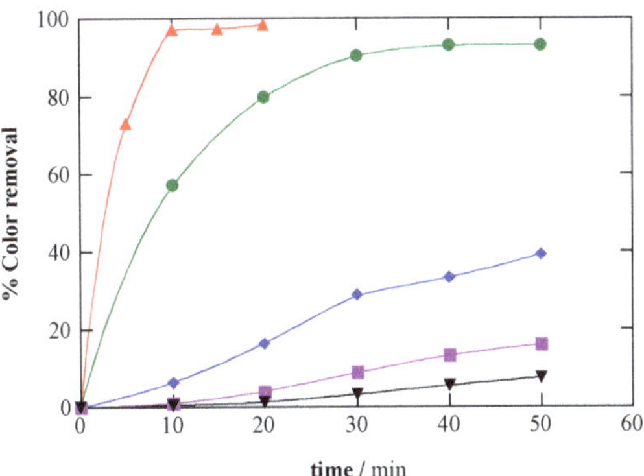

Figure 7. Influence of the initial pH on the percentage of color removal during solar photocatalytic treatment of 100 mL of 5 mg L^{-1} of MO with 1.0 g L^{-1} of Nb_2O_5 0.2 M of H_2O_2 at pH: (▲) 3.0, (●) 5.0, (♦) 7.0, (■) 9.0 and (▼) 11.0.

The nearest working pH to pH_{PZC} = 4.86 was 5.0. Under these conditions, the catalyst surface is practically neutral and has no electro-attractive or electro-repulsive effects that affect the adsorption processes. Thus, as observed in Figure 7, it presents faster decolorization kinetics than higher, more alkaline pH, but slightly slower kinetics than those obtained for pH 3.0. The pH 5.0 was considered the optimum condition, because it is the natural pH of MO solutions and the nearest to the circumneutral pH of water effluents. This would decrease the potential environmental health and safety risks, as well as the costs to small/mid-sized industry, by minimizing the handling and storage of acids and bases. Furthermore, this may have an impact on operational costs, since it would obviate acidification for the treatment and neutralization steps prior to the release of treated effluent.

2.4. Evaluation of Optimal Dosage of Nb_2O_5 to Ensure Maximum Performance

Optimizing the dosage of catalysts is a critical engineering parameter when designing sustainable reactors for solar photocatalytic applications at large scale [16,17]. The impact of Nb_2O_5 dose on decolorization kinetics was evaluated within the range from 0.25 g L^{-1} up to 2.00 g L^{-1} by treating solutions of 5 mg L^{-1} of MO azo dye at pH 5.0 in the presence of 0.20 M of H_2O_2. Figure 8 reports an enhancement in the decolorization rate with increasing dosage of Nb_2O_5. These improved performances can be explained due to the increasing number of active sites resulting from the higher total specific surface of Nb_2O_5 available, thereby accelerating the photocatalytic generation of oxidants in solution by Reactions (1) and (2). The effective reduction of photocatalytic efficiency at excessively high catalyst dosages is commonly reported in the literature, and is attributed to the detrimental effects on light transport in solution caused by: (i) the increase of the solution opacity, which diminishes the radiation penetration and consequently the photogeneration of vacancies [9,40]; (ii) the aggregation of suspended catalyst particles diminishing the specific area [27,36]; and (iii) light scattering effects that also reduce the UV light penetration [22,42]. In addition, the surface available for adsorption processes is also increased, favoring the mechanisms of organic degradation. However, in our experiments, no appreciable loss of efficiency was observed, although the decolorization rate increase became lower from 1 g L^{-1}, resulting in a plateau. The kinetic analysis of MO color removal enabled the estimation of pseudo-first-order rate constants of decolorization (k_{dec}). These analyses showed excellent fits for a pseudo-first-order reaction, assuming that the $^{\bullet}OH$ radicals achieve a pseudo-constant concentration on the Nb_2O_5 surface. Increasing k_{dec} of $3.18 \cdot 10^{-4}$ s^{-1} ($R^2 = 0.996$) with 0.25 g L^{-1} of Nb_2O_5, $6.12 \cdot 10^{-4}$ s^{-1} ($R^2 = 0.997$) with 0.50 g L^{-1} of Nb_2O_5, $1.00 \cdot 10^{-3}$ s^{-1} ($R^2 = 0.997$) with 1.00 g L^{-1} of Nb_2O_5 and $1.07 \cdot 10^{-3}$ s^{-1} ($R^2 = 0.998$) with 0.25 g L^{-1} of Nb_2O_5 were observed. It is important to remark that the slight increase in decolorization rate from 1.0 to 2.0 g L^{-1} is presumably related to the loss of photocatalytic efficiency at excessively high catalyst dosages. Thus, a catalyst dosage of 1.0 g L^{-1} was identified as the optimal condition, because it is the lowest dosage at which a faster decolorization rate can be achieved.

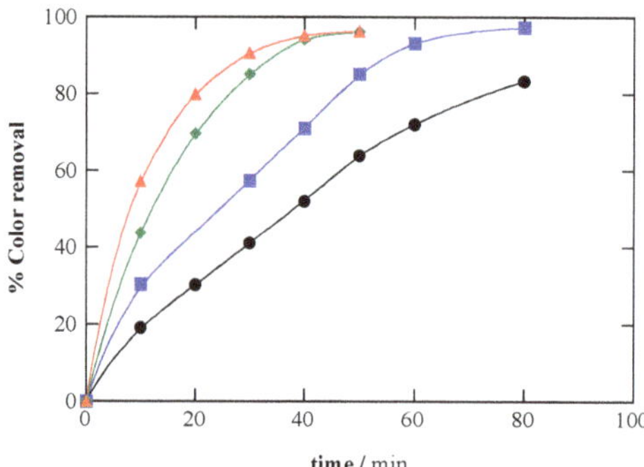

Figure 8. Influence of Nb_2O_5 photocatalyst dose on the photocatalytic performance. Dosing: (●) 0.25 g L^{-1}, (■) 0.5 g L^{-1}, (♦) 1.0 g L^{-1}, and (▲) 2.0 g L^{-1}.

2.5. Effect of Initial Dye Concentration on Nb_2O_5 Performance

The initial dye concentration is a variable of interest that gives invaluable information about the range of pollutant concentration that is efficiently treatable within a reasonable timeframe. For this reason, MO solutions of 5, 10, 15 and 30 mg L^{-1} were treated under the optimum conditions of pH 5.0 with 0.20 M of H_2O_2 and 1.0 g L^{-1} of Nb_2O_5 photocatalyst. Figure 9 depicts the abatement of dye concentration as a function of photocatalytic treatment time for different initial concentrations of MO. The estimation of k_{dec} exhibits variation over an order of magnitude when increasing the dye concentration. Decreasing k_{dec} values from $1.07 \cdot 10^{-3}$ s^{-1} ($R^2 = 0.998$) for 5.0 mg L^{-1} of MO, $4.37 \cdot 10^{4}$ s^{-1} ($R^2 = 0.999$) for 10.0 mg L^{-1} of MO, $2.17 \cdot 10^{-4}$ s^{-1} ($R^2 = 0.996$) for 15.0 mg L^{-1} of MO, down to $1.01 \cdot 10^{-4}$ s^{-1} ($R^2 = 0.993$) for 30.0 mg L^{-1} of MO were estimated as depicted in the inset panel of Figure 9. The greater azo dye concentration implies a greater colorization of the water being treated, thus limiting the light penetration into the solution and diminishing photocatalytic Reactions (1) and (2), which generate oxidant species. Furthermore, the greater accumulation of organic intermediates susceptible adsorption onto the Nb$_2$O$_5$ active sites could inhibit the photocatalytic generation of vacancies and the production of other oxidants, decreasing the organic events and, consequently, the decolorization rate [9,23,43]. Therefore, longer treatment times would be required to completely decolorize MO at higher pollutant loads.

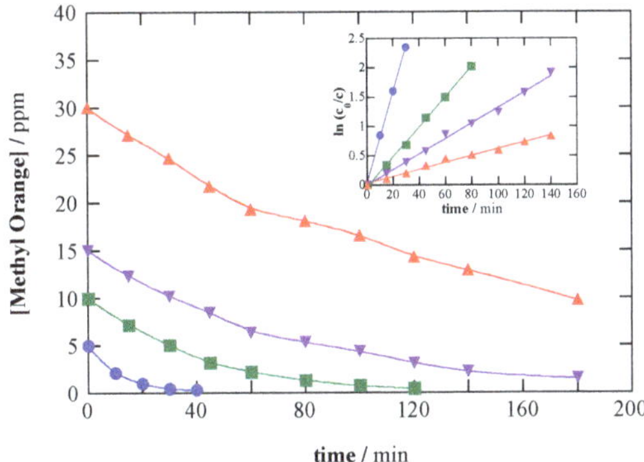

Figure 9. Methyl Orange azo dye abatement vs photoelectrocatalytic treatment time of 100 mL of solution with 1.0 g L^{-1} of Nb$_2$O$_5$, 0.20 M of H$_2$O$_2$ at pH 5.0 and dye concentration of: (●) 5 mg L^{-1}, (■) 10 mg L^{-1}, (▼) 15 mg L^{-1} and (▲) 30 mg L^{-1}. The corresponding kinetic analysis assuming a pseudo-first-order reaction for MO is given in the inset panel.

2.6. Evaluating Reutilization Capabilities of Nb$_2$O$_5$ Catalyst Aiming for Sustainable Implementation and Comparison with Other Catalysts' Performance

Sustainable processes must consider the continuous reutilization of photocatalytic materials within catalytic converters [44]. Reutilization capabilities and stabilities of novel materials should be assessed in order to demonstrate material stability. Therefore, Nb$_2$O$_5$ was submitted to continuous treatment operation. Suspended Nb$_2$O$_5$ photocatalyst was recovered and tested over consecutive decolorization cycles of MO solutions. As depicted in Figure 10, the niobium metal oxide semiconductor presented excellent stability, and retained its catalytic properties over 10 cycles. Please note that consecutive cycles demonstrated reproducible MO removal performance. It is important to remark that previous reports in the literature reported higher stability of niobium oxides than for conventional TiO$_2$ or ZnO catalysts [9,27,28]. In this context, the results reported here support these previous studies, indicating Nb$_2$O$_5$ as an emerging photocatalyst for environmental remediation. However, more studies related to catalyst aging and fouling during continuous operation are required to ensure the lifetime of these emerging catalysts.

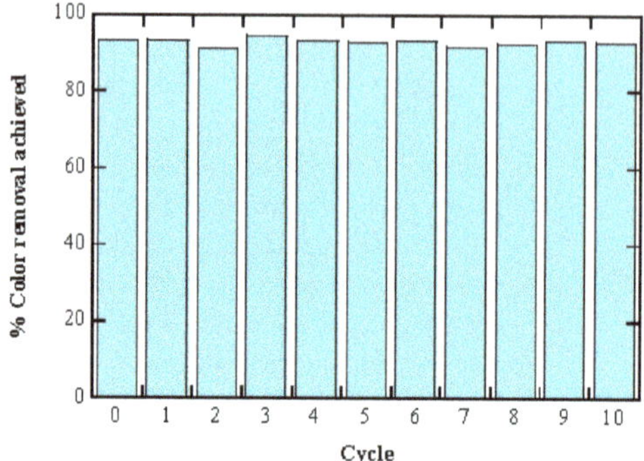

Figure 10. Percentage of color removal achieved after 40 min of solar photoelectrocatalytic treatment of 5 mg L^{-1} of MO with 1.0 g L^{-1} of catalyst Nb_2O_5 in slurry at pH 5.0 with 0.20 M of H_2O_2 after consecutive reuse of Nb_2O_5.

A quick comparison with other photocatalytic materials reported in the literature demonstrates the promising performance of novel niobium oxide semiconductors for water treatment applications. As summarized in Table 2, the results reported in this work are highly competitive, since Nb_2O_5 makes it possible to reduce operational times by more than half when compared with doped TiO_2 and other complex mixed oxides.

Table 2. Comparative performance of visible photocatalytic decolorization of Methyl Orange azo dye with different catalysts to attain over 95% color removal.

Photocatalysts	[Methyl Orange]/mg L^{-1}	Decolorization Time/Min	References
Pd-doped TiO_2	20	150	[45]
Cu-doped TiO_2	10	140	[46]
Cu-doped ZnO	13	120	[47]
Ag/TiO_2	20	120	[48]
Bi_3TiNbO_9	10	75	[49]
Nb_2O_5	10	65	This work

3. Materials and Methods

3.1. Chemicals

MO azo dye of 85.0% purity and the H_2O_2 of 33% (*w/w*) were supplied by Sigma-Aldrich. The ammonium salt of the niobium oxalate complex ($NH_4[NbO(C_2O_4)_2(H_2O)]$. $3H_2O$) of 99.0% purity used in the photocatalyst synthesis was purchased from CBMM. The pH was adjusted prior to experiments using H_2SO_4 or NaOH of analytical grade, supplied by Sigma-Aldrich. All solutions were prepared with high-purity water obtained from a Millipore Milli-Q system with resistivity >18 MΩ cm at 25 °C.

3.2. Synthesis of Microparticulated Nb_2O_5 Photocatalyst

The ammonium salt of the niobium oxalate complex was calcined using a temperature ramp of 10 °C min^{-1} to 500 °C, where it remained for 4 h under ambient atmosphere. Under these conditions, the oxalate was completely incinerated, leading to the formation of amorphous Nb_2O_5 photocatalyst.

3.3. Solar Photocatalytic Experiences

The photochemical reactor used during photocatalytic experiments consisted of an undivided open cell directly exposed to sunlight. The photocatalyst was suspended in the solution in slurry and the tests were carried out under vigorous stirring with a magnetic bar at 700 rpm to ensure the homogeneous distribution of the catalyst in the bulk, while also favoring the transport of reactants to/from the catalyst surface. The photochemical cell had a double jacket in which water was circulated to maintain the solution temperature at 25 °C using a LAUDA A100 thermostat, avoiding the evaporation of the treated solution by solar irradiation heating. Prior to the photocatalytic experiments, the solutions were maintained at the defined pH, dye and catalyst concentration conditions for 30 min in the dark.

3.4. Apparatus and Analytical Procedures

Scanning electron microscopy (SEM) micrographies of the synthesized Nb_2O_5 photocatalyst were obtained using a Hitachi TM-3000 system with a frequency of 50/60 Hz and a magnification capacity of up to 3000x. The X-ray diffractogram (XRD) patterns were recorded on a Bruker D2 Phaser diffractometer with Cu-K$_\alpha$ (λ = 1.54 Å) irradiation source using a Ni filter and a Lynxeye detector, performing the analysis in the 2θ range from 2° to 70°. The average crystallite sizes were estimated by applying Scherrer's Formula (13) to the identified crystal planes [50,51]:

$$\tau = K \, \lambda/\beta \, \cos\theta \tag{13}$$

where τ refers to the mean size of the ordered crystalline domains, K is the Scherrer constant, a dimensionless shape factor that usually has values close to unity, λ is the X-ray wavelength, β is the line width at half maximum in radians and θ is the Bragg angle in degrees.

Fourier transform infrared (FTIR) spectra were recorded from a Shimadzu-8400S FTIR spectrometer in the medium IR region (4000–400 cm^{-1}) after 30 scans with a resolution of 4 cm^{-1}. The UV-vis diffuse reflectance spectra between 200–600 nm, used to determine the band gap from solid samples on the basis of Kubelka-Munk and Tauc plots [52], was obtained with a UV-vis spectrometer Cary 500 Scan using the barium sulfate pattern as a reference material. The specific surface area was determined on the basis of nitrogen adsorption-desorption isotherms registered on a Micrometrics ASAP 2020. The Nb_2O_5 catalyst samples were previously degassed at 300 °C for 3 h and subsequently submitted to a nitrogen atmosphere at 77 K. The nitrogen adsorption-desorption curve made it possible to determine the specific surface by area using BET method in the region of low relative pressure (p/p_0 = 0.1–1.0). The Barret–Joyner–Halenda method (BJH) was used to determine the pore size distribution.

The pH of the treated solutions was adjusted using a pH-meter Tecnopon mPA-210. The percentage of color removal for the solution during the photocatalytic treatment was estimated using Equation (14) [12,53]:

$$\%\text{Color Removal} = (A_0 - A_t)/A_0 \times 100 \tag{14}$$

where A_0 is the initial absorbance and A_t the absorbance at the treatment time t. The absorbance was determined at the maximum absorptivity of MO (λ_{max} = 642 nm) using a UV-vis spectrophotometer Analytikjena SPECORD 210 PLUS. The point of zero charge (PZC) of Nb_2O_5 was determined by the pH drift method, as described by Hashemzadeh et al. [40]. Solutions of 50 mL of 0.01 M of NaCl where adjusted to different pH between 1 and 11 by adding HCl or NaOH. After achieving the defined initial pH, 0.05 g of Nb_2O_5 photocatalyst was added to the solution, which was maintained at 25 °C for 48 h under constant stirring at 700 rpm before measuring the solution pH to determine the pH_{final}. The pH_{PZC} was determined from the intersection of the curve pH_{final} vs. $pH_{initial}$ with the straight line $pH_{final} = pH_{initial}$.

4. Conclusions

The potential application of Nb_2O_5 for photocatalytic decontamination and decolorization of wastewater containing azo dyes was proved on basis of the efficient removal of a model pollutant: MO. A novel Nb_2O_5 photocatalyst was successfully synthesized using a facile calcination method from a natural precursor extracted as a natural resource in Brazil and characterized. The photocatalytic assays demonstrated high removal efficiency of MO azo dye in the presence of H_2O_2, which was used as an e_{cb}^- scavenger, and oxidants, which acted as a photogeneration enhancer. The effects of different control parameters were analyzed and optimized to enable the faster decolorization under the mildest conditions. Thereby, a concentration of Nb_2O_5 catalyst of 1.0 g L^{-1} in slurry was identified as the optimal conditions for the complete removal of color at lower catalyst dosages. The tests carried out also defined as optimal the mild conditions of pH 5.0 and 0.20 M of H_2O_2, with treatable concentrations of MO ranging up to 15 mg L^{-1}. It should be noted that the study was developed in order to find potential applications for niobium materials, which represent a key material produced extensively in Brazil. Our results prove the promising applicability of these innocuous and highly re-utilizable photocatalysts in AOPs for wastewater treatment. It is worth mentioning that this technique is emerging as suitable approach for depollution treatment of textile effluent in mid-sized industry in the Northeast region of Brazil, which receives approximately 10 h/day of sunlight irradiation for more than 350 days a year due to its proximity to the equatorial line.

Author Contributions: Conceptualization: A.J.d.S. and S.G.-S.; methodology: A.J.d.S., L.M.B.B. and S.G.-S.; validation: A.J.d.S., C.A.M.-H. and A.P.d.M.A.; formal analysis: A.J.d.S., S.G.-S. and C.A.M.-H.; investigation: A.J.d.S., L.M.B.B. and S.G.-S.; resources: C.A.M.-H. and A.P.d.M.A.; data curation: A.J.d.S., L.M.B.B. and S.G.-S.; writing—original draft preparation: A.J.d.S. and S.G.-S.; writing—review and editing: C.A.M.-H., A.P.d.M.A. and S.G.-S.; visualization; A.J.d.S. and L.M.B.B.; supervision: C.A.M.-H., A.P.d.M.A. and S.G.-S.; project administration: C.A.M.-H. and A.P.d.M.A.; funding acquisition: C.A.M.-H., A.P.d.M.A. and S.G.-S.

Funding: Financial supports from National Council for Scientific and Technological Development (CNPq—465571/2014-0; CNPq—446846/2014-7 and CNPq—401519/2014-7) and FAPESP (2014/50945-4) are gratefully acknowledged. A.J. dos Santos and L.M.B. Batista gratefully acknowledge the grants awarded from CAPES.

Conflicts of Interest: The authors declare no conflict of interest.

References

1. Nakata, K.; Fujishima, A. TiO$_2$ photocatalysis: Design and applications. *J. Photochem. Photobiol. C Photochem. Rev.* **2012**, *13*, 169–189. [CrossRef]
2. Garcia-Segura, S.; Brillas, E. Applied photoelectrocatalysis on the degradation of organic pollutants in wastewaters. *J. Photochem. Photobiol. C Photochem. Rev.* **2017**, *31*, 1–35. [CrossRef]
3. Kirner, J.T.; Finke, R.G. Water-oxidation photoanodes using organic light-harvesting materials: A review. *J. Mater. Chem. A* **2017**, *5*, 19560–19592. [CrossRef]
4. Kawamura, G.; Matsuda, A. Synthesis of plasmonic photocatalysts for water splitting. *Catalysts* **2019**, *9*, 982. [CrossRef]
5. Cerrón-Calle, G.A.; Aranda-Aguirre, A.J.; Luyo, C.; Garcia-Segura, S.; Alarcón, H. Photoelectrocatalytic decolorization of azo dyes with nano-composite oxide layers of ZnO nanorods decorated with Ag nanoparticles. *Chemosphere* **2019**, *219*, 296–304. [CrossRef]
6. Loeb, S.K.; Alvarez, P.J.J.; Brame, J.A.; Cates, E.L.; Choi, W.; Crittenden, J.; Dionysiou, D.D.; Li, Q.; Li-puma, G.; Quan, X.; et al. The Technology Horizon for Photocatalytic Water Treatment: Sunrise or Sunset? *Environ. Sci. Technol.* **2019**, *53*, 2937–2947. [CrossRef]
7. Simioni, T.; Schaeffer, R. Georeferenced operating-efficiency solar potential maps with local weather conditions—An application to Brazil. *Sol. Energy* **2019**, *184*, 345–355. [CrossRef]
8. Prado, N.T.; Oliveira, L.C.A. Nanostructured niobium oxide synthetized by a new route using hydrothermal treatment: High efficiency in oxidation reactions. *Appl. Catal. B Environ.* **2017**, *205*, 481–488. [CrossRef]
9. Hashemzadeh, F.; Gaffarinejad, A.; Rahimi, R. Porous p-NiO/n-Nb$_2$O$_5$ nanocomposites prepared by an EISA route with enhanced photocatalytic activity in simultaneous Cr(VI) reduction and methyl orange decolorization under visible light irradiation. *J. Hazard. Mater.* **2015**, *286*, 64–74. [CrossRef]
10. Gibson, C.E.; Kelebek, S.; Aghamirian, M. Niobium oxide mineral flotation: A review of relevant literature and the current state of industrial operations. *Int. J. Miner. Process.* **2015**, *137*, 82–97. [CrossRef]
11. Garcia, S.; Cordeiro, A.; de Nääs, I.A.; de Neto, P.L.O.C. The sustainability awareness of Brazilian consumers of cotton clothing. *J. Clean. Prod.* **2019**, *215*, 1490–1502. [CrossRef]
12. Brillas, E.; Martínez-Huitle, C.A. Decontamination of wastewaters containing synthetic organic dyes by electrochemical methods. An updated review. *Appl. Catal. B Environ.* **2015**, *166*, 603–643. [CrossRef]
13. Mansour, F.; Alnouri, S.Y.; Al-Hindi, M.; Azizi, F.; Linke, P. Screening and cost assessment strategies for end-of-Pipe Zero Liquid Discharge systems. *J. Clean. Prod.* **2018**, *179*, 460–477. [CrossRef]

14. Soares, P.A.; Silva, T.F.C.V.; Ramos Arcy, A.; Souza, S.M.A.G.U.; Boaventura, R.A.R.; Vilar, V.J.P. Assessment of AOPs as a polishing step in the decolourisation of bio-treated textile wastewater: Technical and economic considerations. *J. Photochem. Photobiol. A Chem.* **2016**, *317*, 26–38. [CrossRef]

15. Al-Mamun, M.R.; Kader, S.; Islam, M.S.; Khan, M.Z.H. Photocatalytic activity improvement and application of UV-TiO$_2$ photocatalysis in textile wastewater treatment: A review. *J. Environ. Chem. Eng.* **2019**, *7*, 103248. [CrossRef]

16. Spasiano, D.; Marotta, R.; Malato, S.; Fernandez-Ibañez, P.; Di Somma, I. Solar photocatalysis: Materials, reactors, some commercial, and pre-industrialized applications. A comprehensive approach. *Appl. Catal. B Environ.* **2015**, *170*, 90–123. [CrossRef]

17. Marcelino, R.B.P.; Amorim, C.C. Towards visible-light photocatalysis for environmental applications: Band-gap engineering versus photons absorption—A review. *Environ. Sci. Pollut. Res.* **2019**, *26*, 4155–4170. [CrossRef]

18. Vaiano, V.; Iervolino, G. Photocatalytic removal of methyl orange azo dye with simultaneous hydrogen production using Ru-modified Zno photocatalyst. *Catalysts* **2019**, *9*, 964. [CrossRef]

19. AlSalka, Y.; Granone, L.I.; Ramadan, W.; Hakki, A.; Dillert, R.; Bahnemann, D.W. Iron-based photocatalytic and photoelectrocatalytic nano-structures: Facts, perspectives, and expectations. *Appl. Catal. B Environ.* **2019**, *244*, 1065–1095. [CrossRef]

20. Montenegro-Ayo, R.; Morales-Gomero, J.C.; Alarcon, H.; Cotillas, S.; Westerho, P.; Garcia-segura, S. Scaling up photoelectrocatalytic Reactors: A TiO$_2$ nanotube-coated disc compound reactor effectively degrades acetaminophen. *Water* **2019**, *11*, 2522. [CrossRef]

21. Deng, F.; Zhang, Q.; Yang, L.; Luo, X.; Wang, A.; Luo, S.; Dionysiou, D.D. Visible-light-responsive graphene-functionalized Bi-bridge Z-scheme black BiOCl/Bi$_2$O$_3$ heterojunction with oxygen vacancy and multiple charge transfer channels for efficient photocatalytic degradation of 2-nitrophenol and industrial wastewater treatment. *Appl. Catal. B Environ.* **2018**, *238*, 61–69. [CrossRef]

22. Tugaoen, H.O.N.; Garcia-Segura, S.; Hristovski, K.; Westerhoff, P. Compact light-emitting diode optical fiber immobilized TiO$_2$ reactor for photocatalytic water treatment. *Sci. Total Environ.* **2018**, *613*, 1331–1338. [CrossRef] [PubMed]

23. Fagan, R.; Mccormack, D.E.; Dionysiou, D.D.; Pillai, S.C. A review of solar and visible light active TiO$_2$ photocatalysis for treating bacteria, cyanotoxins and contaminants of emerging concern. *Mater. Sci. Semicond. Process.* **2016**, *42*, 2–14. [CrossRef]

24. Batista, L.M.B.; dos Santos, A.J.; da Silva, D.R.; Alves, A.P.D.M.; Garcia-Segura, S.; Martínez-Huitle, C.A. Solar photocatalytic application of NbO$_2$OH as alternative photocatalyst for water treatment. *Sci. Total Environ.* **2017**, *596*, 79–86. [CrossRef] [PubMed]

25. Dominguez, S.; Huebra, M.; Han, C.; Campo, P.; Nadagouda, M.N.; Rivero, M.J.; Ortiz, I.; Dionysiou, D. Magnetically recoverable TiO$_2$-WO$_3$ photocatalyst to oxidize bisphenol A from model wastewater under simulated solar light. *Environ. Sci. Pollut. Res.* **2017**, *24*, 12589–12598. [CrossRef] [PubMed]

26. Zuorro, A.; Lavecchia, R.; Monaco, M.M.; Iervolino, G.; Vaiano, V. Photocatalytic degradation of azo dye Reactive Violet 5 on Fe-doped Titania catalysts under visible light irradiation. *Catalysts* **2019**, *9*, 645. [CrossRef]

27. Idrees, F.; Dillert, R.; Bahnemann, D.; Butt, F.K.; Tahir, M. In-situ synthesis of Nb$_2$O$_5$/g-C$_3$N$_4$ heterostructures as highly efficient photocatalysts for molecular h2 evolution under solar illumination. *Catalysts* **2019**, *9*, 169. [CrossRef]

28. Prado, A.G.S.; Bolzon, L.B.; Pedroso, C.P.; Moura, A.O.; Costa, L.L. Nb$_2$O$_5$ as efficient and recyclable photocatalyst for indigo carmine degradation. *Appl. Catal. B Environ.* **2008**, *82*, 219–224. [CrossRef]

29. Leite, E.R.; Vila, C.; Bettini, J.; Longo, E. Synthesis of niobia nanocrystals with controlled morphology. *J. Phys. Chem. B* **2006**, *110*, 18088–18090. [CrossRef]

30. Heitmann, A.P.; Patrício, P.S.O.; Coura, I.R.; Pedroso, E.F.; Souza, P.P.; Mansur, H.S.; Mansur, A.; Oliveira, L.C.A. Nanostructured niobium oxyhydroxide dispersed Poly (3-hydroxybutyrate) (PHB) films: Highly efficient photocatalysts for degradation methylene blue dye. *Appl. Catal. B Environ.* **2016**, *189*, 141–150. [CrossRef]

31. Aegerter, M.A. Sol-gel niobium pentoxide: A promising material for electrochromic coatings, batteries, nanocrystalline solar cells and catalysis. *Sol. Energy Mater. Sol. Cells* **2001**, *68*, 401–422. [CrossRef]

32. Wang, F.; Wu, H.Z.; Liu, C.L.; Yang, R.Z.; Dong, W.S. Catalytic dehydration of fructose to 5-hydroxymethylfurfural over Nb_2O_5 catalyst in organic solvent. *Carbohydr. Res.* **2013**, *368*, 78–83. [CrossRef] [PubMed]

33. Villaluz, F.J.A.; de Luna, M.D.G.; Colades, J.I.; Garcia-Segura, S.; Lu, M. Removal of 4-chlorophenol by visible-light photocatalysis using ammonium iron (II) sulfate-doped nano-titania. *Process Saf. Environ. Prot.* **2019**, *125*, 121–128.

34. Sathasivam, S.; Williamson, B.A.D.; Althabaiti, S.A.; Obaid, A.Y.; Basahel, S.N.; Mokhtar, M.; Scanlon, D.O.; Carmalt, C.; Parkin, I.P. Chemical vapor deposition synthesis and optical properties of Nb_2O_5 thin films with hybrid functional theoretical insight into the band structure and band gaps. *ACS Appl. Mater. Interfaces* **2017**, *9*, 18031–18038. [CrossRef]

35. Corrêa, M.D.P. Solar ultraviolet radiation: Properties, characteristics and amounts observed in Brazil and south America. *An. Bras. Dermatol.* **2015**, *90*, 297–313. [CrossRef]

36. Lin, J.C.; Sopajaree, K.; Jitjanesuwan, T.; Lu, M. Application of visible light on copper-doped titanium dioxide catalyzing degradation of chlorophenols. *Sep. Purif. Technol.* **2018**, *191*, 233–243. [CrossRef]

37. Anotai, J.; Jevprasesphant, A.; Lin, Y.M.; Lu, M.C. Oxidation of aniline by titanium dioxide activated with visible light. *Sep. Purif. Technol.* **2012**, *84*, 132–137. [CrossRef]

38. Brillas, E.; Garcia-Segura, S. Benchmarking recent advances and innovative technology approaches of Fenton, photo-Fenton, electro-Fenton, and related processes: A review on the relevance of phenol as model molecule. *Sep. Purif. Technol.* **2019**, 116337. [CrossRef]

39. Fujishima, A.; Zhang, X.; Tryk, D.A. TiO_2 photocatalysis and related surface phenomena. *Surf. Sci. Rep.* **2008**, *63*, 515–582. [CrossRef]

40. Hashemzadeh, F.; Rahimi, R.; Gaffarinejad, A. Influence of operational key parameters on the photocatalytic decolorization of Rhodamine B dye using $Fe^{2+}/H_2O_2/Nb_2O_5/UV$ system. *Environ. Sci. Pollut. Res.* **2014**, *21*, 5121–5131. [CrossRef]

41. Venkatachalapathy, R.; Davila, G.P.; Prakash, J. Catalytic decomposition of hydrogen peroxide in alkaline solutions. *Electrochem. Commun.* **1999**, *1*, 614–617. [CrossRef]

42. Garcia-Segura, S.; Tugaoen, H.O.N.; Hristovski, K.; Westerhoff, P. Photon flux influence on photoelectrochemical water treatment. *Electrochem. Commun.* **2018**, *87*, 63–65. [CrossRef]

43. da Costa Filho, B.M.; Araujo, A.L.P.; Padrao, S.P.; Boaventura, R.A.R.; Dias, M.M.; Lopes, J.C.B.; Vilar, V.J.P. Effect of catalyst coated surface, illumination mechanism and light source in heterogeneous TiO_2 photocatalysis using a Mili-Photoreactor for n-Decane oxidation at gas phase. *Chem. Eng. J.* **2019**, *366*, 560–568. [CrossRef]

44. Heck, K.N.; Garcia-Segura, S.; Westerhoff, P.; Wong, M.S. Catalytic Converters for Water Treatment. *Acc. Chem. Res.* **2019**, *52*, 906–915. [CrossRef] [PubMed]

45. Nguyen, C.H.; Fu, C.C.; Juang, R.S. Degradation of methylene blue and methyl orange by palladium doped TiO_2 photocatalyst for water reuse: Efficiency and degradation pathways. *J. Clean Prod.* **2018**, *202*, 413–427. [CrossRef]

46. Wu, M.C.; Wu, P.Y.; Lin, T.H.; Lin, T.F. Photocatalytic performance of Cu-doped TiO_2 nanofibers treated by the hydrothermal synthesis and air-thermal treatment. *Appl. Surf. Sci.* **2018**, *430*, 390–398. [CrossRef]

47. Perillo, P.M.; Atia, M.N. Solar-assisted photodegradation of methyl orange using Cu-doped ZnO nanorods. *Mater. Today Commun.* **2018**, *17*, 252–258. [CrossRef]

48. Zheng, X.; Zhang, D.; Gao, Y.; Wu, Y.; Liu, Q.; Zhu, X. Synthesis and characterization of cubic Ag/TiO_2 nanocomposites for the photocatalytic degradation of methyl orange in aqueous solutions. *Inorg. Chem. Commun.* **2019**, *110*, 107589. [CrossRef]

49. Yin, H.; Zhou, A.; Chang, N.; Xu, X. Characterization and photocatalytic activity of Bi_3RiNbO_9 nanocrystallines synthesized by sol-gel process. *Mater. Res. Bull.* **2009**, *44*, 377–380. [CrossRef]

50. Tallapally, V.; Esteves, R.J.A.; Nahar, L.; Arachchige, U. Multivariate synthesis of tin phosphide nanoparticles: Temperature, time, and ligand control of size, shape, and crystal structure. *Chem. Mater.* **2016**, *28*, 5406–5414. [CrossRef]

51. Tallapally, V.; Damma, D.; Darmakkolla, S.R. Facile synthesis of size-tunable tin arsenide nanocrystals. *ChemmComm* **2019**, *55*, 1506–1563. [CrossRef] [PubMed]

52. Tallapally, V.; Nakagawara, T.A.; Demchenko, D.O.; Ozgur, U.; Arachchige, I.U. $Ge_{1-x}Sn_x$ alloy quantum dots with composition-tunable energy gaps and near-infrared photoluminescence. *Nanoscale* **2018**, *10*, 20296–20305. [CrossRef] [PubMed]
53. Dos Santos, A.J.; De Lima, M.D.; Da Silva, D.R.; Garcia-Segura, S.; Martínez-Huitle, C.A. Influence of the water hardness on the performance of electro-Fenton approach: Decolorization and mineralization of Eriochrome Black T. *Electrochim. Acta* **2016**, *208*, 156–163. [CrossRef]

 *catalysts*

Article

Porous Ni Photocathodes Obtained by Selective Corrosion of Ni-Cu Films: Synthesis and Photoelectrochemical Characterization

Laura Mais [1], Simonetta Palmas [1], Michele Mascia [1], Elisa Sechi [1], Maria Francesca Casula [1], Jesus Rodriguez [2] and Annalisa Vacca [1],*

[1] Chimica e dei Materiali, Dipartimento di Ingegneria Meccanica, Università degli studi di Cagliari, via Marengo 2, 09123 Cagliari, Italy; l.mais@dimcm.unica.it (L.M.); simonetta.palmas@dimcm.unica.it (S.P.); michele.mascia@unica.it (M.M.); elisa.sechi85@gmail.com (E.S.); casulaf@unica.it (M.F.C.)

[2] Centro Nacional del Hidrógeno (CNH2). Prolongación Fernando El Santo s/n, Puertollano, 13500 Ciudad Real, Spain; jesus.rodriguez@cnh2.es

* Correspondence: annalisa.vacca@dimcm.unica.it; Tel.: +39-070-6755-059

Received: 5 April 2019; Accepted: 13 May 2019; Published: 16 May 2019

Abstract: In this work, a dealloying technique is proposed as a synthesis method to obtain highly porous Nickel electrodes starting from Ni-Cu co-deposit: pulsed corrosion is applied adopting different corrosion and relaxation times. Different morphologies, pore size distribution and residual copper amount were obtained depending on the corrosion conditions. For the developed electrodes, the surface roughness factor, R_f, was evaluated by electrochemical impedance spectroscopy (EIS). The hydrogen evolution reaction (HER) on these electrodes was evaluated by means of steady-state polarization curves, and the related parameters were derived by Tafel analysis. Finally, a thin layer of NiO on the porous structures was obtained to exploit the semiconductor characteristic of the oxide, so that an extra-photopotential was obtained by the simulated solar light action. Results demonstrate greater apparent activity of the developed electrodes towards HER in comparison with commercial smooth Ni electrode, which can be mainly attributed to the large R_f obtained with the proposed technique.

Keywords: porous nickel; selective corrosion; hydrogen evolution reaction; photoelectrocatalysis

---

1. Introduction

The production of hydrogen by electrocatalytic or photoelectrocatalytic (PEC) water splitting driven by renewable energy, and its subsequent use in a fuel cell, could represent a zero-emission process in which the storage of H_2 could mitigate the spatial and temporal discontinuities of renewable energy resources [1,2]. However, due to the high overpotential losses involved in the gas evolution reactions occurring at the electrodes, the electrolytic H_2 production is still not competitive, at large scale, with the traditional process of H_2 production from fossil fuels [3]. In fact, noble metals such as Pt, Ru, and Pd are ideal electrocatalysts for hydrogen evolution reaction (HER) in terms of overpotential. Nevertheless, high cost and scarcity make them impractical choices, and the quest for finding inexpensive electrocatalysts is indeed an active area of research [3].

Although many earth-abundant HER catalysts such as transition metal chalcogenides [4,5] or carbides [6,7] have exhibited high activities in acid solutions approaching that of Pt, all of them cannot operate satisfactorily in alkaline electrolytes [8]. Recently, nickel-based electrocatalysts, either as monofunctional or bifunctional materials, were proposed as an economical and efficient replacement to these expensive metal precursors, that exhibited very promising electrocatalytic activity and stability toward oxygen and hydrogen evolution reactions [9–12].

Due to its electronic properties, high conductivity, and thermal stability, Ni has been a very frequent choice for designing electrocatalytic materials. Ni and its related oxides and hydroxides have also been proposed as cathode materials which are effective for HER in non-acidic solutions [13]. Moreover, the NiO p-type semiconductor has emerged as the most frequently used material in this field for its low cost, good stability, and suitable band position alignment, for the transfer of photogenerated holes to counter electrode, that is a key efficiency-determining step of PEC water splitting [14–16]. In order to increase the catalytic activity of nickel based materials towards the HER, two basic approaches could be adopted, namely the use of multicomponent catalysts and the increase of the real surface area [17]. Ni-based alloys, together with other transition metals or rare earths, have been widely studied and exhibited better catalytic capability than single Ni catalyst, due to synergistic effects of the elements forming the alloy [18,19]. Among the Ni-based alloy electrodes studied, Ni–Cu alloy has shown potential as cathode for alkaline HER due to the improved electrocatalytic activity [20,21], high corrosion resistance [22] and good stability [23].

As already stated, the second way to increase the catalytic activity is the preparation of an electrode with a high surface area, that represents a crucial point to which attention should be paid to achieve high efficiency [24–26]. However, the synthesis of ordered nanoporous metals faces great challenges, since metals at the nanoscale tend to present low surface area in order to minimize the surface energy [27]. Moreover, the simple increase of the surface area does not always reflect an increase in the electrochemically-active surface area, because the morphology and the dimension of pores may affect the accessible surface area for the electrochemical reactions.

Nickel foam with a nanoporous structure can be obtained by several methods, such as alkaline leaching of the aluminum from Ni-Al Raney nickel [28], chemical vapor deposition [29,30], electrodeposition [31] and template synthesis [32,33]. However, these processes were found to be imperfect due to limitations in controlling pore sizes and relative density which are important for metallic foams [34].

Based on the above results, in this work, the electrochemical corrosion of Cu-Ni co-deposit was selected for the fabrication of porous Ni-based electrodes, which combine both high surface area and good intrinsic catalytic activity in the process of HER. The selective electrochemical corrosion of copper from Ni_xCu_{1-x} alloy was demonstrated for the first time by Sun and co-workers; by taking advantage from the formation of a passive nickel oxide film in sulfamate aqueous solutions, nanoporous nickel was prepared through selective electrochemical dissolution of the more noble copper, rather than the less noble Ni. The authors performed electrochemical etching under potentiostatic conditions; depending on the composition of the deposited alloy, different morphologies and dimensions of the pores were obtained [35].

Similarly, Chang and co-workers obtained nano-hollow tubes starting from Ni-Cu alloys prepared by electrodeposition, which showed a columnar structure and contained separated Cu- and Ni- rich multiple phases; the selective etching of the less reactive copper from the alloys was performed in solutions containing H_3BO_3 0.5 M [36]. A dendritic Ni-Cu alloy foam with high surface area was fabricated by electrodeposition during HER by Jeong and co-workers; then nanoporous dendritic Ni foam was successfully prepared by selective electrochemical dealloying of copper from Ni-Cu alloy using a sulfuric acid solution [37]. In order to control the morphological features of the porous structure, the dealloying of a co-deposited Ni-Cu was performed under pulsed electric field. By tuning voltage and duration of the bias applied, different final composition and degree of porosity were obtained [38].

To the best of our knowledge, few works have been devoted to the systematic study of the effects of the corrosion conditions of copper on the morphology and dimensions of the pores of nickel, as well as to the amount of residual copper. In this work, we studied the effect of the dealloying conditions, with the application of a voltage waveform, on morphology, real area and composition of the resulting samples. Starting from the same deposit containing 30% of copper and 70% of nickel, different values of corrosion and relaxation times were adopted: two classes of samples were prepared with two values of the ratio between the corrosion and relaxation times.

Moreover, also the electrocatalytic performance of the developed electrodes for HER and for photoelectrochemical tests were studied. To distinguish the effect of both surface roughness and intrinsic activity of the material, the real active surface area of the catalysts, in terms of roughness factor (R_f), was determined.

2. Results and Discussion

The electrodeposition of Ni-Cu was realized under potentiostatic conditions at E = −0.8 V for 130 minutes on niobium discs; linear sweep voltammetry was firstly performed in the electrodeposition solution at 5 mV s^{-1}, starting from open circuit potential (OCP) up to −0.8 V, to favor a slow deposition of a thin film onto niobium surface.

Figure 1 displays the scanning electron microscopy (SEM) images and the Auger mappings of the Ni-Cu deposit onto the niobium surface; as can be seen, a homogeneous distribution of both the elements, with some copper agglomerations, characterized the surface. The chemical composition of the film calculated by energy dispersive X-ray (EDX) analysis indicated an average molar fraction of 30% of copper and 70% of nickel.

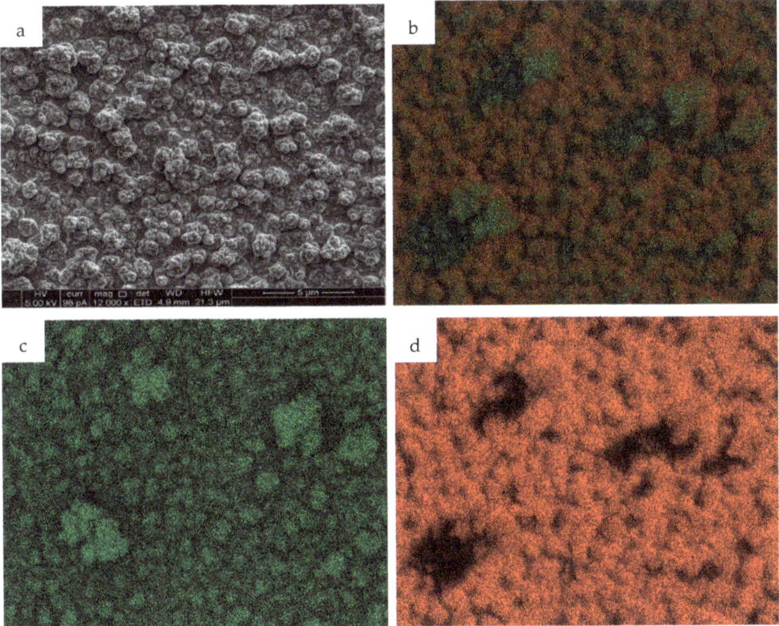

Figure 1. SEM image (**a**) and Auger mappings (**b**) of Ni-Cu deposit onto niobium substrates: Cu in green and Ni in red. The separated Auger mappings of Cu and Ni are reported in (**c**) and (**d**), respectively.

The SEM images of the samples obtained using different corrosion conditions (Figure 2), show the presence of porous structures in each sample. Table 1 reports the conditions adopted for the corrosion of each sample, denoted as S$_{tcorr-trelax}$ where t_{corr} and t_{relax} are the corrosion and relaxation times used in the pulsed corrosion steps, respectively. Table 1 also reports the ratio $\phi = t_{corr}/t_{relax}$ together with the range of pore diameters and molar fractions of Cu and Ni (oxygen was found as third element up to unit).

As can be seen from SEM images, different morphologic features can be recognized depending on the experimental conditions; samples prepared with the highest ϕ as S$_{1-5}$ (Figure 2a) and S$_{0.1-0.5}$ (Figure 2b) were featured by a tubular structure, while the samples prepared using the lowest ϕ (S$_{1-50}$ in Figure 2d, S$_{0.1-5}$ in Figure 2e and S$_{0.01-0.5}$ in Figure 2f) presented a cauliflower-like structure with multi-dimensional pores. Only the sample $S_{0.01-0.05}$ (Figure 2c) is characterized by a poor pore density with very small diameters.

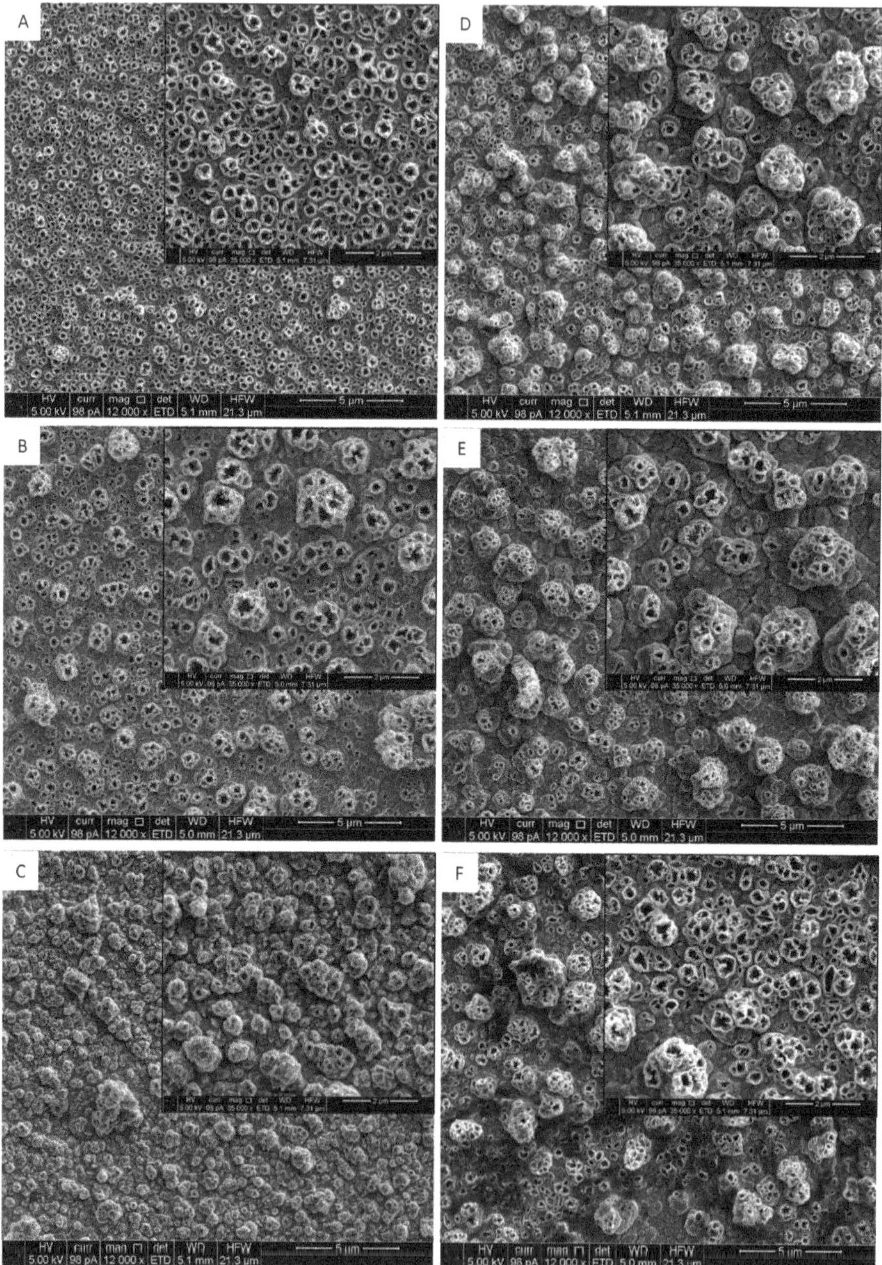

Figure 2. SEM images of samples submitted to different conditions of anodic dissolution: (**A**) S_{1-5}; (**B**) $S_{0.1-0.5}$; (**C**) $S_{0.01-0.05}$; (**D**) S_{1-50}; (**E**) $S_{0.1-5}$; (**F**) $S_{0.01-0.5}$. Insets report the magnifications of the SEM images of the samples.

Data in Table 1 and Figure 2, indicated a combined effect of t_{corr} and t_{relax} on the pore size and morphology of the resulting samples: in fact, t_{relax} being the same, a decrease in t_{corr} leads to lower pores diameter (See Figure 2 for sample S_{1-5} compared with $S_{0.1-5}$, and $S_{0.1-0.5}$ compared with $S_{0.01-0.5}$) and higher residual copper amount.

Table 1. Range of diameters d (nm) and molar fractions of Cu and Ni (X_{Cu} and X_{Ni}) of porous nickel electrodes prepared using different t_{corr} and t_{relax} during the anodic pulsed voltage dissolution experiments.

	t_{corr}/s	t_{relax}/s	ϕ	d/nm	X_{Ni}	X_{Cu}
S_{1-5}	1	5	0.2	200–350	0.90	0.064
$S_{0.1-0.5}$	0.1	0.5	0.2	140–300	0.86	0.086
$S_{0.01-0.05}$	0.01	0.05	0.2	50–150	0.76	0.195
S_{1-50}	1	50	0.02	170–250	0.81	0.14
$S_{0.1-5}$	0.1	5	0.02	170–250	0.89	0.083
$S_{0.01-0.5}$	0.01	0.5	0.02	70–250	0.82	0.11

Conversely, an increase in t_{relax} leads to a decrease of the pores size, t_{corr} being the same, as for samples S_{1-5} and S_{1-50} (t_{corr} of 1 s), or for samples $S_{0.1-0.5}$ and $S_{0.1-5}$ (t_{corr} of 0.1 s) (See Figure 2). Moreover, in the case of t_{corr} of 0.01 s, both samples present the highest copper residual amount.

As it was firstly proposed by Erlebacher et al. [39], these results can be explained considering that the mechanism involved in the formation of porous structures, by dealloying of binary systems under potentiostatic conditions, resulted from two concurrent processes occurring at the alloy/electrolyte interface: the chemical dissolution of the most reactive metal atoms and the atom rearrangement of the most inert atoms, which expose the underlying more reactive metal atoms to further dissolution.

Using pulsed potential, it is possible to influence the interplay between atoms rearrangement and chemical dissolution [38]. Thus, the increase in the ratio ϕ favors the rearrangement of nickel with respect the copper dissolution, which in turn leads to more dendritic samples with high content of residual copper, as observed comparing samples S_{1-5} with $S_{0.1-5}$ or $S_{0.1-0.5}$ with $S_{0.01-0.5}$. Moreover, very short relaxation times, as in the case of sample $S_{0.01-0.05}$, limit the nickel atoms rearrangement thus hindering the exposure of underlying copper and its progressive dissolution: few and very small pores with very high residual copper content was obtained under this condition.

To gain insight into the structural features of the deposited films, X-ray diffraction patterns were acquired, as shown in Figure 3. The pattern of the initial deposit (Figure 3a) shows the occurrence of sharp peaks due to the niobium substrate and additional broader reflections which can be ascribed to the occurrence of nanocrystalline nickel and copper phases, based on comparison with reference PDF cards (Figure 3c). After the electrochemical corrosion, changes in the X-ray pattern can be observed in Figure 3b (sample S_{1-5}): in particular, the relative intensity of the peaks due to Ni and Cu is decreased, suggesting the occurrence of a lower contribution from nanocrystalline phases as compared to the original film. Additional peaks are also observed which can be tentatively ascribed to a copper-rich oxide (Cu_4O_3).

In order to characterise the electrodes, electrochemical impedance spectroscopy (EIS) measurements were performed in KOH 1 M solutions at open circuit potential (OCP). The related Nyquist and Bode plots are reported in Figures 4 and 5: as can be seen, samples S_{1-5}, $S_{0.01-0.05}$ and S_{1-50} are characterized by a wide flattened semicircle in the low frequency (LF) region of Nyquist plot, and by one wave in the phase angle Bode plot. This behavior is like that of smooth electrodes covered with flat pores, indicating that the surface behaves like a flat one, i.e. the pores are well accessible at all the frequency values [40,41]. For the samples $S_{0.1-0.5}$, $S_{0.1-5}$ and $S_{0.01-0.5}$ a small arc is visible at high frequency (HF), followed by a second branch in the LF region of the Nyquist plot, as well as the beginning of a second wave appears in the phase angle Bode plot, indicating a behavior typical of porous and rough electrodes [42], which respond differently, depending on the frequency region.

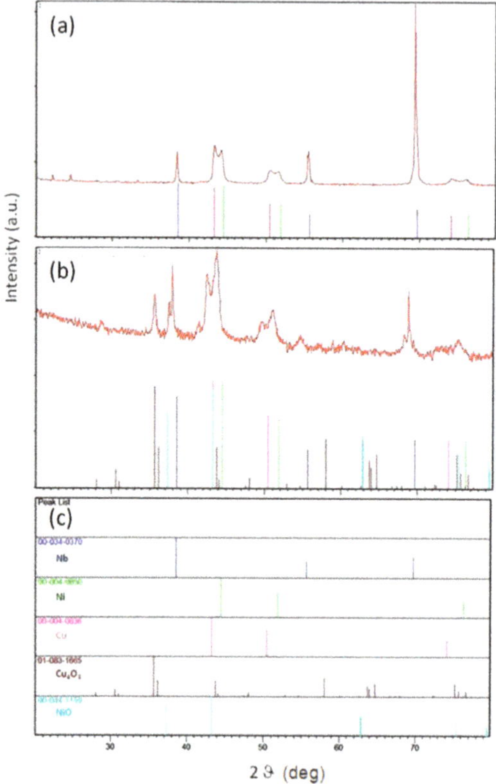

Figure 3. XRD pattern of the films as-deposited (**a**) and of the sample S_{1-5} after the corrosion (**b**) and relevant PDF cards for reference structures.

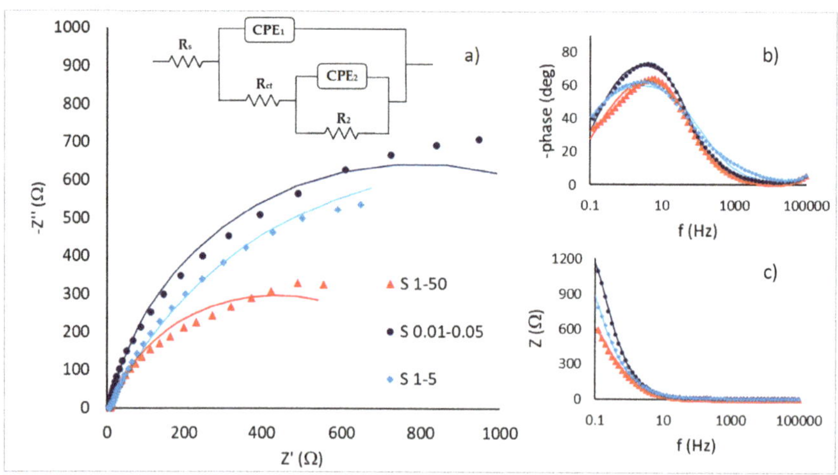

Figure 4. Nyquist diagram (**a**), Bode phase (**b**) and Bode modulus (**c**) recorded in KOH 1 M at open circuit potential. Inset: equivalent circuit used to model the EIS spectra.

To model the experimental data, an electrochemical equivalent circuit, which involves two time constants, was used (inset of Figure 4): the model is a slightly modified version of that originally proposed by Armstrong and Henderson [43], in which the capacitances were replaced by the constant phase elements (CPE) [44–46] which represent a deviation from the purely capacitive behaviour, related to surface in-homogeneity, or to variations of properties in the direction that is normal to the electrode surface. Such variability may be attributed, for example, to changes in the conductivity of oxide layers, or to porosity and surface roughness [47].

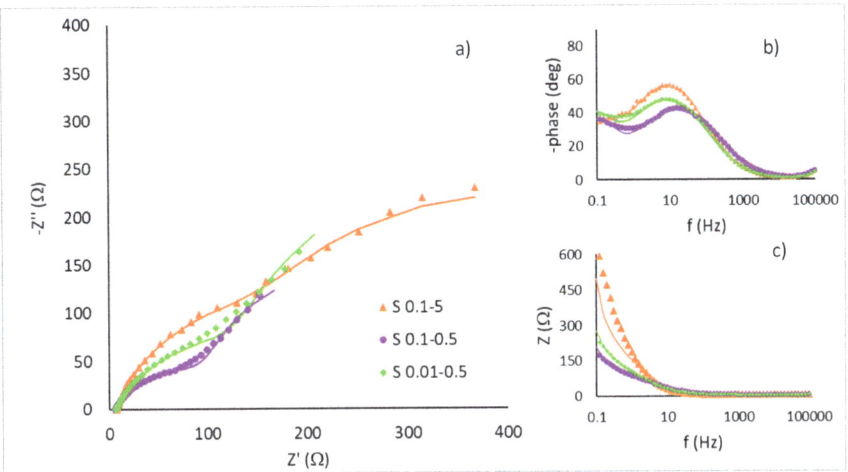

Figure 5. Nyquist diagram (**a**), Bode phase (**b**) and Bode modulus (**c**) recorded in KOH 1 M at open circuit potential.

The impedance Z_{CPE} is described by the following equation:

$$Z_{CPE} = \frac{1}{Q(j\omega)^n} \tag{1}$$

where Q is a capacitance parameter and n is a parameter characterizing the rotation of the complex plane impedance plot [47].

This two time constants model was widely used to describe the response of Ni-based porous electrodes during HER; when the semicircle at HF is potential-independent, it can be related to the electrode surface porosity response, while the potential-dependent LF semicircle can be related to the charge transfer resistance process [44,48,49]. On the other hand, when both semicircles change with overpotential, the LF time constant is associated to the hydrogen adsorption on the electrode surface, while the HF time constant is related to the charge transfer resistance [48,50].

In the present work, the same electrical circuit was used to model the response of the porous electrodes at OCP and the different time constants were correlated to the different pore size distribution. Analogous approach was adopted with hierarchical structures by Abouelamaiem et al. [51]: the smaller the pores, the higher the constant time values.

As can be seen from Figures 4 and 5, in the present case, the proposed model was able to interpret the Nyquist and Bode plots of the selected samples. The fitting parameters are presented in Table 2, along with the values of chi-squared (χ^2), which were always in the order of 10^{-4}.

Table 2. Electrical circuital parameters obtained from the fit of EIS spectra at open circuit potential for the synthesized samples, recorded in KOH 1 M.

Samples	S_{1-5}	$S_{0.1-0.5}$	$S_{0.01-0.05}$	S_{1-50}	$S_{0.1-5}$	$S_{0.01-0.5}$
χ^2	2.06×10^{-4}	8.33×10^{-4}	7.69×10^{-4}	9.51×10^{-4}	5.84×10^{-4}	9.82×10^{-4}
R_s/Ω cm^2	3.16	3.16	3.44	3.70	3.28	3.38
R_{ct}/Ω cm^2	829	65.90	784.5	433.4	140.65	103.60
Q_1/mS s^n cm^{-2}	0.52	3.18	0.242	0.76	1.99	3.24
n_1	0.86	0.68	1.00	1.00	0.80	0.72
R_2/Ω cm^2	6.055	141.4	4.71	3.4	157.7	239.8
Q_2/mS s^n cm^{-2}	1.60	16.66	1.17	2.08	6.82	12.14
n_2	0.71	1.00	0.85	0.73	0.99	1.00
C_{dl}/μF cm^{-2}	454	1522	242	760	1439	2118
C_2/μF cm^{-2}	250	16660	460	335	6821	12140
τ_1 (s)	0.38	0.10	0.19	0.33	0.20	0.22
τ_2 (s)	0.0015	2.36	0.0022	0.0011	1.08	2.91
R_f	22.7	76.1	12.1	38.0	71.9	105.9
A_r (cm^2)	11	38	6	19	36	53

Table 2 reports also the values of capacitance (C) calculated as proposed by Brug et al. [52] for non-Faradaic system:

$$C_i = \frac{(Q_i R_i)^{1/n_i}}{R_i} \qquad (2)$$

and the time constant (τ)

$$\tau_i = C_i R_i \qquad (3)$$

Depending on the samples, different time constant values are calculated from the relevant circuital parameters, that can give information on the processes occurring at the electrode surface.

So, for example, the values of τ_1 between 0.10 and 0.38 s, are in the order of magnitude typically related to the charge transfer between electrode/electrolyte interface. Time constants in the order of half second were reported by Cardona et al [48,53], for porous Ni and Ni-Cu electrodes, and were attributed to the charge transfer kinetics. On the other hand, as suggested by other references [46,50,53], the τ_2 values 10-fold higher, calculated for samples $S_{0.1-0.5}$, $S_{0.1-5}$, $S_{0.01-0.5}$, could be instead associated to the diffusion inside the smaller, less accessible pores. Finally, a very fast charge transfer inside the bulk material should be the reason of the very low values of τ_2 (in the order of ms) calculated for samples S_{1-5}, $S_{0.01-0.05}$, S_{1-50}.

This different behavior can be explained considering the different morphology of the samples: when a tubular structure with little variations of the pore diameter is obtained, as in the case of sample S_{1-5}, the response of the impedance is dominated by accessible surface inside the pores. Conversely, when a multidimensional pores structure, such as a cauliflower-like surface is obtained (see for example sample $S_{0.01-0.5}$), the slow diffusive process in the smaller pores may become important, given that it is revealed when low frequency region is explored.

The values of C_{dl} were used, in turn, to compare the real surface area of the samples accessible to the electrolyte: the surface roughness factor R_f, was determined relating the C_{dl} of the samples with that of a smooth Ni electrode equal to 20 μF/cm^2 [46] an in turn, the values of real active surface area (A_r) were calculated (see Table 2).

The results reported in Table 2 highlight that the group of samples prepared using a lower ϕ presents higher values of C_{dl} and of roughness factor (R_f). Moreover, at the same ϕ values, an increase in the porosity is recorded with the decrease of the corrosion and relaxation time. This is not verified in the case of the sample $S_{0.01-0.05}$, where very low pore density was obtained. Moreover, R_f values of 8 and 13 were obtained by EIS at commercial nickel plate and at Ni-Cu co-deposit, respectively.

In the last part of the work attention was paid on the evaluation of the catalytic activity of the samples toward HER. As is well known, especially when heterogeneous reactions are involved, the achievement of high specific area is of a great concern. However, also the morphology, and then the exploitability of the surface, may affect the material performances, particularly when the reactions involve the production of gas, as in the case of HER. Moreover, a synergistic effect on the catalytic activity was often attributed to the presence of copper, either in alloy, or co-deposited with nickel.

In order to study these effects, two samples were selected, which were characterized by different roughness factor, morphology and residual copper content: in particular, S_{1-5} was selected among those with $\phi = 0.2$, and $S_{0.01-0.5}$ among those with $\phi = 0.02$ (see Table 1). Their electrocatalytic performance and their photoactivity were investigated, and their behavior compared with those of commercial nickel plate and Ni-Cu co-deposit.

Figure 6 reports the results of linear sweep polarization curves, performed in 1 M KOH solution: the cathodic current densities are reported as a function of the overpotential (η) together with the corresponding Tafel linearization, as inset.

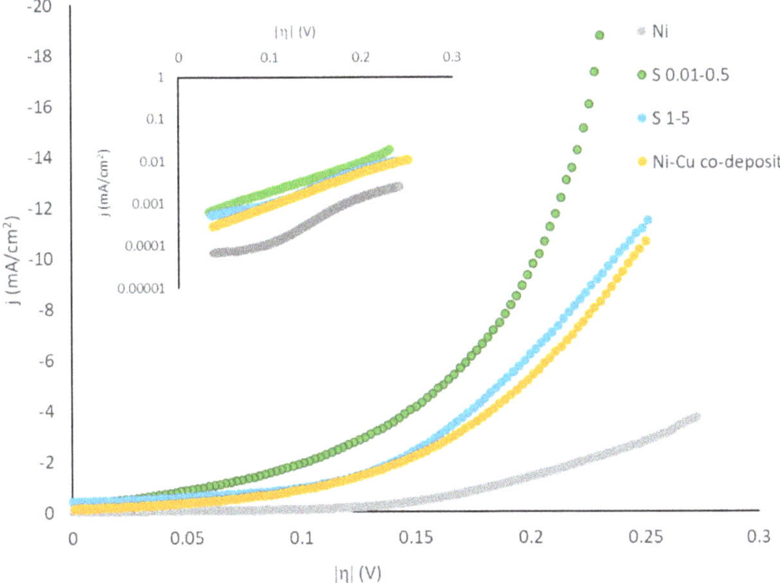

Figure 6. Linear sweep voltammetries in 1 M KOH solution; inset: magnification of the linear Tafel polarization curves.

If compared with commercial nickel plate, the samples S_{1-5} and $S_{0.01-0.5}$ present higher catalytic activity, the highest being obtained for sample $S_{0.01-0.5}$. This enhancement in the catalytic activity can be connected to the increased specific surface area. Moreover, the trend of the linear sweep polarization of the Cu-Ni co-deposit indicates the positive effect of the presence of copper.

The kinetic parameters of the related processes were determined by considering both the geometric and the real surface area of the electrodes. The polarization curves are represented by Tafel equation [54]:

$$\eta = a + b \, log j \tag{4}$$

where η is the overpotential, b is the Tafel slope, j is the current density and a is the intercept of the curve related to the exchange current density j_0 through equation:

$$a = \frac{2.3 \, RT}{\beta n F} log j_0 \tag{5}$$

where n represents the number of electrons exchanged, F is the Faraday constant, β is the symmetry factor and R is the gas constant.

Values of exchange current density j_0, and Tafel slope b, estimated from the linear polarization curves using Equations (4) and (5), are listed in Table 3.

As can be observed, the values of Tafel slope (b) obtained for the synthesized samples range from 126 to 140 mV/dec, indicating that HER proceeds via the Volmer-Heyrosky mechanism [44,55]; similar values were reported in the literature for porous Ni-based samples also in presence of copper [46,48]. Both samples S_{1-5} and $S_{0.01-0.5}$ present exchange current density values considerably higher with respect to the commercial nickel, indicating a considerable improvement of the apparent electrocatalytic properties of the fabricated electrodes.

Table 3. Kinetic parameters for HER obtained from the polarization curves recorded in 1 M KOH.

Sample	Rf	b (mV/dec)	j_0 (mA/cm^2)	j_{0r} (μA/cm^2)
Ni	8	130	0.02	4.2
Cu-Ni co-deposit	13	124	0.13	12
S_{1-5}	23	158	0.29	8.6
$S_{0.01-0.5}$	106	138	0.35	4.6

The value measured for the Ni-Cu co-deposit is instead indicative of the effect of the presence of copper in the deposit. Normalizing the exchange current densities for the active surface area of the electrodes, the real exchange current density j_{0r} was calculated.

The higher j_{0r} value for the sample S_{1-5} with respect to that for $S_{0.01-0.5}$, suggests that the inner porous surface of this sample is not totally exploitable during HER, due to the gas bubbles shielding. This behavior can be explained by considering the more open structure of sample S_{1-5} with respect to $S_{0.01-0.5}$, thus confirming the strong effect of the pores (size, shape and distribution) on the resulting electrocatalytic performance. Comparing the j_{0r} values for sample S_{1-5} with that of commercial nickel, the effect of the composition of the sample can be observed: in fact, the enhancement in the performance is not only related to the increased surface area, but also to the copper amount which affects the overall electroactivity of the electrodes. This effect can be confirmed at Ni-Cu co-deposit, where the enhancement of j_{0r}, compared to a commercial nickel electrode, can be connected essentially to the presence of a rough dual Ni-Cu system.

In order to investigate possible applications of the developed samples as photocathodes, they were submitted to thermal annealing at 500 °C in order to allow the formation of NiO.

Figure 7 shows, as an example, the cathodic photocurrent response measured at sample $S_{0.01-0.5}$: as can be seen, under irradiation a remarkable increase in photocurrent is observed.

In order to compare the photocurrent of samples S_{1-5}, $S_{0.01-0.5}$ and commercial nickel, in Figure 8 the values of photocurrent are reported for three values of applied potential; as can be seen, applied potential being the same, samples S_{1-5} and $S_{0.01-0.5}$ present higher values of photocurrent with respect to the commercial nickel. These values are comparable with those reported in the literature and measured in similar conditions: for NiO photocathodes fabricated by alkaline etching, anodizing nickel foil in an organic-based electrolyte, about 400 μA/cm^2 was measured irradiating the samples with 300 W arc xenon lamp [15]. The values in the order of magnitude of the tens were indicated in a recent review, where sensitized NiO photocathodes for water splitting cells were used: the authors state that photocurrent values varied consistently with the specific surface area of the electrode, suggesting that NiO electrodes made under different conditions should possess comparable photoelectrochemical performance [56].

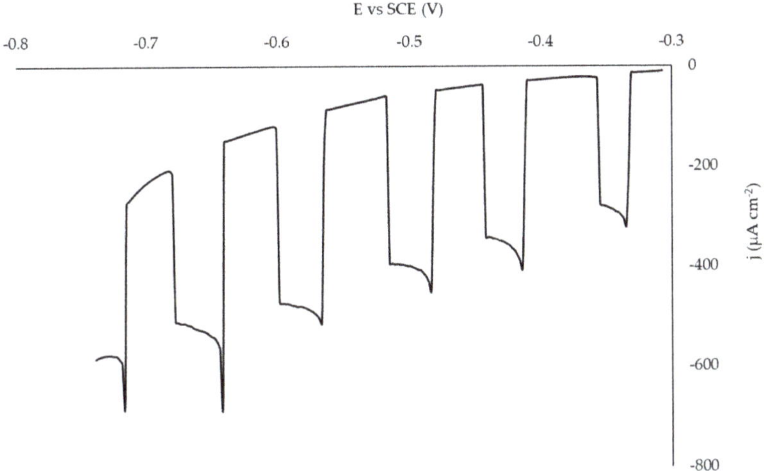

Figure 7. LSV of sample $S_{0.01–0.5}$ in 0.1 M KNO_3. The potential was ramped (5 mV s^{-1}) from the OCP to −0.8 V. Data were recorded under dark and irradiation condition (AM0 filter).

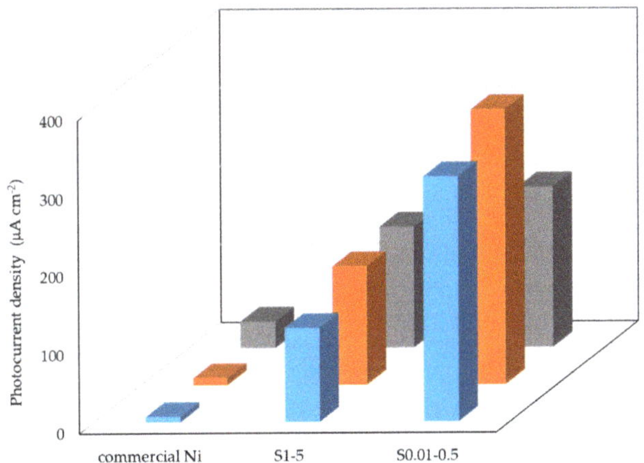

Figure 8. Cathodic photocurrent density recorded in KNO_3 0.1 M solution for different samples, at different applied potential: −0.4 V (blue bars), −0.6 V (orange bars) and −0.8 V (grey bars).

If data at different applied potential are considered, a different behavior of $S_{1–5}$ and $S_{0.01–0.5}$ samples can be observed: as the cathodic potential is increased, while a regular increase in performance is measured at sample $S_{1–5}$, sample $S_{0.01–0.5}$ shows lower value of photocurrent when the most cathodic potentials is applied. This trend may be explained considering that at the highest cathodic potential, gas can be generated which may limit the exploitability of the whole pore structure, especially when small pores are involved in the structure, as at sample $S_{0.01–0.5}$. Of note is that if the photocurrent values are normalized by the real surface area (A_r) of the samples, the highest performance is measured at sample $S_{1–5}$, at all the potentials; moreover, at the highest cathodic potential, the performance of $S_{0.01–0.5}$ sample decreases, and it becomes even worst than the commercial Ni sample.

3. Materials and Methods

Boric acid (H_3BO_3) and potassium hydroxide (KOH) were purchased from Sigma-Aldrich (Sigma-Aldrich Chemie, GmbH, Schnelldorf; Germany). Sodium sulfate anhydrous (Na_2SO_4), nickel sulfate hexahydrate ($NiSO_4 \cdot 6H_2O$), copper sulfate pentahydrate ($CuSO_4 \cdot 5H_2O$) and potassium nitrate (KNO_3) were supplied by Carlo Erba (Carlo Erba Reagents, Cornaredo, Milano, Italy). All electrochemical experiments were performed at room temperature using an AUTOLAB PGSTAT302N (Metrohm, Herisau, Switzerland) potentiostat/galvanostat equipped with a frequency response analyzer controlled with the NOVA software. Two cylindrical hand-made three-electrode cells realized by Teflon were used, in which the working electrode was at the bottom of the cell and the electrical contact consisted of an aluminium disc. A platinized titanium grid, placed in front of the anode at 1 cm distance, and a saturated calomel electrode (SCE) constituted the counter and reference electrodes, respectively. The preparation of the Ni-Cu co-deposits was performed using niobium foils (thickness 0.25 mm, 99.8%, Sigma-Aldrich Chemie, GmbH, Schnelldorf; Germany) as working electrode, cut as discs. Prior to deposition, niobium was mechanically polished with diamond paste (sizes: 3, 1, 0.5 μm) and colloidal silica gel (particles size: 0.05 μm), then submitted to sonication in acetone for 15 minutes and rinsed with distilled water. The cell used for the electrodeposition experiment contained 40 ml of solution (inner diameter = 5 cm, height 4 cm); the exposed geometrical area was 13.5 cm^2. The electrodeposition was performed using solution containing 0.5 M $NiSO_4$, 0.005 M $CuSO_4$ and 0.5 M H_3BO_4 (pH = 4); linear sweep voltammetry was firstly performed in the electrodeposition solution at 5 mV s^{-1}, starting from the OCP up to -0.8 V, after that a constant potential of -0.8 V for a total time of 130 minutes was applied while the solution was stirred. The charge amount recorded during the electrodeposition experiments was about 6 C/cm^2. Considering unit faradaic yield and average atomic weight of 60 g/mol, the amount of deposited metals was roughly estimated equal to 1.8 mg/cm^2, by Faraday's Law.

In order to study the effect of the selective corrosion, starting from the same co-deposit, after the deposition the discs were cut in six slices and subjected to anodic dissolution under different corrosion conditions. The cell used for corrosion experiment contained 10 ml of solution (inner diameter = 1.5 cm, height 3 cm): the exposed geometrical area was 0.5 cm^2.

The runs were performed in aqueous solution, containing 0.5 M H_3BO_4 and 0.5 M Na_2SO_4, under stirring conditions, using pulsed voltage modulated between V_{corr} (E = 0.5 V) and V_{relax} (E = OCP) for time durations of t_{corr} and t_{relax}, respectively. Values of the ratio between t_{corr} and t_{relax} (called φ in the rest of the text) equal to 0.2 and 0.02 were adopted. A total time of 30 minutes were required (typically the dissolution current drops to zero within this period). Prepared samples were denoted as $S_{tcorr\text{-}trelax}$ where t_{corr} and t_{relax} are the corrosion and relaxation times (in seconds), respectively. After oxidation, the electrodes were rinsed in deionized water and dried in a nitrogen stream. The charges amount, recorded during the corrosion test, ranged from 1.2 to 1.6 C/cm^2. Also in this case, applying Faraday's Law, the amount of copper removed was evaluated from 0.4 to 0.55 mg/cm^2.

A scanning electron microscope (SEM) equipped with EDX detector (Zeiss, Oberkochen, Germany) was used to characterize the morphology and the chemical composition of the nanoporous nickel electrodes. Auger electron Spectroscopy (AES) was also used to investigate the distribution of the copper and nickel.

X-ray diffraction (XRD) patterns were recorded in the range of 20–80 (2θ) on a Panalytical Empyrean diffractometer equipped with a Cu Kα radiation and an X'Celerator linear detector. XRD patterns were collected at a grazing incidence of 2 on the films mounted on a flat sample stage. Data were processed by Empyrean X'pert High Score software and phase identification was performed by comparison with the Powder Diffraction Files (PDF-2 JCPDS International Centre for Diffraction Data, Swarthmore, PA, USA) database.

Linear sweep voltammetry (LSV) were performed to study the behavior of the electrodes in the potential range from E = OCP to E = -1.0 V in cathodic direction at a sweep rate of 5 mV/s. The characterization of the electrode/electrolyte interface was also carried out through electrochemical impedance spectroscopy (EIS). The measurements were performed in a frequency range from 100 kHz to 0.1 Hz with excitation amplitude of 10 mV. The impedance spectra were then fitted to an equivalent electrical circuit by using the ZSimpWin 2.0 software (EChem software).

A subsequent annealing treatment was performed in order to convert $Ni(OH)_2$ and NiOOH groups in NiO, which is the semiconductive form of nickel. Thermal treatment was carried out in air atmosphere for 30 min at 500 °C, after ramping 5 °C min^{-1}.

Photoelectrochemical measurements were carried out in a hand-made photoelectrochemical cell (PEC), equipped with a quartz window. The investigated samples were adopted as working electrodes while a platinum wire constituted the counter electrode and a SCE the reference. 0.1 M KNO_3 aqueous solution was used as the supporting electrolyte. The PEC-cell was irradiated with a 300 W Xe lamp (LOT-Quantum Design Europe) equipped with AM 0 optical filter. The incident power density of the light was measured by LP 471 UVU or LP 471 PAR quantum radiometric probes: the recorded value was 138–140 W/m^2. Photocurrent density was calculated with respect the nominal surface area of the samples as the difference between the current recorded under illumination and dark conditions.

4. Conclusions

In this work, a dealloying method was used to obtain porous nickel electrodes as possible alternative cathodic materials in HER. A suitable combination of corrosion and relaxation times in the pulsed potential steps made it possible to obtain different morphologies and pore size distributions. Depending on the samples, roughness factors ranging from 22 and 106 were obtained, but the increased surface area was not always exploitable. In fact, the Tafel analysis revealed that the exchange current density calculated with respect to the real surface area, was higher in sample S_{1-5} rather than sample $S_{0.01-0.5}$, even if a higher surface area was measured at this last sample. The higher j_{0r} value suggests that the inner porous surface area of sample $S_{0.01-0.5}$ is not totally exploitable during HER, due to gas bubbles shielding. This behavior can be explained by considering the more open structure of sample S_{1-5} with respect to $S_{0.01-0.5}$, thus confirming the strong effect of the pores (size, shape and distribution) on the resulting electrocatalytic performance. The presence of thin oxide layer of NiO, as well as of residual copper were indicated as responsible for the photocatalytic activity of the samples. Once again, the different pore size and distribution become crucial in determining the final performance of the samples: thus, for example, at the highest cathodic potential gas can be generated which may limit the exploitability of the whole pore structure, especially when small pores are involved in the structure, as at sample $S_{0.01-0.5}$.

Author Contributions: Conceptualization, S.P. and A.V.; Investigation, L.M., M.M., M.F.C. and E.S.; Writing—original draft, L.M., S.P., E.S. and A.V.; Writing—review & editing, M.M. and J.R.

Funding: This research was funded by Fondazione di Sardegna, grant number project F71I17000280002-2017 and project F71I17000170002.

Conflicts of Interest: The authors declare no conflict of interest.

References

1. Zhao, G.; Rui, K.; Dou, S.X.; Sun, W. Heterostructures for Electrochemical Hydrogen Evolution Reaction: A Review. *Adv. Funct. Mater.* **2018**, *28*, 1803291. [CrossRef]
2. Murthy, A.P.; Madhavan, J.; Murugan, K. Recent advances in hydrogen evolution reaction catalysts on carbon/carbon-based supports in acid media. *J. Power Sources* **2018**, *398*, 9–26. [CrossRef]
3. Eftekhari, A. Electrocatalysts for hydrogen evolution reaction. *Int. J. Hydrog. Energy* **2017**, *42*, 11053–11077. [CrossRef]
4. Kibsgaard, J.; Jaramillo, T.F.; Besenbacher, F. Building an appropriate active-site motif into a hydrogen-evolution catalyst with thiomolybdate $[Mo_3S_{13}]^{2-}$ clusters. *Nat. Chem.* **2014**, *6*, 248–253. [CrossRef] [PubMed]

5. Chang, Y.H.; Lin, C.T.; Chen, T.Y.; Hsu, C.L.; Lee, Y.H.; Zhang, W.; Wei, K.H.; Li, L.J. Highly efficient electrocatalytic hydrogen production by MoSx grown on graphene-protected 3D Ni foams. *Adv. Mater.* **2013**, *25*, 756–760. [CrossRef]

6. Youn, D.H.; Han, S.; Kim, J.Y.; Kim, J.Y.; Park, H.; Choi, S.H.; Lee, J.S. Highly active and stable hydrogen evolution electrocatalysts based on molybdenum compounds on carbon nanotube-graphene hybrid support. *ACS Nano* **2014**, *8*, 5164–5173. [CrossRef] [PubMed]

7. Chen, W.F.; Muckerman, J.T.; Fujita, E. Recent developments in transition metal carbides and nitrides as hydrogen evolution electrocatalysts. *Chem. Commun.* **2013**, *49*, 8896–8909. [CrossRef]

8. Faber, M.S.; Jin, S. Earth-abundant inorganic electrocatalysts and their nanostructures for energy conversion applications. *Energy Environ. Sci.* **2014**, *7*, 3519–3542. [CrossRef]

9. Yu, X.Y.; Feng, Y.; Guan, B.; Lou, X.W.; Paik, U. Carbon coated porous nickel phosphides nanoplates for highly efficient oxygen evolution reaction. *Energy Environ. Sci.* **2016**, *9*, 1246–1250. [CrossRef]

10. Zhao, Z.; Wu, H.; He, H.; Xu, X.; Jin, Y. A high-performance binary Ni-Co hydroxide-based water oxidation electrode with three-dimensional coaxial nanotube array structure. *Adv. Funct. Mater.* **2014**, *24*, 4698–4705. [CrossRef]

11. Zhang, G.; Yuan, J.; Liu, Y.; Lu, W.; Fu, N.; Lia, W.; Huang, H. Boosting the oxygen evolution reaction in non-precious catalysts by structural and electronic engineering. *J. Mater. Chem. A* **2018**, *6*, 10253–10263. [CrossRef]

12. Wei, J.; Zhou, M.; Long, A.; Xue, Y.; Liao, H.; Wei, C.; Xu, Z.J. Heterostructured Electrocatalysts for Hydrogen Evolution Reaction Under Alkaline Conditions. *Nano-Micro Lett.* **2018**, *10*, 75. [CrossRef] [PubMed]

13. Liu, X.; Wang, X.; Yuang, X.; Dong, W.; Huang, F. Rational composition and structural design of in situ grown nickel-based electrocatalysts for efficient water electrolysis. *J. Mater. Chem. A* **2016**, *4*, 167–172. [CrossRef]

14. Macdonald, T.J.; Xu, J.; Elmas, S.; Mange, Y.J.; Skinner, W.M.; Xu, H.; Nann, T. NiO Nanofibers as a Candidate for a Nanophotocathode. *Nanomaterials* **2014**, *4*, 256–266. [CrossRef] [PubMed]

15. Hu, C.; Chu, K.; Zhao, Y.; Teoh, E.Y. Efficient Photoelectrochemical Water Splitting over Anodized p Type NiO Porous Films. *ACS Appl. Mater. Interfaces* **2014**, *6*, 18558–18568. [CrossRef]

16. Li, L.; Gibson, E.A.; Qin, P.; Boschloo, G.; Gorlov, M.; Hagfeldt, A.; Sun, L. Double-Layered NiO Photocathodes for p-Type DSSCs with Record IPCE. *Adv. Mater.* **2010**, *22*, 1759–1762. [CrossRef] [PubMed]

17. Wu, L.; Guo, X.; Xu, Y.; Xiao, Y.; Qian, J.; Xu, Y.; Guan, Z.; He, Y.; Zeng, Y. Electrocatalytic activity of porous Ni–Fe–Mo–C–LaNi$_5$ sintered electrodes for hydrogen evolution reaction in alkaline solution. *RSC Adv.* **2017**, *7*, 32264–32274. [CrossRef]

18. Arce, E.M.; Lopez, V.M.; Martinez, L.; Dorantes, H.J.; Saucedo, M.L.; Hernandez, F. Electrocatalytic properties of mechanically alloyed Co-20 wt % Ni-10 wt % Mo and Co-70 wt % Ni-10 wt % Mo alloy powders. *J. Mater. Sci.* **2003**, *38*, 275–278. [CrossRef]

19. Kedzierzawski, P.; Oleszak, D.; Janik-Czachor, M. Hydrogen evolution on hot and cold consolidated Ni-Mo alloys produced by mechanical alloying. *Mater. Sci. Eng. A* **2001**, *300*, 105–112. [CrossRef]

20. Ngamlerdpokin, K.; Tantavichet, N. Electrodeposition of nickel–copper alloys to use as a cathode for hydrogen evolution in an alkaline media. *Int. J. Hydrog. Energy* **2014**, *39*, 2505–2515. [CrossRef]

21. Solmaz, R.; Döner, A.; Kardaş, G. Electrochemical deposition and characterization of NiCu coatings as cathode materials for hydrogen evolution reaction. *Electrochem. Commun.* **2008**, *10*, 1909–1911. [CrossRef]

22. Friend, W.Z. *Corrosion of Nickel and Nickel Alloys*; Wiley-Interscience: Manhattan, NY, USA, 1980.

23. Solmaz, R.; Döner, A.; Kardaş, G. The stability of hydrogen evolution activity and corrosion behavior of NiCu coatings with long-term electrolysis in alkaline solution. *Int. J. Hydrog. Energy* **2009**, *34*, 2089–2094. [CrossRef]

24. Bae, J.H.; Han, J.-H.; Chung, T.D. Electrochemistry at nanoporous interfaces: New opportunity for electrocatalysis. *Phys. Chem. Chem. Phys.* **2012**, *14*, 448–463. [CrossRef] [PubMed]

25. Dominguez Crespo, M.A.; Plata Torres, M.; Torres Huerta, A.M.; Ortiz Rodriguez, I.A.; Ramirez Rodriguez, C.; Arce Estrada, E.M. Influence of Fe contamination and temperature on mechanically alloyed Co-Ni-Mo electrodes for hydrogen evolution reaction in alkaline water. *Mater. Charact.* **2006**, *56*, 138–146. [CrossRef]

26. Aymard, L.; Dumont, B.; Viau, G. Production of Co-Ni alloys by mechanical-alloying. *J. Alloys Compd.* **1996**, *242*, 108–113. [CrossRef]

27. Zhang, J.; Li, C.M. Nanoporous metals: fabrication strategies and advanced electrochemical applications in catalysis, sensing and energy systems. *Chem. Soc. Rev.* **2012**, *41*, 7016–7031. [CrossRef]

28. Martínez, W.; Fernández, A.; Cano-Castillo, U. Synthesis of nickel-based skeletal catalyst for an alkaline electrolyzer. *Int. J. Hydrog. Energy* **2010**, *35*, 8457–8462. [CrossRef]
29. Olurin, O.B.; Wilkinson, D.S.; Weatherly, G.C.; Paserin, V.; Shu, J. Strength and ductility of as-plated and sintered CVD nickel foams. *Compos. Sci. Technol.* **2003**, *63*, 2317–2329. [CrossRef]
30. Paserin, V.; Marcuson, S.; Shu, J.; Wilkinson, D.S. CVD Technique for Inco Nickel Foam Production. *Adv. Eng. Mater.* **2004**, *6*, 454–459. [CrossRef]
31. Marozzi, C.; Chialvo, A. Development of electrode morphologies of interest in electrocatalysis. Part 1: Electrodeposited porous nickel electrodes. *Electrochim. Acta* **2000**, *45*, 2111–2120. [CrossRef]
32. Li, Y.; Xiao, X.; Zhao, R.; Yu, K.; Hou, C.; Liang, J. Study of the fabrication and characterization of porous ni using polystyrene sphere template. *Synth. React. Inorg. Met.-Org. Nano-Met. Chem.* **2016**, *46*, 286–290. [CrossRef]
33. Böhme, O.; Leidich, F.U.; Salge, H.J.; Wendt, H. Development of materials and production technologies for molten carbonate fuel cells. *Int. J. Hydrog. Energy* **1994**, *19*, 349–355. [CrossRef]
34. Rahman, A.; Zhu, X.; Wen, C. Fabrication of Nanoporous Ni by chemical dealloying Al from Ni-Al alloys for lithium-ion batteries. *Int. J. Electrochem. Sci.* **2015**, *10*, 3767–3783.
35. Sun, L.; Chien, C.L.; Searson, P.C. Fabrication of Nanoporous Nickel by Electrochemical Dealloying. *Chem. Mater.* **2004**, *16*, 3125–3129. [CrossRef]
36. Chang, J.K.; Hsu, S.H.; Sun, I.W.; Tsai, W.T. Formation of Nanoporous Nickel by Selective Anodic Etching of the Nobler Copper Component from Electrodeposited Nickel–Copper Alloys. *J. Phys. Chem. C* **2008**, *112*, 1371–1376. [CrossRef]
37. Jeong, M.G.; Zhuo, K.; Cherevko, S.; Chung, C.H. Formation of nanoporous nickel oxides for supercapacitors prepared by electrodeposition with hydrogen evolution reaction and electrochemical dealloying. *Korean J. Chem. Eng.* **2012**, *29*, 1802–1805. [CrossRef]
38. Zhang, J.; Zhan, Y.; Bian, H.; Li, Z.; Tsang, C.K.; Lee, C.; Cheng, H.; Shu, S.; Li, Y.Y.; Lu, J. Electrochemical dealloying using pulsed voltage waveforms and its application for supercapacitor electrodes. *J. Power Sources* **2014**, *257*, 374–379. [CrossRef]
39. Erlebacher, J.; Aziz, M.J.; Karma, A.; Dimitrov, N.; Sieradzki, K. Evolution of nanoporosity in dealloying. *Nature* **2001**, *410*, 450–453. [CrossRef] [PubMed]
40. Losiewicz, B.; Budniok, A.; Rowinski, E.; Lagiewka, E.; Lasia, A. The structure, morphology and electrochemical impedance study of the hydrogen evolution reaction on the modified nickel electrodes. *Int. J. Hydrog. Energy* **2004**, *29*, 145–157. [CrossRef]
41. Panek, J.; Serek, A.; Budniok, A.; Rowinski, E.; Lagiewka, E. Ni + Ti composite layers as cathode materials for electrolytic hydrogen evolution. *Int. J. Hydrog. Energy* **2003**, *28*, 169–175. [CrossRef]
42. Hiltz, C.; Lasia, A. Experimental study and modeling of impedance of the her on porous Ni electrodes. *J. Electroanal. Chem.* **2001**, *500*, 213–222. [CrossRef]
43. Armstrong, R.D.; Henderson, M. Impedance plane display of a reaction with an adsorbed intermediate. *J. Electroanal. Chem.* **1972**, *39*, 81–90. [CrossRef]
44. Lasia, A.; Rami, A. Kinetics of hydrogen evolution on Nickel electrodes. *J. Electroanal. Chem.* **1990**, *294*, 123–141. [CrossRef]
45. Birry, L.; Lasia, A. Studies of the hydrogen evolution reaction on Raney nickel molybdenum electrodes. *J. Appl. Electrochem.* **2004**, *34*, 735–749. [CrossRef]
46. Herraiz-Cardona, I.; Ortega, E.; Vazquez-Gomez, L.; Perez-Herranz, V. Double-template fabrication of three-dimensional porous nickel electrodes for hydrogen evolution reaction. *Int. J. Hydrog. Energy* **2012**, *37*, 2147–2156. [CrossRef]
47. Orazem, M.E.; Tribollet, B. *Electrochemcial Impedance Spectroscopy*; John Wiley & Sons, Inc.: Hoboken, NJ, USA, 2008.
48. Herraiz-Cardona, I.; Ortega, E.; Pérez-Herranz, V. Impedance study of hydrogen evolution on Ni/Zn and Ni–Co/Zn stainless steel based electrodeposits. *Electrochim. Acta* **2011**, *56*, 1308–1315. [CrossRef]
49. Borresen, B.; Hagen, G.; Tunold, R. Hydrogen evolution on $Ru_xTi_{1-x}O_2$ in 0.5 M H_2SO_4. *Electrochim. Acta* **2002**, *47*, 1819–1827. [CrossRef]
50. Castro, E.B.; De Giz, M.J.; Gonzalez, E.R.; Vilche, J.R. An electrochemical impedance study on the kinetics and mechanism of the hydrogen evolution reaction on nickel molybdenite electrodes. *Electrochim. Acta* **1997**, *42*, 951–959. [CrossRef]

51. Abouelamaiem, D.I.; He, G.; Neville, T.P.; Patel, D.; Ji, S.; Wang, R.; Parkin, I.P.; Jorge, A.B.; Titirici, M.M.; Shearing, P.R.; et al. Correlating electrochemical impedance with hierarchical structure for porous carbon-based supercapacitors using a truncated transmission line model. *Electrochim. Acta* **2018**, *284*, 597–608. [CrossRef]

52. Brug, G.J.; Van den Eeden, A.L.G.; Sluyters-Rehbach, M.; Sluyters, J.H. The Analysis of Electrode Impedances Complicated by the Presence of a Constant Phase Element. *J. Electroanal. Chem.* **1984**, *176*, 275–295. [CrossRef]

53. Valero-Vidal, C.; Herraiz-Cardona, I.; Pérez-Herranz, V.; Igual-Munoz, A. Stability of 3D-porous Ni/Cu cathodes under real alkaline electrolyzer operating conditions and its effect on catalytic activity. *Appl. Catal. B Environm.* **2016**, *198*, 142–153. [CrossRef]

54. Pletcher, D.; Greff, R.; Peat, R.; Peter, L.M.; Robinson, J. *Instrumental Methods in Electrochemistry*; Woodhead Publishing Limited: Philadelphia, PA, USA, 2010.

55. Delgado, D.; Minakshi, M.; Kim, D.J.; Kyeong, C.W. Influence of the Oxide Content in the Catalytic Power of Raney Nickel in Hydrogen Generation. *Anal. Lett.* **2017**, *50*, 2386–2401. [CrossRef]

56. Xu, P.; McCool, N.S.; Mallouk, T.E. Water splitting dye-sensitized solar cells. *Nano Today* **2017**, *14*, 42–58. [CrossRef]

Article

Effect of the Degree of Inversion on the Photoelectrochemical Activity of Spinel ZnFe₂O₄

Luis I. Granone [1,2,*], Konstantin Nikitin [3], Alexei Emeline [3], Ralf Dillert [1,2,***] and Detlef W. Bahnemann [1,2,3]**

[1] Institute of Technical Chemistry, Gottfried Wilhelm Leibniz University Hannover, Callinstrasse 3, D-30167 Hannover, Germany; bahnemann@iftc.uni-hannover.de

[2] Laboratory of Nano- and Quantum-Engineering (LNQE), Gottfried Wilhelm Leibniz University Hannover, Schneiderberg 39, D-30167 Hannover, Germany

[3] Laboratory "Photoactive Nanocomposite Materials", Saint-Petersburg State University, Ulyanovskaya str. 1, 198504 Peterhof, Saint-Petersburg, Russia; konstantin.nikitin@spbu.ru (K.N.); alexei.emeline@spbu.ru (A.E.)

* Correspondence: granone@iftc.uni-hannover.de (L.G.); dillert@iftc.uni-hannover.de (R.D.); Tel.: +49-511-762-16039 (R.D.)

Received: 7 April 2019; Accepted: 7 May 2019; Published: 9 May 2019

Abstract: Physicochemical properties of spinel $ZnFe_2O_4$ (ZFO) are known to be strongly affected by the distribution of the cations within the oxygen lattice. In this work, the correlation between the degree of inversion, the electronic transitions, the work function, and the photoelectrochemical activity of ZFO was investigated. By room-temperature photoluminescence measurements, three electronic transitions at approximately 625, 547, and 464 nm (1.98, 2.27, and 2.67 eV, respectively) were observed for the samples with different cation distributions. The transitions at 625 and 547 nm were assigned to near-band-edge electron-hole recombination processes involving $O^{2-}\,2p$ and $Fe^{3+}\,3d$ levels. The transition at 464 nm, which has a longer lifetime, was assigned to the relaxation of the excited states produced after electron excitations from $O^{2-}\,2p$ to $Zn^{2+}\,4s$ levels. Thus, under illumination with wavelengths shorter than 464 nm, electron-hole pairs are produced in ZFO by two apparently independent mechanisms. Furthermore, the charge carriers generated by the $O^{2-}\,2p$ to $Zn^{2+}\,4s$ electronic transition at 464 nm were found to have a higher incident photon-to-current efficiency than the ones generated by the $O^{2-}\,2p$ to $Fe^{3+}\,3d$ electronic transition. As the degree of inversion of ZFO increases, the probability of a transition involving the $Zn^{2+}\,4s$ levels increases and the probability of a transition involving the $Fe^{3+}\,3d$ levels decreases. This effect contributes to the increase in the photoelectrochemical efficiency observed for the ZFO photoanodes having a larger cation distribution.

Keywords: $ZnFe_2O_4$; degree of inversion; cation distribution; photoelectrochemical activity

---

1. Introduction

During the last few years, there has been increasing interest in the study of spinel $ZnFe_2O_4$ (ZFO) as a photoanode material for photoelectrochemical water oxidation [1–3]. As new scientific investigations in this field are reported, differences concerning the photoelectrochemical activity of ZFO photoanodes prepared by different routes have become evident [4–11]. It is well known that ZFO exhibits a variable structure where the distribution of Zn^{2+} and Fe^{3+} cations between octahedral and tetrahedral sites within the crystal lattice depends on the synthetic conditions [12–18]. Therefore, the reason behind the broad variety of results found in the scientific literature for ZFO photoanodes might be related to the cation distribution. The parameter used to quantify the cation distribution is the degree of inversion, x, which is defined as $^T[Zn_{1-x}Fe_x]^O[Zn_xFe_{2-x}]O_4$ with $0 \le x \le 1$ (where the superscripts T and O denote the tetrahedral and octahedral sites, respectively). When $x = 0$

and $x = 1$, ZFO adopts the so-called normal (^{T}[Zn]^O[Fe$_2$]O$_4$) and inverse (^{T}[Fe]^O[ZnFe]O$_4$) structure, respectively. The degree of inversion of bulk ZFO can be controlled by the calcination of the samples at temperatures higher than 737 K and subsequent quenching [12–15]. Thus, bulk ZFO samples with x ranging from approximately 0.02 to 0.20 can be prepared. Pavese et al. [13] reported degrees of inversion up to $x \approx 0.34$ at 1600 K for bulk ZFO by in situ high-temperature neutron powder diffraction measurements. Nevertheless, as shown by O'Neill [12], degrees of inversion higher than $x \approx 0.20$ cannot be experimentally accessed for bulk ZFO samples prepared by means of a solid-state reaction and subsequent quenching. Calcination temperatures higher than 1200 K are required to increase the degree of inversion above this upper limit and, under these conditions, the rate of re-ordering is too fast to quench the sample. For nanoparticulate ZFO, higher degrees of inversion are likely to be obtained [16–18], and the synthesis of ZFO nanoparticles having an almost completely inverted structure ($x = 0.94$) has been reported [19].

In a recent publication, Zhu et al. [8] reported that the cation distribution in partially reduced ZFO anodes affected the performance of light-induced water oxidation. The authors showed that partially reduced ZFO had a relatively poor crystallinity, but a high degree of inversion exhibited superior photogenerated charge carrier transport when compared to ZFO with a high crystallinity but a low degree of inversion. The research of Zhu et al. pioneered the investigation of the effect of the cation distribution on the photoelectrochemical activity of ZFO. However, a study of the effect of the degree of inversion on the photoelectrochemical activity of pristine ZFO samples exhibiting uniform particle size, crystallinity, and crystallite size is, to the best of our knowledge, missing.

Recently, we reported the preparation of ZFO by a solid-state reaction and its further processing into pellets with varying degrees of inversion [14,20]. The elemental analysis of the ZFO pellets revealed a Fe to Zn ratio of 2:1 within the limit of the experimental error, as expected for ZFO [14]. The absence of non-reacted α-Fe$_2$O$_3$ and ZnO as well as the absence of secondary iron oxide phases, were confirmed by XRD and Raman spectroscopy [14]. Mössbauer spectroscopy confirmed the absence of Fe^{2+} and, hence, of oxygen vacancies for all the ZFO pellets [14]. The crystallite size values deduced from the Rietveld refinements and the particle size distribution obtained from the SEM confirmed that the pellets exhibited similar crystallite and particle sizes independently of the degree of inversion [14,20]. Thus, the degree of inversion was found to be the only independent variable between the different ZFO pellets. These characteristics made it possible to investigate the impact of the degree of inversion on the photoelectrochemical activity of ZFO photoanodes unaffected by other variable parameters such as impurities, the number of oxygen vacancies, the particle size, the crystallite size, and the crystallinity.

In the present work, the photoelectrochemical activity of photoanodes made of pristine ZFO with degrees of inversion increasing from $x \approx 0.07$ to $x \approx 0.20$ is reported. Furthermore, the effect of the cation distribution on electronic properties such as the Fermi level and the electronic structure was investigated for the first time. The electronic structure was studied by means of time-averaged and transient room-temperature photoluminescence spectroscopy. It is well known that the photoluminescence spectrum of a material depends on its particle size, crystallinity, and the presence of point defects [21,22]. Therefore, the crystallinity and crystallite size homogeneity of the synthesized ZFO pellets is of utmost importance in order to access meaningful information concerning the effect of the degree of inversion on the photoluminescence properties. The nature of the observed transitions as well as their lifetime and the impact of the degree of inversion are discussed.

2. Results

ZFO pellets with degrees of inversion of $x \approx 0.07$ (ZFO_773), $x \approx 0.10$ (ZFO_873), $x \approx 0.13$ (ZFO_973), $x \approx 0.16$ (ZFO_1073), and $x \approx 0.20$ (ZFO_1173) were prepared by employing a spinel zinc ferrite synthesized by a solid-state reaction as reported previously [14]. In order to access information concerning the porosity of the ZFO pellets, N$_2$ and Ar physisorption isotherms were measured. Total pore volumes below 10 cm^3 g^{-1} were obtained for all pellets. These values were at the lower limit of quantification, suggesting that the samples did not exhibit a considerable porosity. This is an expected result for dense pellets pressed at high pressure. Other consequences of the low porosity were small BET surface areas below 10 m^2 g^{-1}. The pellets exhibited values ranging from 3.0 to 9.9 m^2 g^{-1} with an average of 5.3 m^2 g^{-1} and no systematic trend concerning the degree of inversion. Thus, independent of the degree of inversion, negligible total pore volumes and sizes were obtained for the different ZFO pellets.

The photoelectrochemical activity of the photoanodes was evaluated by measuring the photocurrent for the methanol oxidation reaction. Figure 1a shows the current density – voltage (j-V) curves measured under chopped solar simulated light for the ZFO photoanodes with increasing degrees of inversion. The light was turned on and off at 20 s intervals. Onset photocurrents for the methanol oxidation were observed at an anodic bias potential of around +0.9 V vs. RHE. At a bias potential of +1.2 V vs. RHE, the dark currents were still negligible and the photocurrents were high enough to allow a direct comparison between the different photoanodes. Figure 1b shows the chopped light chronoamperometry measured at an applied bias of +1.2 V vs. RHE. It was observed that the photocurrent for the methanol oxidation increased as the degree of inversion rose from $x \approx 0.07$ to $x \approx 0.20$. The photoanodes with degrees of inversion of $x \approx 0.07$, $x \approx 0.10$, and $x \approx 0.13$ showed current densities below 0.05 µA cm^{-2}. There was a significant increase in the current density from 0.05 to 0.24 µA cm^{-2} as the degree of inversion increased from $x \approx 0.13$ to $x \approx 0.16$. The current density further increased up to 0.77 µA cm^{-2} as the degree of inversion increased from $x \approx 0.16$ to $x \approx 0.20$.

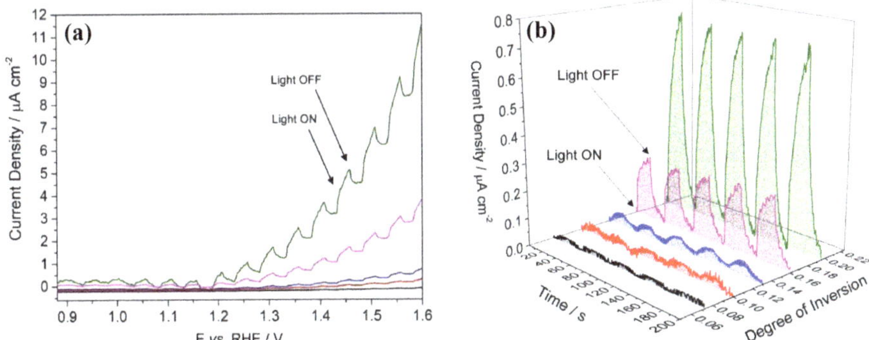

Figure 1. (a) Current density – voltage curves for the photoanodes made from ZFO pellets with different degrees of inversion. The measurements were performed in a 50% v/v methanol aqueous solution containing 0.1 mol L^{-1} KNO$_3$ under chopped solar simulator irradiation (intensity output of 680 W m^{-2}). The light was turned on and off at 20 s intervals. (b) Chopped light chronoamperometry for the ZFO photoanodes measured with an externally applied bias of +1.2 V vs. RHE. $x = 0.074$ (ZFO_773); $x = 0.104$ (ZFO_873); $x = 0.134$ (ZFO_973); $x = 0.159$ (ZFO_1073); and $x = 0.203$ (ZFO_1173).

Figure 2 shows the result of an incident photon-to-current efficiency (IPCE) measurement performed with the ZFO photoanode with a degree of inversion of $x \approx 0.20$ (ZFO_1173). This photoanode showed the highest photocurrent density for the methanol oxidation (0.77 µA cm^{-2}, Figure 1b). Considering the optical properties of the ZFO_1173 pellet reported previously [14], the absorbed photon-to-current efficiency (APCE) was calculated. It can be observed from Figure 2 that the APCE and, therefore, the ratio between the number of photogenerated holes reacting with methanol and the number of absorbed photons increased as the wavelength of the incident light became shorter. Thus, the photoanode converts the incident light into an electrical current more efficiently as the energy of the photons emitted by the excitation source increases.

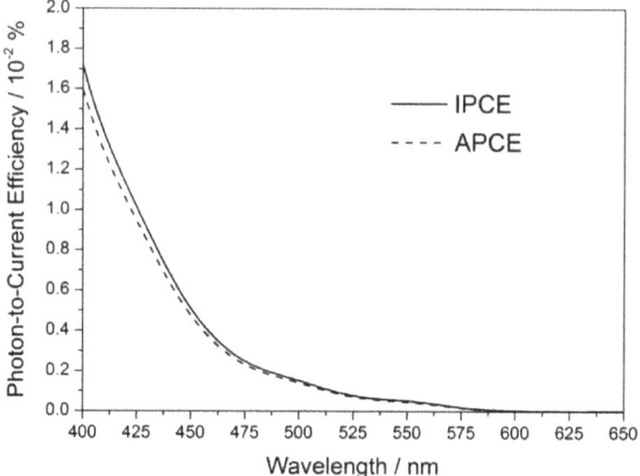

Figure 2. IPCE and APCE of the ZFO photoanode with $x \approx 0.20$ measured under an externally applied bias of +1.2 V vs. RHE in a 50% v/v methanol aqueous solution containing 0.1 mol L^{-1} KNO_3. A monochromator was used for fine-tuning the wavelengths of the analyzing light to a final resolution of 1 nm.

Figure 3 shows the work function measured by the Kelvin probe technique for the ZFO pellet samples with different degrees of inversion. It was observed that the work function exhibited values ranging from 5.28 to 5.47 eV. Considering the experimental uncertainty (±0.13 eV), no significant changes in the Fermi level were observed as the degree of inversion of the ZFO pellets increased.

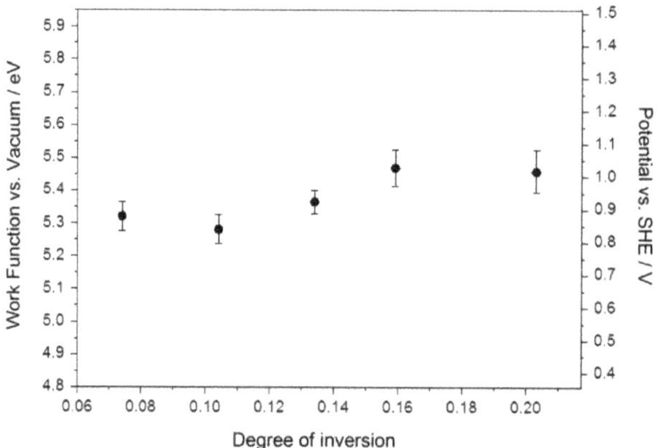

Figure 3. Work function of the ZFO pellet samples with degrees of inversion ranging from $x \approx 0.07$ to $x \approx 0.20$. The values were measured using a scanning Kelvin probe system. Before the measurement, the pellets were calcined at 673 K for 12 h to remove adsorbed water.

Time-averaged room-temperature photoluminescence measurements were carried out to investigate the effect of the degree of inversion on the electronic structure of ZFO. It becomes obvious from Figure 4 that all pellets exhibited three emission signals at approximately 464, 547, and 625 nm. The relative fluorescence quantum yield of the photoluminescence was determined by using quinine hemisulfate as a standard. Fluorescence quantum yields of 0.05% for the ZFO pellets with degrees of inversion of $x \approx 0.07$, $x \approx 0.10$, and $x \approx 0.13$, 0.04% for the pellet with $x \approx 0.15$, and 0.03% for the pellets with $x \approx 0.20$ were obtained.

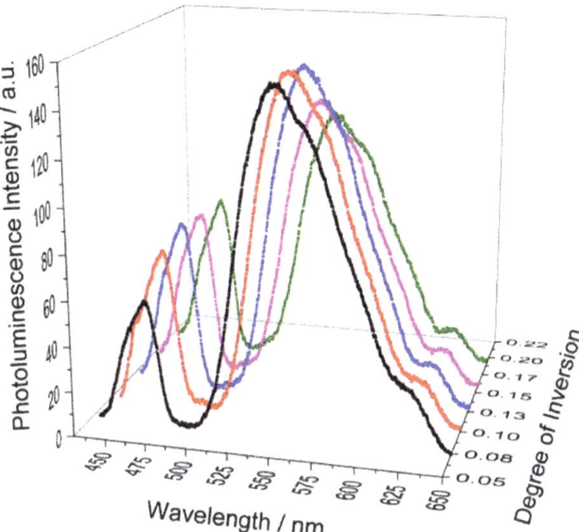

Figure 4. Time-averaged room-temperature photoluminescence of the ZFO pellets as a function of the degree of inversion. An excitation wavelength of 355 nm with an emission slit width of 10 nm was used for the measurement.

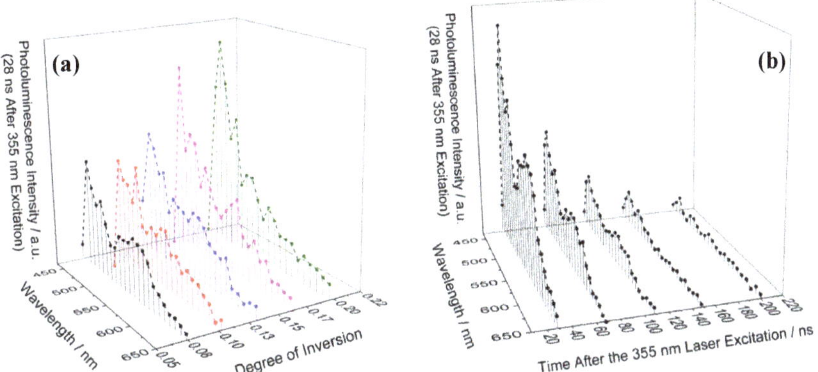

Figure 5. (**a**) Transient room-temperature photoluminescence spectra of the ZFO pellets as a function of the degree of inversion. The third harmonic (355 nm) of a Brilliant Nd:YAG laser with a pulse duration of 6 ns was used as the excitation source. The spectra were measured 28 ns after the excitation. $x = 0.074$ (ZFO_773); $x = 0.104$ (ZFO_873); $x = 0.134$ (ZFO_973); $x = 0.159$ (ZFO_1073); and $x = 0.203$ (ZFO_1173). (**b**) Transient room-temperature photoluminescence spectra of the ZFO pellet with $x = 0.074$ (ZFO_773) measured at different points in time after the 355 nm laser excitation.

Transient room-temperature photoluminescence measurements were conducted to study the lifetime of the electronic transitions. Figure 5a shows the photoluminescence spectra of the ZFO pellets with different degrees of inversion measured at 28 ns after the 355 nm laser excitation. Since 10 nm steps were used to record the transient signals, the spectral resolution was lower in comparison to the time-averaged measurements. Thus, the emission observed at approximately 625 nm in the time-averaged photoluminescence measurements (Figure 4) was not detected in the transient measurements. However, the emissions centered at 547 and 464 nm were clearly observed. The emission centered at 464 nm showed a higher intensity than the emission centered at 547 nm. Figure 5b shows the room-temperature photoluminescence spectra of the ZFO pellet with a degree of inversion of $x \approx 0.07$ (ZFO_773) measured at different times after the laser excitation. The emission signal centered at 464 nm was weakly observed at 208 ns after the excitation, while the emission signal centered at 547 nm was no longer observed at 148 ns after the excitation.

3. Discussion

A series of ZFO photoanodes with different cation distributions between the tetrahedral and octahedral sites of the oxygen lattice were investigated to study the impact of the degree of inversion on the photoelectrochemical activity of ZFO. Due to the high-temperature calcination steps carried out during the preparation of the samples, large crystallite sizes of approximately 300 nm were obtained [14]. Large crystallite sizes imply large diffusion lengths for the photogenerated charge carriers [23,24]. Consequently, the charge carrier recombination rate increases and the photoelectrochemical efficiency of the material decreases [25]. Furthermore, low active surface areas are expected as the crystallite size increases [23,24]. Therefore, the large crystallite size of the ZFO photoanodes were responsible for the low photocurrent values obtained for the photoelectrochemical methanol oxidation (Figure 1). The large thickness of the films (approximately 750 µm) also had a negative impact on the photoelectrochemical activity for the same reasons above-mentioned. Higher photocurrent densities are reported in the literature for ZFO photoelectrodes [4,6,26,27]. In fact, the current benchmark in the performance of a partially reduced ZFO photoanode for solar water oxidation is 1.0 mA cm^{-2} at 1.23 V [8]. However, the aim of this work was to understand the impact of the degree of inversion on both the electronic properties and the photoelectrochemical activity of ZFO. The large particle size and thickness of the prepared ZFO electrodes as well as the absence of oxygen vacancies were the main reasons behind the observed low photocurrents. These drawbacks cannot be avoided when the synthesis of pristine ZFO samples in which the degree of inversion is the only independent variable is intended. However, although the measured photocurrents were below 1.0 µA cm^{-2}, an impact of the degree of inversion on the current density was clearly observed. It becomes obvious from Figure 1 that an increase of the cation distribution results in an increase in the photoelectrochemical activity of ZFO. The electrical conductivity of the pellets used in the present work was reported in a previous study [20]. An increase in the conductivity from 9.28×10^{-9} to 1.82×10^{-5} S cm^{-1} was observed as the degree of inversion increased from $x \approx 0.07$ to $x \approx 0.20$ [20]. Interestingly, a significant increase of two orders of magnitude was observed in the conductivity by increasing the degree of inversion from $x \approx 0.13$ to $x \approx 0.16$. This increase in the conductivity agreed with the large increase in the photocurrent from 0.05 to 0.24 µA cm^{-2} that was observed for methanol oxidation as the degree of inversion increased from $x \approx 0.13$ to $x \approx 0.16$ (Figure 1b). It becomes evident from the preceding discussion that the increase in the electrical conductivity of the pellets with a higher degree of inversion is closely related to the improvement of the photoelectrochemical activity. However, the interrelation between the photoelectrochemical activity and the degree of inversion might also be related to other physicochemical properties of ZFO. Therefore, the impact of the cation distribution on the work function and the electronic transitions of ZFO was investigated.

As mentioned in the previous section, no significant changes in the work function were observed as the degree of inversion of the ZFO pellets increased. In a semiconductor, the work function represents the minimum energy required to remove an electron from the Fermi level into the free space [28]. The work function values obtained for the ZFO pellets corresponded, considering the IUPAC recommended value of -4.44 V for the absolute electrode potential of the hydrogen electrode [29], to Fermi levels around $+0.94$ V vs. SHE. Sun et al. [5] reported work function values ranging from 5.46 to 5.55 eV ($+1.02$ and $+1.11$ V vs. SHE) for mesoporous ZFO nanoparticles obtained via an evaporation-induced self-assembly method and calcined at different temperatures. The results were in good agreement with the values obtained in the present work. The work function of a semiconductor strongly depends on its doping level [30]. As well as for many n-type metal oxide semiconductors [31], the mechanism for n-type doping of ZFO is the formation of oxygen vacancies [32], which results in the reduction of Fe^{3+} to Fe^{2+} [33]. As no Fe^{2+} was detected by Mössbauer spectroscopy for the ZFO pellets used in this work [14], a low donor density was expected. The high work function values obtained and the large anodic potential necessary for the methanol photoelectrochemical oxidation were the consequences of the low donor density. Furthermore, as similar work functions were observed for the different pellets, it was concluded that their donor densities were in the same order of magnitude and did not depend on the degree of inversion. For highly doped ZFO samples, work function values between 4.23 and 4.97 eV (−0.21 and +0.53 V vs. SHE) have been reported in the literature [27,34–36].

The optical band gap and the nature of the optical transitions of the pellets used for the present work were determined by measuring the diffuse reflectance and applying the derivation of absorption spectrum fitting (DASF) method, as was shown in a previous report [14]. Independent of the degree of inversion, the pellets exhibited an indirect band gap transition at approximately 614 nm (2.02 eV), and a direct band gap transition at approximately 532 nm (2.33 eV) [14]. These values were in good agreement with the 1.9 and 2.3 eV reported by Guijarro et al. [27] for the indirect and direct band gap, respectively. From density functional theory calculations, Yao et al. [37] showed that the valence band of ZFO consisted of O^{2-} 2s, Zn^{2+} 3d, Fe^{3+} 3d, and O^{2-} 2p states. However, the valence band edge consisted of Fe^{3+} 3d and mainly O^{2-} 2p states [37]. The conduction band edge consisted of O^{2-} 2p and mainly Fe^{3+} 3d states [37]. At higher energies, the contribution from the Zn^{2+} 4s states was observed in the density of states. Lv et al. [38] claimed that the energy band structures of ZFO were defined by considering the O^{2-} 2p levels as the valence band edge and the Fe^{3+} 3d levels as the conduction band edge. The emissions observed in Figure 4 at approximately 547 and 625 nm were in reasonable agreement with the direct and indirect band gap transitions at 532 and 614 nm, respectively [14]. Thus, these bands were ascribed to near-band-edge emissions due to the electron relaxation from Fe^{3+} 3d levels located in the conduction band edge to O^{2-} 2p levels in the valence band edge [39,40]. The emission centered at 464 nm might be related to transitions involving the Zn^{2+} cations [41,42]. It is well known for Fe-doped ZnO that the near-band-edge emission of ZnO centered at approximately 379 nm becomes red-shifted as the amount of Fe increases [42,43]. If the amount of Fe is high enough and spinel ZFO is formed as a secondary phase, a new emission centered at 464 nm is observed [42]. Therefore, the transition observed at approximately 464 nm could be ascribed to the electron relaxation from Zn^{2+} 4s levels within the conduction band to O^{2-} 2p in the valence band edge. This assignment agrees with the density of states presented by Yao et al. [37] Regardless of the nature of the electronic transitions, it is important to stress that the conduction band of ZFO does not exhibit a continuous density of empty electronic energetic states. Contradicting the semiconductor band theory [44], the conduction band of photoexcited electrons were delocalized in confined densities of states involving either Fe^{3+} 3d levels or Zn^{2+} 4s levels. Whether the photoexcited electron is delocalized in Fe^{3+} 3d levels or Zn^{2+} 4s levels depends on the energy of the excitation source. A scheme of the electronic excitation mechanism of ZFO is shown in Figure 6.

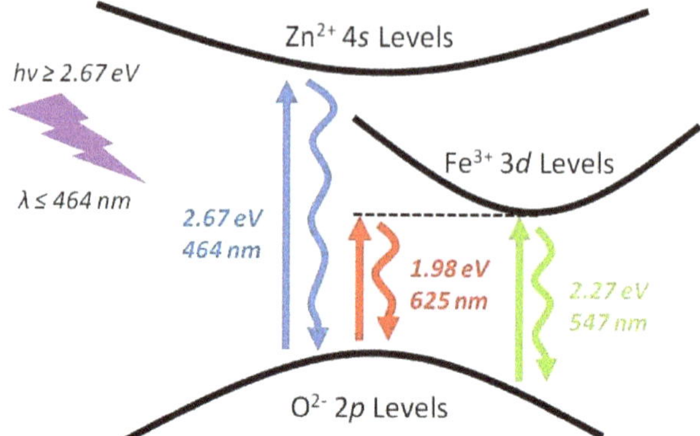

Figure 6. Scheme of the electronic transitions observed for ZFO when photons with wavelengths shorter than 464 nm (energy higher than 2.67 eV) are used for the excitation.

The influence of the O^{2-} 2p to Fe^{3+} 3d (indirect and direct) and the O^{2-} 2p to Zn^{2+} 4s electronic transitions on the photoelectrochemical efficiency of ZFO was deduced from the APCE measurement. According to Figure 2, when the ZFO photoanode is irradiated with light at wavelengths longer than 600 nm, the APCE is approximately zero. Thus, the methanol oxidation efficiency of the charge carriers generated via the indirect O^{2-} 2p to Fe^{3+} 3d transition (625 nm) is negligible. As also reported by Guijarro et al. [27], the indirect transition of ZFO does not effectively drive photoelectrochemical processes. The APCE increases as wavelengths ranging from 475 to 600 nm are used for the excitation of the photoanode. Under these irradiation conditions, the direct O^{2-} 2p to Fe^{3+} 3d transition occurs and the generated charge carriers can convert the incident light into an electrical current. Interestingly, a significant increase in the APCE values can be observed when the photoanode is irradiated with light at wavelengths shorter than 475 nm. This phenomenon could be attributed to two reasons. One is the increase in the absorptivity of the material as is observed from the diffuse reflectance measurements of the ZFO pellets reported elsewhere [14]. The second reason is the contribution of the charge carriers generated by the electronic transition from O^{2-} 2p to Zn^{2+} 4s levels (464 nm). The absorption coefficient of ZFO has approximately the same magnitude for wavelengths ranging from 400 to 500 nm [45,46]. Hence, the significant increase in the APCE should be observed at wavelengths longer than 475 nm if the larger light absorption of the material is responsible for this effect. Therefore, the increase in the APCE values at wavelengths shorter than 475 nm must be mainly due to the contribution to the methanol photooxidation of the charge carriers generated via the O^{2-} 2p to Zn^{2+} 4s electronic transition (464 nm).

The data presented in Figure 4 revealed that the degree of inversion affected the relative intensity of the room-temperature photoluminescence bands of ZFO. As the cation disorder increased, the intensity of the near-band-edge emission centered at 547 nm decreased, and the intensity of the emission centered at 464 nm increased. These emission bands were the result of electron-hole recombination processes and, thus, it is reasonable to assume that the observed increase or decrease in the emission intensity is due to an increase or decrease, respectively, in the number of generated electron-hole pairs. Therefore, as Zn^{2+} cations located in tetrahedral sites are interchanged by Fe^{3+} cations from octahedral sites, the probability of the electronic transition at 464 nm (from O^{2-} 2p to Zn^{2+} 4s levels) increases and the probability of the near-band-edge electronic transitions (from O^{2-} 2p to Fe^{3+} 3d levels) decreases. From the time-averaged photoluminescence measurements, it was observed that the intensity of the emission centered at 547 nm was higher than that of the emission centered at 464 nm (Figure 4). However, transient photoluminescence measurements showed a higher intensity for the emission centered at 464 nm (Figure 5a). This can be explained by a faster decay of the emission centered at 547 nm. In fact, the signal centered at 547 nm was no longer observed at 148 ns after the excitation while the signal centered at 464 nm (due to the relaxation of the O^{2-} $2p$ to Zn^{2+} $4s$ electron excitation) was observed even 208 ns after the excitation (Figure 5b).

As discussed above, the electron-hole pairs generated by the electronic transitions observed at 547 and 464 nm are involved in the photoelectrochemical process occurring at the ZFO electrodes. It was shown that the transition centered at 464 nm had a higher efficiency for photoelectrochemical methanol oxidation than the transition centered at 547 nm (Figure 2). However, the valence band holes generated via both transitions had the same redox potential and, therefore, the same oxidizing activity. The higher efficiency of the transition centered at 464 nm was due to the longer lifetime of the generated charge carriers (Figure 5b). As the degree of inversion of the ZFO pellets increased, both, the amount of O^{2-} $2p$ to Zn^{2+} $4s$ electronic transitions and the photoelectrochemical activity increased.

4. Materials and Methods

Polycrystalline ZFO samples with degrees of inversion increasing from $x \approx 0.07$ to $x \approx 0.20$ were synthesized by means of a solid-state reaction as reported previously [14]. Briefly, stoichiometric amounts of ZnO (60 mmoles, Sigma Aldrich, Taufkirchen, Germany, $\geq$ 99.0%) and α-Fe_2O_3 (60 mmoles, Sigma Aldrich, Taufkirchen, Germany, $\geq$99.0%) powders were ground in an agate mortar. The mixture was calcined in air at 1073 K for 12 h, cooled down to room temperature, and ground once again. Aliquots of 0.500 g were pressed into 13 mm diameter pellets applying a pressure of 55 MPa. The pellets were calcined at 1273 K for 2 h, cooled down to 1073 K and kept at this temperature for 12 h, then cooled down to 773 K and kept at this temperature for 50 h, and finally quenched in cold water. Some of the pellets (referred as ZFO_773) were separated and the rest were divided into four sets of pellets. These pellets were heated up with a rate of 300 K h^{-1} and calcined at 873, 973, 1073, and 1173 K for 25, 20, 12, and 10 h, respectively. As reported by O'Neill [12], these calcination times were sufficiently long for the ZFO pellets to reach a steady-state value of the degree of inversion. After this period of time, the calcined pellets were immediately quenched in cold water. These pellets are referred as ZFO_873, ZFO_973, ZFO_1073, and ZFO_1173. The ZFO pellets were sanded with Al_2O_3 sandpaper (KK114F, grit size P320, VSM Abrasives, Hannover, Germany) to a final thickness of 0.75 mm $\pm$ 0.02 mm.

Molecular nitrogen physisorption isotherms were measured at 77 K on a Quantachrome Autosorb-3MP instrument and argon physisorption isotherms were measured at 87 K on a Quantachrome Autosorb-1 instrument (3P Instruments GmbH & Co. KG, Odelzhausen, Germany). The pellets were outgassed in vacuum at 423 K for 24 h prior to the sorption measurements. Surface areas were estimated by applying the Brunauer–Emmett–Teller (BET) equation [47] and the total pore volumes were estimated by the single-point method at $p/p_0 = 0.95$.

Electrochemical and photoelectrochemical measurements were performed by employing a ZENNIUM Electrochemical Workstation (Zahner Scientific Instruments, Kronach, Germany) equipped with a three-electrode electrochemical cell with a Pt counter electrode and an Ag/AgCl/NaCl (3 mol kg^{-1}) reference electrode. ZFO working electrodes were prepared by attaching a copper wire with silver paint (Ferro GmbH, Frankfurt am Main, Germany) and conductive epoxy (Chemtronics, Kennesaw, GA, USA) to one face of the pellet sample. Photoelectrochemical measurements were performed in a 50% v/v methanol aqueous solution containing 0.1 mol L^{-1} KNO_3. A solar simulator (LOT-Quantum Design GmbH, Darmstadt, Germany) consisting of a 300 W xenon-arc lamp provided with an AM 1.5-global filter was used as the irradiation source. An intensity output of 680 W m^{-2} at the position of the working electrode was measured using a SpectraRad Xpress spectral irradiance meter (B&W Tek, Newark, DE, USA). For the IPCE measurements, a TLS03 tunable light source (Zahner Scientific Instruments, Kronach, Germany) consisting of an array of monochromatic LEDs with emission wavelengths ranging from 400 to 650 nm was used. A monochromator was employed for fine tuning the wavelengths to a final resolution of 1 nm.

A scheme of the set-up used for the transient photoluminescence spectroscopy is shown in Figure 7. The third harmonic (355 nm) of a Brilliant Nd:YAG laser (Quantel, Lannion, France) with a pulse duration of 6 ns was used as the excitation source. A laser intensity of 3.0 mJ per pulse was selected for the measurements. The intensity of the laser was measured using a Maestro laser power meter (Gentec-EO, Québec, Canada). The angle of the laser beam path was adjusted by rotating a Pellin-Broca prism beam steering module. The illumination area of the laser beam was approximately 0.5 cm^2. The light emitted by the pellets after the laser excitation was collected by a Spectrosil lens and directed toward the monochromator by a folding mirror. The monochromator was connected with a PMT R928 photomultiplier detector (Hamamatsu Photonics, Hamamatsu, Japan). To avoid the overloading of the photomultiplier, a 370 nm cut off filter was introduced in front of the monochromator entrance. For the transient photoluminescence measurements, a constant voltage of 700 V was applied to the photomultiplier.

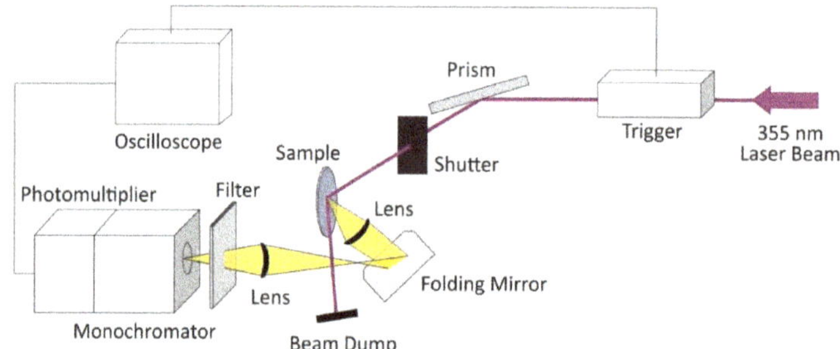

Figure 7. Scheme of the diffuse reflectance setup used for the transient photoluminescence measurements. The 355 nm laser excitation is shown in violet and the radiation emitted by the ZFO pellet is represented in yellow.

Time-averaged photoluminescence measurements were performed in an F-7100 fluorescence spectrophotometer (Hitachi, Tokyo, Japan). A wavelength of 355 nm was selected as the excitation source and the photoluminescence spectra were measured from 440 to 650 nm with a 240 nm min^{-1} scan rate. Excitation and emission slit widths of 10 nm were used. The relative fluorescence quantum yield was determined by employing a reported standard method [48–50]. Briefly, a 5×10^{-3} mol L^{-1} solution of quinine hemisulfate monohydrate ($C_{20}H_{24}N_2O_2 \cdot 0.5H_2SO_4 \cdot H_2O$, Sigma Aldrich, Taufkirchen, Germany, $\geq 98.0\%$) in 1 N H_2SO_4, with an absolute quantum yield efficiency of 0.51 at 25 °C [48], was used to determine the relative photoluminescence quantum yield of the ZFO pellets. For the time-averaged and the transient photoluminescence measurements, suspensions of the ZFO pellets were prepared. ZFO pellets were ground in an agate mortar and dispersed in deionized water by a one-hour ultrasound treatment (340 W L^{-1}). A concentration of 2.5 g L^{-1} was used for the measurements.

Work function measurements were performed with a scanning Kelvin probe system SKP5050 (KP Technology, Wick, Scotland) versus a gold reference probe electrode (probe area of 2 mm^2). The probe oscillation frequency was 74 Hz, and the backing potential was 7000 mV. Work function values were obtained by averaging 1000 data points for two different sites of each pellet. Prior to the measurements, the pellets were heated at 673 K for 12 h to remove adsorbed water. The cation diffusion process in ZFO is kinetically hindered at temperatures lower than 773 K [12]. Thus, the degree of inversion of the pellets did not change during the heat treatment.

5. Conclusions

The effect of the degree of inversion on the photoelectrochemical activity of ZFO was investigated. As the cation distribution changed from $x \approx 0.07$ to $x \approx 0.20$, the photoelectrochemical activity of ZFO increased. In order to study the correlation between this phenomenon and the electronic properties of the material, the work function as well as the time-averaged and transient photoluminescence of the different ZFO samples were studied. No significant effect of the degree of inversion on the Fermi level was observed. Regarding the photoluminescence results, the characteristic near-band-edge emissions (547 and 625 nm for the direct and indirect transition, respectively) due to the electron-hole recombination involving Fe^{3+} $3d$ and O^{2-} $2p$ levels were observed. Furthermore, an emission of energy higher than the band gap was also detected. This emission was assigned to the relaxation of the excited state produced after the electronic transition from O^{2-} $2p$ to Zn^{2+} $4s$ levels (464 nm). Interestingly, the lifetime of the latter emission was observed to be longer than the lifetime of the near-band-edge emission. After excitation with photons with wavelengths shorter than 464 nm (blue and violet region of the visible light spectrum), electron-hole pairs were produced in ZFO by two apparently independent mechanisms (Figure 6). One pathway was the excitation of electrons from O^{2-} $2p$ levels at the valence band maximum to Fe^{3+} $3d$ levels at the conduction band minimum. The other was the excitation of electrons from O^{2-} $2p$ levels at the valence band maximum to Zn^{2+} $4s$ levels located within the conduction band. The charge carriers generated by the latter mechanism showed a longer lifetime and, consequently, a higher efficiency for the photoelectrochemical methanol oxidation. As the degree of inversion of ZFO increased, the transition involving the Zn^{2+} $4s$ levels was favored, thus contributing to the observed increase in the photoelectrochemical activity.

Author Contributions: Conceptualization, L.I.G.; Investigation, L.I.G. and K.N.; Project administration, D.W.B.; Supervision, A.E., R.D., and D.W.B.; Writing—original draft, L.I.G.; Writing—review & editing, R.D.

Funding: This research was funded by the Deutsche Forschungsgemeinschaft (DFG) under the SPP 1613 program (BA 1137/22-1), the Niedersächsisches Ministerium für Wissenschaft und Kultur (NTH-research group "ElektroBak"), and the Korea government (MSIP) through NRF under the Global Research Laboratory program (2014K1A1A2041044).

Acknowledgments: The authors would like to thank Peter Behrens and Malte Schäfer (Institute for Inorganic Chemistry, Gottfried Wilhelm Leibniz University Hannover) for the physisorption measurements. The publication of this article was funded by the Open Access Fund of the Gottfried Wilhelm Leibniz Universität Hannover.

Conflicts of Interest: The authors declare no conflict of interest.

References

1. Dillert, R.; Taffa, D.H.; Wark, M.; Bredow, T.; Bahnemann, D.W. Research update: Photoelectrochemical water splitting and photocatalytic hydrogen production using ferrites (MFe_2O_4) under visible light irradiation. *APL Mater.* **2015**, *3*, 104001. [CrossRef]

2. Taffa, D.H.; Dillert, R.; Ulpe, A.C.; Bauerfeind, K.C.L.; Bredow, T.; Bahnemann, D.W.; Wark, M. Photoelectrochemical and theoretical investigations of spinel type ferrites ($M_xFe_{3-x}O_4$) for water splitting: A mini-review. *J. Photonics Energy* **2016**, *7*, 12009. [CrossRef]

3. Chandrasekaran, S.; Bowen, C.; Zhang, P.; Li, Z.; Yuan, Q.; Ren, X.; Deng, L. Spinel photocatalysts for environmental remediation, hydrogen generation, CO_2 reduction and photoelectrochemical water splitting. *J. Mater. Chem. A* **2018**, *6*, 11078–11104. [CrossRef]

4. Kim, J.H.; Jang, Y.J.; Kim, J.H.; Jang, J.W.; Choi, S.H.; Lee, J.S. Defective $ZnFe_2O_4$ nanorods with oxygen vacancy for photoelectrochemical water splitting. *Nanoscale* **2015**, *7*, 19144–19151. [CrossRef]

5. Sun, M.; Chen, Y.; Tian, G.; Wu, A.; Yan, H.; Fu, H. Stable mesoporous $ZnFe_2O_4$ as an efficient electrocatalyst for hydrogen evolution reaction. *Electrochim. Acta* **2016**, *190*, 186–192. [CrossRef]

6. Hufnagel, A.G.; Peters, K.; Müller, A.; Scheu, C.; Fattakhova-Rohlfing, D.; Bein, T. Zinc ferrite photoanode nanomorphologies with favorable kinetics for water-splitting. *Adv. Funct. Mater.* **2016**, *26*, 4435–4443. [CrossRef]

7. Peeters, D.; Taffa, D.H.; Kerrigan, M.M.; Ney, A.; Jöns, N.; Rogalla, D.; Cwik, S.; Becker, H.W.; Grafen, M.; Ostendorf, A.; et al. Photoactive zinc ferrites fabricated via conventional CVD approach. *ACS Sustain. Chem. Eng.* **2017**, *5*, 2917–2926. [CrossRef]

8. Zhu, X.; Guijarro, N.; Liu, Y.; Schouwink, P.; Wells, R.A.; Le Formal, F.; Sun, S.; Gao, C.; Sivula, K. Spinel structural disorder influences solar-water-splitting performance of $ZnFe_2O_4$ nanorod photoanodes. *Adv. Mater.* **2018**, *30*, 1801612. [CrossRef] [PubMed]

9. Kirchberg, K.; Wang, S.; Wang, L.; Marschall, R. Mesoporous $ZnFe_2O_4$ photoanodes with template-tailored mesopores and temperature-dependent photocurrents. *ChemPhysChem* **2018**, *19*, 2313–2320. [CrossRef]

10. Kim, J.H.; Jang, Y.J.; Choi, S.H.; Lee, B.J.; Kim, J.H.; Park, Y.B.; Nam, C.M.; Kim, H.G.; Lee, J.S. A multitude of modifications strategy of $ZnFe_2O_4$ nanorod photoanodes for enhanced photoelectrochemical water splitting activity. *J. Mater. Chem. A* **2018**, *6*, 12693–12700. [CrossRef]

11. Guo, Y.; Zhang, N.; Wang, X.; Qian, Q.; Zhang, S.; Li, Z.; Zou, Z. A facile spray pyrolysis method to prepare Ti-doped $ZnFe_2O_4$ for boosting photoelectrochemical water splitting. *J. Mater. Chem. A* **2017**, *5*, 7571–7577. [CrossRef]

12. O'Neill, H.S.C. Temperature dependence of the cation distribution in zinc ferrite ($ZnFe_2O_4$) from powder XRD structural refinements. *Eur. J. Miner.* **1992**, *4*, 571–580. [CrossRef]

13. Pavese, A.; Levy, D.; Hoser, A.H. Cation distribution in synthetic zinc ferrite ($Zn_{0.97}Fe_{2.02}O_4$) from in situ high-temperature neutron powder diffraction. *Am. Mineral.* **2000**, *85*, 1497–1502. [CrossRef]

14. Granone, L.I.; Ulpe, A.C.; Robben, L.; Klimke, S.; Jahns, M.; Renz, F.; Gesing, T.M.; Bredow, T.; Dillert, R.; Bahnemann, D.W. Effect of the degree of inversion on optical properties of spinel $ZnFe_2O_4$. *Phys. Chem. Chem. Phys.* **2018**, *20*, 28267–28278. [CrossRef] [PubMed]

15. Bræstrup, F.; Hauback, B.C.; Hansen, K.K. Temperature dependence of the cation distribution in $ZnFe_2O_4$ measured with high temperature neutron diffraction. *J. Solid State Chem.* **2008**, *181*, 2364–2369. [CrossRef]

16. Akhtar, M.J.; Nadeem, M.; Javaid, S.; Atif, M. Cation distribution in nanocrystalline $ZnFe_2O_4$ investigated using x-ray absorption fine structure spectroscopy. *J. Phys. Condens. Matter* **2009**, *21*, 405303. [CrossRef] [PubMed]

17. Kumar, G.S.Y.; Naik, H.S.B.; Roy, A.S.; Harish, K.N.; Viswanath, R. Synthesis, optical and electrical properties of $ZnFe_2O_4$ nanocomposites. *Nanomater. Nanotechnol.* **2012**, *2*. [CrossRef]

18. Nakashima, S.; Fujita, K.; Tanaka, K.; Hirao, K.; Yamamoto, T.; Tanaka, I. First-principles XANES simulations of spinel zinc ferrite with a disordered cation distribution. *Phys. Rev. B* **2007**, *75*, 174443. [CrossRef]

19. Šepelák, V.; Tkáčová, K.; Boldyrev, V.V.; Wigmann, S.; Becker, K.D. Mechanically induced cation redistribution in $ZnFe_2O_4$ and its thermal stability. *Phys. B* **1997**, *234–236*, 617–619. [CrossRef]

20. Granone, L.I.; Dillert, R.; Heitjans, P.; Bahnemann, D.W. Effect of the degree of inversion on the electrical conductivity of spinel $ZnFe_2O_4$. *ChemistrySelect* **2019**, *4*, 1232–1239. [CrossRef]

21. Mitra, J.; Ghosh, M.; Bordia, R.K.; Sharma, A. Photoluminescent electrospun submicron fibers of hybrid organosiloxane and derived silica. *RSC Adv.* **2013**, *3*, 7591–7600. [CrossRef]

22. Vinosha, P.A.; Mely, L.A.; Jeronsia, J.E.; Krishnan, S.; Das, S.J. Synthesis and properties of spinel $ZnFe_2O_4$ nanoparticles by facile co-precipitation route. *Optik* **2017**, *134*, 99–108. [CrossRef]

23. Kočí, K.; Obalová, L.; Matějová, L.; Plachá, D.; Lacný, Z.; Jirkovský, J.; Šolcová, O. Effect of TiO_2 particle size on the photocatalytic reduction of CO_2. *Appl. Catal. B Environ.* **2009**, *89*, 494–502. [CrossRef]

24. Dodd, A.C.; McKinley, A.J.; Saunders, M.; Tsuzuki, T. Effect of particle size on the photocatalytic activity of nanoparticulate zinc oxide. *J. Nanoparticle Res.* **2006**, *8*, 43–51. [CrossRef]

25. Sivula, K.; Le Formal, F.; Grätzel, M. Solar water splitting: Progress using hematite (α-Fe_2O_3) photoelectrodes. *ChemSusChem* **2011**, *4*, 432–449. [CrossRef]

26. Kim, J.H.; Kim, J.H.; Jang, J.W.; Kim, J.Y.; Choi, S.H.; Magesh, G.; Lee, J.; Lee, J.S. Awakening solar water-splitting activity of $ZnFe_2O_4$ nanorods by hybrid microwave annealing. *Adv. Energy Mater.* **2015**, *5*, 1401933. [CrossRef]

27. Guijarro, N.; Bornoz, P.; Prévot, M.S.; Yu, X.; Zhu, X.; Johnson, M.; Jeanbourquin, X.A.; Le Formal, F.; Sivula, K. Evaluating spinel ferrites MFe_2O_4 (M = Cu, Mg, Zn) as photoanodes for solar water oxidation: Prospects and limitations. *Sustain. Energy Fuels* **2018**, *2*, 103–117. [CrossRef]

28. Kahn, A. Fermi level, work function and vacuum level. *Mater. Horizons* **2016**, *3*, 7–10. [CrossRef]

29. Trasatti, S. The absolute electrode potential: An explanatory note. *Pure Appl. Chem.* **1986**, *58*, 955. [CrossRef]

30. Matsumoto, Y. Energy positions of oxide semiconductors and photocatalysis with iron complex oxides. *J. Solid State Chem.* **1996**, *126*, 227–234. [CrossRef]

31. Raebiger, H.; Lany, S.; Zunger, A. Origins of the p-type nature and cation deficiency in Cu_2O and related materials. *Phys. Rev. B* **2007**, *76*, 45209. [CrossRef]

32. Nunome, T.; Irie, H.; Sakamoto, N.; Sakurai, O.; Shinozaki, K.; Suzuki, H.; Wakiya, N. Magnetic and photocatalytic properties of n- and p-type $ZnFe_2O_4$ particles synthesized using ultrasonic spray pyrolysis. *J. Ceram. Soc. Jpn.* **2013**, *121*, 26–30. [CrossRef]

33. Šutka, A.; Pärna, R.; Kleperis, J.; Käämbre, T.; Pavlovska, I.; Korsaks, V.; Malnieks, K.; Grinberga, L.; Kisand, V. Photocatalytic activity of non-stoichiometric $ZnFe_2O_4$ under visible light irradiation. *Phys. Scr.* **2014**, *89*, 44011. [CrossRef]

34. Zheng, X.-L.; Dinh, C.T.; Pelayo García de Arquer, F.; Zhang, B.; Liu, M.; Voznyy, O.; Li, Y.-Y.; Knight, G.; Hoogland, S.; Lu, Z.-H.; et al. $ZnFe_2O_4$ leaves grown on TiO_2 trees enhance photoelectrochemical water splitting. *Small* **2016**, *12*, 3181–3188. [CrossRef]

35. Liu, X.; Zheng, H.; Li, Y.; Zhang, W. Factors on the separation of photogenerated charges and the charge dynamics in oxide/$ZnFe_2O_4$ composites. *J. Mater. Chem. C* **2013**, *1*, 329–337. [CrossRef]

36. Nada, A.A.; Nasr, M.; Viter, R.; Miele, P.; Roualdes, S.; Bechelany, M. Mesoporous $ZnFe_2O_4$@TiO_2 nanofibers prepared by electrospinning coupled to PECVD as highly performing photocatalytic materials. *J. Phys. Chem. C* **2017**, *121*, 24669–24677. [CrossRef]

37. Yao, J.; Li, X.; Li, Y.; Le, S. Density functional theory investigations on the structure and electronic properties of normal spinel $ZnFe_2O_4$. *Integr. Ferroelectr.* **2013**, *145*, 17–23. [CrossRef]

38. Lv, H.; Ma, L.; Zeng, P.; Ke, D.; Peng, T. Synthesis of floriated $ZnFe_2O_4$ with porous nanorod structures and its photocatalytic hydrogen production under visible light. *J. Mater. Chem.* **2010**, *20*, 3665–3672. [CrossRef]

39. Lemine, O.M.; Bououdina, M.; Sajieddine, M.; Al-Saie, A.M.; Shafi, M.; Khatab, A.; Al-Hilali, M.; Henini, M. Synthesis, structural, magnetic and optical properties of nanocrystalline $ZnFe_2O_4$. *Phys. B Condens. Matter* **2011**, *406*, 1989–1994. [CrossRef]

40. Zhu, X.; Zhang, F.; Wang, M.; Ding, J.; Sun, S.; Bao, J.; Gao, C. Facile synthesis, structure and visible light photocatalytic activity of recyclable $ZnFe_2O_4/TiO_2$. *Appl. Surf. Sci.* **2014**, *319*, 83–89. [CrossRef]

41. Fang, Z.; Wang, Y.; Xu, D.; Tan, Y.; Liu, X. Blue luminescent center in ZnO films deposited on silicon substrates. *Opt. Mater.* **2004**, *26*, 239–242. [CrossRef]

42. Srivastava, A.K.; Deepa, M.; Bahadur, N.; Goyat, M.S. Influence of Fe doping on nanostructures and photoluminescence of sol-gel derived ZnO. *Mater. Chem. Phys.* **2009**, *114*, 194–198. [CrossRef]

43. Chen, A.J.; Wu, X.M.; Sha, Z.D.; Zhuge, L.J.; Meng, Y.D. Structure and photoluminescence properties of Fe-doped ZnO thin films. *J. Phys. D. Appl. Phys.* **2006**, *39*, 4762–4765. [CrossRef]

44. Kittel, C. *Introduction to Solid State Physics*, 8th ed.; John Wiley & Sons: New York, NY, USA, 2005; ISBN 0-471-41526-X.

45. Sun, S.; Yang, X.; Zhang, Y.; Zhang, F.; Ding, J.; Bao, J.; Gao, C. Enhanced photocatalytic activity of sponge-like $ZnFe_2O_4$ synthesized by solution combustion method. *Prog. Nat. Sci. Mater. Int.* **2012**, *22*, 639–643. [CrossRef]

46. Xie, T.; Xu, L.; Liu, C.; Wang, Y. Magnetic composite $ZnFe_2O_4/SrFe_{12}O_{19}$: Preparation, characterization, and photocatalytic activity under visible light. *Appl. Surf. Sci.* **2013**, *273*, 684–691. [CrossRef]

47. Brunauer, S.; Emmett, P.H.; Teller, E. Adsorption of gases in multimolecular layers. *J. Am. Chem. Soc.* **1938**, *60*, 309–319. [CrossRef]

48. Melhuish, W.H. Quantum efficiencies of fluorescence of organic substances: Effect of solvent and concentration of the fluorescent solute. *J. Phys. Chem.* **1961**, *65*, 229–235. [CrossRef]

49. Fletcher, A.N. Quinine sulfate as a fluorescence quantum yield standard. *Photochem. Photobiol.* **1969**, *9*, 439–444. [CrossRef]

50. Würth, C.; Grabolle, M.; Pauli, J.; Spieles, M.; Resch-Genger, U. Relative and absolute determination of fluorescence quantum yields of transparent samples. *Nat. Protoc.* **2013**, *8*, 1535–1550. [CrossRef]

Review

Advanced Design and Synthesis of Composite Photocatalysts for the Remediation of Wastewater: A Review

Jianlong Ge, Yifan Zhang, Young-Jung Heo and Soo-Jin Park *

Department of Chemistry and Chemical Engineering, Inha University, 100 Inharo, Incheon 22212, Korea; gejianlong1121@126.com (J.G.); zyf910626@inhaian.net (Y.Z.); heoyj1211@inhaian.net (Y.-J.H.)
* Correspondence: sjpark@inha.ac.kr; Tel.: +82-32-860-7234; Fax: +82-32-860-5604

Received: 29 December 2018; Accepted: 28 January 2019; Published: 30 January 2019

Abstract: Serious water pollution and the exhausting of fossil resources have become worldwide urgent issues yet to be solved. Solar energy driving photocatalysis processes based on semiconductor catalysts is considered to be the most promising technique for the remediation of wastewater. However, the relatively low photocatalytic efficiency remains a critical limitation for the practical use of the photocatalysts. To solve this problem, numerous strategies have been developed for the preparation of advanced photocatalysts. Particularly, incorporating a semiconductor with various functional components from atoms to individual semiconductors or metals to form a composite catalyst have become a facile approach for the design of high-efficiency catalysts. Herein, the recent progress in the development of novel photocatalysts for wastewater treatment via various methods in the sight of composite techniques are systematically discussed. Moreover, a brief summary of the current challenges and an outlook for the development of composite photocatalysts in the area of wastewater treatment are provided.

Keywords: Composite catalysts; photocatalysis; synergy effect; solar energy; wastewater remediation

1. Introduction

In the past several decades, with the booming of industry, the ever-increasing consumption of natural resources, especially fresh water and fossil resources, have caused alarming damage to the environment and seriously threaten the sustainability of human society [1–3]. As a worldwide concern, freshwater pollution drives people to seek for an effective approach to repair the polluted water environment. In general, the contaminants in water are mainly derived from the sewage effluent of industries (e.g., textile industry, paper industry, the pharmaceutical industry, etc.), and domestic contaminants (e.g., pharmaceuticals, pesticide, detergent, etc.) [4]. Until now, numerous contaminants have been detected and are classified as inorganic ions, organic chemicals, and pathogens; most of those contaminants are toxic to organisms [4–7]. Up to now, a variety of strategies including chemical or physical coagulation [8], sedimentation [9], adsorption [10], membrane filtration [11], and biological degradation method [12] have been invented to treat wastewater. However, due to the complex compositions and different physico-chemical properties of the contaminants, there are still several limitations of these traditional techniques, such as the low efficiency, high energy consumption, and the risk of secondary pollution [13–15]. Consequently, a promoted technique with high efficiency, low energy consumption, and being environmentally friendly is highly desired for the remediation of wastewater.

Nowadays, the advanced oxidation processes (AOPs) have been extensively explored to remove the non-biodegradable and highly stable compounds in water [16,17]. In fact, the AOPs are chemical processes that can generate highly reactive hydroxyl radicals ($\cdot$OH) in situ. The $\cdot$OH in water exhibits an extremely strong oxidizing property with a high oxidation potential of 2.80 V/SHE ($\cdot$OH/H$_2$O), such that it can non-selectively oxidize the contaminants and finally convert them to CO$_2$, H$_2$O, or small inorganic ions in a short time [17,18]. In most cases, the $\cdot$OH could be produced with the presence of one or more primary oxidants, and/or energy sources or catalysts. Therefore, the typical AOPs could be classified as Fenton reactions, the electrochemical advanced oxidation processes, and the heterogeneous photocatalysis [17]. Compared with the traditional water remediation techniques, the AOPs exhibit many advantages, which include: (1) the contaminants are directly destroyed or reduced in the water body, rather than simply coagulated or filtrated from the water, thus the secondary pollution could be avoided; (2) the AOPs are suitable for a wide range of contaminants including some inorganics and pathogens because of their robust non-selectively oxidizability; and (3) no hazardous byproducts will be generated due to the final reduction products of the AOPs being just CO$_2$, H$_2$O, or small inorganic ions. With the abovementioned merits, the AOPs have attracted significant attention from both scientific research and industrial processing [19].

Solar energy is a green, costless, and inexhaustible energy resource. Effective utilization of solar energy is of vital importance for enhancing the sustainability of industry, reducing pollution, and retarding global warming. Consequently, solar energy has been widely used in a range of applications, such as solar heating, photovoltaics, solar thermal energy, solar architecture, artificial photosynthesis, photocatalysis, etc. [20] Among which, photocatalysis is one of the most effective strategies for the AOPs, which just rely on the light radiation on the photocatalysts to drive the oxidization reaction at the ambient condition, and during the whole reaction process, no additional energy is needed and no toxic byproduct will be generated; therefore, it is a green chemical technique [21,22]. Actually, the core of photocatalytic AOPs are photocatalysts; semiconductors as the most employed heterogeneous photocatalysis for the AOPs have attained considerable development since Fujishima et al. [23] carried out the first photo-catalyzed AOP based on the titanium-oxide (TiO$_2$) in 1972. Up to now, a myriad of photocatalytic AOPs have been designed for water treatment based on various semiconductors. In general, semiconductors are light-sensitive because of their unique electronic structure with a filled valence band (VB) and an empty conduction band (CB) [18,21]. Figure 1 and Equations (1)–(6) demonstrate the basic reaction process of a semiconductor to generate the photocatalytic radicals, which could be decomposed in the following steps: (i) photons with a certain energy are absorbed by the semiconductor; (ii) the absorbed photons with energy greater than the band gap energy (E$_b$) of semiconductors lead to the formation of electrons in the CB and corresponding holes in the VB; and (iii) the generated electron–hole pairs will migrate to the surface of semiconductors for redox reactions, and fast recombination in nanoseconds will happen at the same time (it should be mentioned that this process is negative for the AOPs, which shall be suppressed [21,22]).

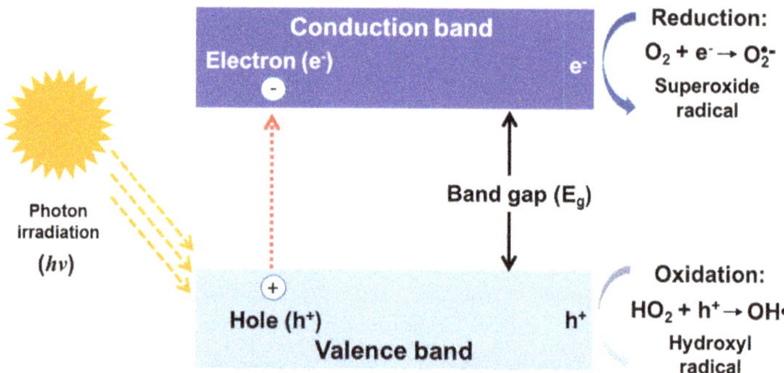

Figure 1. Schematic illustration of the photocatalytic reaction process of a semiconductor. Adapted with permission from Reference [18]. Copyright (2012) Elsevier.

$$\text{Excitation: Photon } (h\nu) + \text{Semiconductor} \rightarrow e^-_{CB} + h^+_{VB} \tag{1}$$

$$\text{Recombination: } e^- + h^+ \rightarrow \text{energy} \tag{2}$$

$$\text{Oxidation of } H_2O: H_2O + h^+_{VB} \rightarrow \bullet OH + H^+ \tag{3}$$

$$\text{Reduction of adsorbed } O_2: O_2 + e^- \rightarrow O_2\bullet^- \tag{4}$$

$$\text{Reaction with } H^+: O_2\bullet^- + H^+ \rightarrow \bullet OOH \tag{5}$$

$$\text{Electrochemical reduction: } \bullet OOH + \bullet OOH \rightarrow H_2O_2 + O_2 \tag{6}$$

However, it remains a significant challenge to fabricate a high-efficiency visible light photocatalyst solely based on an individual semiconductor photocatalyst. For example, the TiO_2, as the most used photocatalyst, possesses various advantages with excellent chemical stability, large surface area, non-toxicity, and low cost [24]; however, its wide energy band gap (3.0–3.2 eV) means it can only be excited by the UV light ($\lambda < 400$ nm), such that less than 5% of the irradiated solar energy can be effectively used [25]. Moreover, the fast recombination speed of electron–hole pairs seriously limits the further improvement of its photocatalytic activity [18,22]. On the other hand, although the recently developed visible light response semiconductors have a lower energy band gap (<3 eV), such as BiOX (X = I or Br) [26], they still suffer from serious photo-corrosion problems in aqueous media via redox reactions and the fast recombination of electron–hole pairs during the reaction process. Therefore, it is highly urgent to find an effect strategy to further improve the performance of semiconductor photocatalysts.

From ancient times, people have recognized that the incorporation of two or more constituent materials could obtain various composite materials with intriguing properties superior to the individual components. Nowadays, a myriad of functional composite materials have been developed for different applications [27,28]. Actually, the enhanced performance of a composite material is mainly attributed to the synergistic effect of its individual constituent materials; meanwhile, this principle is also appropriate to the design of semiconductor photocatalysts. Up to now, there have been numerous pioneering studies reporting the design and fabrication of composite semiconductor photocatalysts via various methods, such as doping heteroatoms or constructing heterojunctions via directly compositing with individual semiconductors or carbonaceous nanomaterials, among others. Therefore, as shown in Scheme 1, in this review, we aim to provide a systematic appraisal of the recent development in the design and fabrication of various composite photocatalysts for the application of wastewater treatment. Meanwhile, some representative photocatalysts with composite structures and morphologies from the atomic scale to macroscopic scale are reviewed. Finally, the current developing status, challenges, and evolution trend of the composite semiconductor photocatalysts for wastewater remediation are briefly proposed.

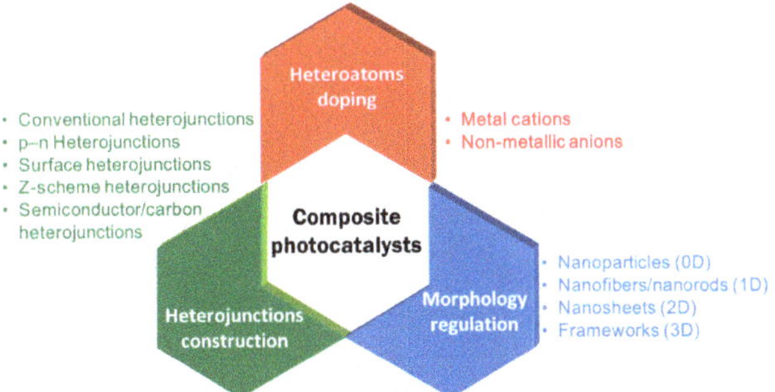

Scheme 1. The schematic illustration demonstrating the design and synthesis strategies for composite photocatalysts.

2. Principle of the Semiconductor Photocatalysts for Wastewater Remediation

As mentioned above, the trace contaminants (e.g., phenol, chlorophenol, oxalic acid) derived from the dyeing industry, petrochemical industry, and the agricultural chemicals are quite difficult to remove from the water due to the low concentration and complex compositions [4]. A photocatalytic degradation method is considered as the most promising strategy to deal with this problem. According to the previous studies [18,29,30], as shown in Figure 2, the basic mechanism of the photocatalytic degradation process of a contaminant could be characterized as the following steps: (i) the target contaminants transfer from the water body to the surface of the photocatalysts, in which the migration rate of corresponding contaminants may be influenced by the morphology and surface properties of the catalysts (e.g., surface area, porosity, and surface charges); (ii) the contaminants are adsorbed on the surface of catalysts with photon excited reaction sites, therefore a high surface area of the catalysts can provide more active sites for the reaction; (iii) the redox reactions of the photon activated sites with the adsorbed contaminants and the degraded intermediates are produced, which are finally degraded to CO_2 and H_2O; (iv) part of the generated intermediates and the resultant mineralization products (CO_2 and H_2O) desorb from the surface of catalysts to expose the active sites for the subsequent reactions; and (v) the desorbed intermediates transfer from the interface of catalysts and water to the bulk liquid, and part of the intermediates will repeat the procedure i–v until they are completely degraded to CO_2 and H_2O. Based on the abovementioned principles of the semiconductor photocatalysts for water contaminants degradation, five main criteria for the design of an effective photocatalyst could be proposed as follow: (1) a semiconductor with a lower E_g is preferred so that the electron–hole pair could be excited easier; (2) the photon absorption capacity of the catalysts shall be as high as possible to generate more electron–hole pairs; (3) the recombination process of electron–hole pairs must be prevented as much as possible to enhance the quantum efficiency of the photo-generated electron–hole pairs; (4) the surface area of the catalysts shall be large to provide more reaction sites; and (5) the chemical and physical structures of photocatalysts must be stable and be beneficial for the mass transfer in water. To meet the abovementioned requirements, a variety of strategies have been developed for the design, some of the most-used strategies will be summarized in this review.

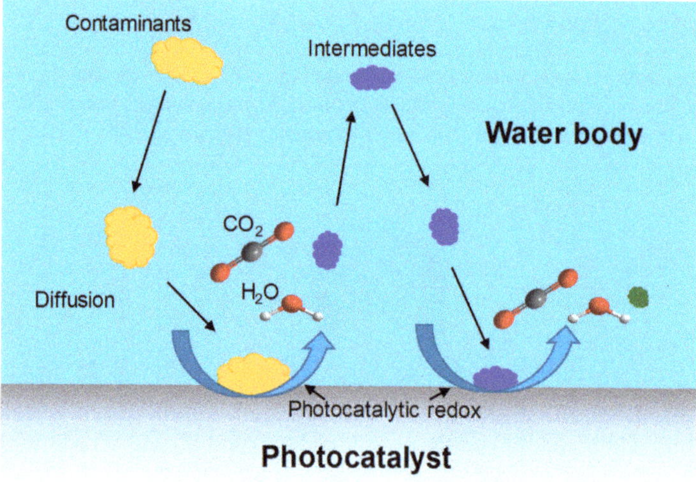

Figure 2. Schematic diagram demonstrating the removal of contaminants in water with the presence of photocatalysts [18,29,30].

3. Heteroatoms Doping

Recently, the strategy of introducing heteroatoms into the lattice of corresponding semiconductors has been widely employed to regulate the band gap of the semiconductor photocatalysts so as to improve their absorption capacity for visible lights, which takes up almost 45% in the solar light spectrum [31]. In general, the most commonly used dopants in semiconductors (e.g., TiO_2) could be classified as the metal cations and the non-metallic elements [32,33].

3.1. Metal Cations Doping

The most-used metal cation dopants for semiconductors mainly involve transition metal ions, such as Fe^{3+}, Co^{3+}, Mo^{5+}, Ru^{3+}, Ag^+, Cu^{2+}, Rb^+, Cr^{3+}, V^{4+}, etc. [32,34–37]. In most cases, the redox energy states of those employed metal cations lie within the band gap states of corresponding semiconductors (e.g., TiO_2); therefore, the introduction of those metal ions will result in an intraband state near the CB or VB edge of a semiconductor. Consequently, the red shift in band gap absorption of a metal-cation-doped semiconductor is mainly contributed by the charge migration between the d electrons of the doped cations and the CB (or VB) of the corresponding semiconductors. In addition, the doped metal cations could act as an electron–hole trap, regulating the charge carrier equilibrium concentration [38–40]. Although some transition metal cations could provide new energy levels as electron donors or acceptors, and virtually improved the visible light absorption capacity of corresponding semiconductors, this approach is also known to suffer from many disadvantages, such as bad thermal stability, significant increase in the carrier-recombination centers, and the high cost for an expensive facility, which are critical limitations for the generalization of this strategy.

3.2. Non-Metallic Anions Doping

Alternatively, doping the semiconductors with appropriate non-metallic anions has been proven to be a facile way to regulate the intrinsic electronic structure of semiconductors and could construct various heteroatomic surface structures such that the resultant non-metallic-anion-doped semiconductors exhibit improved photocatalytic performances under solar light irradiation [33,41]. In general, the chemical states and locations are key factors for the regulation of the electronic state of the dopant and the corresponding heteroatomic surface structures of the composite semiconductor catalysts. According to a previous study [18], three requirements needed to be satisfied for the doping of a semiconductor: (i) the doping process should construct states in the band gap of corresponding semiconductors with an enhanced visible light absorption capacity, (ii) the CB minimum including the doped states should be equal to that of the semiconductor's or higher than that of the H_2/H_2O level such that the photoreduction can be conducted, and (iii) the states in the gap should sufficiently overlap with the band states of semiconductors to ensure that the photoexcited carriers could migrate to the surface of catalysts within their lifetime. Based on the abovementioned principles, various elements, including C, N, F, P, and S, were employed to substitute for the O in TiO_2 [42], and the results showed that N was the most effective dopant for the improvement of visible-light photocatalysis of TiO_2 because the *p* states of N can narrow the band gap of N-doped TiO_2 via mixing with the O *2p* states [43]. Moreover, owing to the comparable atomic size with oxygen, small ionization energy, and high stability, the nitrogen has been one of the most promising elements for promoting the photocatalysis performance of the semiconductors. In general, the doped N in the TiO_2 could be classified as the substitutional type and interstitial type (Figure 3), the substitutional type N-doped TiO_2 is attributed to the oxygen replacement, while the interstitial type is attributed to the additional N element in the lattice of TiO_2 [41]. Up to now, the N-doping of semiconductors can be realized via several strategies, and the most-used techniques with certain industrial application prospects could be mentioned as the magnetron sputtering, ion implantation, chemical vapor deposition, atomic layer deposition, and sol-gel and combustion method, which will be discussed as follows.

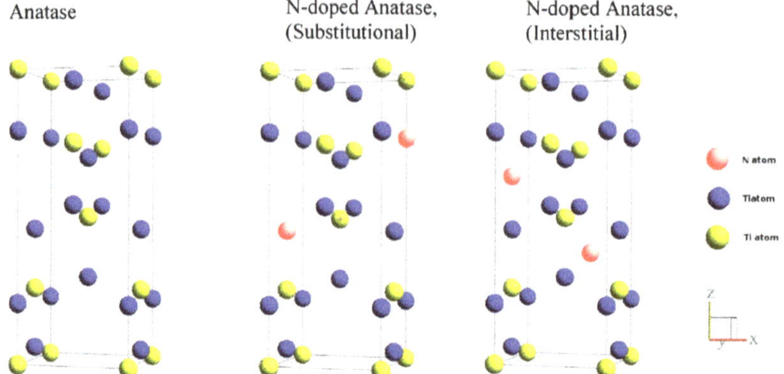

Figure 3. Schematic diagram demonstrating the N doping in the lattice of TiO_2. Adapted with permission from Reference [41]. Copyright (2011) Royal Society of Chemistry.

3.2.1. Magnetron Sputtering Method

The magnetron sputtering method is widely used for the preparation of various hybrid semiconductors. For example, Kitano et al. [44] fabricated nitrogen-substituted TiO_2 thin films by using a radio frequency magnetron sputtering (RF-MS) method. The N_2/Ar gas mixtures with different concentration of N_2 was used as the sputtering gas. They systematically investigated the influence of nitrogen content on the properties of the obtained N-TiO_2 thin films via regulating the concentration of N_2 in the sputtering gases. Meanwhile, they proved that the extent of substitution of oxygen positions with N in the lattice of TiO_2 as well as the surface morphologies of TiO_2 could be controlled well. As a result, the visible light absorption capacity of the obtained N-TiO_2 was obviously enhanced with bands up to 550 nm, and it was found that the band red shift extent was closely related to the content of the substituted N element in the TiO_2 lattice. Moreover, they found that the as-prepared N-TiO_2 photocatalyst exhibited an optimized photocatalysis reactivity with the N content of 6%. This result was because of the excessive substituted N, which causes the formation of undesirable Ti^{3+} species and acts as the recombination centers to decrease the photocatalytic activity [44]. Apart from the TiO_2, some other N-doped semiconductors could also be prepared based on the RF-MS method. Recently, Salah et al. [45] fabricated a series of N-doped ZnO nanoparticles films by employing the RF-MS method. As shown in Figure 4, the obtained N-doped ZnO films exhibited an improved response to the visible light, and possessed significantly enhanced degradation/mineralization performance for 2-chlorophenol (2-CP), 4-chlorophenol (4-CP), and 2,4-dichlorophenoxyaceticacid (2,4-D) solely under the drive of natural sunlight.

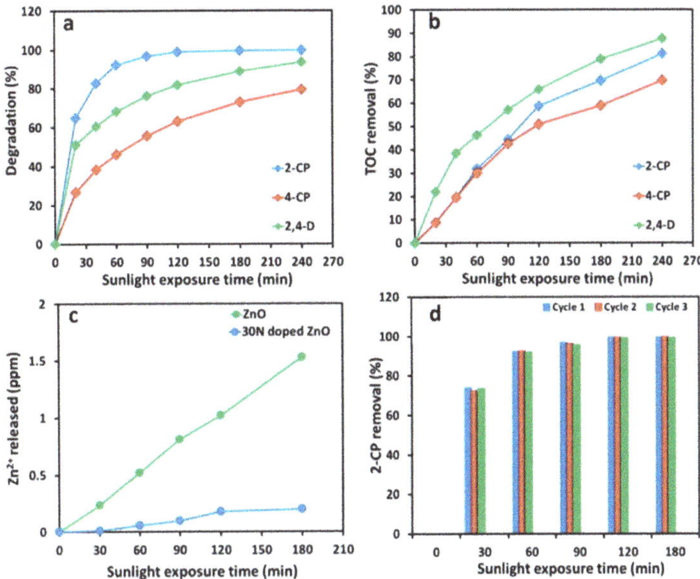

Figure 4. (**a**) The degradation and (**b**) mineralization of N-doped ZnO films for 2-CP, 4-CP, and 2,4-D. (**c**) The stability of pristine ZnO and N-doped ZnO film. (**d**) The stability and reusability of an N-doped ZnO film for the degradation of 2-CP. Adapted with permission from Reference [45]. Copyright (2016) Elsevier.

3.2.2. Ion Implantation Method

The ion implantation method as a typical materials engineering strategy that can effectively regulate the physical, chemical, and electronic properties of semiconductors, and the operation process does not involve any other elements except the selected element, which ensures the purity of the dopant [46]. Moreover, owing to the controllable parameters of ion beam implantation, such as ion element, ion energy, ion density, uniformity of ion beam, and the doping efficiency, ion beam implantation is a powerful approach for the heteroatom doping of semiconductors. For example, Tang et al. [47] fabricated an N-doped TiO_2 layer with macrospores on a titanium substrate by using the plasma-based ion implantation method. The fabrication process involves four steps: (i) a helium plasma was employed to generate He bubbles in the substrate, (ii) an oxygen plasma treatment and a followed annealing in air were used to obtain rutile and anatase phases of TiO_2, (iii) an Ar ion sputtering method was used to exposure the He bubbles on the surface; and (iv) the pre-treated samples were doped by nitrogen though the nitrogen beam ion implantation method. Moreover, co-doping of two or more non-metallic anions into a semiconductor photocatalyst (e.g., TiO_2) could also be realized using the ion implantation method. For example, Song et al. [48] prepared C/N-implanted single-crystalline rutile TiO_2 nanowire arrays by using carbon and nitrogen ions beam to treat the as-prepared TiO_2 nanowire arrays. After an annealing treatment, the obtained C/N-doped TiO_2 nanowire arrays exhibited a superior visible light response activity, which was attributed to the synergistic effect between the doped C and N atoms. Their work proved that the co-doped C and N in the lattice of TiO_2 not only greatly improves the visible light absorption capability, but also enhances the separating and transferring property of photo-generated electron–hole pairs (Figure 5).

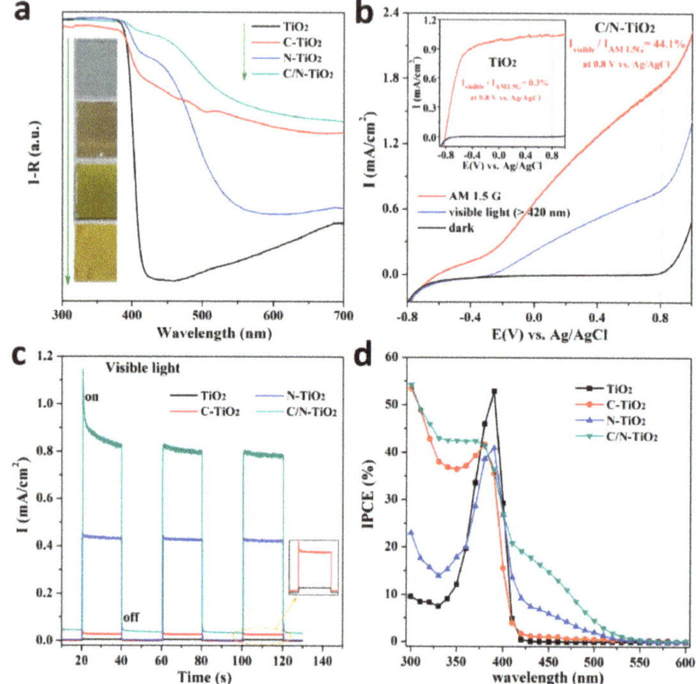

Figure 5. (**a**) UV–vis absorption spectra of TiO$_2$ and various doped TiO$_2$. (**b**) Linear sweep voltammograms of C/N-TiO$_2$ and TiO$_2$. (**c**) Photo-response of TiO$_2$ and the various doped TiO$_2$ samples under visible light. (**d**) Incident photon-to-current conversion efficiency spectra of TiO$_2$ and various doped TiO$_2$. Adapted with permission from Reference [48]. Copyright (2018) Wiley.

3.2.3. Chemical Vapor Deposition Method

Chemical vapor deposition (CVD) is a low-cost and scalable technique, which can directly grow a solid-phase material from a gas phase containing specific precursors. The CVD method has been widely used for the fabrication of semiconductors and the corresponding composite of oxides, sulfides, nitrides, and other mixed anion materials [49]. For example, Lee et al. [50] prepared TiO$_2$ composite materials doped by C (TiOC) and N (TiON) with the titanium tetraisopropoxide (TTIP), oxygen, and NH$_3$ as the precursors via combing the CVD method with a fluidized bed. The results demonstrated that the visible light photocatalysis performance of the composite TiO$_2$ (e.g., TiON) was significantly improved compared to the commercial TiO$_2$ catalyst (P25, Degussa). Similarly, Kafizas et al. [51] employed a combinatorial atmospheric pressure chemical vapor deposition (cAPCVD) method to prepare an anatase TiO$_2$ film with a gradating N content. The obtained TiO$_2$ film exhibited a gradating substitutional (N$_s$) and interstitial (N$_i$) nitrogen concentration, and the transition process from predominantly N$_s$-doped TiO$_2$ to N$_s$/N$_i$ mixtures, and finally to purely N$_i$-doped TiO$_2$ was precisely characterized. In addition, the UV and visible light photocatalytic activities of the obtained N-doped TiO$_2$ were evaluated. As a result, this work demonstrated that N$_i$-doped anatase TiO$_2$ results in a better visible light photocatalytic activity than that of predominantly N$_s$-doping. They proved that the different influences of substitutional and interstitial nitrogen doping on the photocatalytic activity of TiO$_2$ were due to that the greater stability of electron–holes in N$_i$-doped TiO$_2$ compare with that of N$_s$-doped TiO$_2$, while the propensity of the N$_s$-doped TiO$_2$ for recombination is greater. This result indicated that the doped structures is well-deigned to improve the photocatalytic activity of a semiconductor. Additionally, the CVD could also be combined with other materials synthesis strategy; for example, as shown in Figure 6, Youssef et al. [52] prepared the N-doped anatase films via a one-step low-frequency plasma enhanced chemical vapor deposition (PECVD) process. Furthermore, they demonstrated that this method did not need the subsequential annealing step or post-incorporation of the doping agent, and the as–prepared N-TiO$_2$ film exhibited good visible-light-induced photocatalytic performance.

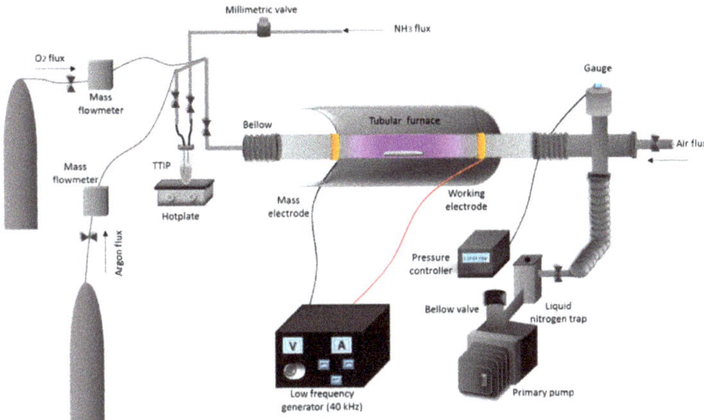

Figure 6. Schematic view of the capacitively-coupled low frequency PECVD reactor. Adapted with permission from Reference [52]. Copyright (2017) Elsevier.

3.2.4. Atomic Layer Deposition

The atomic layer deposition (ALD) method is a recently developed and facile strategy for the element doping of semiconductors. Actually, ALD is a gas-phase deposition process based on alternate surface reactions of the substrates, and the ALD method possesses several advantages, such as good reproducibility, considerable conformality, and excellent uniformity [53]. Consequently, the ALD method is considered as a promising strategy for the preparation of doped and composite photocatalysts [54]. For example, Pore et al. [55] prepared a series of N-TiO$_2$ films via employing the ALD processes. In this study, TiCl$_4$ was used as the titanium precursor and there were two ALD cycles during the fabrication process: (i) a thin layer of TiN was grown from the TiCl$_4$ and NH$_3$; and (ii) TiO$_2$ was deposited on the surface of TiN layer from TiCl$_4$ and H$_2$O, meanwhile the as-prepared TiN layer was part-oxidized to TiO$_2$, thus resulting in the TiO$_{2-x}$N$_x$. Moreover, the nitrogen concentration of the obtained TiO$_{2-x}$N$_x$ could be well controlled via changing the ratio of TiN and TiO$_2$ deposition cycles. Similarly, Lee et al. [56] reported a facile and effective vapor-phase synthesis strategy to prepare a conformal N-TiO$_2$ thin film based on the ALD process. As shown in Figure 7, the fabrication process of the corresponding N-TiO$_2$ film involved four main steps: (i) pulse the TiCl$_4$ vapor on the surfaces of a substrate to produce a monolayer of chemisorbed TiCl$_x$ species; (ii) remove the remaining unreacted TiCl$_4$ and corresponding HCl byproducts using nitrogen gas; (iii) NH$_4$OH as the nitrogen source was subsequently pulsed to generate a mixture of gaseous H$_2$O and NH$_3$, which react with the as-prepared TiCl$_x$ species to obtain the N-TiO$_2$; and (iv) remove the unreacted precursors and HCl byproducts again. This cycle could be repeated to achieve the N-TiO$_2$ film with the desired thickness. The as-prepared N-TiO$_2$ exhibited significantly enhanced photocatalytic degradation performance for organic pollutants solely driven by the solar irradiation.

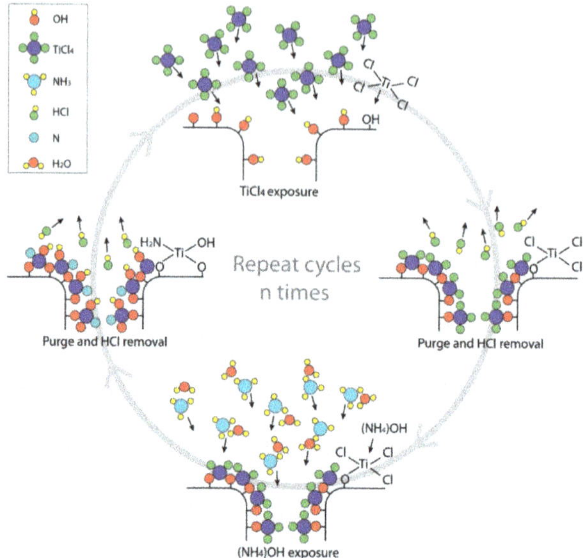

Figure 7. Schematic illustration demonstrating the synthesis process of the N-doped TiO_2 conformal film via the ALD method. Adapted with permission from Reference [56]. Copyright (2017) Wiley.

3.2.5. Sol-Gel and Combustion Method

Compared with the abovementioned synthesis approaches, the sol-gel and combustion method is a facile and low-cost strategy for the preparation of various semiconductors and the corresponding hybrid semiconductors. With the merits of simplicity and the possibility of controlling the synthesis conditions, the sol-gel methods have been well developed and several extended sol-gel techniques have been invented to fabricate new types of semiconductor photocatalysts. For example, Albrbar et al. [57] reported the synthesis of a series of mesoporous anatase TiO_2 powders doped by N, and S, as well as the N,S co-doped anatase TiO_2 powder using a non-hydrolytic sol-gel process. During the gel synthesis process, titaniumtetrachloride and titaniumisopropoxide were used as the precursor of Ti, dimethylsulfoxide (DMSO) was used as the precursor of S, and NH_3 was used as the precursor of N. For the preparation of S-doped TiO_2, the obtained gel derived from the solvent of DMSO was calcined in air, while N and S co-doped TiO_2 was obtained when the gel was annealed in the atmosphere of NH_3. In addition, the pristine TiO_2 and corresponding N-doped TiO_2 was further obtained via calcining the gel derived from the solvent of cyclohexane in air and NH_3, respectively. In their studies, the photocatalytic activities of the samples were evaluated via the degradation of dye C.I. Reactive Orange16 in water under the irradiation of visible light. The obtained results showed that the N-doped TiO_2 exhibited better visible-light photocatalytic activity compared with the pristine TiO_2 and S-doped TiO_2. Similarly, the sol-gel method is also versatile enough to be combined with other materials synthesis techniques. Most recently, Rajoriya et al. [58] successfully fabricated a samarium (Sm) and nitrogen (N) co-doped TiO_2 photocatalyst through an ultrasound-assisted sol-gel process (Figure 8), where they found that after doping TiO_2 with Sm and N, the photocatalytic degradation performance of the TiO_2 for 4-acetamidophenol was greatly improved owing to the significantly improved separation efficiency of the photo-generated electron–hole pair.

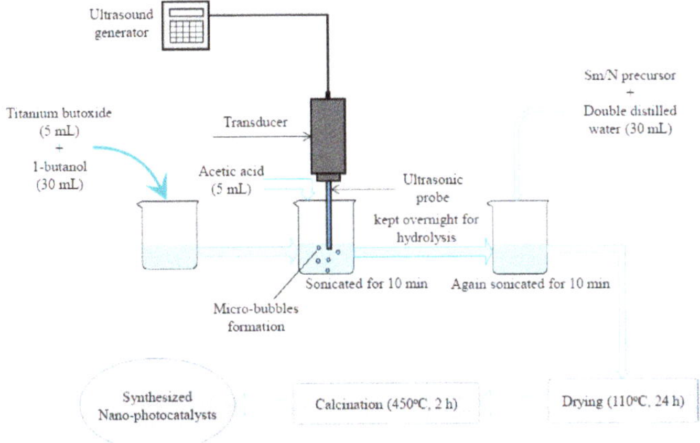

Figure 8. Schematic illustrating the ultrasound assisted sol-gel synthesis process of the Sm/N doped TiO$_2$. Adapted with permission from Reference [58]. Copyright (2019) Elsevier.

4. Heterojunctions Construction

Besides the abovementioned heteroatoms doping strategy, constructing heterojunctions in photocatalysts is also considered as one of the most promising approaches for improving the photocatalysis performance of semiconductors due to its feasibility and effectiveness for the spatial separation of electron–hole pairs. More specifically, the heterojunction is defined as the formed interface between two semiconductors with the unequal band structure, which can form band alignments [59,60]. In fact, there have been several types of heterojunction structures, which could be considered as the conventional heterojunction structures, and the new generation of heterojunction structures.

4.1. Conventional Heterojunctions

In general, the conventional heterojunctions can be classified as three types depending on the different band gaps of the composite semiconductors, which are type I with a straddling gap, type II with a staggered gap, and type III with a broken gap (Figure 9) [59]. As for the type I heterojunction, the VB and CB of semiconductor A are lower and higher than the corresponding VB and CB of semiconductor B, respectively. As a result, the photo-generated electrons and holes transfer to the CB and VB of semiconductor B, which is negative for the separation of electron–hole pairs. Moreover, the redox reaction of the composite semiconductors with a type I heterojunction will conduct on the surface of semiconductor B with a lower redox potential, therefore the redox ability of the whole photocatalyst may be suppressed. Meanwhile, in the composite semiconductor system with type II heterojunctions, the VB and CB of semiconductor A are higher than that of semiconductor B, thus the photo-generated electrons will migrate from the CB of semiconductor A to that of semiconductor B with a lower reduction potential, and the corresponding holes in the VB of semiconductor B will migrate to semiconductor A with a lower oxidation potential, thus a spatial separation of electron–hole pairs will be completed. However, the band gap of the two semiconductors will not overlap in the type III heterojunctions, and as a result, there is no transmission or separation of electrons and holes between semiconductor A and semiconductor B. Consequently, the type II heterojunction is the most effective structure for improving the photocatalysis performance of semiconductors, and has received a great deal of research attention.

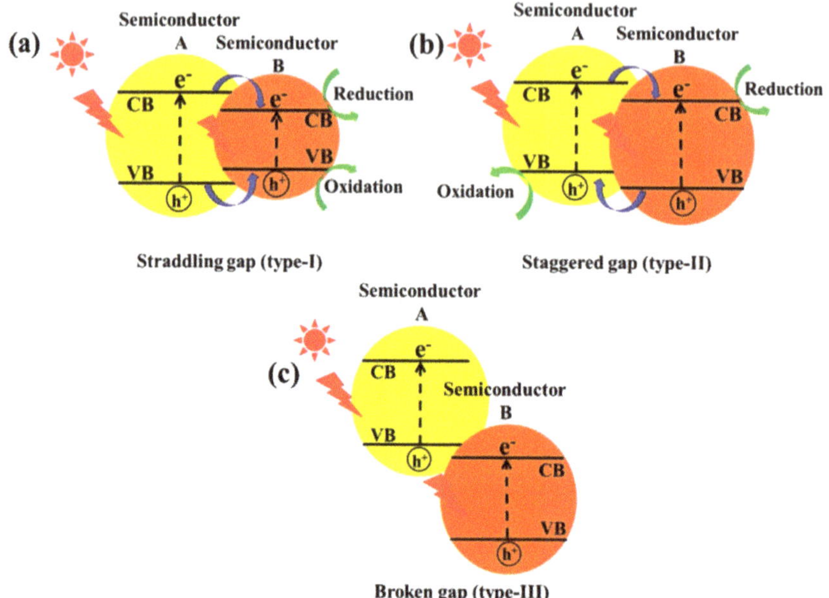

Figure 9. Schematic illustrating the photocatalysis mechanism of the three different types of heterojunction photocatalysts: (**a**) type-I, (**b**) type-II, and (**c**) type-III. Adapted with permission from Reference [59]. Copyright (2017) Wiley.

Up to now, several type-II heterojunction photocatalysts have been developed by creating two different phases in the same semiconductor, or directly compositing different semiconductors together [60,61]. For example, Yu et al. [62] once created the anatase-brookite dual-phase in a TiO_2 photocatalyst to form a type-II heterojunction via hydrolyzing the titanium tetraisopropoxide in water and an ethanol-H_2O mixture solution. They found that the co-presence of brookite and anatase phases in the TiO_2 significantly enhanced the photocatalysis performance. After that, Uddin et al. [63] successfully fabricated the mesoporous SnO_2-ZnO heterojunction photocatalysts using a two-step synthesis strategy. Furthermore, they had carefully examined the band alignment, the results showed that the obtained SnO_2-ZnO heterojunction photocatalyst possessed a type-II band alignment and exhibited higher photocatalytic activity for the degradation of methyl blue in water than that of the individual SnO_2 and ZnO nanocatalysts (Figure 10). Apart from the inorganic semiconductors, organic semiconductors could also be incorporated with the semiconductors to form the type-II heterojunction. For example, Shirmardi et al. [64] used polyaniline (PANI) as the organic semiconductor combined with ZnSe nanoparticles via a simple and cost-effective co-precipitation method in the ambient conditions. The obtained ZnSe/PANI nanocomposites exhibited obvious enhancement in the photocatalytic performance compared to that of the pristine ZnSe nanoparticles.

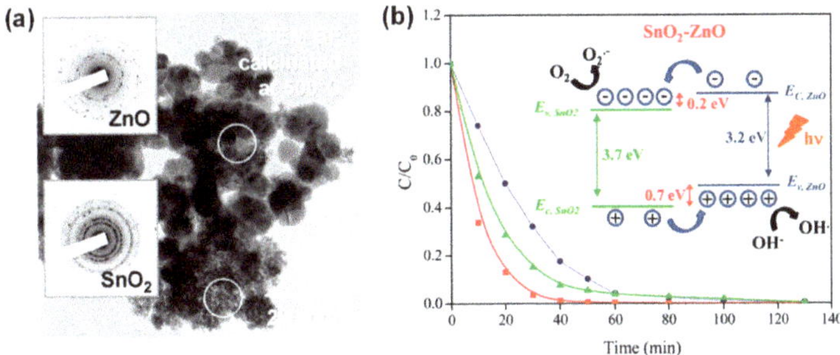

Figure 10. (**a**) Nanostructures of SnO_2-ZnO composite photocatalysts. (**b**) The corresponding photocatalytic performances of SnO_2–ZnO (red line with square dots), SnO_2 (green line with triangle dots), and ZnO (blue line with circle dots). Adapted with permission from Reference [63]. Copyright (2012) American Chemical Society.

4.2. New Generation of Heterojunctions

Although the conventional type-II heterojunctions are capable of spatially separating the photo-generated electron–hole pairs, there remain several critical limitations, such as the relatively weak redox ability due to the lower reduction and oxidation potentials, and the suppressed migration of electrons and holes due to the electrostatic repulsion [59]. Recently, in order to overcome the abovementioned limitations, a new generation of heterojunctions have been developed, including the p-n heterojunctions, the surface heterojunctions, the Z-scheme heterojunctions, and the semiconductor/carbon heterojunctions. Here we will give a brief introduction of each kind of these newly developed heterojunctions.

4.2.1. p–n Heterojunctions

The p-n heterojunctions could be obtained by incorporating a p-type semiconductor with an n-type semiconductor, and it has been proved that the formation of p-n heterojunctions are effective for improving the photocatalytic performance of composite catalysts [65,66]. In general, before the irradiation of light, there is an internal electric field in the region closed to the p-n interface due to the electron–hole diffusion tendency of the composite semiconductors system with unequal Fermi levels [59,67]. Alternatively, when the composite semiconductors are irradiated by a light, and the energy state of the photon is beyond the band gaps of both p-type and n-type semiconductors, electron–hole pairs will be generated in the corresponding semiconductors. However, due to the presence of an internal electric field, the photo-generated electrons and holes will transfer to the CB of the n-type semiconductor and the VB of p-type semiconductor, respectively. Furthermore, it has been proved that this spatial separation of the photo-generated electron–hole pairs is much more efficient compared with that of conventional type-II heterojunction because of the synergy of the internal electric field and band alignment [59,68]. As a result, a variety of composite semiconductors with the p-n heterojunctions have been created for the application of photocatalysis. For example, Wen et al. [69] reported the fabrication of a $BiOI/CeO_2$ p-n junction using a facile in situ chemical bath method. The result demonstrated that the $BiOI/CeO_2$ composite with a mole ratio of 1:1 exhibited a superior photocatalytic performance for the decomposition of bisphenol A (BPA) and methylene orange under visible light irradiation. Most recently, as shown in Figure 11, our group reported a facile method for the preparation of SnS_2/MoO_3 hollow nanotubes based on the hydrothermal method [70]. The obtained SnS_2/MoO_3 hollow nanotubes exhibit a typical p-n heterojunction structure, and a synergistic effect between MoO_3 and SnS_2 was proven to yield an optimal hydrogen peroxide production performance.

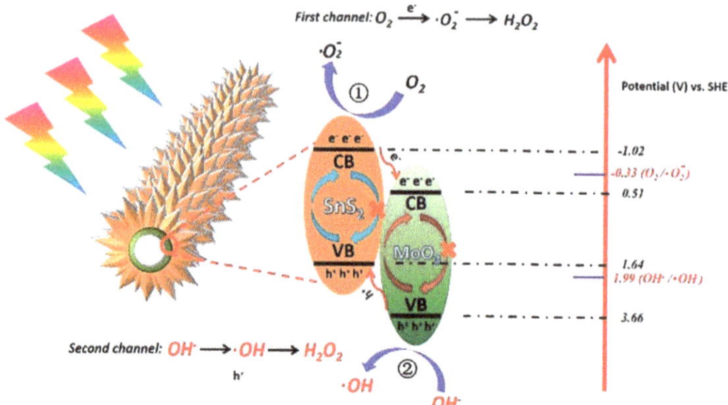

Figure 11. Schematic illustration of the SnS$_2$/MoO$_3$ hollow nanotubes and its photocatalysis mechanism with a two-channel pathway. Adapted with permission from Reference [70]. Copyright (2018) Royal Society of Chemistry.

4.2.2. Surface Heterojunctions

As reported before, a surface heterojunction can be created between two crystal facets of a single semiconductor [59,71]. For example, Yu et al. [72] proved that the formation of a heterojunction between the (001) and (101) facets in TiO$_2$ contribute significantly toward the enhancement of photocatalytic activity. This method enables the construction of a heterojunction on the surface of a single semiconductor, which is less costly because only one semiconductor is used. They also demonstrate that there is an optimal ratio for the (001) and (101) facets in the anatase TiO$_2$ for the improvement of its photocatalysis performance. Subsequently, Gao et al. [73] found that the surface heterojunction of TiO$_2$ could be self-adjusted, and its photocatalytic activity could be significantly improved via combining a proper surface heterojunction with the Schottky junction. Apart from the TiO$_2$, Bi-based semiconductors could also be employed for the design of photocatalysts with surface heterojunctions. Most recently, as shown in Figure 12, Lu et al. [74] synthesized a tetragonal BiOI photocatalyst by regulating the amount of water in the hydrolysis process at room temperature. The as-prepared photocatalyst possessed a typical surface heterojunction structure between (001) facets and (110) facets, and exhibited a promoted photocatalytic performance for the degradation of organic contaminants in water under visible light.

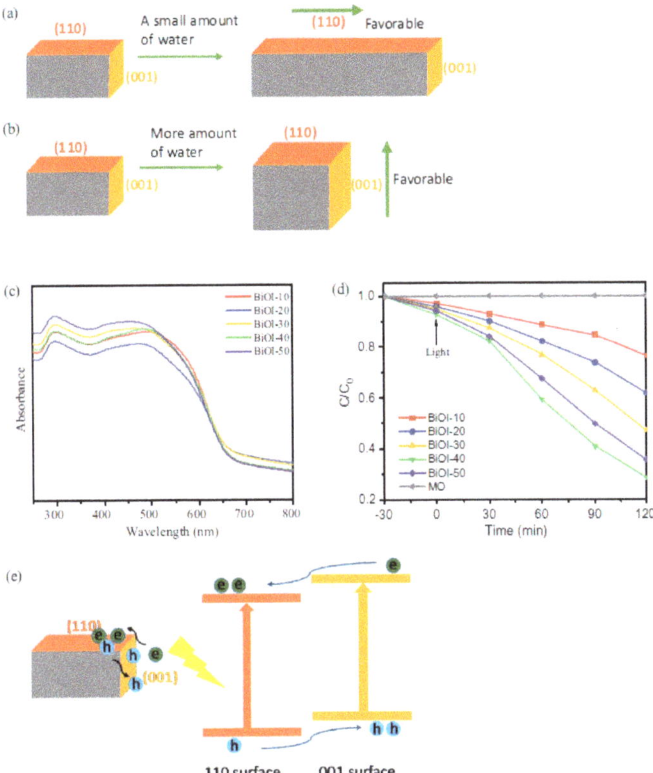

Figure 12. (**a**,**b**) Schematic illustrating the growth of TiO$_2$ nanosheets at different conditions. (**c**) UV-vis images of the related samples. (**d**) Photocatalytic degradation efficiency of different catalysts for methyl orange. (**e**) Schematic demonstrating the migration of electrons and holes in the surface heterojunction. Adapted with permission from Reference [74]. Copyright (2018) Elsevier.

4.2.3. Z-Scheme Heterojunctions

Z-scheme heterojunctions were constructed to overcome the limitation of the lower redox potential of the heterojunction systems. [59,75] In general, the Z-scheme heterojunction is composed of two different semiconductors and an electron acceptor/donor pair. During the photocatalysis process, the photo-generated electrons/holes will transfer from the matrix semiconductor to the coupled semiconductor through the electron acceptor/donor pair or an electron mediator. As a result, the electrons/holes will accumulate on different semiconductors with higher redox potentiasl, and an effective spatial separation of electron–hole pairs is also realized. Up to now, the Z-scheme heterojunctions have been well developed, and various photocatalysts with well-designed Z-scheme heterojunctions have been invented for the wastewater treatment. [75] For example, Wu et al. [76] reported the fabrication of the Ag$_2$CO$_3$/Ag/AgNCO composite photocatalyst via a simple in situ ion exchange method. The obtained composite photocatalyst possessed the Z-scheme heterojunction and exhibited a highly efficient degradation ratio of rhodamine B and the reduction of Cr (VI) under the driving of visible light. They proved that the significantly enhanced photocatalytic activity could be attributed to the low resistance for the interfacial charge transfer and desirable absorption capability. Recently, considering the relative high cost of the common used electron mediators (e.g., Pt, Ag, and Au), a new generation of Z-heterojunctions without the electron mediators have been invented for wastewater treatment, which is named as the direct Z-scheme system [59]. For example, Lu et al. [77] synthesized a $CuInS_2/Bi_2WO_6$ composite catalyst with a direct Z-scheme heterojunction via the in ·situ hydrothermal growth of Bi_2WO_6 on the surface of $CuInS_2$ networks. The obtained composite photocatalysts with an optimal Z-scheme exhibited a superior visible light degradation performance of the tetracycline hydrochloride in water than that of the pristine $CuInS_2$ and Bi_2WO_6. The improved photocatalytic activity was attributed to the formed intimate interface contact, which ensured a good interfacial charge transfer ability (Figure 13).

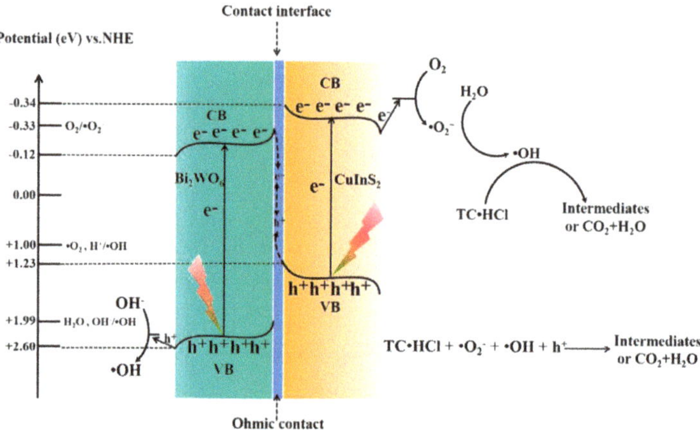

Figure 13. Schematic illustrating the interfacial electron transfer process and possible photocatalytic mechanism of $CuInS_2/Bi_2WO_6$ with the Z-scheme heterojunction. Adapted with permission from Reference [77]. Copyright (2019) Elsevier.

4.2.4. Semiconductor/Carbon Heterojunctions

Carbonaceous nanomaterials have been widely employed for the design of novel photocatalysts due to their advantages of high surface area, good conductivity, and chemical stability. In general, the most commonly used carbonaceous materials for combining with semiconductors involves the carbon dots (CDs), carbon nanotubes (CNTs), and graphene [78].

The CDs as typical nanocarbon materials have been widely used to enhance the photocatalytic activity of semiconductors owing to their intriguing optical and electronic properties, low chemical toxicity, adjustable photoluminescence, and the distinct quantum effect [79]. For example, Long et al. [80] used carbon dots (CDs) to couple with the BiOI with highly exposed (001) facets to form a composition of CDs/BiOI. Furthermore, the obtained CDs/BiOI composite exhibited a greatly improved photocatalytic activity for the degradation of organic dyes in water. It has been proved that the incorporated CDs in the semiconductor formed a CDs/BiOI heterojunction, which was able to construct numerous electron surface trap sites and was beneficial for enhancing the visible light absorption range as well as the charge separation. Recently, Zhao et al. [81] reported the fabrication of carbon quantum dots (CQDs)/TiO_2 nanotubes (TNTs) composite via an improved hydrothermal method. The CQDs were incorporated on the surface of the TNTs, and played a vital role in improving the visible light photocatalytic performance of the composite. As shown in Figure 14, they demonstrated that there were three advantages for the formation of CQDs/TiO_2: (i) the CQDs could effectively trap the photo-generated electrons from TNTs and suppress the recombination of electron-hole pairs, (ii) the up-conversion photoluminescence property of CQDs could further improve the visible light utilization efficiency of CQDs/TNTs, and (iii) the hetero-structure formed between the CQDs and the TNTs could prolong the life of the photogenerated electron and hole pairs.

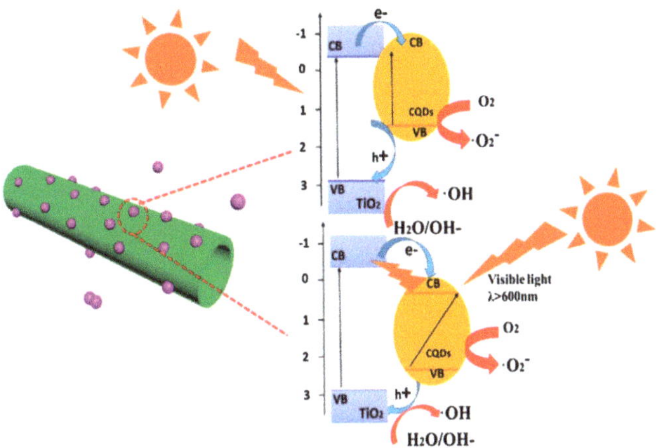

Figure 14. Schematic illustration indicating the photocatalysis mechanism of the CQDs/TNTs photocatalyst. Adapted with permission from Reference [81]. Copyright (2018) Elsevier.

CNTs are typical nanocarbon materials with highly sp^2-ordered structures, and thus exhibit an excellent metallic conductivity, which could form a Schottky barrier junction between the CNT and semiconductors; as reported before, the Schottky barrier junction could effectively increase the recombination time of electron–hole pairs [78,82]. Moreover, CNTs could accept electrons in the composite system with semiconductors due to its large electron-storage capacity, which is beneficial for retarding or hindering the electron–hole recombination. As a result, a variety of semiconductor-CNT composite photocatalysts have been developed. For example, Miribangul et al. [83] prepared a TiO_2/CNT composite via a simple hydrothermal method. The influence of the CNT concentration in the TiO_2-CNT composites on their photocatalytic activity was investigated and the 0.3 wt % CNT content in TiO_2/CNT composite could offer the highest photocatalytic degradation of Sudan (I) in UV–vis light. Apart from the TiO_2, some of the other semiconductors can also be employed to composite with CNT, such as the $CNT/LaVO_4$ composite photocatalyst developed by Xu et al. [84]. As shown in Figure 15, with the presence of CNT, the photocatalytic activity of a $CNT/LaVO_4$ composite was greatly improved due to the synergistic effect between CNT and $LaVO_4$, therefore the corresponding photocatalytic degradation rate of $CNT/LaVO_4$ composite for organic contaminant is 2 times that of pure $LaVO_4$.

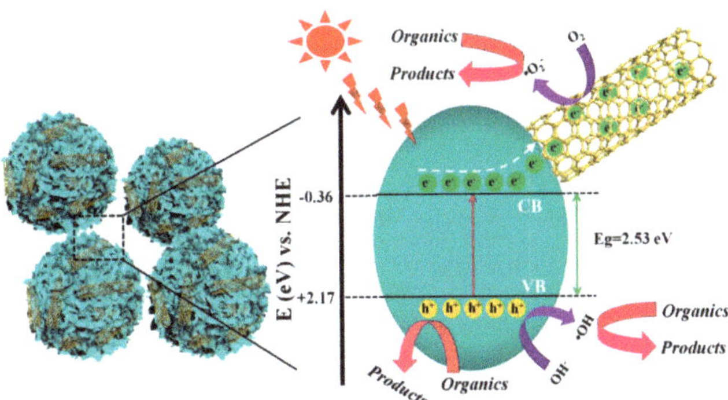

Figure 15. Schematic illustration indicating the reaction mechanism of the photocatalytic procedure of $CNT/LaVO_4$ composite catalyst. Adapted with permission from Reference [84]. Copyright (2019) Elsevier.

Recently, graphene as a newly developed nanocarbon material has attracted numerous research attention in the area of photocatalysts due to its extraordinary physical properties, including superior charge transport ability, unique optical properties, high thermal conductivity, large specific surface area, and good mechanical strength [78,85,86]. Up to now, a myriad of attempts has been carried out to couple the graphene with various semiconductors to further improve their photocatalytic activity. According to the previous reports, the first graphene composite semiconductor for photocatalysis was prepared by Zhang and co-workers [87]. They fabricated a TiO_2 (P25)-graphene composite with a chemically bonding structure via a one-step hydrothermal reaction. As reported, there are three contributions of the graphene for the photocatalytic activity: (i) enhancing the adsorption capacity of pollutants, (ii) extending light absorption range, and (iii) improving charge transportation and separation efficiency. As a result, the photodegradation of the obtained TiO_2 (P25)-graphene composite for methylene blue was significantly improved, and was superior to that of the bare P25 and the commonly reported P25-CNTs composite. After that, numerous photocatalysts based on the composite graphene-semiconductors have been invented for the treatment of wastewater. Most recently, in order to overcome the limitation of the poor homogenous dispersion of graphene, as shown in Figure 16, Isari et al. [88] created a ternary nanocomposite catalyst (Fe-doped TiO_2/rGO) derived from Fe-doped TiO_2 and reduced graphene oxide via a simple sol-gel method. They proved that the band gap of Fe-TiO_2/rGO could be significantly decreased compared with that of the pristine TiO_2, and the obtained Fe-doped TiO_2/rGO exhibited an effective decontamination performance for rhodamine B in water.

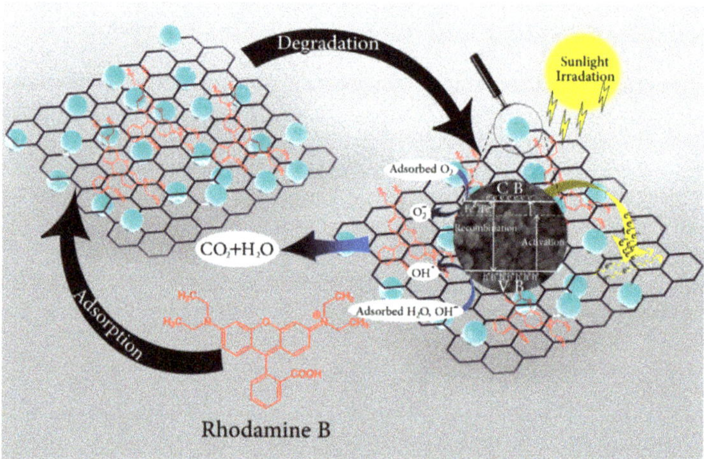

Figure 16. Schematic illustration demonstrating the photocatalytic mechanism of Fe-TiO_2/rGO catalyst for contaminant under solar irradiation. Adapted with permission from Reference [88]. Copyright (2018) Elsevier.

5. Morphology Regulation of the Composite Photocatalysts

Apart from the intrinsic characteristics of semiconductors, the corresponding catalysts with different morphologies may result in different photocatalytic activities and different application processes [89]. Recently, morphology modification of the photocatalysts have attracted more and more attention owing to further improvements in their application performance, not only for the photocatalytic performance but also for the application techniques. With this in mind, in this section we will briefly summarize the recent achievements of the composite photocatalysts with different morphologies of 0D, 1D, 2D, and 3D materials.

5.1. Nanoparticles (0D)

Generally, the 0D materials are characterized as spherically shaped with nano-scaled dimensions. Nanoparticles as a typical 0D material have been widely used in the area of photocatalysis with the merits of large surface area, simple synthesis methods, and easy to be functionalized [90]. Up to now, several synthesis approaches have been invented, among which the sol-gel method, hydrothermal method, and solvothermal method could be the most-used techniques for the fabrication of 0D composite photocatalysts.

5.1.1. Sol-Gel Method

The sol-gel process is a commonly used and effective strategy for the preparation of various inorganic materials, especially for the metal oxides based on the corresponding precursors, and it has several merits including being low cost, processed at low-temperature, and the fine control of the product's chemical composition. Therefore, the sol-gel process is one of the most-used techniques for the preparation of composite semiconductor photocatalysts [58]. For example, Vaiano et al. [91] immobilized the N-doped TiO_2 nanoparticles (NPs) on glass spheres via the sol-gel method. Through regulating the synthesis conditions, and employing the Triton X-100 as the surface active agent, the obtained N-doped TiO_2 NPs/glass spheres exhibited a good photocatalytic activity for methylene blue and eriochrome black-T in water under UV and visible light irradiation. Moreover, the composite catalyst was easy to be separated from the reaction mixture with a good durability. Recently, Chen et al. [92] prepared a Ni-Cu-Zn ferrite@SiO_2@TiO_2 composite via a simple sol-gel method. With the immobilization of Ag and magnetic ferrite, the composite photocatalysts exhibited comparatively good photodegradation performance for the methylene blue under a visible light source with lower power. Moreover, the composite catalysts can be easily separated using a magnet and can be reused well without significant loss of photocatalytic activity (Figure 17).

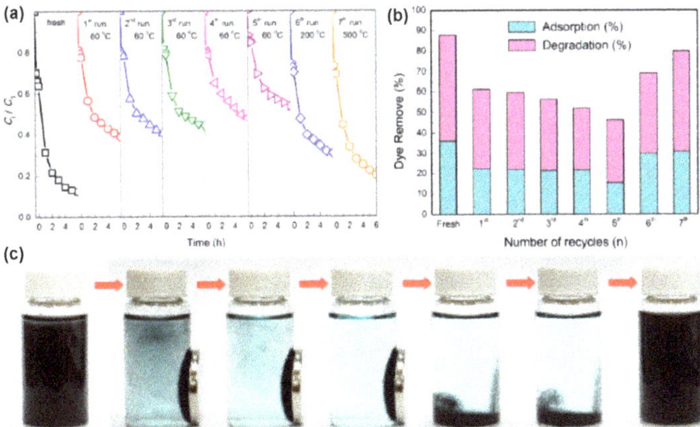

Figure 17. (**a**) The cycling test of the catalytic performance. (**b**) Dye removal capacity by adsorption and degradation for different cycles. (**c**) Digital photos demonstrating the recyclability by using an external magnetic field. Adapted with permission from Reference [92]. Copyright (2017) Elsevier.

5.1.2. Hydrothermal Methods

The hydrothermal method is a wet-chemistry method for synthesizing single crystals. Through the hydrothermal method, a great deal of crystalline phases that are not stable at the melting point can be obtained [93]. As a result, numerous semiconductor nanoparticles with different surface morphologies and compositions can also be prepared using the hydrothermal method. As a representative work, Wu et al. [94] fabricated the F-doped flower-like TiO_2 nanoparticle on the surface of Ti via a low-temperature hydrothermal process. They reported that the presence of HF in water and the hydrothermal reaction time play an important role in the formation of the F-doped flower-like TiO_2 nanostructures. Through regulating the synthesis parameters, the obtained F-doped TiO_2 flower-like nanomaterials exhibited a superior photoelectrochemical activity for the photodegradation of organic pollutants compared with P-25. They also demonstrated that the improved photoelectrochemical activity of the F-doped TiO_2 flower-like nanomaterials was mainly due to the larger surface area and the enhanced visible light harvest capacity. Additionally, magnetic composite photocatalysts can also be synthesized using the hydrothermal method, such as the magnetic $CoFe_2O_4/Ag/Ag_3VO_4$ photocatalysts fabricated by Jing and co-workers [95]. During this study, the as-prepared $CoFe_2O_4$ nanoparticles were dispersed in the solutions with $AgNO_3$ and Na_3VO_4, and the mixture suspensions were hydrothermally treated to prepare $CoFe_2O_4/Ag/Ag_3VO_4$ composites. Through controlling the weight ratios of $CoFe_2O_4$ in the composite system, the optimal $CoFe_2O_4/Ag/Ag_3VO_4$ composite exhibited significantly improved photocatalytic activity toward the degradation of various contaminants including methyl orange, tetracycline, and could even kill *Escherichia coli* solely under the driving of visible light. Moreover, with the advantage of having a good magnetic response property, the corresponding $CoFe_2O_4/Ag/Ag_3VO_4$ composite could be facilely collected from the water by applying an extra magnetic field (Figure 18).

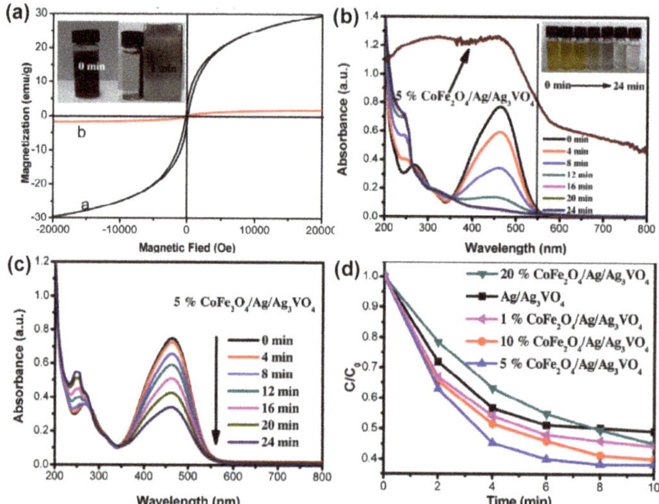

Figure 18. (**a**) Magnetic separation performance of the $CoFe_2O_4/Ag/Ag_3VO_4$ composite. (**b**) The absorption spectra of methyl orange solutions over time in the presence of $CoFe_2O_4/Ag/Ag_3VO_4$ (under visible light $\lambda \geq 440$ nm) and the UV–vis absorption spectrum of $CoFe_2O_4/Ag/Ag_3VO_4$. (**c**) The evolution of the absorption spectra of methyl orange solutions over time in the presence of $CoFe_2O_4/Ag/Ag_3VO_4$ (under visible light $\lambda \geq 550$ nm). (**d**) Photocatalytic degradation of tetracycline with different samples under visible light irradiation. Adapted with permission from Reference [95] Copyright (2016) Elsevier.

5.2. Nanofibers/Nanorods (1D)

Recently, nanofibrous photocatalysts have been intensively studied owing to their unique long aspect ratio, large surface area, and being easily functionalized. Up to now, various strategies had been developed to synthesize the 1D materials with different morphology like: wires, belts, rods, tubes, and rings [61,89,96], among which, the hydrothermal method and electrospinning are the most-used techniques. Consequently, in this part, we present the development of composite semiconductors with nanofibrous morphology derived from the electrospinning method and hydrothermal method for the treatment of wastewater.

5.2.1. Hydrothermal Method

As mentioned above, the hydrothermal method is capable of synthesizing various inorganics with different morphologies, including the nanofibrous materials. For example, Yang et al. [97] once reported the fabrication of a novel TiO_2 nanofibers with a shell of anatase nanocrystals based on the hydrothermal process. Actually, the whole fabrication process included three steps: First, the $H_2Ti_3O_7$ nanofibers were obtained from the anatase TiO_2 particles and NaOH solutions via hydrothermal method. After that, the as-prepared $H_2Ti_3O_7$ nanofibers were treated using a dilute acid solution under certain hydrothermal condition to generate the anatase nanocrystal shell on the outside. Finally, the $H_2Ti_3O_7$ phase was converted to $TiO_2(B)$ phase after a heat treatment while the anatase nanocrystal shell remained unchanged. Owing to the well-matched phase interfaces, which ensures the charge transfer across the interfaces, the recombination of electron-hole pairs was effectively suppressed and the corresponding photoactivity was significantly enhanced. Most importantly, they demonstrated that these nanofibrous photocatalysts possess specific surface areas similar to the commercial P25 powder, and the fibril morphology endowed them with a good recyclability from water, which is critically important in practical applications. Recently, our group also carried out a series of studies on the synthesis of nanofibrous photocatalysts via employing the hydrothermal method such as the bimetallic AuPd alloy nanoparticles deposited on MoO_3 nanowires [98]. As shown in Figure 19, MoO_3 nanowires were firstly prepared from Mo powder and H_2O_2 via the hydrothermal method. Then, the as-prepared MoO_3 nanowires were used as the substrates to synthesize the MoO_3/Au-Pd bimetallic alloy nanowires via a simple chemical reduction method. As expected, the MoO_3/Au–Pd bimetallic alloy nanowires exhibited a good photocatalytic degradation performance for trichloroethylene (TCE) under the driving of visible light. Similarly, the composite nanowires could be easily separated from the reaction slurry in a short time after the reaction.

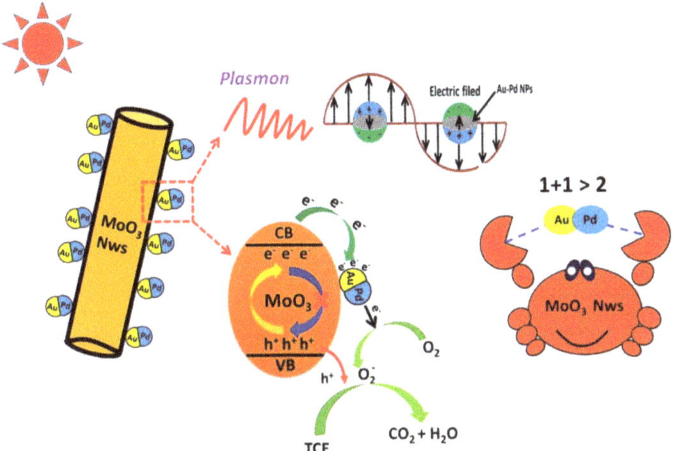

Figure 19. Schematic illustrating the band structure of MoO_3/Au-Pd composite photocatalyst and the possible reaction mechanism. Adapted with permission from Reference [98]. Copyright (2018) Elsevier.

5.2.2. Electrospinning Method

Electrospinning is considered as a promising way to synthesize nanofibers with several advantages, such as easy operation, low cost, and scalable [99–101]. In general, there are four major parts in an electrospinning device: (i) an electrical power supplier, (ii) a metallic needle, (iii) syringes with the polymer solution, and (iv) a conductive collector. Meanwhile, several process parameters, such as the polymer-based solution concentration, the viscosity of solution, the flow rate of the syringe driver, and the electric field power, could also be well-regulated to manipulate the morphology of fibers. During the electrospinning process, the solution is injected through a metallic needle via a syringe with a constant pump speed. At the same time, a voltage is applied on the metallic needle; therefore, the solution droplet will be charged, and then a Taylor cone will be generated when the electronic force is enough to overcome the surface tension. Following this, a liquid jet is formed between the grounded collector and the needle. The generated jets will be stretched by an electrostatic repulsion force until it reaches the collector; meanwhile, the solvent will rapidly evaporate during this process. Finally, the jets are solidified and the corresponding nanofibers are collected on the collector [100].

As for the applications of photocatalysis, high specific surface area is required to provide more active sites for the redox reaction. More specifically, electrospun nanofibers as forefront fibrous materials have attracted considerable research attention in the area of photocatalysis due to its several advantages of large surface area, extremely high aspect ratio, and ease of functionalization [102,103]. For example, Zhang et al. [104] reported the fabrication of a flexible and hierarchical mesoporous TiO_2 nanoparticle (TiO_2 NP) modified TiO_2 nanofiber composites via the combination of electrospinning and in situ polymerization method. At first, flexible TiO_2 nanofibers were prepared via the electrospinning and the subsequently consuming process with the dopant of yttrium. After that, the as-prepared TiO_2 nanofibers were used as a template for the incorporation of TiO_2 NPs by utilizing a bifunctional benzoxazine as the carrier through a calcination process in the N_2 atmosphere. The as-prepared membranes exhibited remarkable photocatalytic activity towards organic dyes in water; moreover, it could be reused well via simply rinsing with water, and without time-consuming separation procedures owing to the long aspect ratio and good mechanical property of the composite nanofibers. In recent years, our group has carried out several works on the design of electrospun nanofibrous photocatalysts [105–109]. As a representative sample, a $BiOCl_{0.3}/BiOBr_{0.3}/BiOI_{0.4}/PAN$ composite fibrous catalyst was fabricated via combining the electrospinning and sol-gel method [109]. As shown in Figure 20, the obtained composite photocatalyst exhibited a typical fibril structure with a good uniformity, and the corresponding field emission transmission electron microscope (FE-TEM) image demonstrated a highly crystalline structure in the composite fibers with a clear lattice spacing relating to the (112) plane of BiOCl, the (110) plane of BiOBr, and the (200) plane of BiOI; therefore, a heterojunction structure was generated via a close contact of the composite semiconductors. After a visible-light driven photocatalysis performance evaluation, it was found that the obtained $BiOCl_{0.3}/BiOBr_{0.3}/BiOI_{0.4}/PAN$ fiber displayed the highest photocatalytic degradation performance of TCE. Moreover, it was concluded that the improved visible-light driven photocatalytic activity is caused by the interfacial contact of a heterojunction and the inhibition of the recombination rate of the electron–hole pairs.

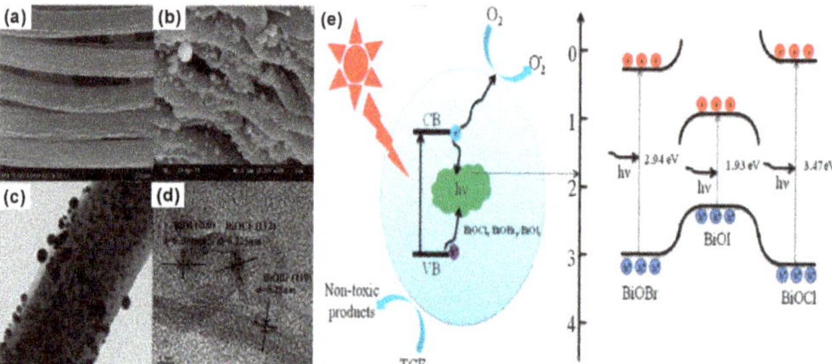

Figure 20. SEM image of pristine PAN fibers (**a**) and $BiOCl_x/BiOBr_y/BiOI_z$ composite fibers (**b**). (**c**) FE-TEM image of $BiOCl_x/BiOBr_y/BiOI_z$ fibers. (**d**) Lattice-resolved image for $BiOCl_x/BiOBr_y/BiOI_z$ nanofibers. (**e**) Schematic indicating the photocatalytic degradation of TCE. Adapted with permission from Reference [109]. Copyright (2016) Elsevier.

5.3. Nanosheets (2D)

Semiconductor nanosheets are typical 2D nanomaterials and have attracted significant attention in the research area of photocatalysis for their larger surface area and tunable structures. Up to now, a great deal of semiconductor nanosheets have been synthesized via various strategies for different applications. The hydrothermal process is one of the most used strategies for the preparation of 2D semiconductor photocatalysts for the application of wastewater treatment [110]. Through the hydrothermal process, various nanosheets derived from a single semiconductor or multi-semiconductors could be synthesized. For example, Chen et al. [111] prepared TiO_2-based nanosheets (TNS) via the alkaline hydrothermal treatment of commercial P25. They reported that the as-prepared TNS exhibited much higher specific surface area and much stronger adsorption for crystal violet molecules than the raw P25. Furthermore, the TNS could be effectively regenerated using a H_2O_2-assisted photocatalysis process, showing great potential for dealing with the high-chroma dye wastewater. Besides TiO_2, various nanosheets derived from different semiconductors could also be fabricated using a hydrothermal process, such as the WO_3 nanosheet/$K^+Ca_2Nb_3O_{10}^-$ ultrathin nanosheet synthesized by Ma et al. [112] via a facile hydrothermal assembly of WO_3 nanosheets and ultrathin $K^+Ca_2Nb_3O_{10}^-$ nanosheets. They demonstrated that the composite nanosheets possess 2D/2D heterojunctions and display remarkably enhanced photocatalytic activity compared to the pristine WO_3 and $K^+Ca_2Nb_3O_{10}^-$ nanosheets, which were mainly caused by the strongly coupled hetero-interfaces that provided more active sites for reactions and band structure. Additionally, some other methods, such as the solvothermal or photo-reduction methods, could also be employed for the preparation of composite nanosheets. For example, Wang et al. [113] fabricated the $Bi_{12}O_{15}Cl_6$ nanosheets with a narrowed band gap via a simple and facile solvothermal method followed by a simple thermal treatment. The obtained $Bi_{12}O_{15}Cl_6$ nanosheets exhibited a good photocatalytic degradation performance of bisphenol A solely under the driving of visible light, and the reaction rate of the composite nanosheets was 13.6 and 8.7 times faster than those of BiOCl and TiO_2 (P25), respectively. In addition, the as-prepared $Bi_{12}O_{15}Cl_6$ nanosheets possessed good stability and recyclability during the photocatalytic process. As shown in Figure 21, our group recently developed a Pt/BiOI composite nanosheet via a photo-reduction method in ambient conditions [114], where the as-prepared Pt/BiOI composites exhibited a flower-like structure and could effectively photocatalytically degrade the rhodamine B and phenol under visible-light irradiation ($\lambda > 420$ nm), where the degradation rate was superior to that of pure BiOI. Moreover, it was found that the content of Pt in the composite plays a vital important role on the photoactivity, and it was found that the optimal ratio of Pt to BiOI in the composite was 3%. It was concluded that the enhanced photocatalytic activity of the Pt/BiOI composite was caused by the superior electron transfer ability with the presence of an appropriate amount of Pt.

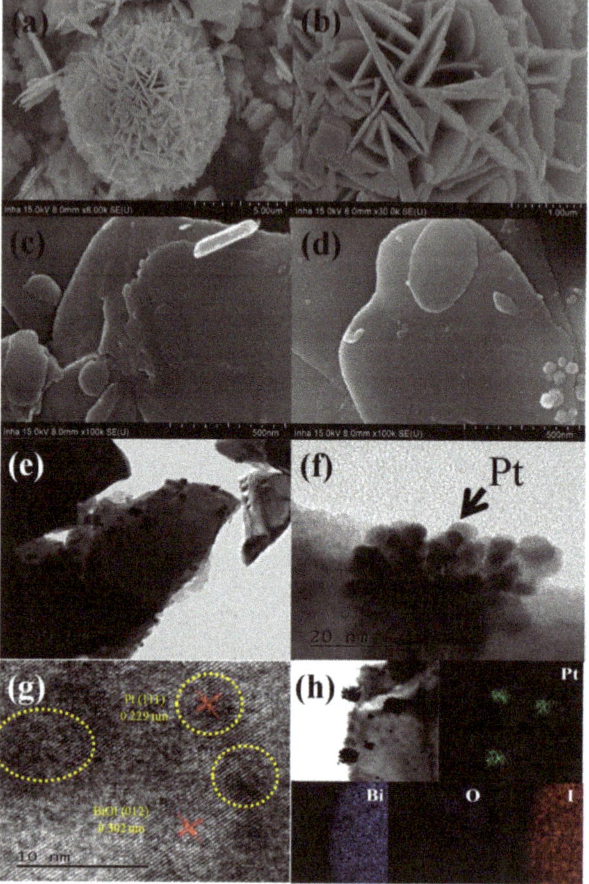

Figure 21. (**a**,**b**) SEM images of the obtained Pt/BiOI composite nanosheets. (**c**,**d**) HR-SEM images of BiOI nanosheets and Pt/BiOI composite nanosheets, respectively. (**e**–**g**) HR-TEM images of Pt/BiOI composite nanosheets. (**h**) Element mappings of Pt, Bi, O, and I for the Pt/BiOI composite nanosheets. Adapted with permission from Reference [114]. Copyright (2017) Elsevier.

5.4. Frameworks (3D)

In recent years, there have been a great number of works reporting a new generation of composite photocatalysts with 3D frameworks [115]. Actually, the 3D photocatalysts with well-deigned frameworks show great advantages such as a large specific surface area, high adsorptive capacity, good structure stability, good mass transfer ability, and a large number of exposed active sites, which make them promising candidates for the highly efficient photodegradation of contaminants in water. In general, the 3D composite photocatalysts could be obtained via two approaches, which are to directly construct a photocatalyst with 3D frameworks (type I), or compositing the photocatalysts with a template with 3D frameworks (type II).

Through employing the commonly reported synthesis methods of various 3D frameworks, such as the sol–gel process, in situ assembly, and template methods, various 3D photocatalysts with different characteristics have be fabricated [116]. The sol-gel process is a well-developed strategy to synthesize aerogels, and some of the produced aerogels, such as the SiO_2 aerogels, have been commercialized [117].

In general, there are two stages for the sol-gel process method: first, a precursor (e.g., metal alkoxide) is subjected to the hydrolysis and condensation reactions to form a wet gel, during which time, numerous networks are generated between the alkoxide groups; subsequently, the formed wet gels are sufficiently dried to obtain aerogels. As a matter of course, a photocatalytic aerogel can be obtained using a metal alkoxide precursor with an appropriate photocatalytic activity. For example, Dagan et al. [118] prepared a series of highly porous TiO_2 aerogels via the sol-gel method and they also proved that the photocatalytic degradation performance of the TiO_2 aerogels for organic contaminants is much better than that of a commercial TiO_2 (P25). Besides, various photoactive metal oxides, metal silylamide, or their composite aerogels have been developed. However, due to the limitation of sol-gel processes, some metal oxides or metal chalcogenides are not able to be synthesized into aerogels, and the obtained aerogel photocatalysts usually suffer from low crystallinity. Therefore, a new generated strategy, namely an assembly method, has been invented to construct aerogels based on various nanoscale units with different morphologies and chemical properties. As reported before [116], there are three typical steps for the assembly process: (i) fabrication of the building blocks, (ii) preparing the dispersion of the building blocks with appropriated concentration, and (iii) solidified the suspension of building blocks to form a 3D monolith. Based on this principle, Heiligtag et al. [119] developed a 3D framework Au-TiO_2 photocatalysts with a preformed TiO_2 nanoparticles as the blocking units without using any templates. Through modifying the surface of anatase TiO_2 nanoparticles with trizma, the nanoparticles undergo an oriented attachment process during gelation and finally result in well-bonded networks. Moreover, based on the above-mentioned aerogel synthesis methods, various phototcatalytic aerogels can also be prepared via employing the preformed aerogels as the templates, such as a C_3N_4 aerogel that was fabricated by Kailasam and co-workers [120] via preparing a C_3N_4/SiO_2 composite aerogel based on the sol-gel method at first, and then remove the SiO_2 via treating the composite in 4 M NH_4HF_2 (Figure 22).

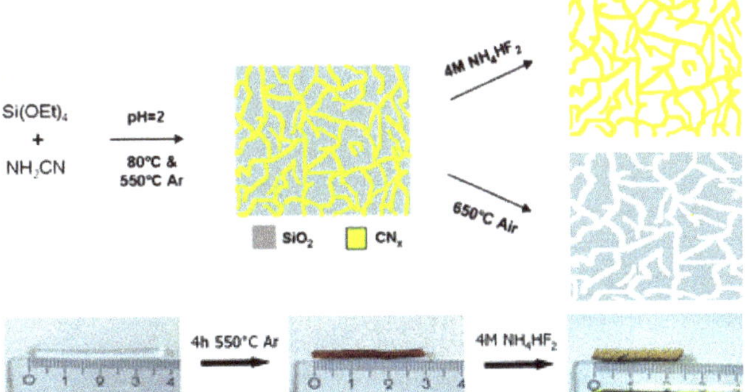

Figure 22. Schematic illustration indicating the synthesis process of porous carbon nitride and silica aerogels based on the sol-gel method and the digital photos of corresponding aerogels. Adapted with permission from Reference [120]. Copyright (2011) Royal Society of Chemistry.

As for the fabrication of type II 3D photocatalysts, an appropriate 3D porous substrate should be prepared before loading the active substance on its frameworks. Considering the requirements for high photoreactivity and good service performance, the aerogels/hydrogels derived from ceramics or carbon are mostly preferred. For example, Li et al. [121] fabricated a ternary magnetic composite of Fe_3O_4@TiO_2/SiO_2 aerogel via combining the sol-gel process and a hydrothermal treatment. During the fabrication process, Fe_3O_4 microspheres were first synthesized via the hydrothermal method; after that, Fe_3O_4@TiO_2 core shell microspheres were fabricated via an in situ reaction method. The used SiO_2 aerogel was derived from the industrial fly ash via a common sol-gel method. Finally, the as-prepared

Fe$_3$O$_4$@TiO$_2$ core shell microspheres and SiO$_2$ aerogel were combined via the hydrothermal method. According to their report, the obtained Fe$_3$O$_4$@TiO$_2$/SiO$_2$ aerogel exhibited an enhanced photocatalytic activity for the degradation of rhodamine B dye under visible light irradiation, and the aerogel could be facilely collected after the reaction due to its good magnetic separation performance. Interestingly, as shown in Figure 23, Jiang et al. [122] recently developed a separation-free PANI/TiO$_2$ 3D hydrogel for the continuous photocatalytic degradation of various contaminants in water. In their studies, the PANI hydrogel with 3D frameworks was synthesized via the polymerization of aniline. During the gelling process, the TiO$_2$ nanoparticles (P25) were incapsulated in the hydrogels. As a result, the obtained PANI/TiO$_2$ composite hydrogel exhibited an intriguing capacity for removing organic contaminants from water, which was mainly caused by the synergistic effect of adsorption enrichment of hydrogel and the in situ photocatalytic degradation of TiO$_2$. Moreover, the presented separation-free characteristics in the obtained bulk materials indicate a good recyclability of the composite hydrogel.

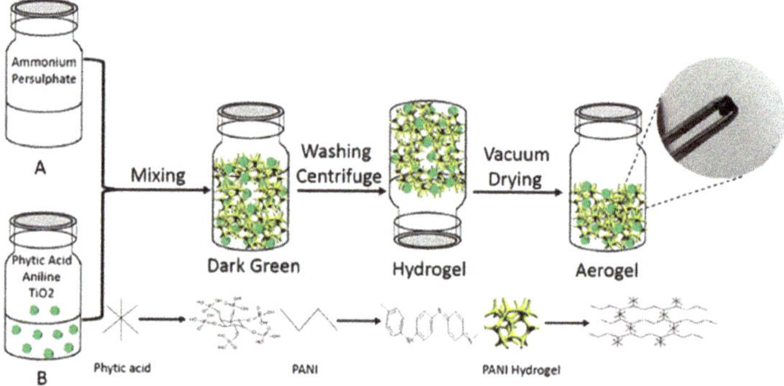

Figure 23. Schematic illustration demonstrating the synthesis process of the 3D PANI/TiO$_2$ composite hydrogel. Adapted with permission from Reference [122]. Copyright (2015) Wiley.

6. Summary and Perspectives

In summary, in order to address the worldwide concerned issues of water pollutions, various photocatalysis processes based on different photocatalysts have been developed; meanwhile, numerous efforts have been made to further improve the photocatalytic activity of the catalysts based on the semiconducors. In this review, the recent progress in the development of composite semiconductor photocatalysts for wastewater treatment is presented, including the most-used strategies to narrow the band gap of semiconductors, to retard the recombination of the photo-generated electron-hole pairs, to enhance the visible light adsorption capacity, as well as to increase the reaction ratio between the photocatalysts and contaminants. Moreover, the composite catalyts with different morphologies and the corresponding photocatlytic performance were also summarized.

Although great development of the photocatalysis process has been obtained, there are still several problems yet to be addressed to further improve the practical application performance of the photocatalysis. Therefore, some plausible perspectives for the developing trend of composite photocatalysts for the wasterwater treatment are proposed based on the presented studies: (i) the existing synthesis methods are relative complex, high cost, and harmful to the environment to some degree, thus a more facile, highly efficent, and green method is anticipated; (ii) the mechanism of the composite semicondutor photocatalysts are still confusing and some of them are unpersuasive, therefore much more effort is needed for the basic studies of the catalytic mechanisms; and (iii) the practical use are limited because the collection and reuse of the catalysts in water are still inconvenient due to their small size and poor mechanical property, therefore novel photocatalysts with easy collection property or new hybrid devices based on the composite of photocatalysts with selected substrates (e.g., polymers, metals) are proposed. Finally, we anticipate that this review can provide some useful guidance for the design of next generation of photocatlysts for the wastewater remediation.

Funding: This work was supported by the Technology Development Program (S2598148) funded by the Ministry of SMEs and Startups (MSS, Korea) and the Commercialization Promotion Agency for R&D Outcomes (COMPA) funded by the Ministry of Science and ICT (MSIT) [2018_RND_002_0064, Development of 800 mAh/g pitch carbon coating.

Acknowledgments: This work was supported by the Technology Development Program (S2598148) funded by the Ministry of SMEs and Startups (MSS, Korea) and the Commercialization Promotion Agency for R&D Outcomes (COMPA) funded by the Ministry of Science and ICT (MSIT) [2018_RND_002_0064, Development of 800 mAh/g pitch carbon coating.

Conflicts of Interest: The authors declare no conflict of interest.

References

1. Carpenter, S.R.; Caraco, N.F.; Correll, D.L.; Howarth, R.W.; Sharpley, A.N.; Smith, V.H. Nonpoint pollution of surface waters with phosphorus and nitrogen. *Ecol. Appl.* **1998**, *8*, 559–568. [CrossRef]
2. Jarup, L. Hazards of heavy metal contamination. *Br. Med. Bull.* **2003**, *68*, 167–182. [CrossRef] [PubMed]
3. Dudgeon, D.; Arthington, A.H.; Gessner, M.O.; Kawabata, Z.I.; Knowler, D.J.; Leveque, C.; Naiman, R.J.; Prieur-Richard, A.H.; Soto, D.; Stiassny, M.L.J.; et al. Freshwater biodiversity: Importance, threats, status and conservation challenges. *Biol. Rev.* **2006**, *81*, 163–182. [CrossRef] [PubMed]
4. Ribeiro, A.R.; Nunes, O.C.; Pereira, M.F.R.; Silva, A.M.T. An overview on the advanced oxidation processes applied for the treatment of water pollutants defined in the recently launched Directive 2013/39/EU. *Environ. Int.* **2015**, *75*, 33–51. [CrossRef] [PubMed]
5. Kolpin, D.W.; Furlong, E.T.; Meyer, M.T.; Thurman, E.M.; Zaugg, S.D.; Barber, L.B.; Buxton, H.T. Pharmaceuticals, hormones, and other organic wastewater contaminants in US streams, 1999-2000: A national reconnaissance. *Environ. Sci. Technol.* **2002**, *36*, 1202–1211. [CrossRef] [PubMed]
6. Schwarzenbach, R.P.; Escher, B.I.; Fenner, K.; Hofstetter, T.B.; Johnson, C.A.; von Gunten, U.; Wehrli, B. The challenge of micropollutants in aquatic systems. *Science* **2006**, *313*, 1072–1077. [CrossRef]
7. Radjenovic, J.; Sedlak, D.L. Challenges and opportunities for electrochemical processes as next-generation technologies for the treatment of contaminated water. *Environ. Sci. Technol.* **2015**, *49*, 11292–11302. [CrossRef]
8. Jiang, J.Q. The role of coagulation in water treatment. *Curr. Opin. Chem. Eng.* **2015**, *8*, 36–44. [CrossRef]
9. Lekang, O.I.; Bomo, A.M.; Svendsen, I. Biological lamella sedimentation used for wastewater treatment. *Aquac. Eng.* **2001**, *24*, 115–127. [CrossRef]
10. Ali, I. Water treatment by adsorption columns: Evaluation at ground level. *Sep. Purif. Rev.* **2014**, *43*, 175–205. [CrossRef]
11. Oe, T.; Koide, H.; Hirokawa, H.; Okukawa, K. Performance of membrane filtration system used for water treatment. *Desalination* **1996**, *106*, 107–113. [CrossRef]
12. Joss, A.; Zabczynski, S.; Gobel, A.; Hoffmann, B.; Loffler, D.; McArdell, C.S.; Ternes, T.A.; Thomsen, A.; Siegrist, H. Biological degradation of pharmaceuticals in municipal wastewater treatment: Proposing a classification scheme. *Water Res.* **2006**, *40*, 1686–1696. [CrossRef] [PubMed]
13. Mamba, G.; Mishra, A. Advances in magnetically separable photocatalysts: Smart, recyclable materials for water pollution mitigation. *Catalysts* **2016**, *6*, 34. [CrossRef]
14. Lee, S.Y.; Park, S.J. TiO$_2$ photocatalyst for water treatment applications. *J. Ind. Eng. Chem.* **2013**, *19*, 1761–1769. [CrossRef]
15. Van der Hoek, J.P.; Bertelkamp, C.; Verliefde, A.R.D.; Singhal, N. Drinking water treatment technologies in Europe: State of the art-challenges-research needs. *J. Water Supply Res. Technol.-Aquac.* **2014**, *63*, 124–130. [CrossRef]
16. Gogate, P.R.; Pandit, A.B. A review of imperative technologies for wastewater treatment I: Oxidation technologies at ambient conditions. *Adv. Environ. Res.* **2004**, *8*, 501–551. [CrossRef]
17. Oturan, M.A.; Aaron, J.J. Advanced oxidation processes in water/wastewater treatment: Principles and applications. a review. *Crit. Rev. Environ. Sci. Technol.* **2014**, *44*, 2577–2641. [CrossRef]

18. Pelaez, M.; Nolan, N.T.; Pillai, S.C.; Seery, M.K.; Falaras, P.; Kontos, A.G.; Dunlop, P.S.M.; Hamilton, J.W.J.; Byrne, J.A.; O'Shea, K.; et al. A review on the visible light active titanium dioxide photocatalysts for environmental applications. *Appl. Catal. B-Environ.* **2012**, *125*, 331–349. [CrossRef]

19. Comninellis, C.; Kapalka, A.; Malato, S.; Parsons, S.A.; Poulios, L.; Mantzavinos, D. Advanced oxidation processes for water treatment: Advances and trends for R&D. *J. Chem. Technol. Biot.* **2008**, *83*, 769–776.

20. Sansaniwal, S.K.; Sharma, V.; Mathur, J. Energy and exergy analyses of various typical solar energy applications: A comprehensive review. *Renew. Sustain. Energy Rev.* **2018**, *82*, 1576–1601. [CrossRef]

21. Mills, A.; Davies, R.H.; Worsley, D. Water-purification by semiconductor photocatalysis. *Chem. Soc. Rev.* **1993**, *22*, 417–425. [CrossRef]

22. Mills, A.; LeHunte, S. An overview of semiconductor photocatalysis. *J. Photochem. Photobiol. A Chem.* **1997**, *108*, 1–35. [CrossRef]

23. Fujishima, A.; Honda, K. Electrochemical photolysis of water at a semiconductor electrode. *Nature* **1972**, *238*, 37. [CrossRef] [PubMed]

24. Schneider, J.; Matsuoka, M.; Takeuchi, M.; Zhang, J.L.; Horiuchi, Y.; Anpo, M.; Bahnemann, D.W. Understanding TiO_2 photocatalysis: Mechanisms and materials. *Chem. Rev.* **2014**, *114*, 9919–9986. [CrossRef] [PubMed]

25. Tong, H.; Ouyang, S.X.; Bi, Y.P.; Umezawa, N.; Oshikiri, M.; Ye, J.H. Nano-photocatalytic materials: Possibilities and challenges. *Adv. Mater.* **2012**, *24*, 229–251. [CrossRef] [PubMed]

26. An, H.Z.; Du, Y.; Wang, T.M.; Wang, C.; Hao, W.C.; Zhang, J.Y. Photocatalytic properties of BiOX (X = Cl, Br, and I). *Rare Metals* **2008**, *27*, 243–250. [CrossRef]

27. Azeez, A.A.; Rhee, K.Y.; Park, S.J.; Hui, D. Epoxy clay nanocomposites—Processing, properties and applications: A review. *Compos. Part B-Eng.* **2013**, *45*, 308–320. [CrossRef]

28. Dhand, V.; Mittal, G.; Rhee, K.Y.; Park, S.J.; Hui, D. A short review on basalt fiber reinforced polymer composites. *Compos. Part B-Eng.* **2015**, *73*, 166–180. [CrossRef]

29. Chong, M.N.; Jin, B.; Chow, C.W.K.; Saint, C. Recent developments in photocatalytic water treatment technology: A review. *Water Res.* **2010**, *44*, 2997–3027. [CrossRef] [PubMed]

30. Ani, I.J.; Akpan, U.G.; Olutoye, M.A.; Hameed, B.H. Photocatalytic degradation of pollutants in petroleum refinery wastewater by TiO_2- and ZnO-based photocatalysts: Recent development. *J. Clean. Prod.* **2018**, *205*, 930–954. [CrossRef]

31. Yu, Z.B.; Chen, X.Q.; Kang, X.D.; Xie, Y.P.; Zhu, H.Z.; Wang, S.L.; Ullah, S.; Ma, H.; Wang, L.Z.; Liu, G.; et al. Noninvasively modifying band structures of wide-bandgap metal oxides to boost photocatalytic activity. *Adv. Mater.* **2018**, *30*, 1706259. [CrossRef] [PubMed]

32. Choi, W.Y.; Termin, A.; Hoffmann, M.R. The role of metal ion dopants in quantum-sized TiO_2: Correlation between photoreativity and charge carrier recombination dynamics. *J. Phys. Chem.* **1994**, *98*, 13669–13679. [CrossRef]

33. Liu, G.; Wang, L.Z.; Yang, H.G.; Cheng, H.M.; Lu, G.Q. Titania-based photocatalysts-crystal growth, doping and heterostructuring. *J. Mater. Chem.* **2010**, *20*, 831–843. [CrossRef]

34. Zhu, J.F.; Deng, Z.G.; Chen, F.; Zhang, J.L.; Chen, H.J.; Anpo, M.; Huang, J.Z.; Zhang, L.Z. Hydrothermal doping method for preparation of Cr^{3+}-TiO_2 photocatalysts with concentration gradient distribution of Cr^{3+}. *Appl. Catal. B-Environ.* **2006**, *62*, 329–335. [CrossRef]

35. Dvoranova, D.; Brezova, V.; Mazur, M.; Malati, M.A. Investigations of metal-doped titanium dioxide photocatalysts. *Appl. Catal. B-Environ.* **2002**, *37*, 91–105. [CrossRef]

36. Bouras, P.; Stathatos, E.; Lianos, P. Pure versus metal-ion-doped nanocrystalline titania for photocatalysis. *Appl. Catal. B-Environ.* **2007**, *73*, 51–59. [CrossRef]

37. Devi, L.G.; Kottam, N.; Murthy, B.N.; Kumar, S.G. Enhanced photocatalytic activity of transition metal ions Mn^{2+}, Ni^{2+} and Zn^{2+} doped polycrystalline titania for the degradation of Aniline Blue under UV/solar light. *J. Mol. Catal. A Chem.* **2010**, *328*, 44–52. [CrossRef]

38. Serpone, N. Is the band gap of pristine TiO_2 narrowed by anion- and cation-doping of titanium dioxide in second-generation photocatalysts? *J. Phys. Chem. B* **2006**, *110*, 24287–24293. [CrossRef]

39. Kudo, A.; Niishiro, R.; Iwase, A.; Kato, H. Effects of doping of metal cations on morphology, activity, and visible light response of photocatalysts. *Chem. Phys.* **2007**, *339*, 104–110. [CrossRef]

40. Chen, J.H.; Yao, M.S.; Wang, X.L. Investigation of transition metal ion doping behaviors on TiO_2 nanoparticles. *J. Nanopart. Res.* **2008**, *10*, 163–171. [CrossRef]

41. Dunnill, C.W.; Parkin, I.P. Nitrogen-doped TiO_2 thin films: Photocatalytic applications for healthcare environments. *Dalton Trans.* **2011**, *40*, 1635–1640. [CrossRef] [PubMed]

42. Asahi, R.; Morikawa, T.; Irie, H.; Ohwaki, T. Nitrogen-doped titanium dioxide as visible-light-sensitive photocatalyst: Designs, developments, and prospects. *Chem. Rev.* **2014**, *114*, 9824–9852. [CrossRef] [PubMed]

43. Asahi, R.; Morikawa, T.; Ohwaki, T.; Aoki, K.; Taga, Y. Visible-light photocatalysis in nitrogen-doped titanium oxides. *Science* **2001**, *293*, 269–271. [CrossRef] [PubMed]

44. Kitano, M.; Funatsu, K.; Matsuoka, M.; Ueshima, M.; Anpo, M. Preparation of nitrogen-substituted TiO_2 thin film photocatalysts by the radio frequency magnetron sputtering deposition method and their photocatalytic reactivity under visible light irradiation. *J. Phys. Chem. B* **2006**, *110*, 25266–25272. [CrossRef] [PubMed]

45. Salah, N.; Hameed, A.; Aslam, M.; Abdel-Wahab, M.S.; Babkair, S.S.; Bahabri, F.S. Flow controlled fabrication of N doped ZnO thin films and estimation of their performance for sunlight photocatalytic decontamination of water. *Chem. Eng. J.* **2016**, *291*, 115–127. [CrossRef]

46. Mikkelsen, N.J.; Pedersen, J.; Straede, C.A. Ion implantation—The job coater's supplement to coating techniques. *Surf. Coat. Technol.* **2002**, *158*, 42–47. [CrossRef]

47. Tang, G.Z.; Li, J.L.; Sun, M.R.; Ma, X.X. Fabrication of nitrogen-doped TiO_2 layer on titanium substrate. *Appl. Surf. Sci.* **2009**, *255*, 9224–9229. [CrossRef]

48. Song, X.Y.; Li, W.Q.; He, D.; Wu, H.Y.; Ke, Z.J.; Jiang, C.Z.; Wang, G.M.; Xiao, X.H. The "Midas Touch" transformation of TiO_2 nanowire arrays during visible light photoelectrochemical performance by carbon/nitrogen coimplantation. *Adv. Energy Mater.* **2018**, *8*, 1800165. [CrossRef]

49. Chen, X.; Mao, S.S. Titanium dioxide nanomaterials: Synthesis, properties, modifications, and applications. *Chem. Rev.* **2007**, *107*, 2891–2959. [CrossRef]

50. Lee, S.Y.; Park, J.; Joo, H. Visible light-sensitized photocatalyst immobilized on beads by CVD in a fluidizing bed. *Sol. Energy Mater. Sol. Cell* **2006**, *90*, 1905–1914. [CrossRef]

51. Kafizas, A.; Crick, C.; Parkin, I.P. The combinatorial atmospheric pressure chemical vapour deposition (cAPCVD) of a gradating substitutional/interstitial N-doped anatase TiO_2 thin-film; UVA and visible light photocatalytic activities. *J. Photochem. Photobiol. A Chem.* **2010**, *216*, 156–166. [CrossRef]

52. Youssef, L.; Leoga, A.J.K.; Roualdes, S.; Bassil, J.; Zakhour, M.; Rouessac, V.; Ayral, A.; Nakhl, M. Optimization of N-doped TiO_2 multifunctional thin layers by low frequency PECVD process. *J. Eur. Ceram. Soc.* **2017**, *37*, 5289–5303. [CrossRef]

53. George, S.M. Atomic layer deposition: An Overview. *Chem. Rev.* **2010**, *110*, 111–131. [CrossRef]

54. Vilhunen, S.H.; Sillanpaa, M.E.T. Atomic layer deposited (ALD) TiO_2 and $TiO_{2-x}N_x$ thin film photocatalysts in salicylic acid decomposition. *Water Sci. Technol.* **2009**, *60*, 2471–2475. [CrossRef]

55. Pore, V.; Heikkila, M.; Ritala, M.; Leskela, M.; Areva, S. Atomic layer deposition of $TiO_{2-x}N_x$ thin films for photocatalytic applications. *J. Photochem. Photobiol. A Chem.* **2006**, *177*, 68–75. [CrossRef]

56. Lee, A.; Libera, J.A.; Waldman, R.Z.; Ahmed, A.; Avila, J.R.; Elam, J.W.; Darling, S.B. Conformal nitrogen-doped TiO_2 photocatalytic coatings for sunlight-activated membranes. *Adv. Sustain. Syst.* **2017**, *1*, 1600041. [CrossRef]

57. Albrbar, A.J.; Djokic, V.; Bjelajac, A.; Kovac, J.; Cirkovic, J.; Mitric, M.; Janackovic, D.; Petrovic, R. Visible-light active mesoporous, nanocrystalline N,S-doped and co-doped titania photocatalysts synthesized by non-hydrolytic sol-gel route. *Ceram. Int.* **2016**, *42*, 16718–16728. [CrossRef]

58. Rajoriya, S.; Bargole, S.; George, S.; Saharan, V.K.; Gogate, P.R.; Pandit, A.B. Synthesis and characterization of samarium and nitrogen doped TiO_2 photocatalysts for photo-degradation of 4-acetamidophenol in combination with hydrodynamic and acoustic cavitation. *Sep. Purif. Technol.* **2019**, *209*, 254–269. [CrossRef]

59. Low, J.; Yu, J.; Jaroniec, M.; Wageh, S.; Al-Ghamdi, A.A. Heterojunction photocatalysts. *Adv. Mater.* **2017**, *29*, 1601694. [CrossRef]

60. Baek, J.H.; Kim, B.J.; Han, G.S.; Hwang, S.W.; Kim, D.R.; Cho, I.S.; Jung, H.S. $BiVO_4/WO_3/SnO_2$ Double-heterojunction photoanode with enhanced charge separation and visible-transparency for bias-free solar water-splitting with a perovskite solar cell. *ACS Appl. Mater. Interfaces* **2017**, *9*, 1479–1487. [CrossRef]

61. Han, H.S.; Han, G.S.; Kim, J.S.; Kim, D.H.; Hong, J.S.; Caliskan, S.; Jung, H.S.; Cho, I.S.; Lee, J.K. Indium-tin-oxide nanowire array based $CdSe/CdS/TiO_2$ one-dimensional heterojunction photoelectrode for enhanced solar hydrogen production. *ACS Sustain. Chem. Eng.* **2016**, *4*, 1161–1168. [CrossRef]

62. Yu, J.C.; Yu, J.G.; Ho, W.K.; Zhang, L.Z. Preparation of highly photocatalytic active nano-sized TiO_2 particles via ultrasonic irradiation. *Chem. Commun.* **2001**, 1942–1943. [CrossRef]

63. Uddin, M.T.; Nicolas, Y.; Olivier, C.; Toupance, T.; Servant, L.; Muller, M.M.; Kleebe, H.J.; Ziegler, J.; Jaegermann, W. Nanostructured SnO_2-ZnO heterojunction photocatalysts showing enhanced photocatalytic activity for the degradation of organic dyes. *Inorg. Chem.* **2012**, *51*, 7764–7773. [CrossRef]
64. Shirmardi, A.; Teridi, M.A.M.; Azimi, H.R.; Basirun, W.J.; Jamali-Sheini, F.; Yousefi, R. Enhanced photocatalytic performance of ZnSe/PANI nanocomposites for degradation of organic and inorganic pollutants. *Appl. Surf. Sci.* **2018**, *462*, 730–738. [CrossRef]
65. Lee, C.H.; Lee, G.H.; van der Zande, A.M.; Chen, W.C.; Li, Y.L.; Han, M.Y.; Cui, X.; Arefe, G.; Nuckolls, C.; Heinz, T.F.; et al. Atomically thin p-n junctions with van der Waals heterointerfaces. *Nat. Nanotechnol.* **2014**, *9*, 676–681. [CrossRef]
66. Kawazoe, H.; Yanagi, H.; Ueda, K.; Hosono, H. Transparent p-type conducting oxides: Design and fabrication of p-n heterojunctions. *MRS Bull.* **2000**, *25*, 28–36. [CrossRef]
67. Lu, M.X.; Shao, C.L.; Wang, K.X.; Lu, N.; Zhang, X.; Zhang, P.; Zhang, M.Y.; Li, X.H.; Liu, Y.C. p-MoO_3 Nanostructures/n-TiO_2 nanofiber heterojunctions: Controlled fabrication and enhanced photocatalytic properties. *ACS Appl. Mater. Interfaces* **2014**, *6*, 9004–9012. [CrossRef]
68. Zhang, L.P.; Jaroniec, M. Toward designing semiconductor-semiconductor heterojunctions for photocatalytic applications. *Appl. Surf. Sci.* **2018**, *430*, 2–17. [CrossRef]
69. Wen, X.J.; Niu, C.G.; Zhang, L.; Zeng, G.M. Novel p-n heterojunction $BiOI/CeO_2$ photocatalyst for wider spectrum visible-light photocatalytic degradation of refractory pollutants. *Dalton Trans.* **2017**, *46*, 4982–4993. [CrossRef]
70. Zhang, Y.; Park, S.J. Formation of hollow MoO_3/SnS_2 heterostructured nanotubes for efficient light-driven hydrogen peroxide production. *J. Mater. Chem. A* **2018**, *6*, 20304–20312. [CrossRef]
71. Wei, Z.D.; Zhao, Y.; Fan, F.T.; Li, C. The property of surface heterojunction performed by crystal facets for photogenerated charge separation. *Comp. Mater. Sci.* **2018**, *153*, 28–35. [CrossRef]
72. Yu, J.G.; Low, J.X.; Xiao, W.; Zhou, P.; Jaroniec, M. Enhanced Photocatalytic CO_2-Reduction Activity of Anatase TiO_2 by Coexposed {001} and {101} Facets. *J. Am. Chem. Soc.* **2014**, *136*, 8839–8842. [CrossRef]
73. Gao, S.J.; Wang, W.; Ni, Y.R.; Lu, C.H.; Xu, Z.Z. Facet-dependent photocatalytic mechanisms of anatase TiO_2: A new sight on the self-adjusted surface heterojunction. *J. Alloys Compd.* **2015**, *647*, 981–988. [CrossRef]
74. Lu, J.; Wu, J.; Xu, W.X.; Cheng, H.Q.; Qi, X.M.; Li, Q.W.; Zhang, Y.A.; Guan, Y.; Ling, Y.; Zhang, Z. Room temperature synthesis of tetragonal BiOI photocatalyst with surface heterojunction between (001) facets and (110) facets. *Mater. Lett.* **2018**, *219*, 260–264. [CrossRef]
75. Li, H.J.; Tu, W.G.; Zhou, Y.; Zou, Z.G. Z-scheme photocatalytic systems for promoting photocatalytic performance: Recent progress and future challenges. *Adv. Sci.* **2016**, *3*, 12. [CrossRef]
76. Wu, X.S.; Hu, Y.D.; Wang, Y.; Zhou, Y.S.; Han, Z.H.; Jin, X.L.; Chen, G. In-situ synthesis of Z-scheme $Ag_2CO_3/Ag/AgNCO$ heterojunction photocatalyst with enhanced stability and photocatalytic activity. *Appl. Surf. Sci.* **2019**, *464*, 108–114. [CrossRef]
77. Lu, X.Y.; Che, W.J.; Hu, X.F.; Wang, Y.; Zhang, A.T.; Deng, F.; Luo, S.L.; Dionysiou, D.D. The facile fabrication of novel visible-light-driven Z-scheme $CuInS_2/Bi_2WO_6$ heterojunction with intimate interface contact by in situ hydrothermal growth strategy for extraordinary photocatalytic performance. *Chem. Eng. J.* **2019**, *356*, 819–829. [CrossRef]
78. Leary, R.; Westwood, A. Carbonaceous nanomaterials for the enhancement of TiO_2 photocatalysis. *Carbon* **2011**, *49*, 741–772. [CrossRef]
79. Li, J.; Liu, K.; Xue, J.; Xue, G.; Sheng, X.; Wang, H.; Huo, P.; Yan, Y. CQDS preluded carbon-incorporated 3D burger-like hybrid ZnO enhanced visible-light-driven photocatalytic activity and mechanism implication. *J. Catal.* **2019**, *369*, 450–461. [CrossRef]
80. Long, B.; Huang, Y.C.; Li, H.B.; Zhao, F.Y.; Rui, Z.B.; Liu, Z.L.; Tong, Y.X.; Ji, H.B. Carbon Dots Sensitized BiOI with Dominant {001} Facets for superior photocatalytic performance. *Ind. Eng. Chem. Res.* **2015**, *54*, 12788–12794. [CrossRef]
81. Zhao, F.F.; Rong, Y.F.; Wan, J.M.; Hu, Z.W.; Peng, Z.Q.; Wang, B. High photocatalytic performance of carbon quantum dots/TNTs composites for enhanced photogenerated charges separation under visible light. *Catal. Today* **2018**, *315*, 162–170. [CrossRef]
82. Di, J.; Li, S.X.; Zhao, Z.F.; Huang, Y.C.; Jia, Y.; Zheng, H.J. Biomimetic $CNT@TiO_2$ composite with enhanced photocatalytic properties. *Chem. Eng. J.* **2015**, *281*, 60–68. [CrossRef]

83. Miribangul, A.; Ma, X.L.; Zeng, C.; Zou, H.; Wu, Y.H.; Fan, T.P.; Su, Z. Synthesis of TiO_2/CNT Composites and its photocatalytic activity toward sudan (I) degradation. *Photochem. Photobiol.* **2016**, *92*, 523–527. [CrossRef]

84. Xu, Y.G.; Liu, J.; Xie, M.; Jing, L.Q.; Xu, H.; She, X.J.; Li, H.M.; Xie, J.M. Construction of novel $CNT/LaVO_4$ nanostructures for efficient antibiotic photodegradation. *Chem. Eng. J.* **2019**, *357*, 487–497. [CrossRef]

85. Wang, Y.J.; Shi, R.; Lin, J.; Zhu, Y.F. Significant photocatalytic enhancement in methylene blue degradation of TiO_2 photocatalysts via graphene-like carbon in situ hybridization. *Appl. Catal. B-Environ.* **2010**, *100*, 179–183. [CrossRef]

86. Mamaghani, A.H.; Haghighat, F.; Lee, C.-S. Hydrothermal/solvothermal synthesis and treatment of TiO_2 for photocatalytic degradation of air pollutants: Preparation, characterization, properties, and performance. *Chemosphere* **2019**, *219*, 804–825. [CrossRef]

87. Zhang, H.; Lv, X.J.; Li, Y.M.; Wang, Y.; Li, J.H. P25-Graphene composite as a high performance photocatalyst. *ACS Nano* **2010**, *4*, 380–386. [CrossRef]

88. Isari, A.A.; Payan, A.; Fattahi, M.; Jorfi, S.; Kakavandi, B. Photocatalytic degradation of rhodamine B and real textile wastewater using Fe-doped TiO_2 anchored on reduced graphene oxide (Fe-TiO_2/rGO): Characterization and feasibility, mechanism and pathway studies. *Appl. Surf. Sci.* **2018**, *462*, 549–564. [CrossRef]

89. Nasr, M.; Eid, C.; Habchi, R.; Miele, P.; Bechelany, M. Recent progress on titanium dioxide nanomaterials for photocatalytic applications. *ChemSusChem* **2018**, *11*, 3023–3047. [CrossRef]

90. Giberman, D. Against zero-dimensional material objects (and other bare particulars). *Philos. Stud.* **2012**, *160*, 305–321. [CrossRef]

91. Vaiano, V.; Sacco, O.; Sannino, D.; Ciambelli, P. Nanostructured N-doped TiO_2 coated on glass spheres for the photocatalytic removal of organic dyes under UV or visible light irradiation. *Appl. Catal. B-Environ.* **2015**, *170*, 153–161. [CrossRef]

92. Chen, C.C.; Jaihindh, D.; Hu, S.H.; Fu, Y.P. Magnetic recyclable photocatalysts of Ni-Cu-Zn ferrite@SiO_2@TiO_2@Ag and their photocatalytic activities. *J. Photochem. Photobiol. A Chem.* **2017**, *334*, 74–85. [CrossRef]

93. Somiya, S.; Roy, R. Hydrothermal synthesis of fine oxide powders. *Bull. Mater. Sci.* **2000**, *23*, 453–460. [CrossRef]

94. Wu, G.S.; Wang, J.P.; Thomas, D.F.; Chen, A.C. Synthesis of F-doped flower-like TiO_2 nanostructures with high photoelectrochemical activity. *Langmuir* **2008**, *24*, 3503–3509. [CrossRef]

95. Jing, L.Q.; Xu, Y.G.; Huang, S.Q.; Xie, M.; He, M.Q.; Xu, H.; Li, H.M.; Zhang, Q. Novel magnetic $CoFe_2O_4$/Ag/Ag_3VO_4 composites: Highly efficient visible light photocatalytic and antibacterial activity. *Appl. Catal. B-Environ.* **2016**, *199*, 11–22. [CrossRef]

96. Cho, I.S.; Lee, C.H.; Feng, Y.Z.; Logar, M.; Rao, P.M.; Cai, L.L.; Kim, D.R.; Sinclair, R.; Zheng, X.L. Codoping titanium dioxide nanowires with tungsten and carbon for enhanced photoelectrochemical performance. *Nat. Commun.* **2013**, *4*, 8. [CrossRef]

97. Yang, D.J.; Liu, H.W.; Zheng, Z.F.; Yuan, Y.; Zhao, J.C.; Waclawik, E.R.; Ke, X.B.; Zhu, H.Y. An Efficient Photocatalyst Structure: TiO_2(B) Nanofibers with a Shell of Anatase Nanocrystals. *J. Am. Chem. Soc.* **2009**, *131*, 17885–17893. [CrossRef]

98. Zhang, Y.; Park, S.J. Bimetallic AuPd alloy nanoparticles deposited on MoO_3 nanowires for enhanced visible-light driven trichloroethylene degradation. *J. Catal.* **2018**, *361*, 238–247. [CrossRef]

99. Li, D.; Xia, Y.N. Electrospinning of nanofibers: Reinventing the wheel? *Adv. Mater.* **2004**, *16*, 1151–1170. [CrossRef]

100. Greiner, A.; Wendorff, J.H. Electrospinning: A fascinating method for the preparation of ultrathin fibres. *Angew. Chem. Int. Edit.* **2007**, *46*, 5670–5703. [CrossRef]

101. Reneker, D.H.; Chun, I. Nanometre diameter fibres of polymer, produced by electrospinning. *Nanotechnology* **1996**, *7*, 216–223. [CrossRef]

102. Wu, X.H.; Si, Y.; Yu, J.Y.; Ding, B. Titania-based electrospun nanofibrous materials: A new model for organic pollutants degradation. *MRS Commun.* **2018**, *8*, 765–781. [CrossRef]

103. Agarwal, S.; Greiner, A.; Wendorff, J.H. Functional materials by electrospinning of polymers. *Prog. Polym. Sci.* **2013**, *38*, 963–991. [CrossRef]

104. Zhang, R.Z.; Wang, X.Q.; Song, J.; Si, Y.; Zhuang, X.M.; Yu, J.Y.; Ding, B. In situ synthesis of flexible hierarchical TiO_2 nanofibrous membranes with enhanced photocatalytic activity. *J. Mater. Chem. A* **2015**, *3*, 22136–22144. [CrossRef]

105. Panthi, G.; Park, M.; Kim, H.Y.; Park, S.J. Electrospun Ag-CoF doped PU nanofibers: Effective visible light catalyst for photodegradation of organic dyes. *Macromol. Res.* **2014**, *22*, 895–900. [CrossRef]

106. Panthi, G.; Park, M.; Park, S.J.; Kim, H.Y. PAN Electrospun nanofibers reinforced with Ag_2CO_3 nanoparticles: Highly efficient visible light photocatalyst for photodegradation of organic contaminants in waste water. *Macromol. Res.* **2015**, *23*, 149–155. [CrossRef]

107. Saud, P.S.; Pant, B.; Park, M.; Chae, S.H.; Park, S.J.; El-Newehy, M.; Al-Deyab, S.S.; Kim, H.Y. Preparation and photocatalytic activity of fly ash incorporated TiO_2 nanofibers for effective removal of organic pollutants. *Ceram. Int.* **2015**, *41*, 1771–1777. [CrossRef]

108. Seong, D.B.; Son, Y.R.; Park, S.J. A study of reduced graphene oxide/leaf-shaped TiO_2 nanofibers for enhanced photocatalytic performance via electrospinning. *J. Solid State Chem.* **2018**, *266*, 196–204. [CrossRef]

109. Zhang, Y.F.; Park, M.; Kim, H.Y.; Ding, B.; Park, S.J. In-situ synthesis of nanofibers with various ratios of $BiOCl_x/BiOBr_y/BiOI_z$ for effective trichloroethylene photocatalytic degradation. *Appl. Surf. Sci.* **2016**, *384*, 192–199. [CrossRef]

110. Zhang, D.Q.; Li, G.S.; Li, H.X.; Lu, Y.F. The development of better photocatalysts through composition- and structure-engineering. *Chem. Asian J.* **2013**, *8*, 26–40. [CrossRef]

111. Chen, F.T.; Fang, P.F.; Gao, Y.P.; Liu, Z.; Liu, Y.; Dai, Y.Q. Effective removal of high-chroma crystal violet over TiO_2-based nanosheet by adsorption-photocatalytic degradation. *Chem. Eng. J.* **2012**, *204*, 107–113. [CrossRef]

112. Ma, X.D.; Jiang, D.L.; Xiao, P.; Jin, Y.; Meng, S.C.; Chen, M. 2D/2D heterojunctions of WO_3 nanosheet/$K^+Ca_2Nb_3O_{10}^-$ ultrathin nanosheet with improved charge separation efficiency for significantly boosting photocatalysis. *Catal. Sci. Technol.* **2017**, *7*, 3481–3491. [CrossRef]

113. Wang, C.Y.; Zhang, X.; Song, X.N.; Wang, W.K.; Yu, H.Q. Novel $Bi_{12}O_{15}Cl_6$ photocatalyst for the degradation of bisphenol a under visible-light irradiation. *ACS Appl. Mater. Interfaces* **2016**, *8*, 5320–5326. [CrossRef]

114. Zhang, Y.F.; Park, S.J. Fabrication and characterization of flower-like BiOI/Pt heterostructure with enhanced photocatalytic activity under visible light irradiation. *J. Solid State Chem.* **2017**, *253*, 421–429. [CrossRef]

115. Jiang, W.J.; Zhu, Y.F.; Zhu, G.X.; Zhang, Z.J.; Chen, X.J.; Yao, W.Q. Three-dimensional photocatalysts with a network structure. *J. Mater. Chem. A* **2017**, *5*, 5661–5679. [CrossRef]

116. Wan, W.C.; Zhang, R.Y.; Ma, M.Z.; Zhou, Y. Monolithic aerogel photocatalysts: A review. *J. Mater. Chem. A* **2018**, *6*, 754–775. [CrossRef]

117. Pierre, A.C.; Pajonk, G.M. Chemistry of aerogels and their applications. *Chem. Rev.* **2002**, *102*, 4243–4265. [CrossRef]

118. Dagan, G.; Tomkiewicz, M. Titanium dioxide aerogels for photocatalytic decontamination of aquatic environments. *J. Phys. Chem.* **1993**, *97*, 12651–12655. [CrossRef]

119. Heiligtag, F.J.; Rossell, M.D.; Suess, M.J.; Niederberger, M. Template-free co-assembly of preformed Au and TiO_2 nanoparticles into multicomponent 3D aerogels. *J. Mater. Chem.* **2011**, *21*, 16893–16899. [CrossRef]

120. Kailasam, K.; Epping, J.D.; Thomas, A.; Losse, S.; Junge, H. Mesoporous carbon nitride-silica composites by a combined sol-gel/thermal condensation approach and their application as photocatalysts. *Energy Environ. Sci.* **2011**, *4*, 4668–4674. [CrossRef]

121. Li, Z.D.; Wang, H.L.; Wei, X.N.; Liu, X.Y.; Yang, Y.F.; Jiang, W.F. Preparation and photocatalytic performance of magnetic Fe_3O_4@TiO_2 core-shell microspheres supported by silica aerogels from industrial fly ash. *J. Alloys Compd.* **2016**, *659*, 240–247. [CrossRef]

122. Jiang, W.J.; Liu, Y.F.; Wang, J.; Zhang, M.; Luo, W.J.; Zhu, Y.F. Separation-free polyaniline/TiO_2 3D hydrogel with high photocatalytic activity. *Adv. Mater. Interfaces* **2016**, *3*, 1500502. [CrossRef]

Article

The Role of Mediated Oxidation on the Electro-irradiated Treatment of Amoxicillin and Ampicillin Polluted Wastewater

Fernando L. Silva [1], Cristina Sáez [2,*], Marcos R.V. Lanza [1], Pablo Cañizares [2] and Manuel A. Rodrigo [2]

[1] Institute of Chemistry of São Carlos, University of São Paulo, P.O. Box 780, 13560-970 São Carlos-SP, Brazil; fernandolindosilva@gmail.com (F.L.S.); marcoslanza@iqsc.usp.br (M.R.V.L.)
[2] Chemical Engineering Department, University of Castilla-La Mancha, Edificio Enrique Costa Novella. Campus Universitario s/n, 13005 Ciudad Real, Spain; pablo.canizares@uclm.es (P.C.); manuel.rodrigo@uclm.es (M.A.R)
* Correspondence: Cristina.Saez@uclm.es; Tel.: +34-902-204-100

Received: 2 November 2018; Accepted: 18 December 2018; Published: 24 December 2018

Abstract: In this work, the electrolysis, photoelectrolysis and sonoelectrolysis with diamond electrodes of amoxicillin (AMX) and ampicillin (AMP) solutions were studied in the context of the search for technologies capable of removing antibiotics from liquid wastes. Single-irradiation processes (sonolysis and photolysis) were also evaluated for comparison. Results showed that AMX and AMP are completely degraded and mineralized by electrolysis in both chloride and sulfate media, although the efficiency is higher in the presence of chloride. The effect of the current density on mineralization efficiency is not relevant and this may be related to the role of mediated oxidation. Irradiation by ultraviolet light or ultrasound (US) waves does not produce a synergistic effect on the mineralization of AMX and AMP solutions. This indicates that the massive formation of radicals during the combined processes can favor their recombination to form stable and less reactive species.

Keywords: Anodic oxidation; diamond electrodes; UV irradiation; ultrasounds; amoxicillin; ampicillin

---

1. Introduction

The development of modern society is providing continuous improvements to quality of life, increasing food production through the use of agrochemicals and improving health with the use of biologically active substances for the control of diseases. These chemicals may pose some environmental risks and as a result of this, they have recently been the focus of a good deal of research globally [1,2]. Regarding medicines, antibiotics are worth highlighting because of their extensive consumption and extremely high potential environmental risks [3–8], reflected by the occurrence of super-bacteria, which are becoming a very serious health problem. Often, these chemicals are not efficiently removed in conventional wastewater treatment facilities, and consequently they are discharged into the environment where they are accumulated [9], altering the biological cycle of many types of organisms.

The use of beta-lactam antibiotics, such as amoxicillin (AMX) and ampicillin (AMP), represented an important contribution to medical science from the end of World War II. These medicines are still widely used because of their high efficiency, low cost and few side effects in humans. While ampicillin is more suitable for the treatment of respiratory tract infections, amoxicillin has several applications in infections of the skin and soft tissues, odontogenic infection, lower respiratory tract infection or urinary tract infection and otitis. Studies related to these molecules have developed because they have been detected in different sites. For example, monitoring performed at sewage treatment stations located in different regions of Italy showed concentrations of amoxicillin in the final effluent in the range of 1.8 ng L^{-1} to 120 ng L^{-1}.

The removal of antibiotics from water is neither easy nor efficient with conventional technologies. For this reason, many advanced oxidation processes (AOPs) have been suggested and proven over the last years [10]. Among them, photochemical oxidation, ozonation, photolysis with H_2O_2 and O_3, photocatalysis, and Fenton processes are worthy of mention. The main characteristic of these processes is the generation of very reactive and oxidizing radicals, such as hydroxyl radicals. Presently, electrolysis is one of the most interesting AOPs, and has been shown to be a very capable technology for the elimination of organic pollutants [11–16], including antibiotics. The efficiency of this technology depends not only on the operating conditions of the system but also on the material used as an anode. In the case of using conductive diamond electrodes, the massive generation of hydroxyl radicals is reported. This is a very powerful oxidant (E^0: 2.80 V vs SHE) that can not only react with organics but also promotes the generation of other oxidants, such as peroxophosphates, peroxosulfates, ozone, and hydrogen peroxide [17–19].

The electrolysis of antibiotics has been studied previously with successful results, which suggests that electrolysis is a technology worth evaluating [20–27]. However, these molecules are very complex and there is a need for further investigations in order to determine the conditions in which these processes can be optimized. Electrolysis is very efficient when the concentration of organics is in the range of 10^3–10^4 ppm, and the concentrations of antibiotics in urine are much lower; thus, improved knowledge about the mechanisms and possible synergies with other technologies (such as photo- or sono-processes) could help the development of future processes which are capable of solving the problem efficiently [28–31].

In this context, the goal of the present study is to evaluate the applicability of electrochemical oxidation with diamond anodes to remove waste consisting of mixtures of amoxicillin and ampicillin, clarifying the effects of the current density and supporting electrolyte on the performance of the process. In order to improve this performance, irradiation by UV light and ultrasounds during electrolysis is also evaluated, with the final aim of explaining the way in which the process mediated by electrogenerated oxidants affects amoxicillin/ampicillin oxidation. As seen in Table 1, the chemical structure of both molecules is quite similar. Hence, initially it may be expected that both molecules will be oxidized at similar rates and as a result of similar mechanisms. Similarities and differences found are expected to clarify mechanisms.

Table 1. Chemical structure of amoxicillin and ampicillin.

Amoxicillin 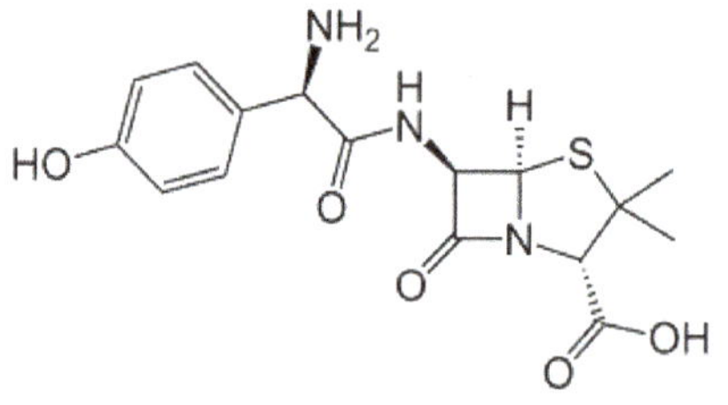	Ampicillin 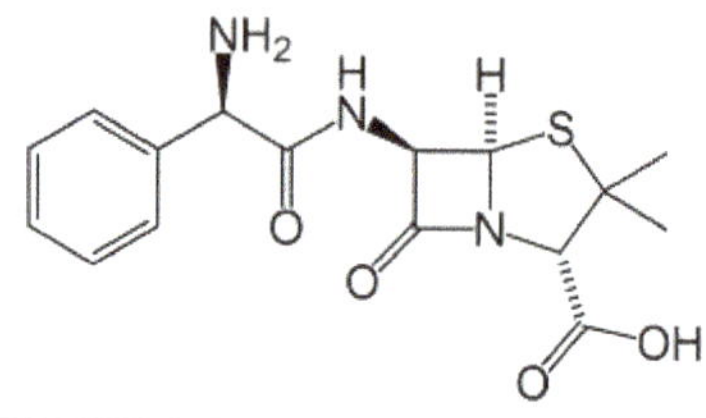

2. Results and Discussion

Figure 1 shows the removal of antibiotics and total organic carbon (TOC), as a function of the current charge passed during the electrolysis of amoxicillin (AMX) and ampicillin (AMP) solutions, carried out in sulphate media at three current densities (15, 30 and 60 mA cm^{-2}). As can be observed in Figure 1a, both AMX and AMP are completely oxidized during the electrolyses, and their decay trends lie over almost the same curve. Thus, they disappeared completely after passing similar electric charges. This similar trend can be explained in terms of their similar chemical structure. Regarding the effect of current density, the electric charges required to attain a given percentage removal depends on the following parameter: 8, 6, and 5.5. Ah dm^{-3} at 15, 30 and 60 mA cm^{-2}, respectively. Moreover, as it is known, the slope of this representation (d[concentration] / dQ) refers to the oxidation efficiency. There is an evident improvement of the degradation when the current density increases from 15 to 30 mA cm^{-2}. From 30 to 60 mA cm^{-2}, this improvement is less evident and it is only observed during the initial stages of the electrolysis. Thus, the higher the current density, the more efficient the partial degradation of the raw material. It is important to bear in mind that the antibiotics were measured by high performance liquid chromatography (HPLC), and any change in functional group causes the transformation of the raw molecule into an intermediate; hence its disappearance in the HPLC chromatogram. However, this trend observed in the raw antibiotic molecules is not reflected for TOC removal (Figure 1b). In fact, the electric charge required to attain the complete mineralization of the waste increases with current density: 12, 14, and 24 Ah dm^{-3} in the electrolyses carried out at 15, 30 and 60 mA cm^{-2}, respectively. Taking into account the resultant cell voltage in each case (4.8, 5.7 and 6.8 V, respectively), these differences are even more relevant in terms of energy consumption. Additionally, the electric current charges required for mineralization are higher than those passed to attain the complete depletion of the initial compound. This observation confirms that the oxidation of AMP and AMX occurs gradually and organic intermediates should be formed (see inset of Figure 1b). Finally, these intermediates are further oxidized to carbon dioxide, being more efficiently degraded at lower current densities, perhaps because of the major influence of the processes occurring in close proximity to the anode surface under those conditions.

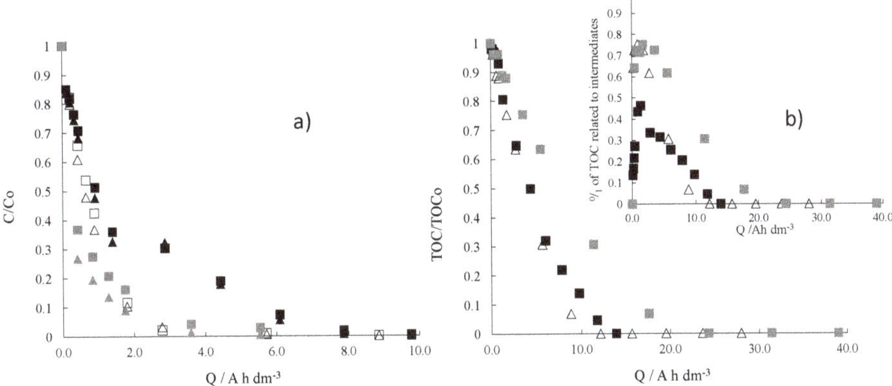

Figure 1. Oxidation trends of the antibiotics amoxicillin (AMX) (▲, △, ▲) and ampicillin (AMP) (■, □, ▨) (**a**) and of total organic carbon (TOC) (**b**) with the electric charge passed during the electrolysis carried out in sulphate media (3 g L^{-1} of Na$_2$SO$_4$) at three current densities: 15 (black symbols), 30 (white symbols), and 60 (grey symbols) mA cm^{-2}. Initial concentration was 100 mg L^{-1} of AMX and 100 mg L^{-1} of AMP.

At this point, it is important to take into consideration that in the electrolysis with conductive diamond electrodes, both direct and mediated oxidation (thanks to oxidants electrogenerated in the system) can contribute to the overall degradation process [32]. In the tests shown in Figure 1, sulphate was used as a supporting electrolyte, and it is well known that electrolysis can lead to the formation of peroxo-sulphates (Equation (1)) [17,33]. Additionally, this type of electrode also promotes the production of free hydroxyl radicals (Equation (2)) that recombine among them or react with oxygen to form O_3 (Equation (3)) and H_2O_2 (Equation (4)). These oxidants can also contribute to the oxidation process.

$$2SO_4{}^{2-} \rightarrow S_2O_8{}^{2-} + 2\,e^- \tag{1}$$

$$H_2O \rightarrow (\bullet OH) + H^+ + e^- \tag{2}$$

$$O_2 + 2\,(\bullet OH) \rightarrow O_3 + H_2O \tag{3}$$

$$2\,(\bullet OH) \rightarrow H_2O_2 \tag{4}$$

The formation of these species is promoted at high current density [26,34,35], but it is not always reflected by a more efficient degradation. In fact, in light of the results obtained in this study, the transformation of AMP and AMX is favored at high current densities, but not their full mineralization. This observation implies that although more oxidants are generated, the mineralization of intermediates becomes less effective under these conditions. This can be explained by the low reactivity of the intermediates with the oxidants, which made the direct oxidation mechanisms the most important in their degradation. This may explain the typical behaviour of diffusion-controlled systems, in which a mass transfer from the bulk to the electrode surface controls the rate of oxidation processes in spite of the presence of large amounts of oxidants in the reaction system. Thus, if the electrogenerated oxidants are not able to oxidize the reaction intermediates, the efficiency of the overall process decreases.

To determine the kinetics of the mineralization process, in Figure 2 the TOC removals are plotted on a semi-log scale. As can be observed, it is possible to distinguish two zones in which the slope of the trend changes, regardless of the current density. In each zone, the data show linear trends, indicating that electrolysis fits the kinetic of the pseudo-first order.

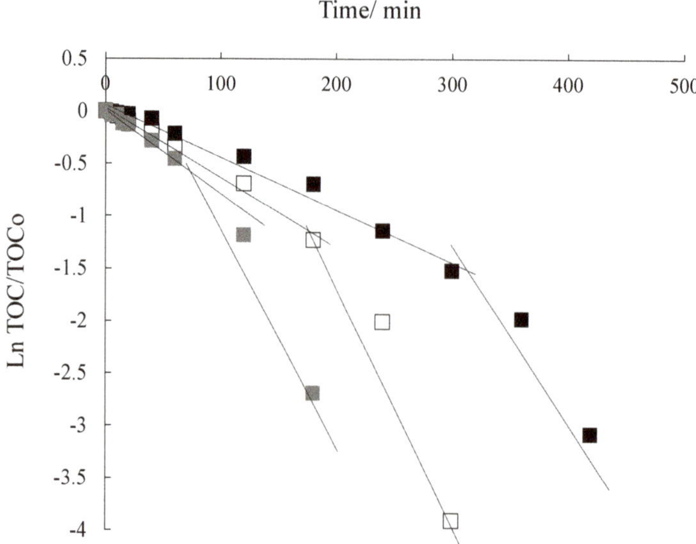

Figure 2. TOC removal in semi-log scale during electrolysis of AMX and AMP solutions carried out in sulphate media (3 g L^{-1} of Na_2SO_4) at three current densities: 15 (■), 30 (□) and 60 (■) mA cm^{-2}. Initial concentration was 100 mg L^{-1} of AMX and 100 mg L^{-1} of AMP.

In the first zone, the kinetic is slower and the current density shows a higher influence (k_1 = 0.0062, 0.0103 and 0.0115 min^{-1} at 15, 30 and 60 mA cm^{-2}, respectively). In this zone, AMX and AMP are the main organics present in the solution and, as shown in Figure 1, their attack is favored with the increase in electron current. These electrons are used to transform the initial pollutants to intermediates by attacking functional groups or double bonds (partial oxidation). Thus, in terms of mineralization, the electric charge supplied is not efficiently used because it is not used to form carbon dioxide but only intermediates.

In the second zone, the mineralization rate increases and the electrons supplied seem to be efficiently used. Before this, a transition region can be observed, in which the change of trends starts to become evident. Here, around 80–90% of the antibiotics have been transformed and around 30% of TOC has been mineralized; therefore, it may be assumed that small amounts of the raw pollutants coexist with both aromatic and aliphatic intermediates. This behaviour has been observed in previous works [32] and can be explained by the presence of short-chain carboxylic acids during the final stages of the degradation process, whose oxidation may be favored by the cocktail of oxidants formed and leads almost directly to carbon dioxide. Consequently, the slope increases significantly until 0.2197, 0.2084 and 0.2236 min^{-1}. Additionally, it is observed that the kinetic constants are similar at the three current densities, confirming the role of the mass transfer limitations, particularly in low-concentration solutions (TOC concentrations below 25 mg L^{-1})

At this point, in the literature it has been reported that electrogenerated oxidants should be activated to promote mediated oxidation efficiently [36–43]. It is well known that the photo-activation of stable oxidants in the bulk can occur when the electrolytic system is irradiated with UV light. Commonly, the radical oxidizing agents formed (Equations (5)–(7)) react faster than the corresponding oxidant-anions, and this enhances the degradation process.

$$S_2O_8{}^{2-} + UV\ light \rightarrow 2\ (SO_4{}^-) \tag{5}$$

$$H_2O_2 + UV\ light \rightarrow 2\ {}^\bullet OH \tag{6}$$

$$H_2O + O_3 + UV\ light \rightarrow 2\ {}^\bullet OH + O_2 \tag{7}$$

In the search for better efficiencies, AMP and AMX solutions undergoing the electrochemical process at 30 mA cm^{-2} were irradiated with 254 nm of UV light. Figure 3 shows the degradation trends of AMX and AMP during the combined process of photoelectrolysis in sulphate media. For comparison purposes, the role of the supporting electrolyte was also evaluated. It is well known that hypochlorite can be formed during electrolysis with conductive diamond electrodes in chlorine media (Equation (8)) and that, under UV light irradiation, hypochlorite is also decomposed into chlorine radicals (Equation (9)). Therefore, attending to the main mediator formed in each case (sulphate or chloride media), differences are expected. Additionally, to evaluate the existence of possible synergisms in the combined process, the results are compared with those obtained in the single processes of photolysis (without applying current) and electrolysis (without irradiating UV light).

$$Cl^- + H_2O \rightarrow HClO + H^+ + 2e^- \tag{8}$$

$$ClO^- + UV\ light \rightarrow Cl^\bullet + O^{-\bullet} \tag{9}$$

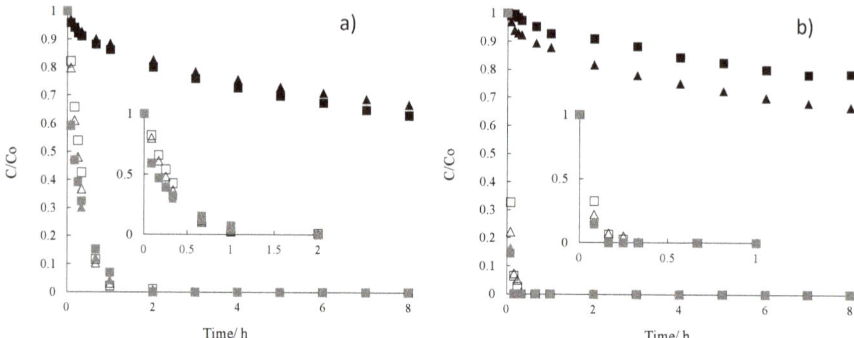

Figure 3. Degradation trends of AMX (triangles) and AMP (squares) during the electrolysis (Δ, □), photolysis (▲, ■) and photoelectrolysis (▲, ▦) tests in sulphate media (**a**) and chlorine (**b**). Initial concentration: 100 mg L^{-1} of AMX and 100 mg L^{-1} of AMP. Electrolyte concentrations: 3 g L^{-1} of Na$_2$SO$_4$ or 3.7 g L^{-1} of NaCl. Current density: 30 mA cm^{-2}: UV light: 254 nm.

As can be observed, the partial oxidation of AMX and AMP depends on the technology used. As expected [26,34,35], photolysis is not able to completely remove the antibiotics, and around 70% of AMX and AMP remains unaltered in the solutions after 8 h of treatment. Moreover, in sulphate media, the removal trends of both compounds are almost overlapping, but they differ significantly in chlorine media, in which the degradation of AMP is slower. As the difference between both antibiotic is just one hydroxyl group in the aromatic ring, it means that the oxidation obtained by photolysis is focused on this group primarily and that mediators formed in chloride medium are not as effective in the oxidation of other groups. On the contrary, less than 1 h of treatment is required to attain the complete depletion of AMP and AMX in the electrolysis tests. In fact, the partial oxidation of AMX and AMP is even faster in chlorine media, and after 0.25 h of electrolysis at 30 mA cm^{-2}, both antibiotics have completely disappeared from the reaction system. According to the literature [26,34,35], in chlorine media, the rapid chlorination of aromatic compounds occurs by the attack of electrogenerated Cl$_2$/HClO/ClO$^-$, although in many cases it is not accompanied by a rapid TOC removal.

Another important observation is that, in both evaluated scenarios, the contribution of the irradiation does not seem to be as relevant as expected: a slight improvement is observed in the first stages of the process (see insets of Figure 3), but this changes when around 70% of the raw organic has been transformed. It may be related to the competitive oxidation of reaction intermediates by activated oxidants.

To evaluate the effect of the irradiation on the mineralization, Figure 4 shows TOC removal in a semi-log scale of the electrolysis, photoelectrolysis and photolysis tests in both sulphate and chlorine media. As can be seen, the slight depletion of AMP and AMX shown in Figure 3 during the photolysis tests is not reflected by the TOC values, indicating that UV light irradiation at 254 nm is not able to convert these organics directly to carbon dioxide, and only a soft transformation of the molecule occurs. On the other hand, the effect of combining UV light irradiation with electrolysis does not seem to be particularly remarkable as regards mineralization and the data of electrolysis and photoelectrolysis lie over the same line, regardless of the supporting electrolyte used.

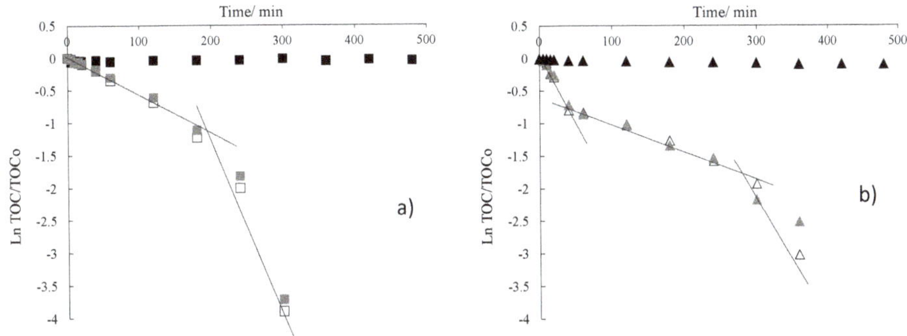

Figure 4. TOC removal in semi-log scale during the electrolysis (white symbols), photoelectrolysis (grey symbols) and photolysis (black symbols) of AMX and AMP solutions carried out in sulphate media (**a**) and chlorine media (**b**). Initial concentrations were 100 mg L^{-1} of AMX and 100 mg L^{-1} of AMP. Current density: 30 mA cm^{-2}. UV light: 254 nm.

More remarkable differences are observed when the results obtained in sulphate and chlorine media are compared, indicating that the oxidation pathways involved in each case should be different. The electrolysis and photoelectrolysis in chlorine media show a very different profile, with three clear regions of oxidation. Initially, a fast mineralization occurs, then the kinetic becomes slower, and at the end of the treatment it increases abruptly. Initially, antibiotics are the majority of pollutants. As was observed in Figure 3, they rapidly react with active chlorine species, leading to the formation of chlorinated intermediates [44]. Additionally, the decrease of TOC concentration seems to indicate that the depletion of any organic may occur—most likely, carboxylic or methyl groups present in the antibiotic molecule and split up in the first stages. These short-chain organics can be further oxidized to carbon dioxide, justifying the mineralization in this first zone. Additionally, the depletion of the initial compound leads also to the appearance of a large amount of chlorinated aromatic intermediates [12], whose reactivity differs from the initial AMX and AMP (chlorinated compounds generally show higher stability to oxidation) and whose further oxidation can lead to both chlorinated and non-chlorinated aliphatic intermediates, and not directly to carbon dioxide. This different reactivity can explain the change observed in the kinetics in the second zone. After that, the rapid and efficient mineralization of short-chain organics may explain the fast kinetic observed in the third zone.

The different profiles observed in sulphate and chlorine media can be related to the oxidants formed in each case, and to the extent of the oxidation reaction from close proximity to the electrode surface to the bulk. In both cases, ozone (Equation (3)) and hydrogen peroxide (Equation (4)) can be formed. Thus, the differences may be related mainly to the different reactivity of persulfate (Equation (1)) and hypochlorite (Equation (7)).

Following this, the effect of ultrasound (US) irradiation on electrolysis was evaluated. As is known [45–49], US irradiation produces a cavitation phenomenon. This releases a large amount of energy that promotes the generation of radicals [32] and the formation of activated oxidant species (Equations (10–13)). Additionally, by coupling sonolysis and electrolysis, the transfer of pollutant from the bulk to the closer to the electrode surface can also be favored.

$$H_2O + US\ wave \rightarrow H^\bullet + {}^\bullet OH \tag{10}$$

$$S_2O_8{}^{-2} + US\ wave \rightarrow 2\,(SO_4{}^-)^\bullet \tag{11}$$

$$S_2O_8{}^{-2} + {}^\bullet OH \rightarrow 2\,HSO_4{}^- + (SO_4{}^-)^\bullet + \frac{1}{2}\,O_2 \tag{12}$$

$$S_2O_8{}^{-2} + {}^\bullet H \rightarrow HSO_4{}^- + (SO_4{}^-)^\bullet \tag{13}$$

Figure 5 shows TOC removal in the semi-log scale of the combined process of sonoelectrolysis, together with the results of the single processes of electrolysis and sonolysis. Two acoustic frequencies were studied: 20 kHz (low-frequency ultrasound) and 10 MHz (high-frequency ultrasound). As expected, sonolysis at both frequencies shows no mineralization. However, when it is coupled to electrolysis, a slight improvement is observed during the first stages of the process, mainly in the case of low-frequency US. Thus, the trend changes and kinetics of electrolysis are higher than those of the combined processes. At low frequencies, ultrasound promotes the violent collapse of bubbles and, consequently, a large amount of radicals is produced, favoring the activation of oxidants. This is not observed at high frequencies [50], where the activation of persulfate is not promoted and an inhibition of its oxidative power is observed.

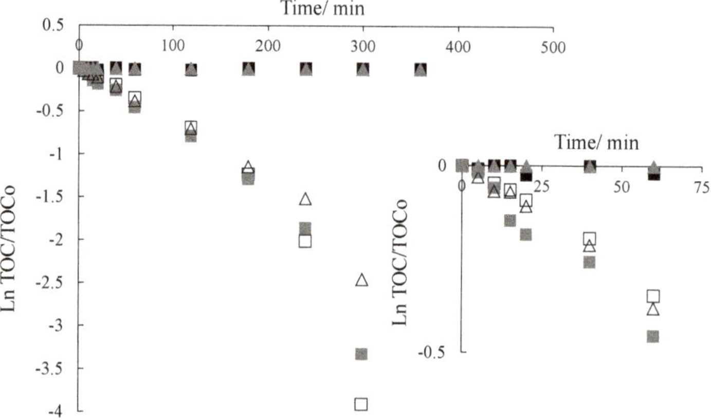

Figure 5. TOC removal in the semi-log scale during the electrolysis (□), sonoelectrolysis at 20 kHz (▨), sonoelectrolysis at 10 MHz (Δ) and sonolysis at 20 kHz (■) sonolysis at 10 MHz (▲) of AMX and AMP solutions carried out in sulphate media (square symbols, 3 g L^{-1} of Na$_2$SO$_4$). Initial concentration: 100 mg L^{-1} of AMX and 100 mg L^{-1} of AMP. Current density: 30 mA cm^{-2}.

In light of the obtained results, it can be said that although irradiation by UV light or US generates SO$_4{}^{-\bullet}$ and Cl$^\bullet$ species, and also promotes the formation of $^\bullet$OH, this is not reflected in an improvement of the degradation process. Therefore, the small effect observed in this study may be indicative of the existence of other non-irradiated activation processes. In this way, the interaction between the oxidants (including ozone, hydrogen peroxide and oxidant formed from the oxidation of the supporting electrolyte) that coexist in a region close to the electrode surface may be the main factor responsible for the typically high efficiencies of conductive diamond electrolysis.

In order to clarify this point, Figure 6 compares the synergistic effect (estimated according to Equation (14)) of the electro-irradiated treatment of AMX and AMP solutions and of other compounds previously studied in the literature [32,51] under similar operation conditions.

$$SI = \frac{k_{electro-irradiated\ process}}{k_{electrolysis} + k_{irradiated\ process}} \tag{14}$$

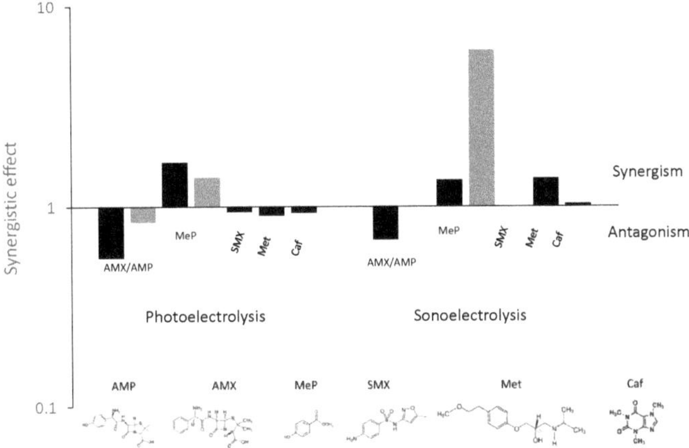

Figure 6. Synergistic/antagonistic effect of coupling irradiation techniques (UV light or ultrasound (US)) to the electrolysis of different organics: a mixture of amoxicillin and ampicillin (AMX/AMP), methylparaben (MeP), sulfomethoxazole (SMX), metroprolol (Met) and caffeine (Caf). Supporting electrolyte: (■) sulphate and (▨) chlorine.

As can be observed, the effect depends on the nature of the organic and supporting media. As reported before, the irradiation of US and/or UV during the electrolysis process leads to an excessive formation of radicals, which can recombine, forming less powerful and more stable oxidants, or can decompose to oxygen (Equations (15–18)). This means that these reactions may compete with organic oxidation and that active radicals are not available to oxidize organics. Thus, irradiation may cause an antagonistic effect on the degradation process.

$$2\,{}^{\bullet}OH \rightarrow H_2O_2 \tag{15}$$

$${}^{\bullet}OH + (SO_4{}^-)^{\bullet} \rightarrow HSO_5{}^- \tag{16}$$

$$HSO_5{}^- \rightarrow HSO_4{}^- + 0.5\,O_2 \tag{17}$$

$$HSO_4{}^- + {}^{\bullet}OH \rightarrow SO_4{}^{2-} + H_2O \tag{18}$$

This negative effect is not observed in all cases and cannot be explained by attending to mass transfer limitations; however, it may indicate the important role of chemical reactivity on the organic molecules. Thus, the synergism or antagonism of coupling irradiation techniques to electrochemical oxidation is difficult to predict at present. Nevertheless, these results shed light on the importance of the cocktail of oxidants that coexist in close proximity to the electrode surface during conductive diamond electrolysis, which may explain the typically high efficiencies of conductive diamond electrolysis.

3. Materials and Methods

3.1. Chemicals

All chemicals were of analytical grade and used as received. Anhydrous sodium sulphate and sodium chloride were purchased from Fluka (Bucharest, Romania), and amoxicillin (90% purity), ampicillin (96% purity) and acetonitrile (HPLCE grade) were purchased from Sigma-Aldrich (St. Louis, MO, USA), and sulphuric acid and hydrochloric acid were purchased from Merck (Darmstadt, Germany). Solutions were prepared using double deionized water (Millipore Milli-Q system, Merck Millipore, Madrid, Spain).

3.2. Electrochemical Cell

The electrolytic essays were carried out in a single-compartment electrochemical flow cell. Synthetic wastewater contain the electrolyte (3 g L^{-1} of Na_2SO_4 or 3.7 g L^{-1} of NaCl) and antibiotics (100 mg dm^{-3} of each). Details of the cell and of the auxiliary equipment can be found elsewhere [52]. Irradiated experiments were carried out using the same fluid-dynamic conditions as the electrolysis setup. For irradiated-electrolysis tests, the UV lamp (Vilber Lourmat filtered lamp, VL-215.MC, with a power of 4 W, Vilber, Marne, France) or the ultrasound (acoustic frequency: 20 kHz (UP200S, Hielscher Ultrasonics GmbH, Hielscher, Berlin, Germany) and 10 MHz (EPOCH 650, Olympus, Barcelona, Spain) were immersed in the reservoir tank.

3.3. Analysis Procedures

The concentration of antibiotics and the mineralization of the solutions were monitored by liquid chromatography at 220 nm wavelength (HPLC, Agilent1260 series, Agilent, Madrid, Spain) and total organic carbon (TOC, Multi N/C 3100 Analytik Jena analyser, Analytik Jena AG, Jena, Germany) analysis, respectively. For the chromatographic analysis the column used was an Elipse Plus 35 μm C18, a mixture of acetonitrile (A) and water with phosphoric acid (pH 2) was used as a mobile phase, the volume injection was set to 20 μL, and the temperature was fixed at 35 °C. Before analysis, all samples were filtered using 0.45 μm nylon filters. The concentration of inorganic species was determined by titration or by ion chromatography (IC, Shimadzu LC-20A (Shimadzu, Duisburg, Germany). The IC chromatograph was equipped with a ShodexIC I-524A column; mobile phase, 2.5 mM phthalic acid at pH 4.0; flow rate, 1×10^{-3} dm^3 min^{-1}. Hypochlorite ($HClO^-$) was determined by titration with 0.001 mol L^{-1} As_2O_3 in 2.0 mol L^{-1} NaOH. Peroxosulfate was quantified iodometrically

4. Conclusions

The following conclusions have been drawn:

- The conductive diamond electrolysis is able to attain the complete mineralization of amoxicillin and ampicillin solutions. The removal efficiency depends on the current density and supporting media, with the antibiotic degradation rate and mineralization favored in the presence of chloride.
- The mineralization rate of AMX and AMP solutions during electrolysis, photoelectrolysis and sonoelectrolysis fits well to pseudo-first order kinetics, although two or three reaction zones are distinguished in sulphate and chloride media, respectively. This may indicate the existence of complex mechanisms in which indirect oxidation is very important.
- The effect of irradiating UV light and/or US waves during the electrolysis of AMX and AMP is not very relevant. This may be explained in terms of the high contribution of mediated oxidation in the single electrolysis process or of the massive formation of radicals during electro-irradiated processes, which recombine to form stable oxidants.

Author Contributions: Investigation, F.L.S.; Methodology, F.L.S., C.S. and M.A.R.; Project administration, M.R.V.L., P.C. and M.A.R.; Supervision, M.R.V.L., P.C. and M.A.R.; Writing—original draft, F.L.S., C.S. and M.A.R.

Funding: Financial support from the Regional Government of Castilla-La Mancha (Spain) and European Union (EU, FEDER) through project SBPLY/17/180501/000396 are gratefully acknowledged. CNPq (grants 301492/2013-1 and 465571/2014-0 (INCT-DATREM)) and Capes and São Paulo Research Foundation (FAPESP) (grants #2016/01937-4, #2014/50945-4, #2013/16690-6, #2016/11672-8 and #2017/10118-0) are also acknowledge.

Conflicts of Interest: The authors declare no conflict of interest.

References

1. Feng, L.; van Hullebusch, E.D.; Rodrigo, M.A.; Esposito, G.; Oturan, M.A. Removal of residual anti-inflammatory and analgesic pharmaceuticals from aqueous systems by electrochemical advanced oxidation processes. A review. *Chem. Eng. J.* **2013**, *228*, 944–964. [CrossRef]
2. Rodrigo, M.A.; Oturan, N.; Oturan, M.A. Electrochemically Assisted Remediation of Pesticides in Soils and Water: A Review. *Chem. Rev.* **2014**, *114*, 8720–8745. [CrossRef] [PubMed]
3. Vieno, N.; Tuhkanen, T.; Kronberg, L. Elimination of pharmaceuticals in sewage treatment plants in Finland. *Water Res.* **2007**, *41*, 1001–1012. [CrossRef] [PubMed]
4. Radjenovic, J.; Jelic, A.; Petrovic, M.; Barcelo, D. Determination of pharmaceuticals in sewage sludge by pressurized liquid extraction (PLE) coupled to liquid chromatography-tandem mass spectrometry (LC-MS/MS). *Anal. Bioanal. Chem.* **2009**, *393*, 1685–1695. [CrossRef] [PubMed]
5. Klavarioti, M.; Mantzavinos, D.; Kassinos, D. Removal of residual pharmaceuticals from aqueous systems by advanced oxidation processes. *Environ. Int.* **2009**, *35*, 402–417. [CrossRef] [PubMed]
6. Urtiaga, A.M.; Perez, G.; Ibanez, R.; Ortiz, I. Removal of pharmaceuticals from a WWTP secondary effluent by ultrafiltration/reverse osmosis followed by electrochemical oxidation of the RO concentrate. *Desalination* **2013**, *331*, 26–34. [CrossRef]
7. Liu, X.; Lu, S.; Guo, W.; Xi, B.; Wang, W. Antibiotics in the aquatic environments: A review of lakes, China. *Sci. Total Environ.* **2018**, *627*, 1195–1208. [CrossRef]
8. Carvalho, I.T.; Santos, L. Antibiotics in the aquatic environments: A review of the European scenario. *Environ. Int.* **2016**, *94*, 736–757. [CrossRef]
9. Al Aukidy, M.; Verlicchi, P.; Voulvoulis, N. A framework for the assessment of the environmental risk posed by pharmaceuticals originating from hospital effluents. *Sci. Total Environ.* **2014**, *493*, 54–64. [CrossRef]
10. Dewil, R.; Mantzavinos, D.; Poulios, I.; Rodrigo, M.A. New perspectives for Advanced Oxidation Processes. *J. Environ. Manag.* **2017**, *195*, 93–99. [CrossRef]
11. Sires, I.; Brillas, E.; Oturan, M.A.; Rodrigo, M.A.; Panizza, M. Electrochemical advanced oxidation processes: Today and tomorrow. A review. *Environ. Sci. Pollut. Res.* **2014**, *21*, 8336–8367. [CrossRef] [PubMed]
12. Martinez-Huitle, C.A.; Rodrigo, M.A.; Sires, I.; Scialdone, O. Single and Coupled Electrochemical Processes and Reactors for the Abatement of Organic Water Pollutants: A Critical Review. *Chem. Rev.* **2015**, *115*, 13362–13407. [CrossRef] [PubMed]
13. Brillas, E. Electro-Fenton, UVA Photoelectro-Fenton and Solar Photoelectro-Fenton Treatments of Organics in Waters Using a Boron-Doped Diamond Anode: A Review. *J. Mex. Chem. Soc.* **2014**, *58*, 239–255. [CrossRef]
14. Brillas, E.; Martínez-Huitle, C.A. Decontamination of wastewaters containing synthetic organic dyes by electrochemical methods. An updated review. *Appl. Catal. B Environ.* **2015**, *166–167*, 603–643. [CrossRef]
15. Martinez-Huitle, C.A.; Brillas, E. Decontamination of wastewaters containing synthetic organic dyes by electrochemical methods: A general review. *Appl. Catal. B Environ.* **2009**, *87*, 105–145. [CrossRef]
16. Sires, I.; Brillas, E. Remediation of water pollution caused by pharmaceutical residues based on electrochemical separation and degradation technologies: A review. *Environ. Int.* **2012**, *40*, 212–229. [CrossRef] [PubMed]
17. Canizares, P.; Saez, C.; Sanchez-Carretero, A.; Rodrigo, M.A. Synthesis of novel oxidants by electrochemical technology. *J. Appl. Electrochem.* **2009**, *39*, 2143–2149. [CrossRef]
18. De Araujo, D.M.; Cotillas, S.; Saez, C.; Canizares, P.; Martinez-Huitle, C.A.; Andres Rodrigo, M. Activation by light irradiation of oxidants electrochemically generated during Rhodamine B elimination. *J. Electroanal. Chem.* **2015**, *757*, 144–149. [CrossRef]
19. Yoon, Y.; Cho, E.; Jung, Y.; Kwon, M.; Yoon, J.; Kang, J.-W. Evaluation of the formation of oxidants and by-products using Pt/Ti, RuO2/Ti, and IrO2/Ti electrodes in the electrochemical process. *Environ. Technol.* **2015**, *36*, 317–326. [CrossRef] [PubMed]
20. Dirany, A.; Sires, I.; Oturan, N.; Oturan, M.A. Electrochemical abatement of the antibiotic sulfamethoxazole from water. *Chemosphere* **2010**, *81*, 594–602. [CrossRef] [PubMed]
21. Garcia-Segura, S.; Cavalcanti, E.B.; Brillas, E. Mineralization of the antibiotic chloramphenicol by solar photoelectro-Fenton. From stirred tank reactor to solar pre-pilot plant. *Appl. Catal. B Environ.* **2014**, *144*, 588–598. [CrossRef]

22. Haidar, M.; Dirany, A.; Sires, I.; Oturan, N.; Oturan, M.A. Electrochemical degradation of the antibiotic sulfachloropyridazine by hydroxyl radicals generated at a BDD anode. *Chemosphere* **2013**, *91*, 1304–1309. [CrossRef] [PubMed]

23. Ji, Y.; Fan, Y.; Liu, K.; Kong, D.; Lu, J. Thermo activated persulfate oxidation of antibiotic sulfamethoxazole and structurally related compounds. *Water Res.* **2015**, *87*, 1–9. [CrossRef] [PubMed]

24. Wu, J.; Zhang, H.; Oturan, N.; Wang, Y.; Chen, L.; Oturan, M.A. Application of response surface methodology to the removal of the antibiotic tetracycline by electrochemical process using carbon-felt cathode and DSA (Ti/RuO2-IrO2) anode. *Chemosphere* **2012**, *87*, 614–620. [CrossRef] [PubMed]

25. Xiao, S.; Song, Y.; Tian, Z.; Tu, X.; Hu, X.; Liu, R. Enhanced mineralization of antibiotic berberine by the photoelectrochemical process in presence of chlorides and its optimization by response surface methodology. *Environ. Earth Sci.* **2015**, *73*, 4947–4955. [CrossRef]

26. Cotillas, S.; Lacasa, E.; Saez, C.; Canizares, P.; Rodrigo, M.A. Electrolytic and electro-irradiated technologies for the removal of chloramphenicol in synthetic urine with diamond anodes. *Water Res.* **2018**, *128*, 383–392. [CrossRef]

27. Cotillas, S.; Lacasa, E.; Saez, C.; Canizares, P.; Rodrigo, M.A. Removal of pharmaceuticals from the urine of polymedicated patients: A first approach. *Chem. Eng. J.* **2018**, *331*, 606–614. [CrossRef]

28. Stackelberg, P.E.; Furlong, E.T.; Meyer, M.T.; Zaugg, S.D.; Henderson, A.K.; Reissman, D.B. Persistence of pharmaceutical compounds and other organic wastewater contaminants in a conventional drinking-watertreatment plant. *Sci. Total Environ.* **2004**, *329*, 99–113. [CrossRef] [PubMed]

29. Dominguez, J.R.; Gonzalez, T.; Palo, P.; Sanchez-Martin, J.; Rodrigo, M.A.; Saez, C. Electrochemical Degradation of a Real Pharmaceutical Effluent. *Water Air Soil Pollut.* **2012**, *223*, 2685–2694. [CrossRef]

30. Sopaj, F.; Rodrigo, M.A.; Oturan, N.; Podvorica, F.I.; Pinson, J.; Oturan, M.A. Influence of the anode materials on the electrochemical oxidation efficiency. Application to oxidative degradation of the pharmaceutical amoxicillin. *Chem. Eng. J.* **2015**, *262*, 286–294. [CrossRef]

31. Perez, J.F.; Llanos, J.; Saez, C.; Lopez, C.; Canizares, P.; Rodrigo, M.A. Treatment of real effluents from the pharmaceutical industry: A comparison between Fenton oxidation and conductive-diamond electro-oxidation. *J. Environ. Manag.* **2017**, *195*, 216–223. [CrossRef] [PubMed]

32. Dionisio, D.; Motheo, A.J.; Sáez, C.; Rodrigo, M.A. Effect of the electrolyte on the electrolysis and photoelectrolysis of synthetic methyl paraben polluted wastewater. *Sep. Purif. Technol.* **2019**, *208*, 201–207. [CrossRef]

33. Serrano, K.; Michaud, P.A.; Comninellis, C.; Savall, A. Electrochemical preparation of peroxodisulfuric acid using boron doped diamond thin film electrodes. *Electrochim. Acta* **2002**, *48*, 431–436. [CrossRef]

34. Cotillas, S.; de Vidales, M.J.M.; Llanos, J.; Saez, C.; Canizares, P.; Rodrigo, M.A. Electrolytic and electro-irradiated processes with diamond anodes for the oxidation of persistent pollutants and disinfection of urban treated wastewater. *J. Hazard. Mater.* **2016**, *319*, 93–101. [CrossRef] [PubMed]

35. Martin de Vidales, M.J.; Cotillas, S.; Perez-Serrano, J.F.; Llanos, J.; Saez, C.; Canizares, P.; Rodrigo, M.A. Scale-up of electrolytic and photoelectrolytic processes for water reclaiming: A preliminary study. *Environ. Sci. Pollut. Res. Int.* **2016**, *23*, 19713–19722. [CrossRef] [PubMed]

36. Boye, B.; Dieng, M.M.; Brillas, E. Electrochemical degradation of 2,4,5-trichlorophenoxyacetic acid in aqueous medium by peroxi-coagulation. Effect of pH and UV light. *Electrochim. Acta* **2003**, *48*, 781–790. [CrossRef]

37. Diagne, M.; Oturan, N.; Oturan, M.A.; Sires, I. UV-C light-enhanced photo-Fenton oxidation of methyl parathion. *Environ. Chem. Lett.* **2009**, *7*, 261–265. [CrossRef]

38. Shih, Y.J.; Chen, K.H.; Huang, Y.H. Mineralization of organic acids by the photo-electrochemical process in the presence of chloride ions. *J. Taiwan Inst. Chem. Eng.* **2014**, *45*, 962–966. [CrossRef]

39. Matafonova, G.; Batoev, V. Recent advances in application of UV light-emitting diodes for degrading organic pollutants in water through advanced oxidation processes: A review. *Water Res.* **2018**, *132*, 177–189. [CrossRef]

40. Souza, F.L.; Sáez, C.; Cañizares, P.; Motheo, A.J.; Rodrigo, M.A. Coupling photo and sono technologies to improve efficiencies in conductive diamond electrochemical oxidation. *Appl. Catal. B Environ.* **2013**, *144*, 121–128. [CrossRef]

41. Rubi-Juarez, H.; Cotillas, S.; Saez, C.; Canizares, P.; Barrera-Diaz, C.; Rodrigo, M.A. Use of conductive diamond photo-electrochemical oxidation for the removal of pesticide glyphosate. *Sep. Purif. Technol.* **2016**, *167*, 127–135. [CrossRef]

42. Aquino, J.M.; Miwa, D.W.; Rodrigo, M.A.; Motheo, A.J. Treatment of actual effluents produced in the manufacturing of atrazine by a photo-electrolytic process. *Chemosphere* **2017**, *172*, 185–192. [CrossRef] [PubMed]

43. Vieira dos Santos, E.; Saez, C.; Canizares, P.; Rodrigo, M.A.; Martinez-Huitle, C.A. Coupling Photo and Sono Technologies with BDD Anodic Oxidation for Treating Soil-Washing Effluent Polluted with Atrazine. *J. Electrochem. Soc.* **2018**, *165*, E262–E267. [CrossRef]

44. Comninellis, C.; Nerini, A. Anodic-oxidation of phenol in the presence of nacl for waste-water treatment. *J. Appl. Electrochem.* **1995**, *25*, 23–28. [CrossRef]

45. Garbellini, G.S.; Salazar-Banda, G.R.; Avaca, L.A. Effects of ultrasound on the degradation of pentachlorophenol by Boron-Doped diamond electrodes. *Port. Electrochim. Acta* **2010**, *28*, 405–415. [CrossRef]

46. Ma, Y.S.; Sung, C.F.; Lin, J.G. Degradation of carbofuran in aqueous solution by ultrasound and Fenton processes: Effect of system parameters and kinetic study. *J. Hazard. Mater.* **2010**, *178*, 320–325. [CrossRef] [PubMed]

47. Bringas, E.; Saiz, J.; Ortiz, I. Kinetics of ultrasound-enhanced electrochemical oxidation of diuron on boron-doped diamond electrodes. *Chem. Eng. J.* **2011**, *172*, 1016–1022. [CrossRef]

48. Rooze, J.; Rebrov, E.V.; Schouten, J.C.; Keurentjes, J.T.F. Effect of resonance frequency, power input, and saturation gas type on the oxidation efficiency of an ultrasound horn. *Ultrason. Sonochem.* **2011**, *18*, 209–215. [CrossRef]

49. Martin de Vidales, M.J.; Barba, S.; Saez, C.; Canizares, P.; Rodrigo, M.A. Coupling ultraviolet light and ultrasound irradiation with Conductive-Diamond Electrochemical Oxidation for the removal of progesterone. *Electrochim. Acta* **2014**, *140*, 20–26. [CrossRef]

50. Steter, J.R.; Dionisio, D.; Lanza, M.R.V.; Motheo, A.J. Electrochemical and sonoelectrochemical processes applied to the degradation of the endocrine disruptor methyl paraben. *J. Appl. Electrochem.* **2014**, *44*, 1317–1325. [CrossRef]

51. Martin de Vidales, M.J.; Saez, C.; Perez, J.F.; Cotillas, S.; Llanos, J.; Canizares, P.; Rodrigo, M.A. Irradiation-assisted electrochemical processes for the removal of persistent organic pollutants from wastewater. *J. Appl. Electrochem.* **2015**, *45*, 799–808. [CrossRef]

52. Cotillas, S.; Lacasa, E.; Saez, C.; Canizares, P.; Rodrigo, M.A. Disinfection of urine by conductive-diamond electrochemical oxidation. *Appl. Catal. B Environ.* **2018**, *229*, 63–70. [CrossRef]

Article

Integrated Au/TiO$_2$ Nanostructured Photoanodes for Photoelectrochemical Organics Degradation

Roberto Matarrese [1,*], Michele Mascia [2], Annalisa Vacca [2], Laura Mais [2], Elisabetta M. Usai [2], Matteo Ghidelli [3,4], Luca Mascaretti [3,5], Beatrice R. Bricchi [3], Valeria Russo [3,6], Carlo S. Casari [3,6], Andrea Li Bassi [3,6], Isabella Nova [1] and Simonetta Palmas [2,*]

[1] Dipartimento di Energia, Laboratorio di Catalisi e Processi Catalitici, Politecnico di Milano, via La Masa 34, I-20156 Milano, Italy; isabella.nova@polimi.it

[2] Università degli studi di Cagliari, Dipartimento di Ingegneria Meccanica, Chimica e dei Materiali, via Marengo 2, 09123 Cagliari, Italy; michele.mascia@unica.it (M.M.); annalisa.vacca@dimcm.unica.it (A.V.); l.mais@dimcm.unica.it (L.M.); elisabetta.usai@unica.it (E.M.U.)

[3] Dipartimento di Energia, Laboratorio Materiali Micro e Nanostrutturati, Politecnico di Milano, via Ponzio 34/3, I-20133 Milano, Italy; m.ghidelli@mpie.de (M.G.); luca.mascaretti@upol.cz (L.M.); beatriceroberta.bricchi@polimi.it (B.R.B.); valeria.russo@polimi.it (V.R.); carlo.casari@polimi.it (C.S.C.); andrea.libassi@polimi.it (A.L.B.)

[4] Department of Structure and Nano/-Micromechanics of Materials, Max-Planck-Institut für Eisenforschung GmbH, Max-Planck-straße 1, 40237 Düsseldorf, Germany

[5] Regional Centre of Advanced Technologies and Materials, Faculty of Science, Palacký University Olomouc, Šlechtitelů 27, 783 71 Olomouc, Czech Republic

[6] Center for Nanoscience and Technology – IIT@Polimi, via Giovanni Pascoli 70/3, 20133 Milano, Italy

[*] Correspondence: roberto.matarrese@polimi.it (R.M.); simonetta.palmas@dimcm.unica.it (S.P.)

Received: 7 March 2019; Accepted: 2 April 2019; Published: 5 April 2019

Abstract: In this work, hierarchical Au/TiO$_2$ nanostructures were studied as possible photoanodes for water splitting and bisphenol A (BPA) oxidation. TiO$_2$ samples were synthetized by Pulsed Laser Deposition (PLD), while Au nanoparticles (NPs) were differently dispersed (i.e., NPs at the bottom or at the top of the TiO$_2$, as well as integrated TiO$_2$/Au-NPs assemblies). Voltammetric scans and electrochemical impedance spectroscopy analysis were used to correlate the morphology of samples with their electrochemical properties; the working mechanism was investigated in the dark and in the presence of a light radiation, under neutral pH conditions towards the possible oxidation of both bisphenol A (BPA) and water molecules. Different behavior of the samples was observed, which may be attributed mainly to the distributions of Au NPs and to their dimension as well. In particular, the presence of NPs at the bottom seems to be the crucial point for the working mechanism of the structure, thanks to scattering effects that likely allow to better exploit the radiation.

Keywords: photoelectrocatalysis; TiO$_2$ nanostructures; Au nanoparticles; water splitting; bisphenol A oxidation

---

1. Introduction

This work focuses on two very important aspects in the field of photoelectrocatalysis: on one side, the synthesis of efficient photocatalysts to be used in a wide range of wavelengths (e.g., also under solar radiation), and on the other side the development of effective methods for the purification of industrial and domestic wastewater, with particular attention to those substances dangerous to human health that are not easily degradable by conventional methods. Among the others, bisphenol A (2,2-bis(4-hydroxyphenyl) propane or BPA), which is composed of two phenol molecules bonded by a methyl bridge and two methyl groups, deserves attention. This compound is used as an intermediate (binding, plasticizing, and hardening) in plastics, paints/lacquers, binding materials, and filling materials, as well as an additive for flame-retardants, brake fluids, and thermal papers [1]. However, the most important field of usage of this compound is represented by polycarbonates and epoxy resins, that determined a huge increasing in the BPA demand, in the last few years: the global demand, which was 5.0 million tons in 2010 and 8 million tons in 2016, is projected to reach 10.6 million tons by 2022 [2]. As a result, BPA has been frequently detected in water and soil and its impact on both environment and human health is a major point of concern [2–5]. Accordingly, the removal of BPA from wastewater has become a priority for the scientific community and several methods for its degradation have been proposed in the literature (e.g., biological [6,7], catalytic [8], photocatalytic [9–11], and photolelectrocatalytic methods [12,13]), most of which reported high yields, up to 90–100%, at least in terms of BPA removal. However, the high number of papers that are still present in the recent literature indicate that the problem is far from resolved and a lot of interest still remains in finding effective and competitive methods for BPA degradation [14–17].

Focusing on photoelectrocatalytic processes, the present work proposes the use of nanostructured TiO_2 electrodes integrated with Au nanoparticles (NPs), and it investigates their electrochemical behavior during electrolysis both in KNO_3 solution and in the presence of BPA.

For a long time, our research group investigated nanostructured TiO_2 materials for photoelectrocatalytic applications [18–21], mostly facing the well-known major problems of TiO_2 which are the low quantum efficiency and the poor activation by visible light (i.e., solar light). More recently, some of us proposed the use of Pulsed Laser Deposition (PLD) for the synthesis of plasmonic Au NPs and their integration in hierarchical TiO_2 nanostructures [22], which have been preliminarily tested for the photodegradation of methyl orange under simulated natural sunlight [23]. Indeed, the use of hierarchical 3D nanostructures, with high roughness, can promote light absorption due the scattering effect over a large angular range. However, such an effect is not always exploitable, for example in the case of thin-film cells where the surface roughness would exceed the film thickness, and because the greater surface area increases minority carrier recombination in the surface and junction regions. Particularly, in these cases, the use of metallic nanostructures that support surface plasmons, could be effective [24]. For a given photoelectrode, various plasmonic mechanisms may be exploited to boost its photoactivity depending on the size, morphology, and chemical nature of the plasmonic unit [25]. For example, the mechanism of hot electron injection from the plasmonic unit to the semiconductor conduction band may allow the use of visible light that is not absorbed by a wide-bandgap semiconductor, as widely reported for the TiO_2 photoelectrodes combined with noble metals [26]. Among them, Au and Ag (i.e., alone or alloys), have been mostly considered as possible additives to exploit the plasmonic effects, or to favor charge carrier separation to inhibit recombination, directly contributing to the production of long-lived charges. For instance, Naseri at al. [27] proposed TiO_2 photoanodes decorated with Au–Ag alloy NPs for photoelectrochemical water splitting applications. In particular, photocurrent measurements showed a 30% increase in the presence of alloy NPs as well as a 50% reduction in charge transfer resistance of the electrodes. Other studies reported the use of Au NPs specifically for the treatment of BPA solutions [27–30], focusing on the photoactivity of Au/TiO_2 films, on the plasmonic effect of Au, and on the nature of the support. Very recently, Sreedhar et al. [31] investigated the photoelectrochemical behavior of Au clusters functionalized TiO_2 thin films to explore the role of Au clusters position on charge carrier generation and incident visible light harvesting.

The results demonstrated the great importance of the structure engineering that represents a key point to maximize the light capture and its concentration even in thin semiconductor layers, by increasing the absorption. In fact, the literature, initially focused on solar cells applications, clearly shows that the effect of the noble metal NPs insertion and the consequent operation mechanism of the final structures strongly depend on the particle size and their dispersion within the structure. For instance, the plasmon resonance energy transfer process and the production of hot electrons as well, more likely occur on small particles, while a simple radiation scattering effect is expected on large particles (>100 nm). Similarly, it was reported that for organic solar cells, the plasmonic effect of small metal NPs can be exploited if the NPs are placed at the interface between two phases where charges separation takes place [32]. Instead, in the case of inorganic solar cells the scattering effect of NPs located away from the p-n junction is exploited [33], even though plasmon effects are also reported in similar cases [34].

As a consequence, the importance of a careful design of the catalyst is evident, which however cannot allow to neglect the analysis of the working mechanism of the structure. In this context, the aim of this work is to investigate the performance of TiO_2 samples with a hierarchical nanostructure, where Au NPs were differently dispersed (i.e., NPs at the bottom or at the top of the TiO_2, as well as integrated TiO_2/Au-NPs assemblies).

The activity of the samples was analyzed by means of photoelectrolysis experiments carried out under neutral pH conditions both towards the possible oxidation of BPA and water molecules. In fact, even if water splitting can not to be particularly favored under such pH conditions, it can be competitive or concomitant with the BPA oxidation.

2. Results

2.1. Structural, Optical, and Electrochemiecal Properties

Table 1 lists all the investigated samples and their description (i.e., the different distribution of Au NPs).

Table 1. Description and schematic representation of the investigated structures. All samples have a nominal surface of 1 cm × 1 cm and the thickness of the TiO_2 layer was 1000 nm.

Sample	Description		Au% Vol.
TiO_2	TiO_2 film on Ti substrate (Reference)		0
TiO_2/Au	Au nanoparticles (NPs) at the bottom of the TiO_2 film		1
Au/TiO_2	Au NPs on the top of the TiO_2 film		1.5
Au/TiO_2/Au	Au NPs both at the bottom and on top of the TiO_2 film		2.5
TiO_2–Au	Au-integrated TiO_2 films		~3.3

Figure 1 shows the morphology of the TiO_2 sample, used as reference (without Au), as well as of the Au loaded samples. All TiO_2 films feature a nanoscale porosity and a hierarchical organization of the nanostructures in 'nanotrees' (Figure 1a), which is beneficial for light scattering, electron transport and to obtain large specific surface area values. This kind of structure and the relative properties have been widely discussed in previous works [23,35,36].

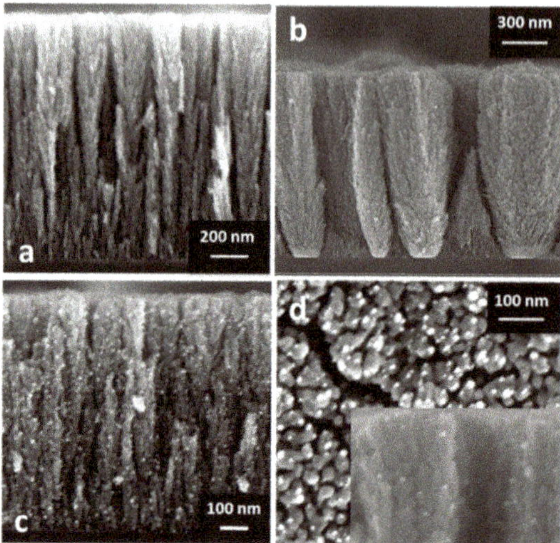

Figure 1. SEM images of (**a**) reference TiO$_2$ film (cross section); (**b**) TiO$_2$/Au film (cross section); (**c**) TiO$_2$–Au film (cross section); (**d**) the Au/TiO$_2$ film (top surface; in the inset the film cross section close to the surface shows the penetration of Au NPs).

In detail, the TiO$_2$/Au sample is characterized by Au NPs at the bottom of the film (Figure 1b) with an average size of 112 ± 44 nm, meant to act as scattering centers in order to induce light diffusion/trapping in the TiO$_2$ layer [25]. The Au/TiO$_2$ sample is characterized by NPs deposited on the top surface of the TiO$_2$ layer, meant to implement plasmonic effects; their average size is 4 ± 1 nm, however nanoparticles as large as ~10 nm are present (Figure 1d) [25]. Moreover, the scanning electron microscopy (SEM) image shows that the Au NPs penetrate in the film for a depth of ~200 nm. The Au/TiO$_2$/Au sample combines the features of the previous two films. The TiO$_2$–Au film is characterized by a nanoscale structure, which is decorated throughout the thickness by Au NPs with an average size of 3 ± 1.5 nm, even though particles as large as 15 nm are present (Figure 1c), as reported in [23]. The amount of Au in the investigated samples is listed in Table 1.

The Au-integrated TiO$_2$ films are characterized by a strong capability of light scattering, as measured by the haze factor defined in the Experimental section (ratio between diffuse and total transmitted radiation). Table 2 reports the average haze factor for the investigated samples in the visible (wavelength 400–800 nm) and in the near infrared (NIR) region (800–2000 nm). It is clear that the large Au NPs at the bottom of the film (sample TiO$_2$/Au) act as scattering centers and, while the Au/TiO$_2$ sample is not characterized by increased light diffusion, the combination of the two Au NP layers has a synergetic effect (sample Au/TiO$_2$/Au with the highest haze). Small NPs on top or distributed inside the film are instead characterized by a plasmonic absorption feature centered at about 650–700 nm for the TiO$_2$–Au sample (as reported in [23]), which is also characterized by a large absorption in the whole visible-near infrared (vis-NIR) range (transmittance <40% in the visible range and <60% in the NIR); Au NPs at the bottom of the film (~100 nm size) are, instead, characterized by plasmonic absorption centered at about 700–800 nm [22].

Table 2. Haze factor % of the samples in two wavelengths.

Sample	Haze Factor (%) 400–800 nm Range	Haze Factor (%) 800–2000 nm Range
$Au/TiO_2/Au$	80	32
TiO_2/Au	58	22
TiO_2	23	Negligible (< 10%)
Au/TiO_2	16	Negligible (< 10%)
TiO_2–Au	11	Negligible (< 10%)

Synthesized samples are submitted to electrochemical characterization in order to investigate their behavior in dark and irradiated conditions. Table 3 reports the values of open circuit voltage (OCV), which give an indication on the equilibrium at the electrode/electrolyte interface. The values measured in the dark at the different samples in neutral and basic solutions are reported. All the values of potential in the text are referred to saturated calomel electrode (SCE).

Table 3. Open circuit voltage (OCV) values measured at different samples, under different pH. The values of OCV at pH 13, calculated by $E = E^0 - 2.3\ RT/F\ pH$, supposing a Nernstian behavior of the surface, are also reported as a comparison.

Sample	OCV KNO_3 V/vs SCE	OCV pH 13 V/vs SCE	OCV pH 13 Nernstian V/vs SCE
TiO_2	−0.14	−0.42	−0.49
TiO_2/Au	0.033	−0.11	−0.32
Au/TiO_2	−0.115	−0.26	−0.47
$Au/TiO_2/Au$	0.022	−0.12	−0.33
TiO_2–Au	−0.001	−0.13	−0.35

If the behavior of the reference sample (TiO_2) is considered, a value of −0.14V is measured as OCV in neutral pH, which becomes −0.42 V in basic solution, with a variation very close to the theoretical one, valuable by a linear Nernstian behavior of the surface (last column in Table 3), as it is expected for oxide electrodes [37].

As it can be observed, OCV measured at neutral pH is higher than that expected for TiO_2: lower values, in the range from −0.5 up to −0.85 V, are generally reported as flat band (FB) potential, for TiO_2 bulk, at this pH [38]. Actually, the form of the electrode material (i.e., thin film, single crystal, polycrystalline), its morphology and phase distribution (anatase/rutile) or possible doping are also decisive factors as far as the OCV or FB is concerned.

Regarding the effect of the metal (M), data shown in Table 3 indicate that the presence of Au in the samples, determines a shift of OCV to more positive values, as might be expected, due to the noble character of Au. However, a direct correlation between M loading and OCV is not observed, the extent of the shift being also dependent on the distribution of the NPs in the structure. In fact, the maximum shift (173 mV) is measured for the less M loaded sample (i.e., TiO_2/Au sample) while, rather than at the highest M loaded sample (i.e., TiO_2–Au sample), the minimum shift (25 mV) is measured at Au/TiO_2 sample.

Variation of OCV is also measured (see Figure 2) when samples were submitted to light irradiation, and in the presence of hole scavengers in the electrolyte solution (BPA 50 ppm, in the specific).

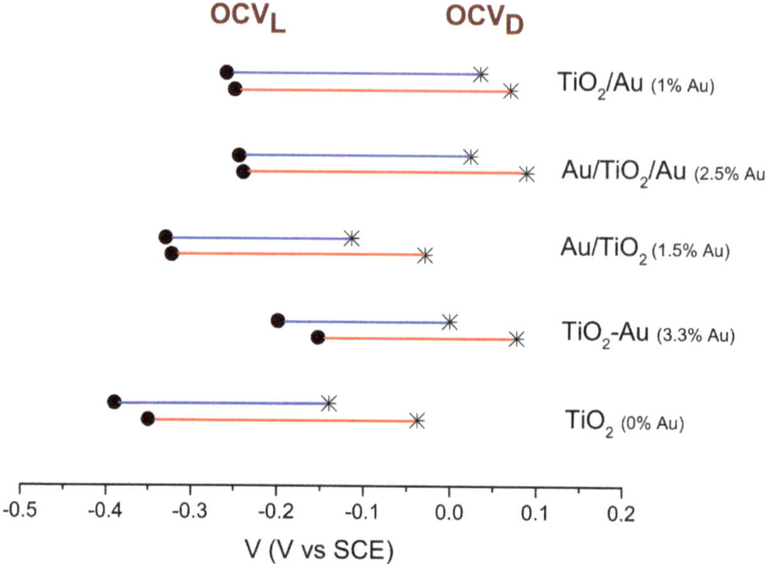

Figure 2. OCV of different samples, under dark (OCV$_D$: stars) and irradiated (OCV$_L$: circles) conditions, in supporting electrolyte (blue lines) or in the presence of 50 ppm BPA (red lines).

The n-type semiconductor (SC) behavior is here highlighted: the light determines a shift of the OCV to more negative values. However, as pointed out above, along with the metal loading in the sample, also the NPs distribution and the accessibility of both solution and light, inside the structure, must be considered. To this aim, the open-circuit photopotential (OCP) may give a better indication on the photo-activity of the sample: OCP represents the band-bending change from dark to light irradiation, resulting from photoexcited carriers in n-type SC, flattening the band bending in the depletion region [39]. The final OCP value also depends on the redox couple which is present in the solution. In Figure 3, the trend of OCP measured in supporting electrolyte is compared to that in two different concentrations of BPA: for all the samples, it is a matter of oxidative OCP, whose absolute values are reported. If the effect of the M is concerned, samples Au/TiO$_2$/Au and TiO$_2$/Au show a slight increase in OCP in KNO$_3$, if compared with that measured at TiO$_2$ sample, which could be in apparent contrast with the relevant values of the OCV. To note, as a consequence of the more positive OCV values measured for the Au/TiO$_2$/Au and TiO$_2$/Au samples, a lower efficiency could be expected in terms of OCP, as it is generally obtained in single-crystal systems. However, when the deposited metal film is in the form of small islands or in nanoparticulate form, an enhancement in the photopotential is possible [40,41].

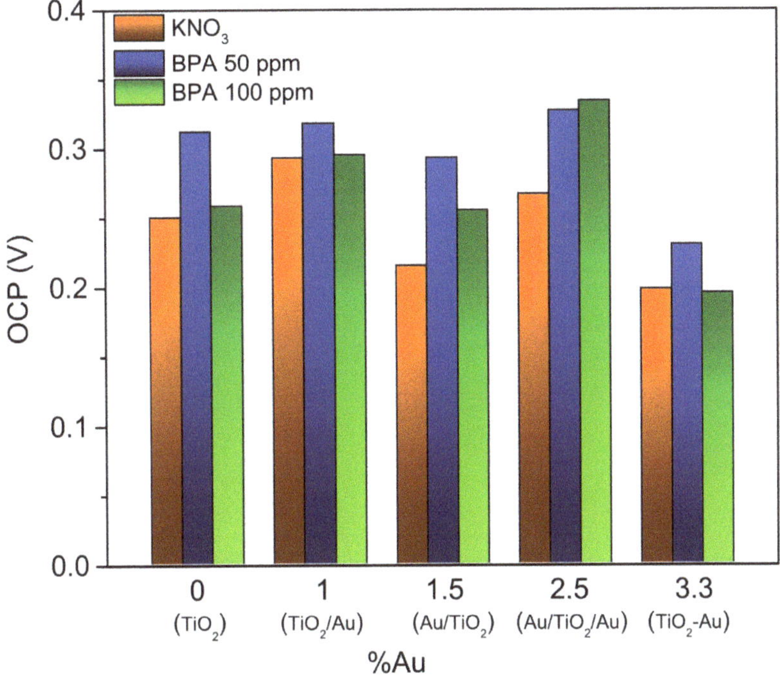

Figure 3. OCP values measured in supporting electrolyte and in two different BPA concentrations at the different samples.

This trend is not observed for the TiO$_2$–Au and Au /TiO$_2$ samples, where the M loading makes OCP lower than in TiO$_2$ sample. At the same time, a not straightforward effect is played by the BPA concentration: on one hand, except for sample TiO$_2$–Au (OCP = 0.23 V), when 50 ppm BPA are present in the solution the OCP is about 0.3 V, regardless of the M load; on the other hand, lower OCP values are measured at the highest concentration, except for the Au/TiO$_2$/Au sample.

Table 4 resumes the values of the photocurrent density measured at constant applied potential of 0.5 V, in KNO$_3$ and in different concentrated BPA solutions, while Figure 4 illustrates the trend of the photocurrents as a function of the different overpotential, evaluated as difference between the applied potential and the OCV$_L$ of the sample.

Table 4. Photocurrent density (μA/cm^2) measured at constant applied potential of 0.5 V vs SCE, in supporting electrolyte and in differently concentrated BPA solutions for the different samples.

	0.1 M KNO$_3$	25 ppm BPA	50 ppm BPA	100 ppm BPA
TiO$_2$	18	36	50	36
TiO$_2$/Au	32	40	55	36
Au/TiO$_2$	25	34	27	39
Au/TiO$_2$/Au	48	44	58	52
TiO$_2$–Au	1.4	1.9	3	1.9

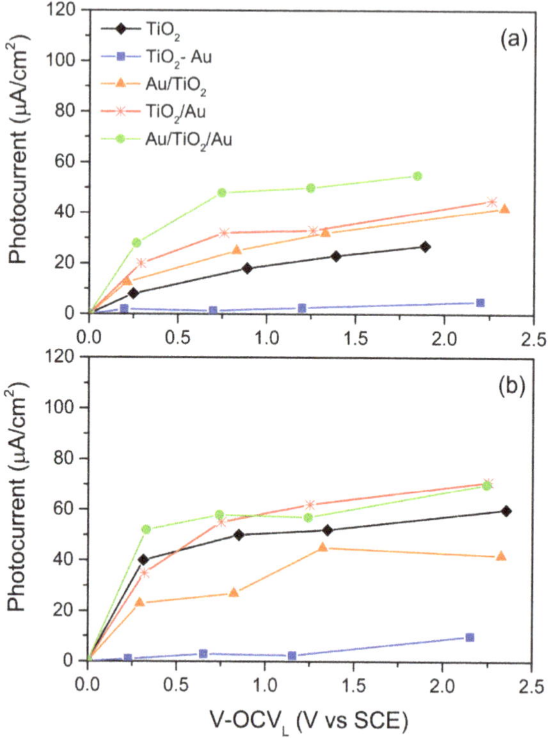

Figure 4. Trend of photocurrent densities as a function of the overpotential with respect to OCV$_L$, measured in (**a**) supporting electrolyte and in (**b**) 50 ppm BPA solution.

Depending on the samples, data appear to be comparable, or even better than those reported in the literature, for analogous electrode materials. As an example, value of photocurrent of 8 µA was measured in K_2SO_4 solution at Au–TiO_2/ITO [30], while 6 and 30 µA were measured at TiO_2/Ti and Au–TiO_2/Ti electrodes, respectively, in [29]. More recent work, on nanoroad TiO_2 modified arrays, reports photocurrent of 3.5 µA/cm^2 in Na_2SO_4 in the presence of 100 µmol/L BPA [42]. Of note those values were obtained under ultraviolet (UV) irradiation, at which the performance of TiO_2 is expected to be higher than in our irradiation conditions.

In the present case, data indicate a complex effect of NPs inside the structure: except for the TiO_2–Au sample, which demonstrated a clearly lower performance than the reference sample, in KNO_3 the positive effect of Au NPs can be assessed in the whole investigated potential range: sample Au/TiO_2/Au appears the most performing in catalyzing the water splitting process. A positive effect of NPs is also measured in the presence of BPA 50 ppm for the Au/TiO_2/Au and TiO_2/Au samples: overpotential being the same, the photocurrents measured on these samples are higher than those measured for the TiO_2 sample. However, in this case, also the performance of Au/TiO_2 is lower than that of the TiO_2 sample. One of the main reasons for the lowest performances of sample TiO_2–Au may be the very high load of M, which reduces the porosity and, in turn, the surface area of the sample. Regarding the performance of the Au/TiO_2 sample, SEM analyses, repeated at the end of those experiments, demonstrated a low stability of the sample. Actually, a certain extent of corrosion was detected for all the samples, but for sample Au/TiO_2, the residual Au content, measured at the end of the experimental campaign, was under the detection limits of the instrument: the low stability of this sample did not allow to perform all the experiments on it.

Among the other possible effects, a conductivity enhancement is expected due to SC/M coupling: the presence of metal NPs should enhance the charge transfer in the structure. The extent of such increase may be deduced by Electrochemical Impedance Spectroscopy (EIS) measurements. Figure 5 shows the Nyquist diagrams obtained by experiments in dark conditions, under −0.5 V of applied potential. This potential was selected in such a way that the space charge of all the n-type semiconductor (SC) samples was in the accumulation regime, so that the majority charge carriers, electrons in the conduction band, could be involved in the charge transfer.

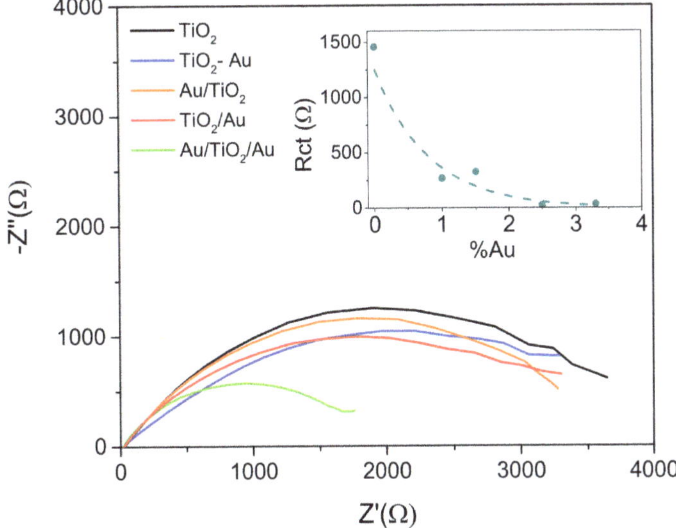

Figure 5. Nyquist plots obtained at V = −0.5 V vs SCE, in supporting electrolyte, at different samples. Inset: trend of R_{ct} evaluated by fitting the EIS data with a Randle circuit.

As it can be observed, all the semicircles in Figure 5 tend to be closed on the x-axis: all samples show a good reactivity at this potential, and all the curves related to samples with Au NPs, are under that related to sample TiO$_2$, indicating that Au catalyzes the charge transfer process to the solution. The interpretation of these curves with a simple Randle (Rs(C-R_{ct})) equivalent circuit, allowed the evaluation of the charge transfer resistance (R_{ct}) for the different samples. The inset in Figure 5 shows that the decrease in R_{ct} is about exponential with the Au load in the samples.

These data are in a nice agreement with the trend of the cyclic voltammetries (CV) recorded in the dark (Figure 6). In the range of negative potential, similar trends are obtained for the different samples, with a first peak (P1) around −0.7 ± −0.8 V, which should correspond to the Ti(IV) → Ti(III) transformation [43], followed by the increase in the negative current due to the H$_2$ evolution.

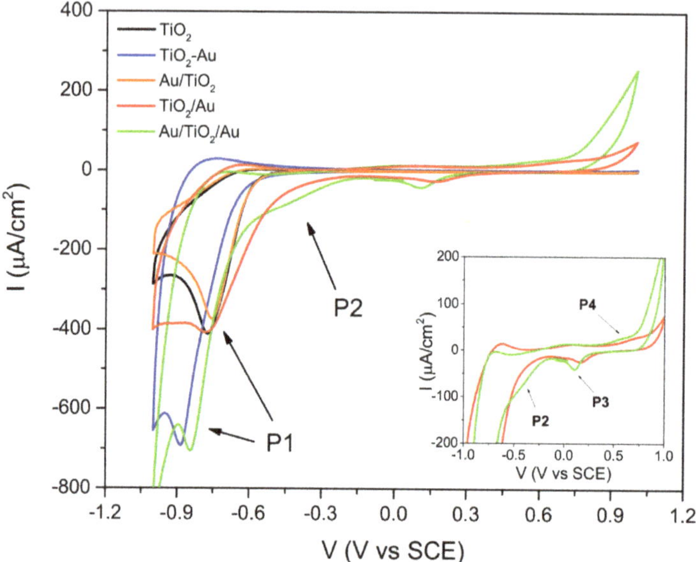

Figure 6. Cyclic Voltammetries (CV) in supporting electrolyte in dark conditions; the typical peaks of Ti(III)/T(IV) (P1), and of Au redox behavior (P3 and P4) are evidenced in the inset, as well as the wave related to O_2 reduction (P2).

The highest values of current peak P1 are obtained in TiO_2–Au and Au/TiO_2/Au samples: these are the most M loaded and the most conductive samples. On samples Au/TiO_2 and TiO_2/Au, which have a lower M loading, the height of P1 is about half.

In the range of positive potentials, Au/TiO_2/Au is the most active sample for O_2 evolution; the corresponding "wave" related to the O_2 reduction (P2 in the Figure 6) appears around −0.4 V. This wave is also visible at TiO_2/Au sample, but in minor extent, because the O_2 evolution at this sample is of minor extent too. The inset of Figure 6 highlights the redox behavior of Au that is well evident at samples Au/TiO_2/Au and TiO_2/Au: in particular, at these samples the reduction peak of Au at about 0.2 V is coupled with the corresponding oxidation wave (P4), that is observed in the range 0.6 to 1 V, always just before the increase in current, due to O_2 evolution. This redox behavior could result quite different from that generally observed for polycrystalline Au, where two oxidative peaks may be detected. However, as reported in the literature, the voltammetric profiles of the supported nanoparticles differ from those associated with the Au polycrystalline electrode, as the nanoparticles exhibit a single, broad oxidation wave, shifted to more positive potentials, with respect to the two peaks present in the voltammetry of the polycrystalline electrode [44].

However, the redox behavior of Au NPs was not evident in all the samples: for Au–TiO_2 and TiO_2/Au, in which there were no NPs on the bottom, the main effect of NPs was only evidenced in the increased conductivity of the sample (higher P1), while there was no evidence of P3 and P4. Analogous results were obtained by other authors in the literature [31], who found that the voltammetric behavior was very different when Au NPs were deposited under or up the TiO_2 layer: only in the former case CV reported evident peaks related to Au redox reactions, both in the dark and under radiation. This was attributed to the fact that Au clusters grown under the TiO_2 exhibited superior charge carrier generation, separation and transportation than that of TiO_2/Au under visible light.

2.2. Behavior of the Irradiated Samples

Among the several factors determining the photoactivity of a SC material, the concentration of defects plays a crucial role. In the case of reduced TiO_2 materials, oxygen vacancies (VO) and Ti^{3+} sites have been intentionally introduced to obtain a higher photocatalytic/photoelectrochemical activity [45]. Nevertheless, such defects may also act as recombination centers for the photoexcited electron–hole pairs. Thus, to investigate the recombination process in a particular SC structure, the kinetics of the photocurrent decay may be analyzed when, after stabilization in dark conditions, the sample is submitted to irradiation. In such conditions, after an initial spike, in absence of applied potential the current tends to a stationary state: the process follows a first-order kinetics if the decay is only due to a surface recombination [46]. A typical example of the trend in time of the current measured at our samples, as effect of the related photopotential is shown in the inset of Figure 7.

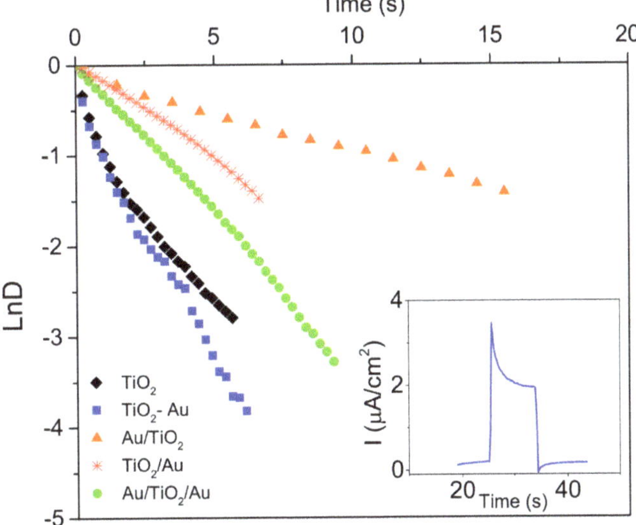

Figure 7. Example of the kinetics of the photocurrent, at TiO$_2$–Au sample, when the light is sudden switched on and off.

Two different current transients are observed. The first one, when the light is switched on, describes the initial increase in photocurrent, caused by a separation of the photogenerated electron–hole pairs at the semiconductor/electrolyte interface, followed by an exponential decrease with time. Then, when the light is switched off, a cathodic spike is observed due to the recombination of the conduction band electrons with the holes trapped at the surface [47]. The process can be described by a first order kinetics in the surface concentration of electrons as:

$$D = \exp\left(-t/\tau\right) \tag{1}$$

where τ represents the transient time constant, and D takes account of the photocurrent relaxation, defined as:

$$D = (I(t) - I_f)/(I_i - I_f) \tag{2}$$

I(t) is the current at a time t, I_i is the current at t = 0, and I_f the stationary current.

Equation (1) can be used to describe both the anodic or the cathodic transients.

In Figure 7 the kinetics of photocurrent relaxation, related to the different samples are compared.

As it can be observed, plots of ln D vs. time did not always show a linear behavior, indicating that the decay mechanism can be complex. However, in the present case, just to make a qualitative comparison between the different samples, τ was always taken where ln D = -1. The calculated values are resumed in Table 5. The obtained values, of the order of seconds, cannot be representative of charge recombination phenomena only. Actually, these characteristic times are probably affected by electrical transport and electrolyte diffusion phenomena, considering that the porosity of the system, the distribution of defectivities, as well as possible diffusive effects in the pores, may complicate the current decay process.

Table 5. τ values for the different samples evaluated at lnD = -1.

	TiO_2	TiO_2/Au	$Au/TiO_2/Au$	TiO_2–Au	Au/TiO_2
τ (s)	0.97	4.88	3.17	0.97	10.5

However, although the exact indication on the charge recombination time is not derivable from τ, its value can be used to compare the different decay trends: the presence of Au NPs makes the rate of photocurrent decay slower, provided that M loading does not exceed a certain limit. At sample TiO_2–Au the decay is the same as in the original sample, and most of the NPs are probably becoming recombination centres. In this context, the best performance is obtained for the sample Au/TiO_2, but, as already mentioned, this sample demonstrated to be highly unstable.

The effect of light on the CV trends is shown in Figure 8, first with regard to TiO_2 sample.

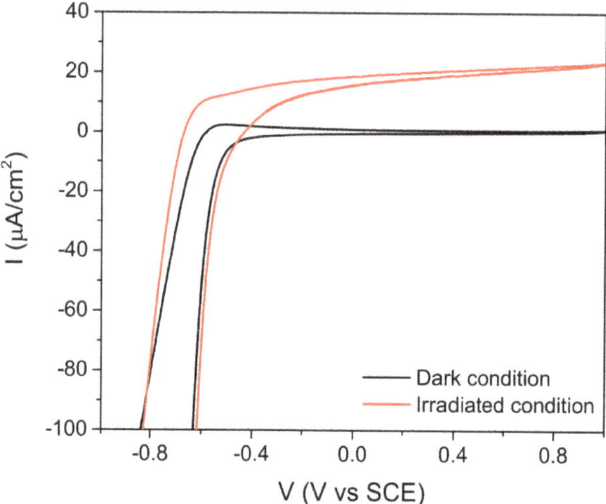

Figure 8. Effect of the irradiation on CV in supporting electrolyte performed at TiO_2 sample.

Attention is focused to the anodic potential range, because, as we expected due to the n-type character of the SC, in this range the effect of the irradiation is more evident.

In Figure 9, the trend of TiO_2 sample is compared with those of the other samples, irradiation conditions being the same.

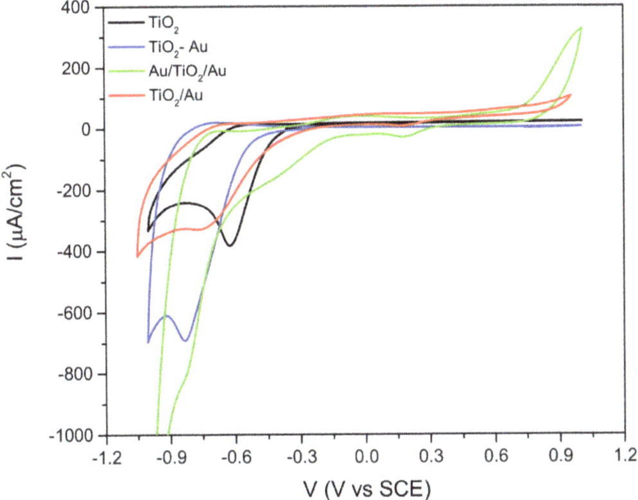

Figure 9. Comparison between CV in supporting electrolyte at different irradiated samples.

Of note is the effect of the irradiation on the O_2 evolution reaction, which is favored for the TiO_2/Au sample and, overall, for the Au/TiO_2/Au sample: for this reason, on this last sample, the corresponding reduction wave P2 is more evident.

2.3. Effect of the Organic Compound on the CV

Specific CV runs performed on the different samples in solutions containing different concentration of BPA, did not show a clear indication of the voltammetric peaks related to the direct oxidation of the organic molecule, at the electrode surface. Nevertheless, the photocurrents recorded in the presence of the organic were higher than those measured in the supporting electrolyte, under almost all the investigated potentials. This may be an indication that oxidation of BPA may occur by means of OH radicals originated from H_2O, by the action of the photogenerated holes. The BPA molecule, acting as OH scavenger, accelerates the separation process of the charges, this in turn gives the increase in the photocurrent.

However, even in the absence of a direct reaction at the electrode surface, the effects of BPA are visible in the CV: the presence of BPA seems to interfere on the redox behavior of the M. For this kind of analysis, we investigated the Au/TiO_2/Au and TiO_2/Au samples, where the redox behavior of the M was more evident: also, they were the most performing systems in terms of photocurrent.

Figure 10a compares the CV obtained in solution with two concentrations of BPA, while Figure 10b compares the trend of CV for the differently M loaded samples in solution with the same concentration of BPA.

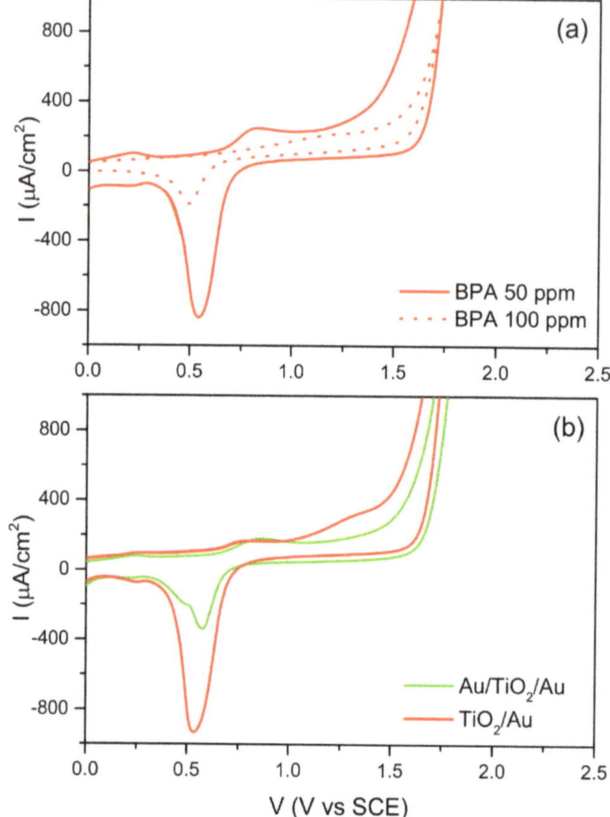

Figure 10. Comparison between the CV obtained at TiO_2/Au sample: (**a**) in solution with two concentrations of BPA and (**b**) at samples differently M loaded, in solution with the same concentration of BPA (50 ppm).

The increase in the BPA concentration has two main effects: on one hand, a decrease in the current of the redox peaks (P3 and P4) it is observed, on the other hand, the peaks are more distant one from another, in terms of potential, this indicating a worst and more irreversible charge transfer process with the substrate.

This behavior may indicate that during the CV, in absence of the organic, the photogenerated charges can be available to activate the redox process of the NPs, under the imposed potentials. In particular, during the oxidative scan, the holes may: (i) generate OH radicals from water, or (ii) oxidise Au NPs. In the presence of the organic, BPA molecules act as OH radical scavengers, thus accelerating the path (i) and subtracting holes to the redox process of Au, which is less evident in the CV.

2.4. Analysis of the Working Mechanism of the Structures

From the results presented up to now, it is clear that the behavior of the investigated structures is rather complex: addition of Au NPs interferes not only with the charge transfer to and from the electrolyte, but also with the equilibrium and non-equilibrium interface energetics.

In order to better understand the working mechanism of the electrodes, it can be useful to recall some of the fundamental concepts on SC, how they behave under illumination, and, finally, how they interact with the electrolyte and the solutes in it contained.

From the structural point of view, we are dealing with coupling of SC and M. In the presence of bulk materials, the SC/M interface is expected to behave according to the well-developed theory of SC/metal Schottky contacts [48]. In the specific case, by considering the electronic affinity of TiO_2 equal to 4,5 eV, and a work function (WF) for Au equal to about 5.3 eV [49], coupling of the two materials should lead to a Schottky barrier of 0.8 eV. However, when we deal with nanostructures, the position of the energetic levels may be different: depending on the conditions, and in particular on the morphology of the coupled phases, different values can be calculated for both the Fermi level of the SC, and the WF of the M, which may be affected also by the excess of charge on the NPs [49].

In our specific case, information on the location of the conduction band-edge (CB) can be derived from CV measurements, and, as suggested by the literature [50], we considered the potential of the peak as representative of the energetic level of the CB edge. Thus, a value of 4.4 eV has been calculated from the value of -0.78 V, at which the main peak of TiO_2 appears in the CV of TiO_2 sample.

Regarding the WF of M, we may consider that the OCV values of the samples with Au, were more positive than that of the reference TiO_2 sample (see Figure 2): this may be an indication that the WF of the M was greater than the Fermi level of the SC. When the two materials are contacted, a spontaneous charge transfer occurred from TiO_2 to Au, up to equilibrium leading to the formation of the Schottky barrier. Thus, the presence of Au can be seen as a reservoir of electrons, which are displaced from the CB of the SC: the Fermi level of the SC is lowered, and its potential made more positive by the presence of metal NPs.

When the effect of the irradiation is considered, we cannot neglect that M is present as NPs. In the specific case, optical analyses indicated that for the different samples, depending on the NP dimensions, and on the deep penetration of the Au deposition, plasmonic or scattering effects are originated. In fact, as pointed out in the Introduction, the presence of NPs, more or less distributed in the bulk of the SC or at the interfaces, may strongly enhance the effectiveness of light absorption.

Finally, also the potential of the donor/acceptor redox couples present in the solution must be considered: the Fermi level of the electrons in the CB should remain higher than the energetic level of the acceptor, otherwise it cannot receive electrons at the cathode. At the same time, the level of the holes (h^+) in the VB of the SC should remain lower than that of the donor to which h^+ will be transferred.

Accordingly, the scheme shown in Figure 11 illustrates the possible working mechanisms of the examined structures, when irradiated with the whole solar simulated light. Depending on the wavelength of the incident light, different effects could be achieved. The final mechanism should be a combination of the response of the TiO_2 nanostructure, which is sensitive only to a fraction of the wavelengths (UV), and of the Au NP effects (both plasmonic and in terms of charge separation).

Under the fraction of UV radiation, effective on TiO_2:

e^- are excited in CB and, assisted by the applied potential, they may:

(1)   go to the cathode (and react with dissolved O_2, or with H_2O to give H_2)
(2)   recombine with h^+

holes that do not recombine are displaced on the surface and they may:

(3)   react with the electrolyte (oxidising it)
(4)   be transferred to the M (depending on the WF)

holes arriving at the M may:

(5)   be transferred to the electrolyte (in this case, not compatible with the redox level)
(6)   oxidise the M (corrosion).

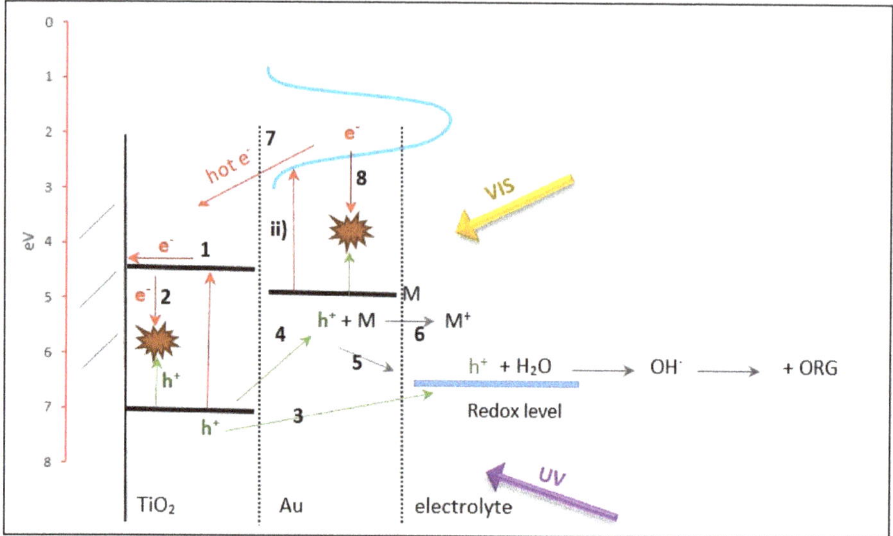

Figure 11. Schematic representation of the energetic levels involved in the charge activation and transfer processes.

The main phenomena related to the illumination of the M nanoparticle can be summarized as:

(i)   scattering effects, able to redistribute radiation in the whole film thickness thus favoring light interaction with the photoanode (this effect is dominated by NPs of great dimensions, not indicated in the scheme),

(ii) resonant absorption by the localized surface plasmon resonance of the NP (centered in the visible light (VIS) range) with subsequent generation of hot electrons which may:

    (7)   be injected into the CB of the SC (then paths 1, 2)

    (8)   decay and thermalize through electron–electron and electron–phonon scattering.

3. Materials and Methods

The investigated TiO_2 thin films were deposited by Pulsed Laser Deposition (PLD) starting from a 99.9% pure TiO_2 target ablated with Nd:YAG ns laser (Continuum) pulses ($\lambda = 532$ nm, pulse duration 5–7 ns, 10 Hz repetition rate, fluence on the target about 3.5 J/cm^2) in a background O_2 atmosphere (P= 8 Pa) and with a fixed target-to-substrate distance of 50 mm, at room temperature. The deposition time was adjusted to obtain a nominal thickness for all the films of about 1000 nm. Thermal annealing in air at 500 °C (4 °C/min heating ramp, dwell time 2 h) was employed to induce crystallization to the anatase phase. Crystallinity was verified by Raman spectroscopy (not shown).

Different distribution of Au NPs was achieved, depending on the samples (see Table 1). The TiO_2/Au sample (i.e., the configuration substrate/Au NPs/TiO_2 film) was obtained depositing a TiO_2 film with the above described PLD procedure on a substrate covered with Au NPs obtained by Au thermal evaporation. An Edwards E306 thermal evaporator (Edwards) was employed to deposit a 10 nm Au film, followed by thermal de-wetting (air annealing at 500 °C with 4 °C/min ramp, 2 h dwell) to induce NP growth. The Au/TiO_2 sample (i.e., the configuration substrate/TiO_2 film/Au NPs) was obtained depositing a TiO_2 film by PLD (details above), followed by deposition of Au NPs by PLD, ablating an Au target in 1000 Pa Ar with a laser fluence of 2 J/cm^2 [22]. The Au/TiO_2/Au sample was obtained by combining the Au/TiO_2 and the TiO_2/Au synthesis procedures. The TiO_2–Au sample, in which Au NPs are embedded in the nanoporous TiO_2 film, was deposited by PLD ablating a composite $Au–TiO_2$ target with same ablation parameters as for the pure TiO_2 film, followed by annealing in air at 500 °C (4 °C/min ramp, 2 h dwell) to induce TiO_2 crystallization and Au NPs formation, as described in detail in [23].

All samples characterized by electrochemical techniques were deposited on Ti plate substrates, while selected samples for SEM and optical measurements were deposited on Si (100) and soda lime glass substrates, respectively.

Optical transmittance spectra were evaluated with a UV-vis-NIR PerkinElmer Lambda 1050 spectrophotometer (PerkinElmer, Waltham, Massachusetts, USA) with a 150 mm diameter integrating sphere in the range 250–2000 nm, illuminating the sample from the glass substrate side and normalizing the spectra with respect to glass transmittance. The haze was obtained as the ratio between the diffuse and the total (i.e., diffuse + direct) transmittance intensity.

SEM images of the synthesized samples were acquired using a field emission scanning electron microscope (FEG-SEM, Zeiss Supra 40, Carl Zeiss Microscopy GmbH, Jena, Germany); measurements were also performed, at the end of the experimental campaign, to check the stability of the samples.

Photo-electrochemical characterization of all the samples was performed in an undivided three-electrode cell; the synthesized materials were used as working electrodes, while platinum constituted the counter electrode and SCE was used as reference. The electrodes were connected to a potentiostat-galvanostat (Metrohm Autolab 302N, Metrohm, Herisau, Switzerland), controlled by NOVA software.

The photo-activity of the samples was tested under a light flux provided by a 300W Xe lamp (Lot Oriel, Darmstadt, Germay) equipped with an AM 1.5G filter.

Aqueous solution 0.1 M KNO_3 was used as supporting electrolyte; depending on the experiments, a fixed amount of BPA (25, 50, or 100 ppm) was added to the electrolyte.

For all the cases, the photocurrent was calculated by subtracting the stable value measured in the dark from that obtained under irradiation. Open circuit voltage (OCV) was also monitored, always after 5 minutes of rest, to allow the sample reaching the equilibrium in the solution under the dark (OCV_D) or irradiation (OCV_L) condition imposed. Open Circuit Photopotential (OCP) was also evaluated as the difference between $OCV_L–OCV_D$ in supporting electrolyte or in the presence of organic.

Chronoamperometric tests were performed at different overpotential from the OCP of each sample in both KNO_3 and different concentrated BPA solutions.

Cyclic voltammograms (CV) were recorded in the potential window between −1.1 V and 1.8 V, at 100 mV/s.

Electrochemical impedance spectroscopy (EIS) were performed in the dark, at negative potential, to investigate on the ability of the samples to possible charge transfer process. The investigated frequency range was from 10^5 Hz to 10^{-1} Hz.

4. Conclusions

Au NPs do not seem to work directly, but we can assess that their presence enhance the ability of the sample to transfer charge with the solvent, and they slow down the decay process of the photocurrent, at least up to a certain load.

If the behavior of the samples is considered in detail, the most performing samples, in both the supporting electrolyte and in the presence of BPA, were the $Au/TiO_2/Au$ and TiO_2/Au samples. On these samples the redox behavior of the M is well evident in the CV, recombination seems slowed down, the current decay is 3–5 times slower than that at the reference sample. On both samples, NPs of great dimensions are present at the bottom under the TiO_2 layer, directly contacting the Ti support. The presence of these NPs at the bottom seems to be the crucial point for the working mechanism of the structure. Thanks to the scattering effects, it is actually possible on these samples to better exploit the radiation, and thus the active sites of the TiO_2 layer.

Measurements for the Au/TiO_2 and $Au/TiO_2/Au$ samples can be useful to interpret the role of the top surface layer of NPs. Considering the smaller dimensions of these NPs, they should be responsible both for possible plasmonic effects and for electron–hole separation, thus leading to reduced recombination. Due to plasmonic effects, localized energetic fields could generate e^-/h^+ couples: however, the low stability evidenced for the Au/TiO_2 sample indicates that holes cannot be transferred neither to the electrolyte, nor to the SC, so that they remain in the NPs and oxidize them. However, corrosion seems to be correlated not only to plasmonic effects, but also to the distribution of the NPs. Of note is that the most important corrosion effects are measured, rather than for the TiO_2–Au system, where the plasmonic effects were the most relevant, for the $Au/TiO_2/Au$ and Au/TiO_2 samples, where an upper Au NPs layer is at direct contact with the electrolyte. Maybe the NPs in TiO_2–Au are somehow protected, as they are embedded in the structure. Other works in the literature on similar SC/M coupling, suggested that a protecting layer could be realized in order to avoid corrosion [51].

Finally, the scarce performance of the TiO_2–Au sample could be attributed mainly to a decreased surface area available for the charge transfer with the electrolyte (see [23]). The large content of Au NPs did not result in enhancement of the global performance of the structure. The photocurrent is lower, because the area is lowered. The low value of τ, and the rapid current decay, could indicate that most of the NPs are behaving as recombination centers for the photogenerated charges.

Author Contributions: Conceptualization, S.P. and A.V.; methodology, A.V.; investigation, L.M., E.M.U. and R.M.; writing—original draft preparation, S.P.; writing—review and editing, E.M.U., L.M., M.G., L.M., A.L.B., R.M. and I.N.; sample preparation, structural/optical characterization: M.G., L.M., B.R.B., V.R., C.S.C. and A.L.B.; supervision, A.V. and A.V.; Funding Acquisition, S.P. All the coauthors contributed to the data discussion.

Funding: This research was funded by Fondazione di Sardegna, CRP project F71I17000280002 – 2017.

Conflicts of Interest: The authors declare no conflict of interest.

References

1. Careghini, A.; Mastorgio, A.F.; Saponaro, S.; Sezenna, E. Bisphenol A, nonylphenols, benzophenones, and benzotriazoles in soils, groundwater, surface water, sediments, and food: A review. *Environ. Sci. Pollut. Res.* **2015**, *22*, 5711–5741. [CrossRef] [PubMed]

2. Industry Experts. Bisphenol A—A global market overview. 2016. Available online: http://industry-experts.com/verticals/files/articles/cp021-bisphenol-a-a-global-market-overview.pdf (accessed on 3 April 2019).

3. Petrie, B.; Lopardo, L.; Proctor, K.; Youdan, J.; Barden, R.; Kasprzyk-Hordern, B. Assessment of bisphenol-A in the urban water cycle. *Sci. Total Environ.* **2019**, *650*, 900–907. [CrossRef] [PubMed]

4. Guerra, P.; Kim, M.; Teslic, S.; Alaee, M.; Smyth, S.A. Bisphenol-A removal in various wastewater treatment processes: Operational conditions, mass balance, and optimization. *J. Environ. Manag.* **2015**, *152*, 192–200. [CrossRef] [PubMed]

5. Flint, S.; Markle, T.; Thompson, S.; Wallace, E. Bisphenol A exposure, effects, and policy: A wildlife perspective. *J. Environ. Manag.* **2012**, *104*, 19–34. [CrossRef] [PubMed]

6. Zhan, M.; Yang, X.; Xian, Q.; Kong, L. Photosensitized degradation of bisphenol A involving reactive oxygen species in the presence of humic substances. *Chemosphere* **2006**, *63*, 378–386. [CrossRef]

7. Zhang, C.; Zeng, G.; Yuan, L.; Yu, J.; Li, J.; Huang, G.; Xi, B.; Liu, H. Aerobic degradation of bisphenol A by Achromobacter xylosoxidans strain B-16 isolated from compost leachate of municipal solid waste. *Chemosphere* **2007**, *68*, 181–190. [CrossRef]

8. Žerjav, G.; Kaplan, R.; Pintar, A. Catalytic wet air oxidation of bisphenol A aqueous solution in trickle-bed reactor over single TiO_2 polymorphs and their mixtures. *J. Environ. Chem. Eng.* **2018**, *6*, 2148–2158. [CrossRef]

9. Davididou, K.; Nelson, R.; Monteagudo, J.M.; Durán, A.; Expósito, A.J.; Chatzisymeon, E. Photocatalytic degradation of bisphenol-A under UV-LED, blacklight and solar irradiation. *J. Clean. Prod.* **2018**, *203*, 13–21. [CrossRef]

10. Blanco-Vega, M.P.; Guzmán-Mar, J.L.; Villanueva-Rodríguez, M.; Maya-Treviño, L.; Garza-Tovar, L.L.; Hernández-Ramírez, A.; Hinojosa-Reyes, L. Photocatalytic elimination of bisphenol A under visible light using Ni-doped TiO_2 synthesized by microwave assisted sol-gel method. *Mater. Sci. Semicond. Process.* **2017**, *71*, 275–282. [CrossRef]

11. Repousi, V.; Petala, A.; Frontistis, Z.; Antonopoulou, M.; Konstantinou, I.; Kondarides, D.I.; Mantzavinos, D. Photocatalytic degradation of bisphenol A over Rh/TiO_2 suspensions in different water matrices. *Catal. Today* **2017**, *284*, 59–66. [CrossRef]

12. Frontistis, Z.; Daskalaki, V.M.; Katsaounis, A.; Poulios, I.; Mantzavinos, D. Electrochemical enhancement of solar photocatalysis: Degradation of endocrine disruptor bisphenol-A on Ti/TiO_2 films. *Water Res.* **2011**, *45*, 2996–3004. [CrossRef] [PubMed]

13. Brugnera, M.F.; Rajeshwar, K.; Cardoso, J.C.; Zanoni, M.V.B. Bisphenol A removal from wastewater using self-organized TiO_2 nanotubular array electrodes. *Chemosphere* **2010**, *78*, 569–575. [CrossRef] [PubMed]

14. Serra-Pérez, E.; Álvarez-Torrellas, S.; Águeda, V.I.; Delgado, J.A.; Ovejero, G.; García, J. Insights into the removal of bisphenol A by catalytic wet air oxidation upon carbon nanospheres-based catalysts: Key operating parameters, degradation intermediates and reaction pathway. *Appl. Surf. Sci.* **2019**, *473*, 726–737. [CrossRef]

15. Nguyen, T.B.; Huang, C.P.; Doong, R. Photocatalytic degradation of bisphenol A over a $ZnFe_2O_4/TiO_2$ nanocomposite under visible light. *Sci. Total Environ.* **2019**, *646*, 745–756. [CrossRef] [PubMed]

16. Xu, L.; Yang, L.; Johansson, E.M.J.; Wang, Y.; Jin, P. Photocatalytic activity and mechanism of bisphenol a removal over TiO_2-x/rGo nanocomposite driven by visible light. *Chem. Eng. J.* **2018**, *350*, 1043–1055. [CrossRef]

17. Abdelraheem, W.H.M.; Patil, M.K.; Nadagouda, M.N.; Dionysiou, D.D. Hydrothermal synthesis of photoactive nitrogen- and boron- codoped TiO_2 nanoparticles for the treatment of bisphenol A in wastewater: Synthesis, photocatalytic activity, degradation byproducts and reaction pathways. *Appl. Catal. B* **2019**, *241*, 598–611. [CrossRef]

18. Mais, L.; Mascia, M.; Palmas, S.; Vacca, A. Photoelectrochemical oxidation of phenol with nanostructured TiO_2-PANI electrodes under solar light irradiation. *Sep. Purif. Technol.* **2019**, *208*, 153–159. [CrossRef]

19. Palmas, S.; Castresana, P.A.; Mais, L.; Vacca, A.; Mascia, M.; Ricci, P.C. TiO_2-WO_3 nanostructured systems for photoelectrochemical applications. *RSC Adv.* **2016**, *6*, 101671–101682. [CrossRef]

20. Palmas, S.; Mascia, M.; Vacca, A.; Llanos, J.; Mena, E. Analysis of photocurrent and capacitance of TiO2 nanotube–Polyaniline hybrid composites synthesized through electroreduction of an aryldiazonium salt. *RSC Adv.* **2014**, *4*, 23957–23965. [CrossRef]

21. Matarrese, R.; Nova, I.; Li Bassi, A.; Casari, C.S.; Russo, V.; Palmas, S. Preparation and optimization of TiO_2 photoanodes fabricated by pulsed laser deposition for photoelectrochemical water splitting. *J. Solid State Electrochem.* **2017**, *21*, 3139–3154. [CrossRef]

22. Ghidelli, M.; Mascaretti, L.; Bricchi, B.R.; Zapelli, A.; Russo, V.; Casari, C.S.; Li Bassi, A. Engineering plasmonic nanostructured surfaces by pulsed laser deposition. *Appl. Surf. Sci.* **2018**, *434*, 1064–1073. [CrossRef]

23. Bricchi, B.R.; Ghidelli, M.; Mascaretti, L.; Zapelli, A.; Russo, V.; Casari, C.S.; Terraneo, G.; Alessandri, I.; Ducati, C.; Li Bassi, A. Integration of plasmonic Au nanoparticles in TiO_2 hierarchical structures in a single-step pulsed laser co-deposition. *Mater. Des.* **2018**, *156*, 311–319. [CrossRef]

24. Atwater, H.A.; Polman, A. Plasmonics for improved photovoltaic devices. *Nat. Mater.* **2010**, *9*, 205–213. [CrossRef] [PubMed]

25. Mascaretti, L.; Dutta, A.; Kment, S.; Shalaev, V.M.; Boltasseva, A.; Zboril, R.; Naldoni, A. Plasmon-Enhanced Photoelectrochemical Water Splitting for Efficient Renewable Energy Storage. *Adv. Mater.* **2019**. [CrossRef] [PubMed]

26. Pu, Y.; Wang, G.; Chang, K.; Ling, Y.; Lin, Y.; Fitzmorris, B.C.; Liu, C.; Lu, X.; Tong, Y.; Zhang, J.Z.; Hsu, Y.; et al. Au Nanostructure-Decorated TiO_2 Nanowires Exhibiting Photoactivity Across Entire UV-visible Region for Photoelectrochemical Water Splitting. *Nano Lett.* **2013**, *13*, 3817–3823. [CrossRef] [PubMed]

27. Naseri, N.; Sangpour, P.; Mousavi, S.H. Applying alloyed metal nanoparticles to enhance solar assisted water splitting. *RSC Adv.* **2014**, *4*, 46697–46703. [CrossRef]

28. Cojocaru, B.; Andrei, V.; Tudorache, M.; Lin, F.; Cadigan, C.; Richards, R.; Parvulescu, V.I. Enhanced photo-degradation of bisphenol pollutants onto gold-modified photocatalysts. *Catal. Today* **2017**, *284*, 153–159. [CrossRef]

29. Fu, P.; Zhang, P. Uniform dispersion of Au nanoparticles on TiO_2 film via electrostatic self-assembly for photocatalytic degradation of bisphenol A. *Appl. Catal. B* **2010**, *96*, 176–184. [CrossRef]

30. Li, X.Z.; He, C.; Graham, N.; Xiong, Y. Photoelectrocatalytic degradation of bisphenol A in aqueous solution using a Au-TiO_2/ITO film. *J. Appl. Electrochem.* **2005**, *35*, 741–750. [CrossRef]

31. Sreedhar, A.; Reddy, I.N.; Kwon, J.H.; Yi, J.; Sohn, Y.; Gwag, J.S.; Noh, J.-S. Charge carrier generation and control on plasmonic Au clusters functionalized TiO_2 thin films for enhanced visible light water splitting activity. *Ceram. Int.* **2018**, *44*, 18978–18986. [CrossRef]

32. Morfa, A.J.; Rowlen, K.L.; Reilly, T.H.; Romero, M.J.; van de Lagemaat, J. Plasmon-enhanced solar energy conversion in organic bulk heterojunction photovoltaics. *Appl. Phys. Lett.* **2008**, *92*, 013504. [CrossRef]

33. Stuart, H.R.; Hall, D.G. Island size effects in nanoparticle-enhanced photodetectors. *Appl. Phys. Lett.* **1998**, *73*, 3815. [CrossRef]

34. Schaadt, D.M.; Feng, B.; Yu, E.T. Enhanced semiconductor optical absorption via surface plasmon excitation in metal nanoparticles. *Appl. Phys. Lett.* **2005**, *86*, 063106. [CrossRef]

35. Di Fonzo, F.; Casari, C.S.; Russo, V.; Brunella, M.F.; Li Bassi, A.; Bottani, C.E. Hierarchically organized nanostructured TiO_2 for photocatalysis applications. *Nanotechnology* **2009**, *20*, 015604. [CrossRef] [PubMed]

36. Passoni, L.; Ghods, F.; Docampo, P.; Abrusci, A.; Martí-Rujas, J.; Ghidelli, M.; Divitini, G.; Ducati, C.; Binda, M.; Guarnera, S.; et al. Hyperbranched Quasi-1D Nanostructures for Solid-State Dye-Sensitized Solar Cells. *ACS Nano* **2013**, *7*, 10023–10031. [CrossRef] [PubMed]

37. Palmas, S.; Polcaro, A.M.; Rodriguez Ruiz, J.; Da Pozzo, A.; Vacca, A.; Mascia, M.; Delogu, F.; Ricci, P.C. Effect of the mechanical activation on the photoelectrochemical properties of anatase powders. *Int. J. Hydrogen Energy* **2009**, *34*, 9662–9670. [CrossRef]

38. Kavan, L.; Grätzel, M. Highly efficient semiconducting TiO_2 photoelectrodes prepared by aerosol pyrolysis. *Electrochim. Acta* **1995**, *40*, 643–652. [CrossRef]

39. Van de Krol, R.; Grätzel, M. *Photoelectrochemical Hydrogen Production*; Springer: New York, NY, USA, 2012; p. 324. Available online: https://www.springer.com/us/book/9781461413790 (accessed on 3 April 2019).

40. Nakato, Y.; Tsubomura, H. The Photoelectrochemical Behavior of an n-TiO_2 Electrode Coated with a Thin Metal Film, as Revealed by Measurements of the Potential of the Metal Film. *Isr. J. Chem.* **1982**, *22*, 180–183. [CrossRef]

41. Nakato, Y.; Ueda, K.; Yano, H.; Tsubomura, H. Effect of Microscopic Discontinuity of Metal Overlayers on the Photovoltages in Metal-Coated Semiconductor-Liquid Junction Photoelectrochemical Cells for Efficient Solar Energy Conversion. *J. Phys. Chem.* **1988**, *92*, 2316–2324. [CrossRef]

42. Fan, Z.; Fan, L.; Shuang, S.; Dong, C. Highly sensitive photoelectrochemical sensing of bisphenol A based on zinc phthalocyanine/TiO_2 nanorod arrays. *Talanta* **2012**, *189*, 16–23. [CrossRef]

43. Palmas, S.; Da Pozzo, A.; Mascia, M.; Vacca, A.; Ricci, P.C.; Matarrese, R. On the redox behaviour of glycerol at TiO_2 electrodes. *J. Solid State Electrochem.* **2012**, *16*, 2493–2502. [CrossRef]

44. Rodriguez, P.; Plana, D.; Fermin, D.J.; Koper, M.T.M. New insights into the catalytic activity of gold nanoparticles for CO oxidation in electrochemical media. *J. Catal.* **2014**, *311*, 182–189. [CrossRef]

45. Naldoni, A.; Altomare, M.; Zoppellaro, G.; Liu, N.; Kment, S.; Zboril, R.; Schmuki, P. Photocatalysis with Reduced TiO_2: From Black TiO_2 to Cocatalyst-Free Hydrogen Production. *ACS Catal.* **2019**, *9*, 354–364. [CrossRef] [PubMed]

46. Radecka, M.; Wierzbicka, M.; Komornicki, S.; Rekas, M. Influence of Cr on photoelectrochemical properties of TiO_2 thin films. *Phys. B* **2004**, *348*, 160–168. [CrossRef]

47. Hagfeldt, A.; Lindström, H.; Södergren, S.; Lindquist, S.-E. Photoelectrochemical studies of colloidal TiO_2 films: The effect of oxygen studied by photocurrent transients. *J. Electroanal. Chem.* **1995**, *381*, 39–46. [CrossRef]

48. Rhoderick, E.H.; Willams, R.H. *Metal-Semiconductor Contacts*, 2nd ed.; Oxford University Press: Oxford, UK, 1988; p. 252.

49. Scanlon, M.D.; Peljo, P.; Méndez, M.A.; Smirnov, E.; Girault, H.H. Charging and discharging at the nanoscale: Fermi level equilibration of metallic nanoparticles. *Chem. Sci.* **2015**, *6*, 2705–2720. [CrossRef] [PubMed]

50. Liu, H.; Tang, J.; Kramer, I.J.; Debnath, R.; Koleilat, G.I.; Wang, X.; Fisher, A.; Li, R.; Brzozowski, L.; Levina, L.; et al. Electron acceptor materials engineering in colloidal quantum dot solar cells. *Adv. Mater.* **2011**, *23*, 3832–3837. [CrossRef] [PubMed]
51. Awazu, K.; Fujimaki, M.; Rockstuhl, C.; Tominaga, J.; Murakami, H.; Ohki, Y.; Yoshida, N.; Watanabe, T. A Plasmonic Photocatalyst Consisting of Silver Nanoparticles Embedded in Titanium Dioxide. *J. Am. Chem. Soc.* **2008**, *130*, 1676–1680. [CrossRef]

 *catalysts*

Article

Ag/Ag$_2$O as a Co-Catalyst in TiO$_2$ Photocatalysis: Effect of the Co-Catalyst/Photocatalyst Mass Ratio

Soukaina Akel [1,2,*], **Ralf Dillert** [1,3], **Narmina O. Balayeva** [1], **Redouan Boughaled** [1], **Julian Koch** [4], **Mohammed El Azzouzi** [2] **and Detlef W. Bahnemann** [1,3,5,*]

[1] Institut für Technische Chemie, Leibniz Universität Hannover, Callinstr. 3, D-30167 Hannover, Germany; dillert@iftc.uni-hannover.de (R.D.); balayeva@iftc.uni-hannover.de (N.O.B.); r.boughaled@gmail.com (R.B.)

[2] Laboratory of Spectroscopy, Molecular Modeling, Materials, Nanomaterials, Water and Environment, (LS3MN2E) Faculty of Sciences, University Mohammed V. BP 1014, Rabat 10000, Morocco; elazzouzim@hotmail.com

[3] Laboratorium für Nano-und Quantenengineering, Leibniz Universität Hannover, Schneiderberg 39, D-30167 Hannover, Germany

[4] Institut für Festkörperphysik, Leibniz Universität Hannover, Appelstraße 2, 30167 Hannover, Germany; koch@fkp.uni-hannover.de

[5] Laboratory "Photoactive Nanocomposite Materials", Saint-Petersburg State University, Ulyanovskaya Street 1, Peterhof, Saint-Petersburg 198504, Russia

* Correspondence: akel@iftc.uni-hannover.de (S.A.); bahnemann@iftc.uni-hannover.de or detlef.bahnemann@spbu.ru (D.W.B.); Tel.: +49-511-762-2773 (S.A.); +49-511-762-5560 (D.W.B.)

Received: 20 October 2018; Accepted: 4 December 2018; Published: 10 December 2018

Abstract: Mixtures and composites of Ag/Ag$_2$O and TiO$_2$ (P25) with varying mass ratios of Ag/Ag$_2$O were prepared, employing two methods. Mechanical mixtures (TM) were obtained by the sonication of a suspension containing TiO$_2$ and Ag/Ag$_2$O. Composites (TC) were prepared by a precipitation method employing TiO$_2$ and AgNO$_3$. Powder X-ray diffraction (XRD) and X-ray photoelectron spectroscopy (XPS) confirmed the presence of Ag(0) and Ag$_2$O. The activity of the materials was determined employing methylene blue (MB) as the probe compound. Bleaching of MB was observed in the presence of all materials. The bleaching rate was found to increase with increasing amounts of TiO$_2$ under UV/vis light. In contrast, the MB bleaching rate decreased with increasing TiO$_2$ content upon visible light illumination. XRD and XPS data indicate that Ag$_2$O acts as an electron acceptor in the light-induced reaction of MB and is transformed by reduction of Ag$^+$, yielding Ag(0). As a second light-induced reaction, the evolution of molecular hydrogen from aqueous methanol was investigated. Significant H$_2$ evolution rates were only determined in the presence of materials containing more than 50 mass% of TiO$_2$. The experimental results suggest that Ag/Ag$_2$O is not stable under the experimental conditions. Therefore, to address Ag/Ag$_2$O as a (photo)catalytically active material does not seem appropriate.

Keywords: photocatalysis; silver(II) oxide; titanium dioxide; mechanical mixture; in situ deposition; hydrogen evolution

1. Introduction

Environmental problems related to water and air contamination, due to increasing world population and the resulting tremendous growth of industry and fuel combustion, have become a major concern of advanced science. In order to deal with this important problem, photocatalytic processes with employment of semiconductors are the most conventional approaches for water and air purification, along with alternative energy storage (e.g., H$_2$) [1–4].

To date, different semiconductor nanoparticles such as TiO$_2$, ZnO, Fe$_2$O$_3$, niobates, tantalates, and metal sulfides, and their underlying working mechanisms, have been investigated with the aim of increasing their photocatalytic activity. It is well known that, besides the ability to decontaminate polluted air and water, a photocatalyst should meet certain requirements such as cost efficiency, stability, non-toxicity, and broad range response towards incident light. TiO_2 is reported as the most durable photocatalyst, responding to all the above-mentioned requirements apart from broad range response to incident solar light due to its wide bandgap energy, (3.2 eV for anatase, 3.0 eV for rutile) which accounts for no more than 5% of the entire solar spectrum [1]. This lack of photocatalytic activity under visible light illumination allows the use of TiO_2 as a UV blocker in sunscreens [5]. The tremendous interest in modification of titanium dioxide with different metals and oxides, to enable absorption of lower energy states and increase stability, has been rising over the last 20 years. Nonetheless, the range of visible-light photocatalysts is still restricted. Thus, it is essential to discover new and efficient photocatalytic materials that are sensitive to visible light.

Ag_2O nanoparticles have been broadly utilized in various manufacturing areas as stabilizers, cleaning agents, electrode supplies, dyes, antioxidants, and catalysts for alkane activation and olefin [6,7]. Several papers have been published reporting the photocatalytic activity of Ag_2O, Ag/Ag_2O, Ag_2O/semiconductors, and Ag/Ag_2O/semiconductor composites, and some reviews are available [8–33]. Ag_2O is reported to be a visible light active photocatalyst. However, due to its photosensitive and labile properties under incident light illumination, Ag_2O is infrequently employed alone as a main photocatalyst rather than as a co-catalyst [8].

Wang et al. investigated the photocatalytic performance of Ag_2O on the photocatalytic decolorization of methyl orange, rhodamine B, and phenol solution under fluorescent light irradiation, and concluded that the stability and high photocatalytic activity of Ag_2O is maintained by the partial formation of metallic Ag on its surface during the photodecomposition of organic compounds [9]. Jiang et al. also reported the decomposition of methyl orange under visible light, ultraviolet light, near-infrared (NIR) light, and sunlight irradiation, using silver oxide nanoparticle aggregation. The superb photo-oxidation performance of Ag_2O is kept almost constant after repeated exposure to light due to its narrow band gap, high surface area, and numerous crystal boundaries supplied by Ag_2O quantum dots [13]. Several authors have claimed that an Ag/Ag_2O structure exhibits 'self-stability' [9,10] during a photocatalytic run, due to rapid electron transfer from the excited Ag_2O to $Ag(0)$ [12,20].

Visible light active nanocomposites of $Ag/Ag_2O/TiO_2$ have been synthesized using different methods, such as a microwave-assisted method [28], a low-temperature hydrothermal method [32], a one-step solution reduction process in the presence of potassium borohydride [22], a simple pH-mediated precipitation [23], and a sol-gel method [27]. Moreover, Su et al. developed a novel multilayer photocatalytic membrane, consisting of an Ag_2O/TiO_2 layer stacked on a chitosan sub-layer immobilized onto a polypropylene [31]. Light-induced hydrogen production via photoreforming of aqueous glycerol has been scrutinized, employing Ag_2O/TiO_2 catalysts prepared by a sol-gel method with varying content of Ag_2O (0.72–6.75 wt %) [30]. Hao et al. have reported that TiO_2/Ag_2O nanowire arrays forming a p-n heterojunction are applicable for enhanced photo-electrochemical water splitting [33]. Hu et al. reported the photocatalytic degradation of tetracycline under UV, visible, NIR, and simulated solar light irradiation with the Z-scheme between visible/NIR light activated Ag_2O and UV light activated TiO_2, using reduced graphene oxide as the electron mediator. They also investigated the stability of Ag_2O, Ag_2O/TiO_2, and Ag_2O/TiO_2 in combination with reduced graphene oxide as an electron mediator. A large amount of $Ag(0)$ was formed into Ag_2O and Ag_2O/TiO_2 after four cycles of tetracycline photodegradation under UV, visible, and NIR illumination [23]. Ren et al. also observed the light-induced reduction of Ag_2O during dye degradation in Ag_2O/TiO_2 suspensions. The authors suggested that the formation of $Ag(0)$ contributed to the high stability of their photocatalyst [29]. The stabilization of Ag_2O/TiO_2 photocatalysts by $Ag(0)$ formed at an initial stage of an experimental run has already been proposed earlier [11]. The photocatalytic stability of Ag-bridged Ag_2O nanowire networks/TiO_2 nanotubes, which were fabricated by a simple electrochemical method, revealed only an insignificant loss in performance, with respect to photocatalytic degradation of the dye acid orange 7, under simulated solar light [15]. On the other hand, Kaur et al. reported a decrease of the degradation efficiency from 81% to 54%, after the third experimental run employing AgO_2/TiO_2 as the photocatalyst and the drug levofloxacin as the probe compound [24]. Very recently, Mandari et al. synthesized plasmonic Ag_2O/TiO_2 photocatalysts, which could absorb visible light by the resonant oscillation of the conduction band electrons under visible light illumination. With this method, they were able to improve the efficiency of TiO_2 as a photocatalyst for hydrogen production by H_2O splitting under natural solar light. The authors observed the formation of Ag(0) by light-induced reduction of Ag_2O [26]. Light-induced reduction of Ag(I) to Ag(0) has also been reported for an Ag(0)/Ag(I) co-doped TiO_2 photocatalyst [34].

The preceding discussion of published experimental results provoked doubt on the stability of Ag_2O-containing photocatalysts under UV/vis illumination. Therefore, visible light harvesting $Ag/Ag_2O/\!/TiO_2$ photocatalysts for water treatment and photocatalytic hydrogen generation were synthesized. To the best of our knowledge, physical $Ag/Ag_2O/\!/TiO_2$ mixtures synthesized by the sonication of a suspension containing TiO_2 (P25) and a self-prepared Ag/Ag_2O were investigated for the first time. $Ag/Ag_2O/\!/TiO_2$ composites, prepared in situ by a simple precipitation method employing TiO_2 and $AgNO_3$, were also prepared, in order to evaluate the effect of the synthesis method on the photocatalytic activity. Additionally, the effect of the mass ratio of Ag/Ag_2O was studied. The as-prepared mixtures and composites showed improved visible light activity for methylene blue (MB) bleaching, compared to blank TiO_2, and high photocatalytic H_2 production from a methanol-water mixture under artificial solar light illumination.

2. Results

2.1. Characterization of the Prepared Materials

The powder X-ray diffraction (XRD) patterns of Ag/Ag_2O, physical mixtures of $Ag/Ag_2O/\!/TiO_2$ with increasing amounts of TiO_2 (20 mass% (TM 41), 50% (TM 11), and 80% (TM 14)), and in situ prepared $Ag/Ag_2O/\!/TiO_2$ composites (20 mass% TiO_2 (TC 41), 50% (TC 11), and 80% (TC 14)) are shown in Figure 1. The XRD peaks for Ag/Ag_2O at 26.7°, 32.8°, 38.1°, 54.9°, 65.4°, and 68.8° perfectly correlate to the (110), (111), (200), (220), (311), and (222) crystal planes of cubic Ag_2O (JCPDS 41–1104). The three peaks at 44.3°, 64.7°, and 77.5° are indexed to the (200), (200), and (311) crystal planes of cubic Ag(0), respectively (JCPDS 04-0783) [35,36].

The TiO_2 containing mixtures (TM) and composites (TC) exhibit diffraction peaks at 25°, 38°, 48°, 54°, 55°, 63°, 69°, 71°, 75°, and 83°, which are attributed to the tetragonal phase of anatase TiO_2, whereas one peak at 27.8° corresponds to the tetragonal phase of rutile TiO_2. Figure 1a presents the patterns of the TM mixtures, where two phases of titania were present. The two strongest peaks of Ag_2O become more prominent, with the Ag_2O mass ratio increasing from TM 14 to TM 41. The small diffraction peaks situated at 44.4°, 64.2°, and 77.5° are indexed to the (200), (200), and (311) plane of metallic Ag(0) (JCPDS 04-0783) [20]. The strongest peak of Ag(111) might likely be masked by the TiO_2 peak at $2\theta = 38°$. The diffraction peaks in the TM mixture patterns correspond to the cubic structure of Ag_2O and the cubic structure of Ag [35,36]. Figure 1b illustrates the XRD patterns of the TC composites. As the figure shows, no significant difference between the two preparations methods was observed, except that in TiO_2-rich composites TC 11 and TC 14 no Ag_2O diffraction peaks were observed, suggesting a complete reduction of Ag_2O to metallic silver Ag(0) during the preparation of these composites. The XRD pattern of TiO_2 is presented for comparison. The diffractogram clearly indicates the presence of two TiO_2 phases with predominance of the anatase phase (JCPDS 21–1272).

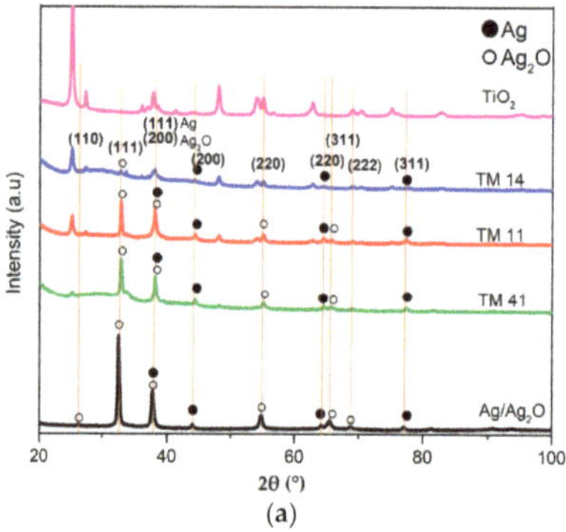
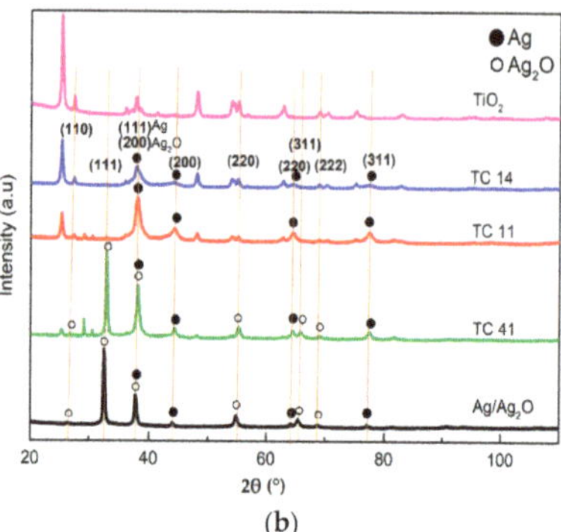

Figure 1. X-ray diffraction (XRD) patterns of (**a**) TiO$_2$ containing mixtures (TM), and (**b**) TiO$_2$ containing composites (TC). The diffractograms of Ag/Ag$_2$O and TiO$_2$ are included in both figures.

In order to investigate the oxidation states of the silver species present on the materials, X-ray photoelectron spectroscopy (XPS) was performed. The results of the XPS analysis for all samples are shown in Figure S3. The deconvolution of the high-resolution spectra for Ag 3d reveals that silver was present in more than one oxidation state in all samples. The binding energies of Ag 3d at 367.5 and 373.5 eV are assigned to the Ag 3d$_{5/2}$ and Ag 3d$_{3/2}$ photoelectrons respectively, indicating the presence of silver in the +1 oxidation state. The other two peaks of Ag 3d$_{5/2}$ and Ag 3d$_{3/2}$, at 368.3 and 374.3 eV respectively, confirm the existence of silver in the Ag(0) state. These binding energies are in good agreement with the values reported for Ag(I) in Ag$_2$O and Ag(0) [16,37,38]. The peaks for O 1s, located in the ranges of 528.9–530.1 eV and 530.5–531.2 eV, are ascribed to O^{2-} in Ag$_2$O and TiO$_2$ respectively (Figure S3). From the Ti 2p core-level spectrum, two peaks at about 464.3 and 458.7 eV can be assigned to the Ti 2p$_{1/2}$ and Ti 2p$_{3/2}$ spin–orbital components respectively, which correspond to the characteristic peaks of Ti^{4+}.

The SEM images of blank TiO$_2$, Ag/Ag$_2$O, TM mixtures, and TC composites are presented in Figure 2. Ag/Ag$_2$O showed well-defined particles with particle sizes ranging from 100 nm to 500 nm (Figure 2a). The small particles that contrast as white spots correspond to the metallic silver Ag(0) distributed on the surface of silver oxide, which is in agreement with the XRD results. The EDX reveals that the sample contained Ag and O without any other impurities (Figure S1).

Figure 2b–d shows SEM images of the physical mixtures of Ag/Ag$_2$O with TiO$_2$. It becomes obvious from these images that Ag/Ag$_2$O changed its shape during preparation of the mixtures by sonification of aqueous suspensions of the oxides. The increasing loading of the Ag$_2$O platelets with TiO$_2$ is also clearly recognizable in these figures. In the Ag/Ag$_2$O$/\!/$TiO$_2$ mixture with the highest mass fraction of TiO$_2$ (TM 14), the appearance was apparently determined by the titanium dioxide distributed over the underlying surface of the Ag$_2$O platelets (Figure 2d). This was also reflected in the specific surface area (SSA) of the materials. The TiO$_2$ (P25) used in this work is known to have an average diameter and specific surface area of 21 nm and about 50 m^2 g^{-1}, respectively [39]. The specific surface area of the Ag/Ag$_2$O synthesized in this work was determined to be 2.7 m^2 g^{-1}. As expected, the specific surface area of the Ag/Ag$_2$O$/\!/$TiO$_2$ mixtures was found to increase with increasing TiO$_2$ content (Table 1), resulting in a SSA of 38.5 m^2 g^{-1} for TM 14.

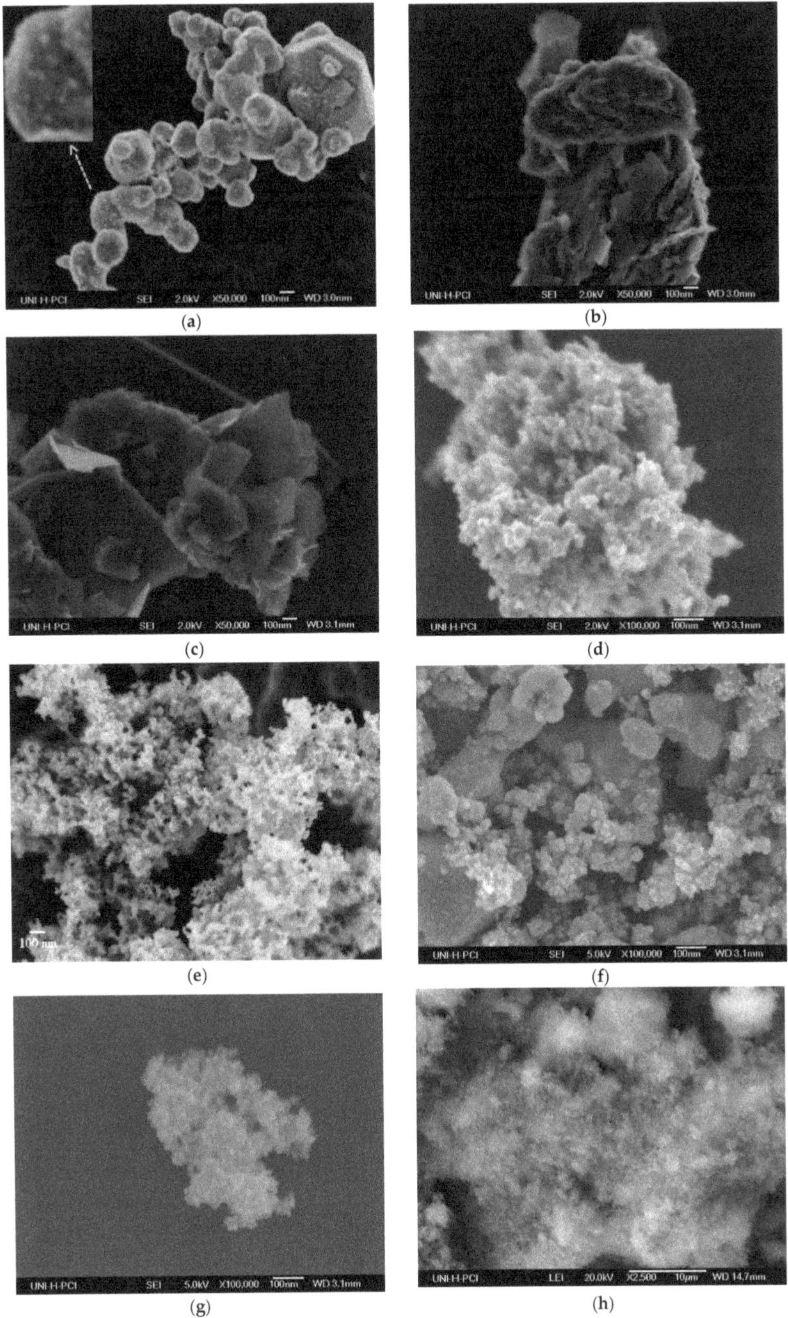

Figure 2. SEM pictures of (**a**) Ag/Ag$_2$O, (**b**) TM 41, (**c**) TM 11, (**d**) TM 14, (**e**) TiO$_2$ (P25), (**f**) TC 41, (**g**) TC 11, and (**h**) TC 14.

SEM images of the TC composites are presented in Figure 2f–h. The image of the TiO$_2$-poor composite TC 41 clearly shows the large Ag/Ag$_2$O particles covered with TiO$_2$ (Figure 2f). The specific surface area of this composite was determined to be 8.4 m^2 g^{-1}, thus being equal within the limits of the experimental error to the surface area of the corresponding physical mixture TC 41 (SSA = 9.7 m^2 g^{-1}). The images of the composites richer in TiO$_2$ (TC 11 and TC 14) seemed to be dominated by aggregates or agglomerates of small TiO$_2$ particles.

The optical properties of TiO$_2$ and the as-prepared Ag-containing mixtures and composites were investigated by UV/vis diffuse reflectance spectroscopy (Figure 3). Ag/Ag$_2$O, as well as the TM, and TC materials, had a dark brown to black color. They displayed strong absorption over the whole UV and visible range (200 nm–800 nm). TiO$_2$ showed only the absorption band below 405 nm, which matches the band gap energy of 3.06 eV calculated from the formula $\lambda = 1239.8/E_{bg}$ due to the charge transfer from O (valence band) to Ti (conduction band).

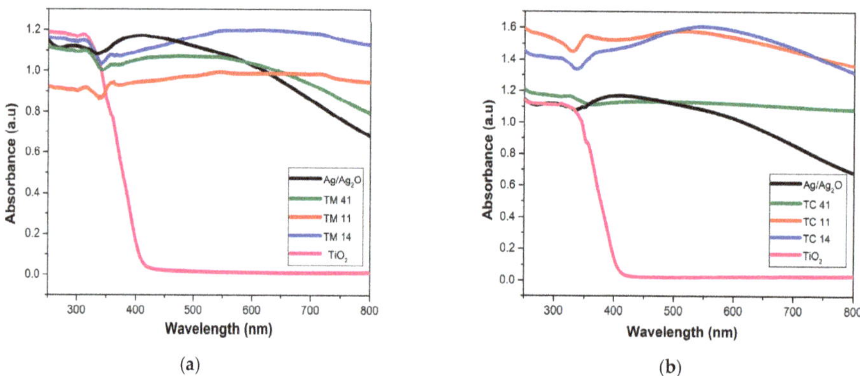

Figure 3. UV/vis diffuse reflectance spectra of (**a**) TiO$_2$, Ag/Ag$_2$O, TM mixtures, and (**b**) TC composites.

Ag/Ag$_2$O exhibited a band gap energy < 1.5 eV, which is in agreement with the reported value of 1.3 ± 0.3 eV [40]. The scattering of the reported values might be due to different particle diameters, as shown for TiO$_2$ [41]. Electrochemical measurements in suspensions yielded flat band potentials of -0.4 V and $+0.3$ V vs. NHE for TiO$_2$ and Ag$_2$O, respectively. The value measured here for the flat band potential of Ag$_2$O is also in reasonably good agreement with published values [42,43].

2.2. Photocatalytic Performance of the Materials

The photocatalytic activity of all materials described above was investigated, employing methylene blue (MB) as the probe compound. The materials in aqueous suspensions were excited by the full output of a xenon lamp (UV/vis illumination), and by Xe light after passing a UV cut-off filter ($\geq$410 nm, vis illumination). Figure 4 illustrates the bleaching of an aqueous solution of MB and the MB-containing suspensions. Photolysis of MB (initiated by the direct excitation of the probe compound) was observed under both UV/vis and visible light illumination. The bleaching of MB was significantly accelerated by the presence of Ag/Ag$_2$O. Under UV/vis illumination, Ag/Ag$_2$O was found to be nearly as active as TiO$_2$ (P25), which is well known to be a very efficient photocatalyst suitable to degrade MB [44] (Figure 4a). In the presence of Ag/Ag$_2$O, MB was bleached very rapidly even when exposed to visible light. As expected, TiO$_2$, having a bandgap energy of 3.1 eV, was found to be inactive under vis illumination (Figure 4c).

In the presence of mixtures of Ag/Ag$_2$O with TiO$_2$, MB was bleached under UV/vis illumination only in the presence of the TiO$_2$-rich TM 14, with a significantly increased rate compared to the rate of MB photolysis. In suspensions containing TM 41 and TM 11, the rate of bleaching was almost the same as the rate of photolysis (Figure 4a). Exposure to visible light in the presence of the Ag/Ag$_2$O-rich TM 41 resulted in bleaching of MB with a slightly increased rate. In contrast, the TiO$_2$-rich mixtures TM 11 and TM 14 were virtually inactive under this illumination condition (Figure 4c).

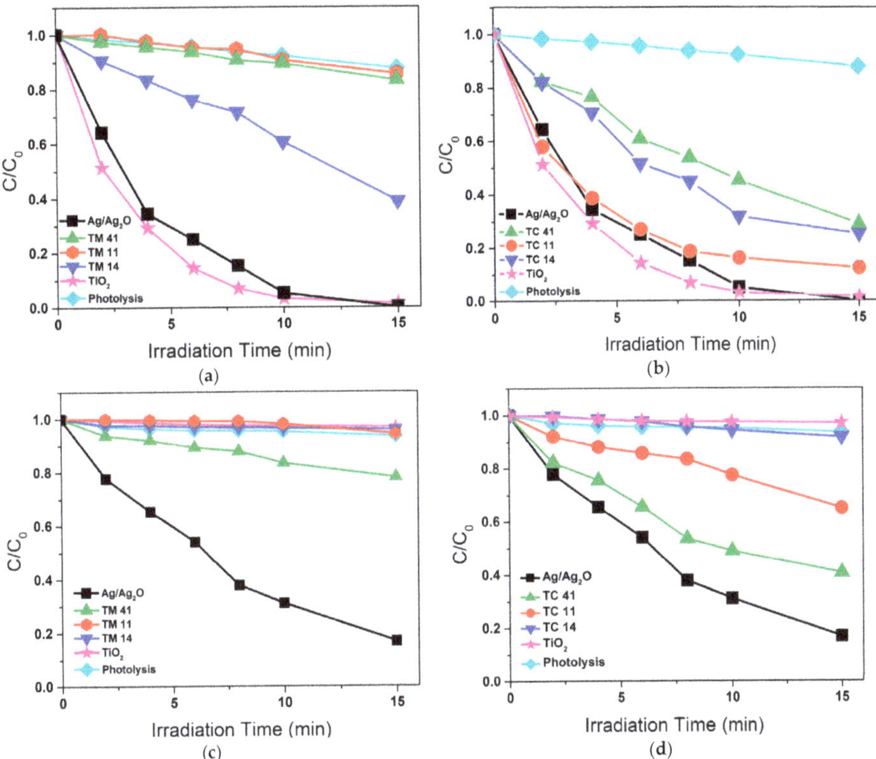

Figure 4. Bleaching of MB in the presence of Ag/Ag$_2$O, TiO$_2$, the TM mixtures and the TC composites under UV/vis (**a,b**) and under vis light only (**c,d**).

In the presence of the composites TC, MB was bleached with significantly faster reaction rates than the rate of photolysis when exposed to UV/vis and visible light. The rates were, however, lower than the rate of bleaching in the presence of the bare TiO$_2$ (Figure 4b,d). Interestingly, while increasing the amount of TiO$_2$ in the TC composites, the visible light activity of the materials seemed to decrease, thus confirming the essential influence of Ag/Ag$_2$O on MB bleaching under illumination with wavelengths $\geq$ 410 nm.

As a second test reaction for the activity of the materials, the UV/vis light-induced evolution of molecular hydrogen by reforming of aqueous methanol was used. Figure 5 shows the amount of H$_2$ vs. illumination time in the presence of TiO$_2$, Ag/Ag$_2$O, and the prepared mixtures and composites. No H$_2$ evolution was observed in the presence of Ag/Ag$_2$O and the Ag/Ag$_2$O-rich TM 41. In the presence of all other materials, the evolution of H$_2$ was detected. However, large amounts of H$_2$ were only evolved with the materials TM 14 (104 µmol/6 h) and TC 11 (174 µmol/6 h).

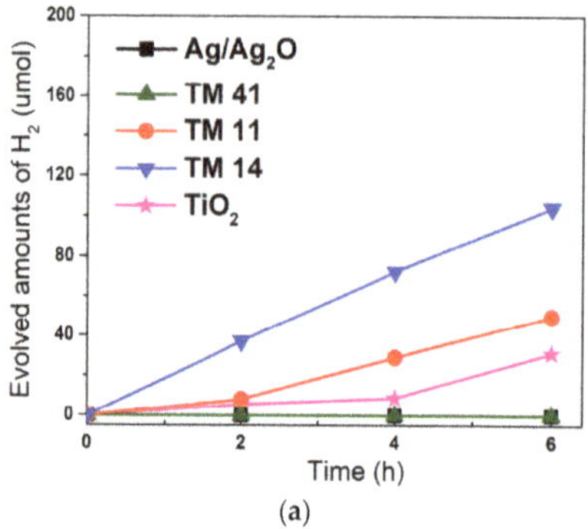
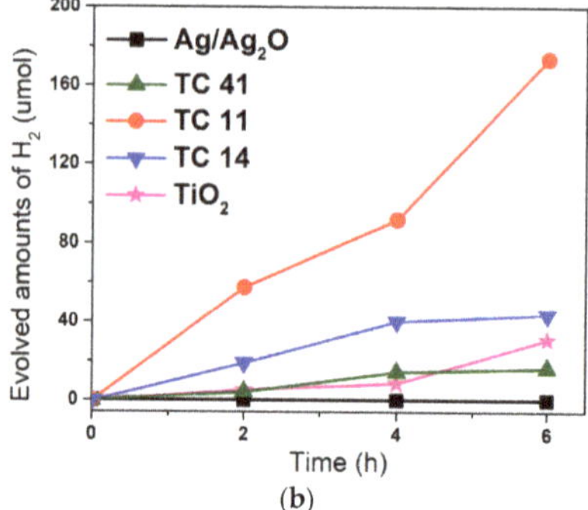

Figure 5. The amount of H_2 evolved from aqueous methanol under UV/vis illumination of Ag/Ag$_2$O, TiO$_2$, (**a**) TM mixtures and (**b**) TC composites vs. illumination time.

Many authors have reported that the kinetic behavior of photocatalytic reactions can be described by a Langmuir–Hinshelwood rate law, with the two limiting cases of zero-order and first-order kinetics [45,46]. To calculate the initial rates r_0 of the bleaching of methylene blue, first-order kinetics have been assumed ($r_0 = kC_0$). To determine the rate constant k, the data given in Figure 4 have therefore been fitted with $C = C_0 \exp(-kt)$. The initial rates are given in Table 1.

Table 1. Brunauer-Emmet-Teller (BET) surface area, initial rates of methylene blue (MB) bleaching and H_2 generation in the presence of Ag/Ag$_2$O, TiO$_2$, the TM mixtures and the TC composites.

Sample	Preparation Method	Composition	SSA $m^2\,g^{-1}$	r_0 (MB) UV/vis $mg\,L^{-1}\,min^{-1}$	r_0 (MB) vis $mg\,L^{-1}\,min^{-1}$	r (H$_2$) UV/vis $\mu mol\,h^{-1}$
Photolysis	-	-	-	0.08	0.05	-
Ag/Ag$_2$O	in situ	Ag/Ag$_2$O	2.7	2.64	1.17	-
TM 41	mechanical mixture	Ag/Ag$_2$O∥TiO$_2$ (20% TiO$_2$)	9.7	0.12	0.17	-
TM 11	mechanical mixture	Ag/Ag$_2$O∥TiO$_2$ (50% TiO$_2$)	22.6	0.09	0.03	9
TM 14	mechanical mixture	Ag/Ag$_2$O∥TiO$_2$ (80% TiO$_2$)	38.5	0.55	0.03	17
TiO$_2$	-	TiO$_2$	50	3.08	0.03	5
TC 41	in situ	Ag/Ag$_2$O∥TiO$_2$ (20% TiO$_2$)	8.4	0.81	0.61	3
TC 11	in situ	Ag/Ag$_2$O∥TiO$_2$ (50% TiO$_2$)	20.1	2.03	0.27	28
TC 14	in situ	Ag/Ag$_2$O∥TiO$_2$ (80% TiO$_2$)	22.1	1.00	0.05	8

3. Discussion

3.1. The Photocatalytic Activity of Ag/Ag$_2$O

It is well known that methylene blue is photocatalytically oxidized in the presence of TiO$_2$ under illumination with photons having an energy equal to or larger than the bandgap energy of the semiconductor. The photocatalytic degradation of methylene blue in the presence of molecular oxygen is reported to follow Equation (1) [44].

$$C_{16}H_{18}N_3SCl + 25.5O_2 \xrightarrow{\text{TiO}_2 + h\nu \geq 3.2\,eV} HCl + H_2SO_4 + 3HNO_3 + 16CO_2 + 6H_2O \quad (1)$$

The energetic positions of the valence and conduction bands of TiO$_2$ and Ag$_2$O, and the reduction potentials of some species (possibly) present in the surrounding electrolyte are shown in Figure 6. As becomes obvious from this Figure, the conduction band electrons generated by UV illumination of TiO$_2$ are able to reduce O$_2$ adsorbed at the semiconductor surface. From a thermodynamic point of view, valence band holes at the TiO$_2$ surface have an energy suitable to oxidize H$_2$O/OH$^-$, yielding OH radicals. These OH radicals are generally assumed to be the oxidizing species in photocatalytic MB degradation.

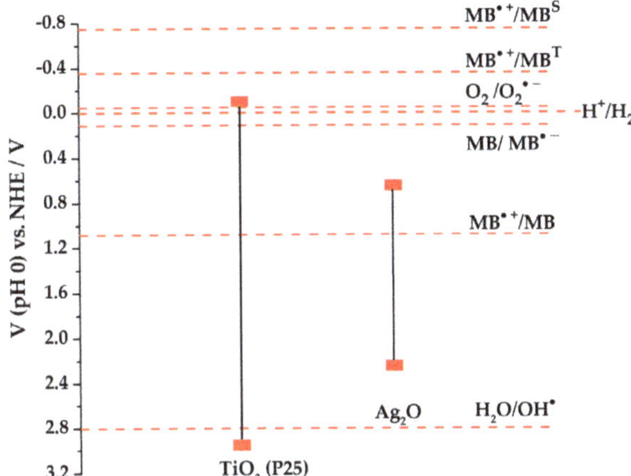

Figure 6. The electrochemical potentials (vs. NHE) of the valence and conduction bands of TiO_2 and Ag_2O, and the reduction potentials of some species (possibly) present in the surrounding electrolyte. MB, MB$^{\bullet-}$, MB$^{\bullet+}$, MB^T, and MB^S denote the MB ground state, the semi-reduced MB, the oxidized MB, the excited triplet state, and the excited singlet state of MB, respectively. The one electron reduction potentials have been calculated with data given in References [44,47,48].

With the assumption that the flat band potential of Ag_2O, which has been determined to be + 0.3 V vs. NHE at pH 7, was equal to the conduction band edge of this semiconductor, and a bandgap energy $E_g = 1.5$ eV, the valence band position was calculated to be +2.0 V vs. NHE. Xu and Schoonen reported a value of +0.2 V vs. NHE for the energy of the Ag_2O conduction band [49]. As becomes obvious from Figure 6, excited Ag_2O was neither able to reduce O_2 nor to oxidize H_2O/OH^-. Consequently, the mechanism of MB bleaching observed in the presence of Ag/Ag_2O (Figure 4 and Table 1) was different from the MB degradation mechanism in the presence of TiO_2. A possible explanation for the decolorization of MB in the presence of Ag/Ag_2O is that MB is excited by light of suitable wavelength (Equation (2), MB* = MB^S and or MB^T), which is subsequently followed by electron injection into the conduction band of Ag_2O (Equation (3)).

$$MB \overset{h\nu}{\to} MB* \tag{2}$$

$$MB* + Ag_2O \to MB^{+\bullet} + Ag_2O\{e^-\} \tag{3}$$

As an alternative to these reactions, the direct oxidation of MB by valence band holes according to

$$Ag_2O \to Ag_2O\{h^+ + e^-\} \tag{4}$$

$$Ag_2O\{h^+ + e^-\} + MB \to Ag_2O\{e^-\} + MB^{+\bullet} \tag{5}$$

has to be considered. Both mechanisms require an electron transfer between Ag_2O and MB. Despite the low surface area available for this reaction, the electron transfer between the solid and the probe compound appears to be very efficient.

It is well known that Ag_2O is sensitive to light and decomposes under illumination. However, it has been suggested that Ag(0) being present in Ag/Ag_2O acts as an electron sink and accepts the conduction band electron of Ag_2O, thus inhibiting the reduction of Ag^+ and stabilizing the Ag_2O [9,10,12,20]. However, the possibility cannot be excluded that Ag^+ is reduced during the processes given in the Equations (2)−(5), yielding Ag(0), since no other suitable electron acceptor is available. Regardless of whether the electrons reduce Ag^+ or become stored in Ag(0), Ag/Ag_2O is not acting as a photocatalyst, because the material changes irreversibly during the reaction.

The potential of the Ag_2O conduction band electron is more positive than the reduction potential of the H^+/H_2 couple (Figure 6). Consequently, light-induced proton reduction yielding H_2 is thermodynamically impossible in suspensions containing only Ag/Ag_2O. This is in accordance with the experimental results reported in Section 2.2.

3.2. The Photocatalytic Activity of Physical $Ag/Ag_2O/\!/TiO_2$ Mixtures

3.2.1. Bleaching of Methylene Blue

When irradiated with light at wavelengths ≥ 410 nm, methylene blue was found to be bleached in the presence of Ag/Ag_2O, and mixtures of this material with TiO_2. The rate of MB bleaching decreased with increasing amounts of TiO_2. Of course, TiO_2 itself was found to be photocatalytically inactive, since it was not excited under this illumination condition (Figure 4c and Table 1). The electron transfer reaction resulting in the observed bleaching of the MB solution occurred at the surface of the Ag_2O, as discussed in Section 3.1. According to the SEM images (Figure 2a–d), the surface of the Ag_2O was increasingly covered by TiO_2 as the content of this oxide in the mixture increased. The interfacial electron transfer was inhibited by this TiO_2 layer (Figure 7). The reaction rates suggest that this inhibition increased with increasing amounts of TiO_2 on the Ag/Ag_2O surface. Consequently, the TiO_2-rich mixtures TM 11 and TM 14 exhibited rates of bleaching almost the same as the rate of photolysis in homogeneous solution (Table 1). Interfacial electron transfer from excited MB to TiO_2 (which is thermodynamically possible; cf. Figure 6) obviously did not contribute significantly, since no MB bleaching was observed under visible light illumination of suspensions containing only this photocatalyst.

Figure 7. Possible mechanism of MB bleaching by Ag_2O and Ag_2O-containing mixtures and composites under visible light illumination.

The situation was different when the TM mixtures were illuminated with UV/vis light. The rate of MB bleaching in the presence of the $Ag/Ag_2O/\!/TiO_2$ mixtures was found to increase with increasing TiO_2 content. However, the rates were always lower than the rates determined for suspensions containing only Ag/Ag_2O or bare TiO_2 (Figure 4a and Table 1). These rates cannot be explained solely by the optical properties of the suspensions. Of course, as the Ag/Ag_2O content increases, more UV photons are absorbed by Ag_2O. They are thus no longer available for the excitation of the TiO_2 that results in decreasing amounts of charge carriers in the TiO_2 and, consequently, decreasing rates of MB degradation. However, the MB bleaching rate calculated for the TiO_2-rich TM 14 mixture suggests that not all photogenerated charge carriers were used in the desired MB bleaching reaction, but some were lost by reactions between excited TiO_2 and Ag/Ag_2O, resulting in the reduction of Ag^+.

XRD measurements revealed the reduction of Ag^+ during the light-induced bleaching of MB under UV/vis illumination. The ratios of the peak intensities corresponding to Ag_2O and TiO_2 of the mixture TM 41 and the composite TC 41 were significantly lower after two experimental runs than before illumination (Figure 8). On the other hand, the ratios of the peak intensities attributed to metallic Ag and TiO_2 obviously increased. In the case of the $Ag/Ag_2O/\!/TiO_2$ mixture TM 11, apart from the TiO_2 peaks, the only visible XRD peaks could be assigned to AgCl and Ag(0) after illumination of a suspension containing MB (Figure S2). The new peaks in the diffractogram, which are indexed to AgCl, were possibly formed by a reaction between Ag^+ and Cl^- known to be present at the surface of TiO_2 P25 [39]. This reaction certainly explains the decrease of the Ag_2O peaks in the diffractogram. However, this explanation does not exclude that Ag_2O is also transformed by a light-induced reduction reaction, yielding Ag(0).

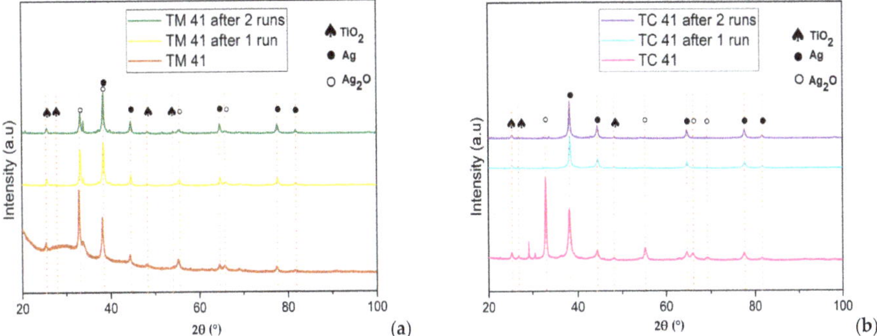

Figure 8. XRD patterns of (**a**) TM 41 and (**b**) TC 41 after two cycles of MB bleaching employing UV/vis light.

The conclusion from the XRD data, that Ag(I) was reduced yielding Ag(0) during the light-induced bleaching of MB in the presence of the mixture TM 41, is supported by the results of the analysis of XPS data taken before and after two experimental runs (Figure 9a,b and Figure S3). It becomes obvious from Figure 9a that the Ag $3d_{5/2}$ and Ag $3d_{3/2}$ peaks of Ag_2O in the mixture TM 41 decreased in intensity and broadened, while the Ag(0) $3d_{5/2}$ and Ag(0) $3d_{3/2}$ peaks increased in intensity after two photocatalytic reactions. Furthermore, the deconvolution of the O 1s peaks denotes that the peak corresponding to the Ag-O bond had a lower intensity compared to the same peak observed before the reaction, indicating significant changes occurred during the light-induced MB bleaching reaction (Figure 9b). These changes were mainly due to the light-induced reduction of Ag^+ yielding Ag(0). Again, the condition of stability of a catalyst was not satisfied.

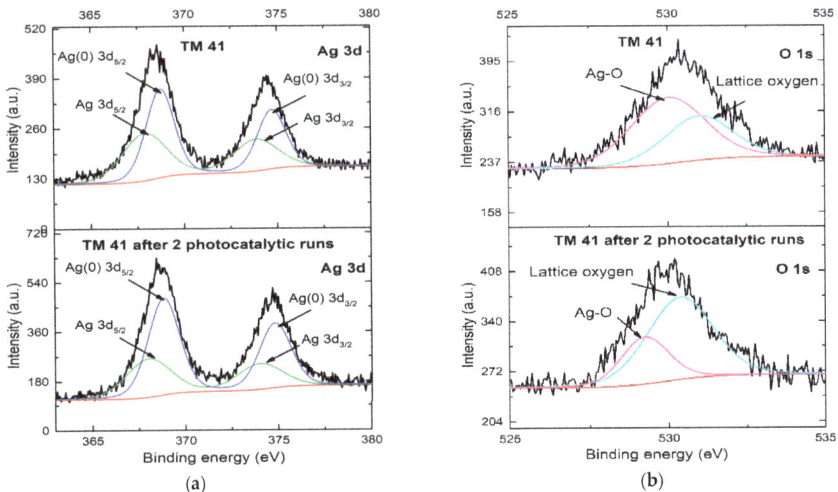

Figure 9. *Cont.*

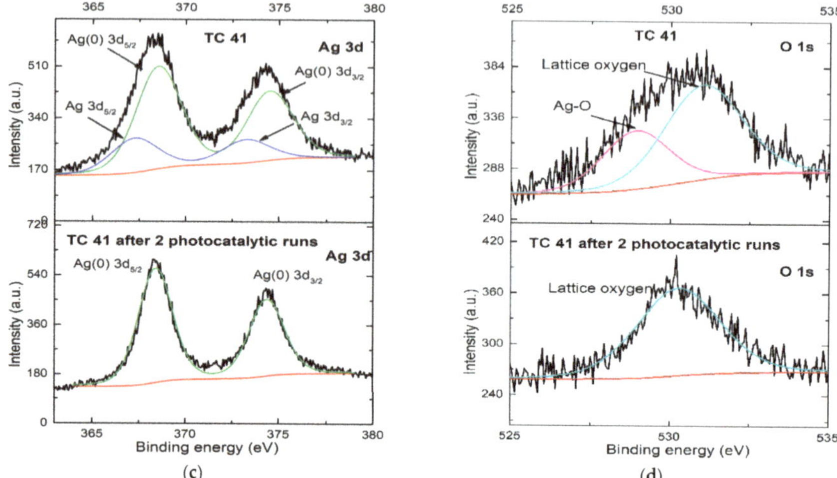

Figure 9. High-resolution XPS spectra of the Ag 3d and O 1s signals of TM 41 (**a,b**) and TC 41 (**c,d**) before and after two experimental runs.

3.2.2. Light-Induced Hydrogen Evolution

From a thermodynamic point of view, excited TiO_2 is able to transfer a conduction band electron to a proton present at the photocatalyst surface (Figure 6). This electron transfer is, however, known to be a kinetically inhibited process. Therefore, it is necessary to deposit an electrocatalyst at the TiO_2 surface, which accelerates the interfacial electron transfer. Ag(0) is known to be a suitable, though relatively inactive, electrocatalyst [50,51]. In this work as well, pure TiO_2 showed only a very low photocatalytic activity with regard to H_2 evolution from aqueous methanol. When using the TM materials, a significant increase in the amount of H_2 evolved (consequently corresponding with an increase in the reaction rate) during six hours of illumination of the mixture was observed with increasing TiO_2 content (Figure 5a and Table 1). On the one hand, this can be explained by the fact that a significant portion of the UV photons was absorbed by Ag_2O being inactive under this illumination condition, and thus was not available for the desired H_2 evolution reaction. However, this portion decreased with increasing TiO_2 amount of the mixture. On the other hand, some of the TiO_2 conduction band electrons were transferred to the Ag_2O, where they were consumed to reduce Ag^+ to Ag(0). These electrons were therefore also not available for the desired reaction. Obviously, these undesired electron losses are lower the higher the mass fraction of TiO_2 in the physical mixture, resulting in increasing H_2 evolution rates with increasing mass fraction of TiO_2.

3.3. The Photocatalytic Activity of Ag/Ag₂O∥TiO₂ Composites

3.3.1. Bleaching of Methylene Blue

When irradiated with light at wavelengths ≥ 410 nm, methylene blue was found to be bleached in the presence of the three TC composites (Figure 4d and Table 1). All TC composites exhibited a higher activity than the corresponding TM mixtures. As in the case of the TM materials, the rate of MB bleaching decreased with increasing amounts of TiO_2. The increased reaction rates for MB bleaching in the presence of Ag_2O containing solids, compared to the rate of photolysis under visible light illumination, were explained in Section 3.2.1 with an interfacial electron transfer from (excited) MB to Ag_2O (cf. Figure 7). However, the experimental result is surprising when it is considered that the surfaces of the composites were smaller than the surfaces of the corresponding TM mixtures. A possible explanation may be due to the preparation method. For the TC materials, the Ag/Ag₂O was prepared in a TiO_2 suspension. Therefore, the Ag/Ag_2O was attached on the surface of the TiO_2 particles. In contrast, in the TM mixtures large Ag/Ag_2O particles were covered by TiO_2, hindering the electron transfer from excited MB to the Ag_2O, as discussed in Section 3.2.2.

The rate of MB bleaching in the presence of TC composite was significantly higher under UV/vis than under visible light illumination. As observed for the TM materials, the bleaching rates were lower in suspensions containing the composites than in suspensions containing only Ag/Ag_2O or TiO_2 (Figure 4b and Table 1).

XRD and XPS data indicate that Ag(I) was reduced, yielding Ag(0), during the light-induced bleaching reaction of MB in the presence of the composite TC 41. A stabilization of Ag_2O by metallic silver, as claimed by several authors [9–12,20,29], was not observed. No XRD peaks that can be attributed Ag_2O, were observed after two experimental runs of the composite. However, the ratios of the peak intensities due to metallic Ag and TiO_2 obviously increased (Figure 8b). No Ag $3d_{5/2}$ and Ag $3d_{3/2}$ peaks, which can be attributed to Ag(I), were present either in the deconvoluted XPS spectra obtained after two experimental runs (Figure 9c and Figure S3). The XPS peak, which was attributed to the presence of Ag-O, also disappeared during the light-induced reaction (Figure 9d and Figure S3).

These observations support the statement made above that Ag/Ag_2O cannot be called a photocatalyst. The XRD pattern shown in Figure 8b as well as the XPS data presented in the Figure 9c,d clearly evince that the $Ag:Ag_2O$ ratio changed during the light-induced bleaching of MB. Thus, the condition for a catalyst to exit a chemical reaction unchanged is not satisfied.

3.3.2. Light-Induced Hydrogen Evolution

The three TC composites were found to be able to promote light-induced H_2 evolution from aqueous methanol. The calculated reaction rates were significantly larger than those of the corresponding TM mixtures. The highest H_2 evolution rate was observed in the presence of TC 11 (Figure 5b and Table 1), which was also characterized by a high MB bleaching rate under UV/vis illumination. A possible mechanistic explanation for the high activity of the TC 11 composite is based on the assumption of synergistic effects, due to the presence of both Ag(0) and Ag_2O at the TiO_2 surface (Figure 10). TiO_2 is excited by UV photons. The photogenerated conduction band electrons migrated to the Ag(0) attached to the TiO_2 surface. In a subsequent step, interfacial electron transfer from Ag(0) to protons present in the surrounding electrolyte occurred, thus yielding molecular hydrogen. The valence band hole inside the TiO_2 particle was filled by an electron from an attached Ag_2O particle. Methanol was oxidized by this hole in the valence band of the Ag_2O. According to this mechanism, Ag(0) acts as an electron sink, thus decreasing the electron-hole recombination, and as electrocatalyst for the hydrogen evolution reaction, while Ag_2O is an electrocatalyst for the oxidation reaction of methanol yielding methanal. The supposition made here, that the methanol oxidation occurs at the Ag_2O surface via electron transfer to the valence band of the excited TiO_2, has already been proclaimed earlier [16,19,23,26]. It should be emphasized again that the energy of an electron in the conduction band of the Ag_2O employed in this study is insufficient to reduce a proton (Figure 6). Consequently, excitation of TiO_2 is a prerequisite for photocatalytic reforming of methanol. TiO_2 is known to be a relatively inactive material for the photocatalytic reduction of protons. High evolution rates of molecular hydrogen are observed only in the presence of a co-catalyst. Ag_2O was found here to be an unsuitable co-catalyst for the hydrogen evolution reaction, since electron transfer from the excited TiO_2 can only occur into the conduction band of this material. The photocatalytic activities of the composites and mixtures discussed here are thus determined to a considerable extent by the competition between interfacial electron transfer to protons in the surrounding electrolyte, and to silver ions in Ag_2O. The mechanism of the photocatalytic hydrogen evolution by reforming of organic compounds in the presence of the mixtures and composites employed in this study does not contradict the mechanism discussed for $Ag/Ag_2O/\!/TiO_2$ samples, which contain Ag_2O with a significantly more negative conduction band energy than TiO_2 [17,24,26,33].

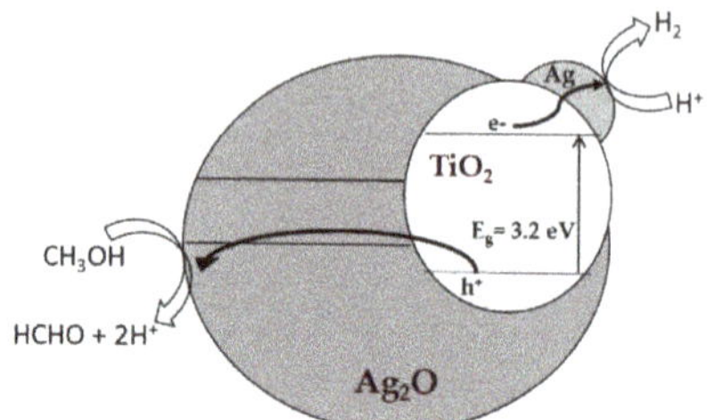

Figure 10. Mechanism of hydrogen evolution from aqueous methanol under UV/vis illumination.

Changes in the respective mass fractions of TiO_2, Ag, and Ag_2O at constant total mass of the solid in suspension may have several impacts on the rate of hydrogen evolution. Increasing mass fractions of UV absorbing and scattering Ag and Ag_2O reduces the number of photons to be absorbed by the TiO_2, thus reducing the H_2 evolution rate. A reduction of the mass fraction of metallic Ag may possibly slow down the interfacial electron transfer to the proton, while a reduction of the mass fraction of Ag_2O might negatively affect the oxidation reaction. It should also be noted that Ag_2O can act as a sink for a TiO_2 conduction band electron (cf. Figure 6). These partially opposing effects may be responsible for the observed differences in the H_2 evolution rates in the presence of the various TC composites (and TM mixtures).

4. Experimental Section

4.1. Materials

Titania P25 (TiO_2) with a mixture of anatase (80%) and rutile (20%) crystal phase, and a specific surface area of 50.1 m^2 g^{-1}, was kindly provided by Evonik, Essen, Germany. Silver nitrate (99%, Sigma Aldrich Chemie GmbH, München, Germany), sodium hydroxide pellets (99%, Carl Roth, Karlsruhe, Germany), methanol (99.9%, Carl Roth), and methylene blue (Sigma Aldrich) were used without further purification. Deionized water with a resistivity of 18.2 MΩ·cm was obtained from a Sartorius Arium 611 device (Sartorius Göttingen, Germany) and used for the preparation of all aqueous solutions and suspensions.

4.2. Synthetic Methods

4.2.1. Preparation of Ag/Ag_2O

An amount of $AgNO_3$ was dissolved in 50 mL of distilled water. The obtained solution was stirred for 30 min. Subsequently, 50 mL NaOH (0.2 M) was added dropwise. The resulting suspension was stirred for another 30 min to promote hydrolysis, and centrifuged, washed with distilled water three times, and dried at 70 °C for 24 h.

4.2.2. Preparation of TM Mixtures

The samples were obtained by mixing the self-prepared Ag_2O with TiO_2 at mass ratios of 4:1 (20 mass% TiO_2), 1:1 (50 mass% TiO_2), and 1:4 (20 mass% TiO_2) with water. The suspensions were sonicated for 1.5 h and dried at 70 °C for 24 h. The Ag/Ag_2O//TiO_2 with 20%, 50%, and 80% of TiO_2 were nominated as TM 41, TM 11, and TM 14, respectively. For purpose of comparison, a TiO_2 sample was prepared by the same procedure without the addition of Ag/Ag_2O.

4.2.3. Preparation of TC Composites

The TC composites were prepared by a published precipitation method [25,29]. A measured amount of TiO_2 was suspended in 50 mL of distilled water, and the calculated amount of $AgNO_3$

corresponding to the desired mass ratio of Ag_2O was added to the solution. The obtained suspension was stirred for 30 min. A volume of 50 mL 0.2 M NaOH was added dropwise. The resulting suspension was stirred for another 30 min to promote hydrolysis and centrifuged, washed with distilled water three times and dried at 70 °C for 24 h. The $Ag/Ag_2O/\!\!/TiO_2$ with 20 mass%, 50 mass%, and 80 mass% of TiO_2 were denoted as TC 41, TC 11, and TC 14, respectively.

4.3. Characterization of the Materials

The crystalline structure of the catalysts was measured by powder X-ray diffraction XRD (D8 Advance system, Bruker, Billerica, MA, USA), using a Cu Kα radiation source with a wavelength of $\lambda = 0.154178$ Å over a 2θ range from 20° to 100°, with a 0.011° step width. The morphology of the prepared materials was determined using a scanning electron microscope (SEM), employing a JEOL JSM-6700F field emission instrument (Tokyo, Japan) with a resolution of 100 nm and 1 µm using an EDXS detector. Measurements of X-ray photoelectron spectra were carried out using a Leybold Heraeus (Cologne, Germany) with X-ray source, Mg & Al anode, nonmonochromatic, hemispherical analyzer, 100 mm radius. Data analysis was performed using XPSPEAK 4.1 software (Hong Kong, China). The energy of the C1s-line was set to 284.8 eV and used as reference for the data correction. Diffuse reflectance UV–Vis spectroscopy was employed using a spectrophotometer (Varian Spectrophotometer Cary-100 Bio, Agilent technologies, Santa Clara, CA, USA) at room temperature. Barium sulfate was used as a standard for 100% reflection. The specific surface area (SSA) of the samples was calculated by N_2 adsorption–desorption measurements, employing the Brunauer-Emmet-Teller (BET) method using a FlowSorb II 2300 apparatus from Micromeritics Instrument Company (Corp., Norcross, GA, USA). Prior to these measurements, the samples were evacuated at 180 °C for 1 h. Measurements of photocurrents and flat band potentials were performed with an electrochemical analyzer using three electrodes employing an Iviumstat potentiostat (Ivium Technologies bv, Eindhoven, The Netherlands). Films of the samples were used as the working electrode, after being coated on cleaned fluorine doped tin oxide (FTO) coated glass using the doctor blade method and calcinated at 400 °C for 2 h. These working electrodes were prepared by grinding 100 mg of the photocatalysts and 50 mg polyethylene glycol with one drop of Triton, followed by addition of 200 µL of deionized water and a sufficient amount of ethanol. An Ag/AgCl electrode (3 M NaCl, +209 mV vs. NHE) and a platinum coil were used as the reference electrode and the counter electrode, respectively. Potassium nitrate aqueous solution (0.1 M) was used as the electrolyte. The impedance spectra were recorded in the range between the chosen potential from -1 V to $+1$ V at frequencies of 10, 100, and 1000 Hz with 20 mV amplitude vs. Ag/AgCl. The capacitance was plotted against V, and the flat band was calculated from the intercept of the plot. (i.e., a plot of C^{-2} vs. V, where C was the capacitance and V was the potential across the space charge layer).

4.4. Photocatalytic Measurements

4.4.1. Methylene Blue Degradation

The apparatus used for carrying out of the photocatalytic degradation reactions consisted of a double jacket cylindrical reactor with a 230 mL volume, which circulated with cold water to maintain the ambient reaction. A volume of 200 mL of aqueous solution of methylene blue (MB, 10 mg L^{-1}) and 200 mg of photocatalysts were used for each reaction experiment. A 300 W Xenon arc lamp (Müller Electronik-Optik, Moosinning, Germany) was used both as the UV/vis light source and as the vis light source by placing a UV cut-off filter ($\geq$410 nm) in the light path. The lamp was started 30 min before the degradation experiments to ensure maximum emission. Aliquots (1.5 mL) of the suspensions were collected at given time intervals (0, 2, 4, 6, 8, 10, 15, and 30 min), centrifuged to remove the solid, and analyzed immediately with the UV–Vis spectrophotometer.

4.4.2. Photocatalytic Hydrogen Formation

The photocatalytic H_2 generation experiments were conducted in quartz vials (capacity of 10 mL) under illumination with a 1000 W Xenon lamp (Hönle UV Technology, Gräfelfing, Germany; Sol 1200 solar). An amount of 6 mg of the photocatalyst was suspended in 6 mL aqueous methanol (10 vol% methanol). The suspension was purged with argon for 20 min to remove the air, and the quartz vial was sealed with a specially made rubber septum degassed for sampling. The amount of H_2 gas evolved during the photocatalytic reaction was quantified every two hours using a gas chromatograph (Shimadzu GC-8A, Shimadzu Deutschland GmbH, Duisburg, Germany) equipped with thermal conductivity detector (TCD) and 60/80 molecular sieve 5 Å column.

5. Conclusions

Ag/Ag_2O was found to enhance the rate of light-induced bleaching of aqueous methylene blue under both UV/vis and vis illumination, in comparison to the bleaching in homogeneous solution. Even in suspensions containing mixtures and composites of Ag/Ag_2O with TiO_2 (P25), with varying mass ratios of Ag/Ag_2O (20%, 50%, and 80%), the reaction rate was slightly increased under these illumination conditions. However, the bleaching rate of methylene blue was lower in the presence of the composites and mixtures than the rate measured for bare Ag/Ag_2O. It is therefore suggested that the bleaching of methylene blue is initiated by an interfacial electron transfer from the excited organic probe compound to Ag_2O. TiO_2 layers covering the Ag_2O seem to inhibit this electron transfer. Since Ag_2O can transfer an electron neither to dissolved molecular oxygen nor to a proton for thermodynamic reasons, it is assumed that Ag^+ is reduced to $Ag(0)$ in the processes investigated here. Results of XRD and XPS measurements support this assumption, and indicate that Ag/Ag_2O is not stable under the experimental conditions employed in this study. A stabilization of Ag_2O by metallic silver, as occasionally claimed, was not observed. Therefore, to address Ag/Ag_2O as a (photo)catalytically active material does not seem appropriate.

Supplementary Materials: The following are available online at http://www.mdpi.com/2073-4344/8/12/647/s1, Figure S1. EDX diagrams of all prepared photocatalysts. Figure S2. XRD patterns of TM 11 photocatalyst after one cycle of MB bleaching employing UV/vis light. Figure S3. XPS spectra of Ag 3d, O 1s and Ti 2p in the Ag/Ag_2O, TM and TC materials. Figure S4. Emission spectrum of the used Xenon lamp. Figure S5. UV/vis spectra of aqueous methylene blue solutions obtained during illumination with visible light in the presence of Ag/Ag_2O. Figure S6. Calibration curve for H_2 experiment.

Author Contributions: Experiments and analysis: S.A.; XPS measurements: J.K.; Redaction of manuscript: S.A., R.D. & R.B.; Review: R.D.; R.B. & N.O.B.; Scientific support: R.D.; Supervision: D.W.B. & M.E.A.

Funding: This research received no external funding.

Acknowledgments: The authors wish to thank Luis Granone for XRD measurements and Stephanie Melchers for SEM/EDAX images.

Conflicts of Interest: The authors declare no conflict of interest.

References

1. Bahnemann, D.W. Mechanisms of Organic Transformations on Semiconductor Particles. In *Photochemical Conversion and Storage of Solar Energy; Proceedings of the Eighth International Conference on Photochemical Conversion and Storage of Solar Energy, IPS-8, Palermo, Italy, 15–20 July 1990*; Pelizzetti, E., Schiavello, M., Eds.; Springer Science/Kluwer Academic Publishers: Dordrecht, The Netherlands; Boston, MA, USA, 1991; pp. 251–276.
2. Hoffmann, M.R.; Martin, S.T.; Choi, W.; Bahnemann, D.W. Environmental Applications of Semiconductor Photocatalysis. *Chem. Rev.* **1995**, *95*, 69–96. [CrossRef]
3. Bahnemann, D.W. Photocatalytic Detoxification of Polluted Waters. In *Environmental Photochemistry*; Boule, P., Ed.; Springer: Berlin/Heidelberg, Germany, 1999; pp. 285–351.
4. Fujishima, A.; Rao, T.N.; Tryk, D.A. Titanium Dioxide Photocatalysis. *J. Photochem. Photobiol. C Photochem. Rev.* **2000**, *1*, 1–21. [CrossRef]

5.   Reinosa, J.J.; Álvarez-Docio, C.M.; Ramírez, V.Z.; Fernández, J.F. Hierarchical Nano ZnO-Micro TiO$_2$ Composites: High UV Protection Yield Lowering Photodegradation in Sunscreens. *Ceram. Int.* **2018**, *44*, 2827–2834. [CrossRef]

6.   Wang, X.; Wu, H.-F.; Kuang, Q.; Huang, R.-B.; Xie, Z.-X.; Zheng, L.-S. Shape-Dependent Antibacterial Activities of Ag$_2$O Polyhedral Particles. *Langmuir* **2010**, *26*, 2774–2778. [CrossRef]

7.   Zou, J.; Xu, Y.; Hou, B.; Wu, D.; Sun, Y. Self-Assembly Ag$_2$O Nanoparticles into Nanowires with the Aid of Amino-Functionalized Silica Nanoparticles. *Powder Technol.* **2008**, *183*, 122–126. [CrossRef]

8.   Lalitha, K.; Reddy, J.K.; Phanikrishna Sharma, M.V.; Kumari, V.D.; Subrahmanyam, M. Continuous Hydrogen Production Activity over Finely Dispersed Ag$_2$O/TiO$_2$ Catalysts from Methanol:Water Mixtures under Solar Irradiation: A Structure-Activity Correlation. *Int. J. Hydrogen Energy* **2010**, *35*, 3991–4001. [CrossRef]

9.   Wang, X.; Li, S.; Yu, H.; Yu, J.; Liu, S. Ag$_2$O as a New Visible-Light Photocatalyst: Self-Stability and High Photocatalytic Activity. *Chem. A Eur. J.* **2011**, *17*, 7777–7780. [CrossRef]

10.   Wang, G.; Ma, X.; Huang, B.; Cheng, H.; Wang, Z.; Zhan, J.; Qin, X.; Zhang, X.; Dai, Y. Controlled Synthesis of Ag$_2$O Microcrystals with Facet-Dependent Photocatalytic Activities. *J. Mater. Chem.* **2012**, *22*, 21189. [CrossRef]

11.   Chen, F.; Liu, Z.; Liu, Y.; Fang, P.; Dai, Y. Enhanced Adsorption and Photocatalytic Degradation of High-Concentration Methylene Blue on Ag$_2$O-modified TiO$_2$-based Nanosheet. *Chem. Eng. J.* **2013**, *221*, 283–291. [CrossRef]

12.   Hu, X.; Hu, C.; Wang, R. Enhanced Solar Photodegradation of Toxic Pollutants by Long-Lived Electrons in Ag–Ag$_2$O Nanocomposites. *Appl. Catal. B Environ.* **2015**, *176–177*, 637–645. [CrossRef]

13.   Jiang, W.; Wang, X.; Wu, Z.; Yue, X.; Yuan, S.; Lu, H.; Liang, B. Silver Oxide as Superb and Stable Photocatalyst under Visible and Near-Infrared Light Irradiation and Its Photocatalytic Mechanism. *Ind. Eng. Chem. Res.* **2015**, *54*, 832–841. [CrossRef]

14.   Li, J.; Fang, W.; Yu, C.; Zhou, W.; Zhu, L.; Xie, Y. Ag-Based Semiconductor Photocatalysts in Environmental Purification. *Appl. Surf. Sci.* **2015**, *358*, 46–56. [CrossRef]

15.   Liu, C.; Cao, C.; Luo, X.; Luo, S. Ag-bridged Ag$_2$O Nanowire Network/TiO$_2$ Nanotube Array p–n Heterojunction as a Highly Efficient and Stable Visible Light Photocatalyst. *J. Hazard. Mater.* **2015**, *285*, 319–324. [CrossRef]

16.   Ren, H.-T.; Jia, S.-Y.; Zou, J.-J.; Wu, S.-H.; Han, X. A facile Preparation of Ag$_2$O/P25 Photocatalyst for Selective Reduction of Nitrate. *Appl. Catal. B Environ.* **2015**, *176–177*, 53–61. [CrossRef]

17.   Kumar, D.P.; Reddy, N.L.; Karthik, M.; Neppolian, B.; Madhavan, J.; Shankar, M.V. Solar Light Sensitized p-Ag$_2$O/n-TiO$_2$ Nanotubes Heterojunction Photocatalysts for Enhanced Hydrogen Production in Aqueous-Glycerol Solution. *Sol. Energy Mater. Sol. Cells* **2016**, *154*, 78–87. [CrossRef]

18.   Kumar, R.; El-Shishtawy, R.M.; Barakat, M.A. Synthesis and Characterization of Ag-Ag$_2$O/TiO$_2$@polypyrrole Heterojunction for Enhanced Photocatalytic Degradation of Methylene Blue. *Catalysts* **2016**, *6*, 76. [CrossRef]

19.   Wei, N.; Cui, H.; Song, Q.; Zhang, L.; Song, X.; Wang, K.; Zhang, Y.; Li, J.; Wen, J.; Tian, J. Ag$_2$O Nanoparticle/TiO$_2$ Nanobelt Heterostructures with Remarkable Photo-Response and Photocatalytic Properties under UV, Visible and Near-Infrared Irradiation. *Appl. Catal. B Environ.* **2016**, *198*, 83–90. [CrossRef]

20.   Yang, H.; Tian, J.; Li, T.; Cui, H. Synthesis of Novel Ag/Ag$_2$O Heterostructures with Solar Full Spectrum (UV, Visible and Near-Infrared) Light-Driven Photocatalytic Activity and Enhanced Photoelectrochemical Performance. *Catal. Commun.* **2016**, *87*, 82–85. [CrossRef]

21.   Zhang, J.; Liu, H.; Ma, Z. Flower-Like Ag$_2$O/Bi$_2$MoO$_6$ p-n Heterojunction with Enhanced Photocatalytic Activity under Visible Light Irradiation. *J. Molec. Catal. A Chem.* **2016**, *424*, 37–44. [CrossRef]

22.   Cui, Y.; Ma, Q.; Deng, X.; Meng, Q.; Cheng, X.; Xie, M.; Li, X.; Cheng, Q.; Liu, H. Fabrication of Ag-Ag$_2$O/reduced TiO$_2$ Nanophotocatalyst and its Enhanced Visible Light Driven Photocatalytic Performance for Degradation of Diclofenac Solution. *Appl. Catal. B Environ.* **2017**, *206*, 136–145. [CrossRef]

23.   Hu, X.; Liu, X.; Tian, J.; Li, Y.; Cui, H. Towards Full-Spectrum (UV, Visible, and Near-Infrared) Photocatalysis: Achieving an All-Solid-State Z-Scheme between Ag$_2$O and TiO$_2$ Using Reduced Graphene Oxide as the Electron Mediator. *Catal. Sci. Technol.* **2017**, *7*, 4193–4205. [CrossRef]

24.   Kaur, A.; Salunke, D.B.; Umar, A.; Mehta, S.K.; Sinha, A.S.K.; Kansal, S.K. Visible Light Driven Photocatalytic Degradation of Fluoroquinolone Levofloxacin Drug Using Ag$_2$O/TiO$_2$ Quantum Dots: A Mechanistic Study and Degradation Pathway. *New J. Chem.* **2017**, *41*, 12079–12090. [CrossRef]

25. Liu, B.; Mu, L.; Han, B.; Zhang, J.; Shi, H. Fabrication of TiO_2/Ag_2O Heterostructure with Enhanced Photocatalytic and Antibacterial Activities under Visible Light Irradiation. *Appl. Surf. Sci.* **2017**, *396*, 1596–1603. [CrossRef]

26. Mandari, K.K.; Kwak, B.S.; Police, A.K.R.; Kang, M. In-Situ Photo-Reduction of Silver Particles and their SPR Effect in Enhancing the Photocatalytic Water Splitting of Ag_2O/TiO_2 Photocatalysts under Solar Light Irradiation: A Case Study. *Mater. Res. Bull.* **2017**, *95*, 515–524. [CrossRef]

27. Olya, M.E.; Vafaee, M.; Jahangiri, M. Modeling of Acid Dye Decolorization by $TiO_2–Ag_2O$ Nano-Photocatalytic Process Using Response Surface Methodology. *J. Saudi Chem. Soc.* **2017**, *21*, 633–642. [CrossRef]

28. Prakoso, S.P.; Taufik, A.; Saleh, R. $Ag/Ag_2O/TiO_2$ Nanocomposites: Microwave-Assisted Synthesis, Characterization, and Photosonocatalytic Activities. *IOP Conf. Ser. Mater. Sci. Eng.* **2017**, *188*, 012029. [CrossRef]

29. Ren, H.-T.; Yang, Q. Fabrication of Ag_2O/TiO_2 with Enhanced Photocatalytic Performances for Dye Pollutants Degradation by a pH-Induced Method. *Appl. Surf. Sci.* **2017**, *396*, 530–538. [CrossRef]

30. Wang, C.; Cai, X.; Chen, Y.; Cheng, Z.; Luo, X.; Mo, S.; Jia, L.; Shu, R.; Lin, P.; Yang, Z.; et al. Efficient Hydrogen Production from Glycerol Photoreforming Over $Ag_2O–TiO_2$ Synthesized by a Sol–Gel Method. *Int. J. Hydrogen Energy* **2017**, *42*, 17063–17074. [CrossRef]

31. Zhao, Y.; Tao, C.; Xiao, G.; Su, H. Controlled Synthesis and Wastewater Treatment of Ag_2O/TiO_2 Modified Chitosan-Based Photocatalytic Film. *RSC Adv.* **2017**, *7*, 11211–11221. [CrossRef]

32. Deng, A.; Zhu, Y. Synthesis of $TiO_2/SiO_2/Ag/Ag_2O$ and $TiO_2/Ag/Ag_2O$ Nanocomposite Spheres with Photocatalytic Performance. *Res. Chem. Intermed.* **2018**, *44*, 4227–4243. [CrossRef]

33. Hao, C.; Wang, W.; Zhang, R.; Zou, B.; Shi, H. Enhanced Photoelectrochemical Water Splitting with $TiO_2@Ag_2O$ Nanowire Arrays via p-n Heterojunction Formation. *Sol. Energy Mater. Sol. Cells* **2018**, *174*, 132–139. [CrossRef]

34. Liu, R.; Wang, P.; Wang, X.; Yu, H.; Yu, J. UV- and Visible-Light Photocatalytic Activity of Simultaneously Deposited and Doped Ag/Ag(I)-TiO_2 Photocatalyst. *J. Phys. Chem. C* **2012**, *116*, 17721–17728. [CrossRef]

35. Kang, J.-G.; Sohn, Y. Interfacial Nature of Ag Nanoparticles Supported on TiO_2 Photocatalysts. *J. Mater. Sci.* **2012**, *47*, 824–832. [CrossRef]

36. Jiang, B.; Hou, Z.; Tian, C.; Zhou, W.; Zhang, X.; Wu, A.; Tian, G.; Pan, K.; Ren, Z.; Fu, H. A Facile and Green Synthesis Route Towards Two-Dimensional $TiO_2@Ag$ Heterojunction Structure with Enhanced Visible Light Photocatalytic Activity. *CrystEngComm* **2013**, *15*, 5821. [CrossRef]

37. Hammond, J.S.; Gaarenstroom, S.W.; Winograd, N. X-ray Photoelectron Spectroscopic Studies of Cadmium- and Silver-Oxygen Surfaces. *Anal. Chem.* **1975**, *47*, 2193–2199. [CrossRef]

38. Zaccheo, B.A.; Crooks, R.M. Stabilization of Alkaline Phosphatase with $Au@Ag_2O$ Nanoparticles. *Langmuir* **2011**, *27*, 11591–11596. [CrossRef]

39. Aeroxide, Aerodisp and Aeroperl. Titanium Dioxide as Photocatalyst. Technical Information. Available online: https://www.aerosil.com/sites/lists/RE/DocumentsSI/TI-1243-Titanium-Dioxide-as-Photocatalyst-EN.pdf (accessed on 16 October 2018).

40. Tjeng, L.H.; Meinders, M.B.J.; van Elp, J.; Ghijsen, J.; Sawatzky, G.A.; Johnson, R.L. Electronic Structure of Ag_2O. *Phys. Rev. B* **1990**, *41*, 3190–3199. [CrossRef]

41. Reinosa, J.J.; Leret, P.; Álvarez-Docio, C.M.; Del Campo, A.; Fernández, J.F. Enhancement of UV Absorption Behavior in $ZnO-TiO_2$ Composites. *Bol. Soc. Esp. Ceram. Vidr.* **2016**, *55*, 55–62. [CrossRef]

42. Jiang, Z.; Huang, S.; Quian, B. Semiconductor Properties of Ag_2O Film Formed on the Silver Electrode in 1 M NaOH Solution. *Electrochim. Acta* **1994**, *39*, 2465–2470. [CrossRef]

43. Vvedenskii, A.; Grushevskaya, S.; Kudryashov, D.; Ganzha, S. The Influence of the Conditions of the Anodic Formation and the Thickness of Ag(I) Oxide Nanofilm on its Semiconductor Properties. *J. Solid State Electrochem.* **2010**, *14*, 1401–1413. [CrossRef]

44. Mills, A.; Wang, J. Photobleaching of Methylene Blue Sensitised by TiO_2: An Ambiguous System? *J. Photochem. Photobiol. A Chem.* **1999**, *127*, 123–134. [CrossRef]

45. Herrmann, J.-M. Fundamentals and Misconceptions in Photocatalysis. *J. Photochem. Photobiol. A Chem.* **2010**, *216*, 85–93. [CrossRef]

46. Herrmann, J.-M. Photocatalysis Fundamentals Revisited to Avoid Several Misconceptions. *Appl. Catal. B Environ.* **2010**, *99*, 461–468. [CrossRef]

47. Bratsch, S.G. Standard Electrode Potentials and Temperature Coefficients in Water at 298.15 K. *J. Phys. Chem. Ref. Data* **1989**, *18*, 1–21. [CrossRef]
48. Arimi, A.; Megatif, L.; Granone, L.I.; Dillert, R.; Bahnemann, D.W. Visible-Light Photocatalytic Activity of Zinc Ferrites. *J. Photochem. Photobiol. A Chem.* **2018**, *366*, 118–126. [CrossRef]
49. Xu, Y.; Schoonen, M.A.A. The Absolute Energy Positions of Conduction and Valence Bands of Selected Semiconducting Minerals. *Am. Mineral.* **2000**, *85*, 543–556. [CrossRef]
50. Harifi, T.; Montazer, M.; Dillert, R.; Bahnemann, D.W. $TiO_2/Fe_3O_4/Ag$ Nanophotocatalysts in Solar Fuel Production: New Approach to Using a Flexible Lightweight Sustainable Textile Fabric. *J. Clean. Prod.* **2018**, *196*, 688–697. [CrossRef]
51. Hamid, S.; Dillert, R.; Bahnemann, D.W. Photocatalytic Reforming of Aqueous Acetic Acid into Molecular Hydrogen and Hydrocarbons over Co-catalyst-Loaded TiO_2: Shifting the Product Distribution. *J. Phys. Chem. C* **2018**, *122*, 12792–12809. [CrossRef]

Article

TiO$_2$-SiO$_2$-PMMA Terpolymer Floating Device for the Photocatalytic Remediation of Water and Gas Phase Pollutants

Valentina Sabatini [1,2,3,*], **Luca Rimoldi** [1,2,*], **Laura Tripaldi** [1,2], **Daniela Meroni** [1,2,*], **Hermes Farina** [1,2,3], **Marco Aldo Ortenzi** [1,2,3] and **Silvia Ardizzone** [1,2,3]

[1] Department of Chemistry, Università degli Studi di Milano, Via Golgi 19, 20133 Milano, Italy; laura.tripaldi@studenti.unimi.it (L.T.); hermes.farina@unimi.it (H.F.); marco.ortenzi@unimi.it (M.A.O.); silvia.ardizzone@unimi.it (S.A.)

[2] Consorzio Interuniversitario Nazionale per la Scienza e la Tecnologia dei Materiali (INSTM), Via Giusti 9, 50121 Firenze, Italy

[3] CRC Materials & Polymers (LaMPo), Università degli Studi di Milano, Via Golgi 19, 20133 Milano, Italy

* Correspondence: valentina.sabatini@unimi.it (V.S.); luca.rimoldi@unimi.it (L.R.); daniela.meroni@unimi.it (D.M.); Tel.: +39-02-5031-4212 (D.M.)

Received: 23 October 2018; Accepted: 16 November 2018; Published: 21 November 2018

Abstract: Floating photocatalytic devices are highly sought-after as they represent good candidates for practical application in pollutant remediation of large water basins. Here, we present a multilayer floating device for the photocatalytic remediation of contaminants present in water as well as of volatile species close to the water surface. The device was prepared on a novel tailored ter-polymer substrate based on methylmethacrylate, α-methylstyrene and perfluoroctyl methacrylate. The ad hoc synthesized support presents optimal characteristics in terms of buoyancy, transparency, gas permeability, mechanical, UV and thermal stability. The adhesion of the TiO$_2$ top layer was favoured by the adopted casting procedure, followed by a corona pre-treatment and by the deposition of an intermediate SiO$_2$ layer, the latter aimed also at protecting the polymer support from photocatalytic oxidation. The device was characterized by contact angle measurement, UV-vis transmittance and scanning electron microscopy. The final device was tested for the photocatalytic degradation of an emerging water pollutant as well as of vapors of a model volatile organic compound. Relevant activity was observed also under simulated solar irradiation and the device showed good stability and recyclability, prospecting its use for the photocatalytic degradation of pollutants in large water basins.

Keywords: composite; polymethylmethacrylate; photocatalytic oxidation; titanium dioxide; tetracycline; ethanol

---

1. Introduction

Polymer/TiO$_2$ micro and nano-composites have raised a great deal of interest in recent years due to their broad range of applications, including the enhancement of thermal, dielectric and mechanical properties of polymers [1–4], water purification [5,6], biomaterials [7] and anti-bacterial surfaces [8], energy conversion and storage such as in solar and fuel cells, lithium batteries and electrochemical capacitors [9–12].

In the field of surface water and wastewater treatment by photocatalytic oxidation [13,14], TiO$_2$/polymer composites benefit from the high durability, light-weight, controlled surface properties and ease-of-processing of the polymeric component [15]. One major challenge in this field is the development of photoactive and durable floating devices for the remediation of large, polluted areas, such as water basins [16]. With respect to powder photocatalysts, floating systems enable an easy retrieval of the photocatalyst as well as a more efficient light usage, since light, especially UV, attenuates rapidly in water (less than 1% of the UV light or ca. 20% of visible light irradiated on the water surface reaches a depth of 0.5 m [17]. The use of inorganic coatings and polymer substrates aims at filling this gap by combining the unique photocatalytic properties deriving from TiO_2 and the excellent polymer processability for an easy scalable technology. Key to the success of such composite devices is the engineering of the fabrication materials and of the device design. Tu et al. described the development of a ternary system made of polypropylene, TiO_2 and activated carbon for the adsorption and degradation of phenol [18]. Sponge-like polyurethane composite foams were adopted by Ni et al. for surface water remediation [19]. Han et al. coated commercial polypropylene with different TiO_2 layers for the degradation of methyl orange [20]. However, an unresolved issue for nano/microcomposites is represented by their poor photochemical, thermal and mechanical stability. As a matter of fact, the mechanical stability of composite devices is limited by the inherent low compatibility between the polymer and the oxide layers [21]. Moreover, the device stability under prolonged irradiation depends on the photostability of the polymer component as well as on the possible occurrence of polymer degradation due to the TiO_2 photocatalytic activity [22]. Overall, a relatively fast loss of photocatalytic performance is often reported [8,18].

In order to increase the mechanical and photochemical stability of the composite, the properties of the device are to be carefully tailored. In particular, the wetting properties of the polymer surface have to be modified to promote the adhesion of the oxide film by increasing the polymer surface hydrophilicity [23]. However, the bottom side of the floating device should present good hydrophobic properties in order to display stable buoyancy. In this respect, the addition of fluorinated chains to enhance hydrophobicity can reduce the photostability of the polymer. The tailoring of the wetting features of the polymer is thus a critical issue for the creation of stable polymer/oxide composites.

Most of the literature uses commercial polymers as substrates for the oxide deposition due to their flexibility, availability and lower cost [21]. However, the poor thermal stability of common commercial polymers (e.g., polyesters and polyacrylates) severely restricts the range of available stabilization treatments that can be used to improve the oxide layer adhesion. The UV resistance and mechanical properties are also critical issues when commercial polymers like polypropylene and polyesters are employed. Moreover, a good transparency in the UV-vis range and high oxygen permeability are also required for the application in open water basins, limiting the applicability of polyurethanes and polyacrylonitriles, respectively.

To solve these issues, in the present work a novel tailored ter-polymer based on methylmethacrylate (MMA), α-methylstyrene and perfluoroctyl methacrylate (POMA) co-monomers was synthesized to be adopted as substrate for the photoactive layer, to achieve good buoyancy, transparency and high mechanical, UV and thermal stability. The TiO_2 layer adhesion was ensured via a surface pre-treatment of the polymer aimed at enhancing its hydrophilicity as well as via addition of an intermediate SiO_2 layer, which also protects the polymer from the TiO_2 generated radicals. The device showed good stability under prolonged irradiation in working conditions. The photocatalytic performance was tested towards the degradation of volatile organic compounds (VOCs) in the gas phase and of an emerging pollutant in water, showing good recyclability.

2. Results and Discussion

2.1. Synthesis and Characterization of MMA_α-Methylstyrene_POMA ter-Polymer

Poly(methyl methacrylate) is the lightweight and shatter-resistant alternative to glass par excellence due to its optimal transparency. It is widely used for outdoor applications thanks to its UV resistance and excellent mechanical properties. However, this polymer suffers from a relatively low thermal stability (glass transition temperature, T_g, 105.0 °C) that makes it unsuitable for the present application.

In this work, a new type of methacrylic ter-polymer was prepared via free radical polymerization among MMA, α-methylstyrene and POMA, according to the reaction scheme represented in Figure 1. The ^{1}H NMR spectra of the synthesized fluorinated comonomer and of the ter-polymer are reported in Figures S1 and S2, respectively. The addition of α-methylstyrene in a molar ratio of 20% with respect to MMA significantly enhanced the thermal properties of the material, leading to a T_g of 123.9 °C (Table 1, 2nd column) and furthermore, the corresponding polymer foils are characterized by mechanical properties (Table 1, 7th–9th column) comparable with industrial films of polyacrylates [24] and excellent oxygen permeability (oxygen transmission rate, OTR: 314 cm^3 m^{-2} d^{-1}) [25] with high homogeneity of the film casted (Figure S3).

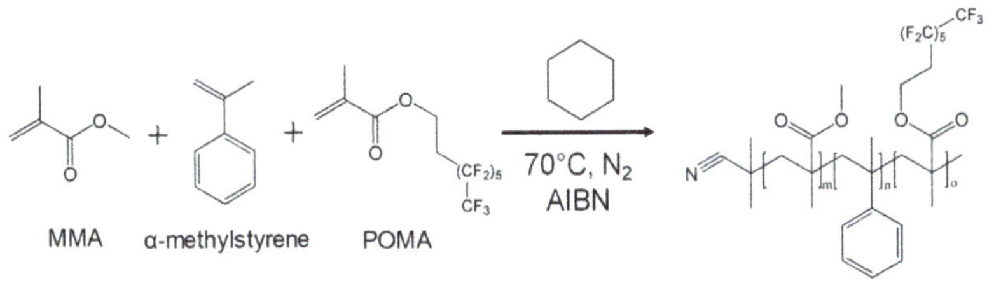

Figure 1. Synthetic route for MMA_α-methylstyrene_POMA ter-polymer.

Table 1. Main physicochemical (glass transition temperature, T_g; number average molecular weight, $\overline{Mn}$; molecular weight distribution, D; water contact angles, wCA) and mechanical properties (elastic Young's tensile modulus, tensile strength, elongation at break) of the MMA_α-methylstyrene_POMA ter-polymer before and after the UV stability test.

	T_g (°C)	$\overline{Mn}$(Da)	D	wCA Air-Side (°)	wCA Mould-Side (°)	Elastic Modulus (GPa)	Tensile Strength (Mpa)	Elongation at Break (%)
before UV test	123.9	28100	2.1	67 ± 2	114 ± 3	3.1	70	2.3
after UV test	124.0	28000	2.1	66 ± 3	116 ± 2	2.9	68	1.9

The wetting features of the polymer films were tailored by the addition of a new fluorinated methacrylic monomer, POMA, which was synthesized via esterification reaction between methacryloyl chloride and 1H,1H,2H,2H-Perfluoro-1-octanol. The fluorinated monomer, together with the adopted solvent casting deposition technique, allowed us to achieve different wetting features on the two sides of the polymer film. As described in previous works for other polymers [26,27], during the drying process the apolar fluorinated chains of the polymer tend to reorganize orienting towards the hydrophobic mould surface due to the higher affinity with polytetrafluoroethylene (PTFE) with respect to the solvent. This gives rise to a polymer film characterized by a hydrophobic side (PTFE-side), and a hydrophilic one (air-side), as appreciable from water contact angle measurements (Table 1, 5th and 6th columns). Surface free energy could not be reliably determined by methods based on contact angle measurements [28] as the polymer was dissolved by some of the most commonly employed solvents used for this purpose (e.g., CH_2I_2).

The POMA content was selected in order to impart the desired hydrophobic properties while preserving the UV stability of the polymer, as determined by stability tests upon prolonged UV irradiation (100 h). FT-IR spectra collected at the air and the PTFE sides of the polymer film before and after UV exposure (Figure S4) show the same features: Peaks in the ~ 3100–2800 cm^{-1} range, which can be attributed to stretching modes of C–H aliphatic bonds [29], the stretching of carbonyl ester groups (C=O) between ~ 1750 cm^{-1} and ~ 1600 cm^{-1} [29], and the characteristic absorption band for the symmetric stretching vibration of C–O conjugated to carbonyl ester groups, which appear between ~ 1350 cm^{-1} and ~ 1100 cm^{-1} [29]. The shape of these peaks does not change upon the UV exposure test, testifying the preservation of the polymeric bonds in correspondence of the aliphatic and carbonyl groups [30,31]. The presence of –CH_3 groups in alpha position to carbonyl groups, which are deriving from the methacrylic co-monomers, inhibits the photo-chemical degradation of the polymer [32]. Moreover, the main physicochemical properties of the polymer do not change upon UV irradiation (Table 1), also in terms of wetting features of the two film sides. Thus, upon UV irradiation the ter-polymer maintains not only its overall structure but also the organization of the fluorinated chains, responsible for the wetting properties of the polymer foils. In their turn, the thermal and mechanical properties remained totally unchanged upon UV stability tests (Table 1).

2.2. Device Preparation and Characterization

Figure 2 shows top SEM images of the different components of the photocatalytic device, i.e., of the device layers in each stage of its assembly. The relative water contact angles are also reported in inset. The air side and the mould side of the as-deposited polymeric foil present notable differences both in terms of wetting and morphological features (Figure 2a,b). While the air side (Figure 2a) appears highly homogeneous and smooth, the mould side (Figure 2b) is characterized by micrometric roughness due to Teflon mould adopted for the deposition. The two sides of the as-deposited polymer foil show different wettability thanks to the orientation of the fluorinated chains of POMA towards the Teflon mould side. The corona treatment increases the hydrophilicity of the polymer surface (reaching a value of $44 \pm 1°$ right after the treatment) surface and imparts a morphological change in the polymer foils (Figure 2c). Micrometric cavities can be detected, in agreement with the literature [33], which are excavated by the energy particles bombardment. These micro pits concur to improve the adhesion of the inorganic layers due to a larger potential bond area [33]. The silica layer spray-deposited onto the polymer surface is crack-free and homogeneously covers the whole foil surface (Figure 2d) and further increases the hydrophilicity of the surface ($32 \pm 6°$). Upon deposition of the TiO_2 layer, the presence of titania particles leads to an appreciable surface roughness (Figure 2e) and to a slight increase of the water contact angle ($63 \pm 2°$) for the unirradiated sample. Even after prolonged irradiation in water, the morphology of the film remains comparable with the one of a pristine sample (Figure 2f).

Cross-sectional SEM images of the bare polymer foil and of the final composite device are reported in Figure 3. Figure 3a shows that the polymer foil has a porous morphology, induced by the choice of the casting solvent, which is at the basis of the lightness of the foil and enhances its floating capabilities. The thickness of the foil was measured to be ca. 200 µm. Figure 3b shows instead the thickness and the morphology of the inorganic layers (SiO_2 and TiO_2) deposited onto the polymer substrate in the complete device. A micrometric silica layer with a very compact morphology favors the protection the organic substrate from the photocatalytically produced radical species. The top titania layer is instead much thinner, in agreement with previous reports [34]; moreover, the active TiO_2 layer displays a rough and porous morphology, as also shown by the top view micrographs, which can be beneficial for the photocatalytic application by enhancing the actual surface area extension and by increasing photon absorption [35].

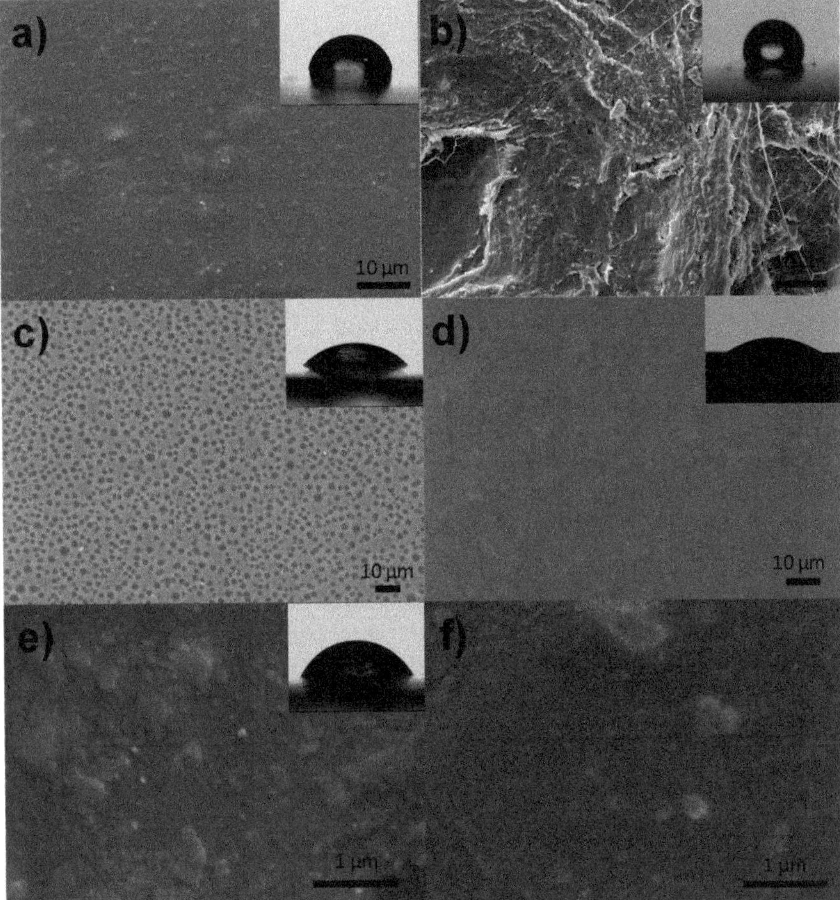

Figure 2. SEM images of the air side (**a**) and the mould side (**b**) of the polymer foil, of the air side after corona treatment (**c**), of the silica layer (**d**) and of the final device before (**e**) and after (**f**) prolonged irradiation under working conditions, together with the relative water contact angles (in insets).

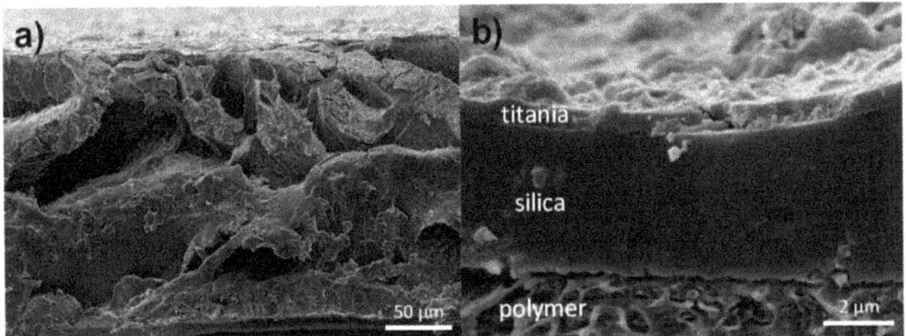

Figure 3. Cross-sectional SEM images of the deposited polymeric foil (**a**) and of the final device (**b**).

XRD patterns of the complete device collected on a Philips PW 3710 Bragg-Brentano goniometer proved unsuccessful in determining the structural composition of the TiO_2 top layer, due to the limited thickness of the film. However, the phase composition of the top layer can be inferred from its synthetic procedure: The addition of Hombikat UV 100 ensures the presence of crystalline anatase particles (see Section 3.2.4). Besides the crystalline commercial particles, the TiO_2 film is expected to show limited crystallinity due to the lack of high temperature post-treatments needed to promote the formation of crystalline anatase [34].

The transmittance spectra of the device before and after the deposition of the TiO_2 layer are reported in Figure 4a together with the spectrum of the bare polymer, for the sake of comparison. The bare polymer foil presents high transparency in the whole visible range (constant transmittance of ca. 80% between 400 and 800 nm). This is a highly desirable feature for application in natural settings as water basins. In the UV region, the transmittance shows a good degree of transparency, presenting a transmittance >60% up to 285 nm, enabling an efficient use of sunlight by the TiO_2 layer in the full device even when the device is capsized. The deposition of the silica layer leads to a further slight enhancement of transmittance in the visible range, owing to the antireflective properties of the silica film [36], while in the UV region a minor decrease of transmittance is observed, due to the characteristic light absorption of SiO_2, remaining however >55% up to 285 nm. The complete photocatalytic device still presents a good transparency (transmittance ca. 70%) in the whole visible region, as also revealed by the photograph reported in Figure 4b. In the UV region, the characteristic absorption of TiO_2 is appreciable, due to the top titania layer. It should be noted that up to 340 nm the device shows substantial transparency (transmittance > 50%), which guarantees the photoactivation of the titania layer also with back illumination. The device can thus be used for both the degradation of water pollutants and gaseous organic compounds present in the atmosphere. In fact, as appreciable from Figure 4c,d, the device revealed high floating capabilities, which remained stable in time, owing to the lightness of the polymer, together with the enhanced hydrophobicity of the bottom side provided by the use of the fluorinated comonomer.

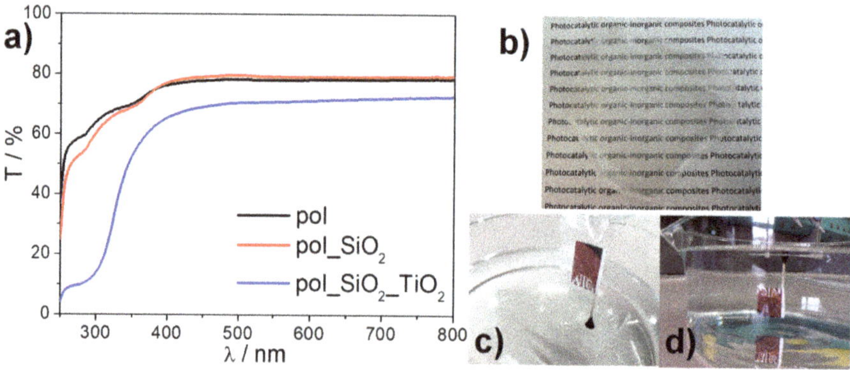

Figure 4. UV-vis transmittance spectra (**a**) and photographs proving the transparency (**b**) and the buoyancy of the composite device on both the mould (**c**) and air sides (**d**); a flag was attached on top of the air side of the transparent device to make it more easily detectable.

2.3. Photocatalytic Activity

The photocatalytic activity of the device was firstly tested in the gas phase degradation of VOCs. In this respect, ethanol was selected as model molecule on the grounds of a previous work [37]. The present device proved to be effective in the degradation of ethanol vapor, achieving complete disappearance of the target molecule after 4 h of irradiation (Figure S5) despite the high pollutant concentration (200 ppm) and the low irradiated TiO_2 amount (ca. 9 mg). Moreover, the main intermediate (acetaldehyde) was almost entirely removed after 6 h, leading to CO_2 and water as final products (Figure S5). Photolysis tests performed in the same conditions but without the device, showed an ethanol disappearance rate of 1.2×10^{-3} min^{-1} and no appreciable formation of intermediates/CO_2.

The present result is thus comparable with previous reports of ethanol gas phase degradation using a similar amount of P25 free powder [38]. However, with respect to previous reports showing substantial deactivation just upon three recycle tests [38], in the present case the device maintained its photocatalytic activity after three consecutive photocatalytic runs (Figure 5a), as appreciable from the pollutant pseudo-first order disappearance rates. In all cases, the mineralization (complete oxidation to CO_2) was larger than 75%, proving the stability and the reusability of the prepared photocatalytic device.

The photocatalytic activity of the device was also evaluated towards the degradation of tetracycline. Tetracyclines are the best-selling antibiotics [39] and have been classified as emerging pollutants [40,41]. Due to their large usage in both humans and animals, tetracyclines are among the most frequently detected micropollutants both in wastewaters [42] and in large water basins as the lakes of Northern Italy [43], leading to increased levels of tetracycline-resistant bacteria [44]. Under simulated solar light with back irradiation, the floating device achieved a tetracycline degradation of 50% after 14 h of irradiation, without any decrease of the performance during the reaction time (Figure 5b), suggesting the possibility to completely degrade the target molecule by prolonging the irradiation. A pseudo-first order kinetics of 4.7 ± 0.1 h^{-1} was observed (Figure 5b). Tests performed in the same conditions in the absence of the device showed a photolysis rate of 1.3 ± 0.1 h^{-1}, in agreement with previous reports [45].

Only few studies reported the photocatalytic degradation of pollutants by floating devices under solar light [21]. Although comparisons are difficult to draw due to the different experimental conditions, the presently reported floating device shows promising performance with respect to previous reports as several literature studies obtain similar degradation rates using much larger TiO_2 actual contents [20,46,47].

The stability of the device after prolonged UV irradiation in working conditions is also testified by cross-sectional SEM images (Figure S6). The rough morphology of the top TiO_2 layer is well appreciable as well as the compact silica layer. It should be noted that the overall thickness of the oxide layer can vary due to the adopted deposition procedure (spray coating).

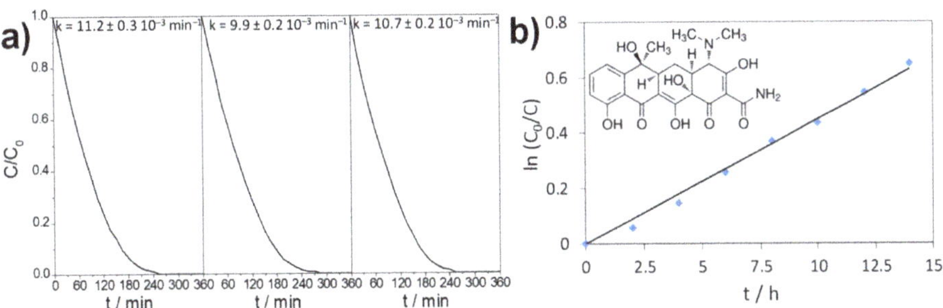

Figure 5. Photocatalytic tests results: (**a**) ethanol disappearance under UV irradiation and the relative rate constant in the recycle tests; (**b**) Determination of the rate constant of tetracycline disappearance in simulated solar photocatalytic tests: logarithmic conversion plot as a function of irradiation time ($R^2 = 0.995$).

3. Materials and Methods

3.1. Materials

Methyl methacrylate (MMA, 99%), methacryloyl chloride (97%), 1H,1H,2H,2H-Perfluoro-1-octanol (97%), α,α'-Azoisobutyronitrile (AIBN, 99%), triethyl amine (TEA, $\geq$99.9%), sodium bicarbonate ($NaHCO_3$, $\geq$99.7%), sodium sulfate (Na_2SO_4, $\geq$99.99%), cyclohexane (99.5% anhydrous), methanol (99.8% anhydrous), α-methylstyrene ($\geq$99% anhydrous), distilled water Chromasolv® ($\geq$99.9%), methylene chloride (CH_2Cl_2, $\geq$99.8% anhydrous), hydrochloric acid (HCl, 37%), tetrahydrofuran (THF, $\geq$99.8% anhydrous) and chloroform-d (CDCl$_3$, 99.96 atom % D) were acquired from Sigma-Aldrich (St. Louis, MI, USA)and used without further purification. Doubly distilled water passed through a Milli-Q apparatus (Sigma-Aldrich, St. Louis, MI, USA) was adopted to prepare solutions and suspensions.

3.2. Preparation Procedures

3.2.1. Synthesis of POMA

The reaction was carried out under inert atmosphere using a 100 mL three-neck round bottom flask equipped with a nitrogen inlet adapter, an internal thermometer adapter, an overhead magnetic stirrer and a reflux condenser. Firstly, 20 mL of methylene chloride, methacryloyl chloride (2.8 g) and 1H,1H,2H,2H-Perfluoro-1-octanol (9.8 g) were mixed. Then, 2.9 g of TEA was added to neutralize HCl formed during the esterification reaction. The solution was carefully cooled down to 0 °C and then stirred for 16 h. Afterwards, it was gradually brought back to room temperature. The solution was washed several times with an aqueous solution of HCl at 5% w/w and then with NaHCO$_3$ 5% w/w to remove traces of TEA and HCl, respectively. Water traces were removed with Na$_2$SO$_4$ and the salt was removed via filtration. The resulting solution was dried under vacuum (ca. 4 mbar) at 40 °C for 1 h (96% yield). The structure of POMA was confirmed via ^{1}H NMR spectroscopy (Figure S1).

3.2.2. Synthesis of MMA_α-Methylstyrene_POMA Ter-Polymer

The reaction was performed under inert atmosphere in a 100 mL two-necked round bottom flask equipped with a nitrogen inlet adapter, a reflux condenser and an overhead magnetic stirrer. 40 mL of cyclohexane was mixed with MMA (14.6 g) and α-methylstyrene (4.3 g) POMA (0.8 g) and AIBN (0.3 g), used as free radical initiator. The resulting molar ratios were 8:2 MMA: α-methylstyrene, 1% mol·mol^{-1} POMA: (MMA and α-methylstyrene), and 1% mol·mol^{-1} AIBN: (MMA, α-methylstyrene and POMA). The solution heated in an oil bath at 70 °C for 24 h, then gradually cooled down to room temperature. A white solid was precipitated by addition of a large excess of methanol. After recovering the solid by filtration, the polymer was washed with methanol for several days under stirring to remove unreacted methacrylic monomers. After washing, the polymer was dried in a vacuum oven (ca. 4 mbar) at 40 °C for 48 h. The structure of the product was confirmed via ^{1}H NMR spectroscopy (Figure S2).

3.2.3. Silica and Titania Sol Preparation

The silica sol was prepared by a modification of the procedure reported by Soliveri et al. [34]. Firstly, 10 g of tetraethyl orthosilicate (TEOS, Sigma-Aldrich, St. Louis, MI, USA) was added to a solution of 4.5 g of 0.1 M HCl and 25 g of ethanol. The mixture was stirred at room temperature for 120 min and then refluxed at 60 °C for 60 min. After cooling down, a solution prepared by dissolving 2 g of Lutensol ON 70 (BASF, Ludwigshafen, Germany) in 25 g of ethanol (>99.8%, Sigma-Aldrich) was added to the reaction mixture and stirred for 1 h. A stable, transparent sol was obtained.

For the preparation of the titania sol [48], 28.4 g of titanium isopropoxide (97%, Sigma-Aldrich) was dissolved in 79 g of ethanol and 0.9 mL of HCl 37% was added while stirring. Then, a solution of 0.47 g of Lutensol ON 70 in 79 g of ethanol was added. The resulting stable and transparent sol was stirred at 1 h at room temperature.

3.2.4. Device Preparation

Polymer films were prepared via solution casting: 1.8 g of polymer was dissolved in 10 mL of CH$_2$Cl$_2$ and the resulting solution was cast onto a PTFE mould (7 cm in diameter). The films were dried overnight at 25 °C and atmospheric pressure. The air side of the polymer film was corona treated (Aslan Machinery, Germantown, MD, USA; voltage: 0.43 kW; exposition time: 5 min) in order to promote the adhesion of the oxide layers. Then, the SiO$_2$ layer was deposited by spray coating the silica sol (nozzle diameter: 0.5 mm, target-nozzle distance ca. 30 cm; spraying time ca. 1 s, 3 layers). After drying at room temperature, the TiO_2 layer was deposited by spray coating a suspension of Hombikat UV100 (Sachtleben Chemie GmbH, Duisburg, Germany) (pure anatase, average crystallite size 10 nm, specific surface area ca. 350 m^2 g^{-1}) in the prepared titania sol (2 mg of Hombikat UV100 + 1 mL of ethanol + 0.3 mL titania sol), in the same conditions adopted for the silica layer. The prepared device was immediately dried in oven at 90 °C for 20 h. A further treatment was performed by immersing the device in water for 10 min at 70 °C and finally in 10^{-4} M HNO_3 for 20 min at 70 °C, according to a previously reported procedure [23]. Finally, the device was dried in an oven and irradiated for 3 h under UV light.

3.3. Characterization Methods

Nuclear magnetic resonance spectroscopy (NMR). ^{1}H NMR spectra were collected at 25 °C with a Bruker 400 MHz spectrometer (Bruker, Billerica, MA, USA). The samples for the analyses were prepared dissolving 10–15 mg of POMA/ter-polymer in 1 mL of $CDCl_3$.

Size Exclusion Chromatography (SEC). The polymer molecular weight before and after accelerated ageing was investigated by SEC using a Waters 1515 Isocratic HPLC pump (Waters, Milford, MA, USA) and a four Phenomenex Phenogel (5×10^{-3} Å–5×10^{-4} Å–5×10^{-5} Å–5×500 Å) column set with a RI detector (Waters 2487, Milford, MA, USA) using a flow rate of 1 mL/min and 40 µL as injection volume. Samples were prepared dissolving 40 mg of polymer in 1 mL of anhydrous THF; before the analysis, the solution was filtered with 0.45 µm filters. Molecular weight data were expressed in polystyrene (PS) equivalents. The calibration was built using monodispersed PS standards having the following nominal peak molecular weight (Mp) and molecular weight distribution (D): Mp = 1,600,000 Da (D ≤ 1.13), Mp = 1,150,000 Da (D ≤ 1.09), Mp = 900,000 Da (D ≤ 1.06), Mp = 400,000 Da (D ≤ 1.06), Mp = 200,000 Da (D ≤ 1.05), Mp =90,000 Da (D ≤ 1.04), Mp = 50,400 Da (D = 1.03), Mp = 30,000 Da (D = 1.06), Mp = 17,800 Da (D = 1.03), Mp = 9730 Da (D = 1.03), Mp = 5460 Da (D = 1.03), Mp = 2032 Da (D = 1.06), Mp = 1241 Da (D = 1.07), Mp = 906 Da (D = 1.12), Mp = 478 Da (D = 1.22); Ethyl benzene (molecular weight = 106 g/mol). For all analyses, 1,2-dichlorobenzene was used as internal reference. The molecular weights of the samples obtained after the UV exposure test were also determined.

Differential Scanning Calorimetry (DSC). The polymer glass transition temperature (T_g) was measured by DSC analyses on a Mettler Toledo DSC1 (Zurich, Switzerland), using a 10 °C/min heating rate and under nitrogen atmosphere. Before measurement, samples were heated at 90 °C to eliminate residual internal stresses from the synthesis.

Fourier Transform Infrared Spectroscopy (FT-IR). FT-IR spectra were collected on a Spectrum 100 spectrophotometer (Perkin Elmer, Waltham, MA, USA) working in attenuated total reflection (ATR) mode. A single-bounce diamond crystal was used with an incidence angle of 45°.

Water Contact Angle (WCA) analyses. Water contact angle measurements were carried out using a Krüss EasyDrop (Krüss, Hamburg, Germany), on at least ten independent measurements using 5 µL water droplets.

UV-vis spectroscopy. UV-vis transmittance spectra were acquired in the 250–800 nm range by using a Shimadzu UV2600 spectrophotometer (Tokyo, Japan).

Scanning electron microscopy (SEM). Top and cross-sectional SEM images were collected on a Zeiss LEO-1430 (Oberkochern, Germany). Samples were sputter coated with Au before measurements.

Mechanical properties. The polymer sample preparation and the determination of their elastic modulus, tensile strength and elongation at break were performed in agreement with the ISO 527-1/2 standard method using a Kistler 9273 dynamometer (Winterthur, Switzerland).

Permeability test. The permeability test was conducted in oxygen atmosphere, at 23 °C and with a relative humidity of 40%, in accordance to the ASTM D3985 standard test method for oxygen gas transmission rate through plastic films and sheeting using a colorimetric sensor.

UV stability test. An accelerated aging test was conducted according to the UNI10925:2001 standard method to evaluate the stability under UV radiation of the prepared polymer foils. The test was conducted for 100 h (T = 25 °C and p = 1 atm), with a Jelosil HG500 lamp (Milan, Italy; 45 mW cm^{-2} in the 280–400 nm range).

3.4. Photocatalytic Activity

The photocatalytic activity of the device was tested under both UV (Jelosil HG500, effective power density: 17 mW cm^{-2} in the 280–400 nm) and simulated solar irradiation (UltraVitalux lamp, Osram, Munich, Germany, effective power density: 4.5 mW cm^{-2} in the 280–400 nm range and 14 mW cm^{-2} in the 400–800 nm range).

Tests were carried out both in the gas phase and in water. The gas phase degradation of ethanol was carried out using a previously reported setup [37]. An active surface area of 38 cm^2 and an initial pollutant concentration of 198 ppm were employed; a gas chromatographic system was adopted for monitoring the pollutant disappearance, and the formation of acetaldehyde (main reaction intermediate) and CO_2 (complete degradation product). Three consecutive photocatalytic tests were performed in the same conditions to test the stability and the reusability of the device. The degradation of a tetracycline hydrochloride (TC) in water was conducted in an open reactor, with a total active surface of 60 cm^2 and an initial pollutant concentration of 12 ppm (V = 300 mL). The reaction was carried out at spontaneous pH and oxygen saturation was maintained via air bubbling. Before irradiation, the device was kept in the dark for 30 min in order to achieve adsorption-desorption equilibrium. The molecule disappearance was monitored by UV-vis spectrophotometry, by recording the intensity of the characteristic absorption peak of tetracycline at 357 nm, in agreement with previous literature reports [41,49,50].

4. Conclusions

In this work, we presented a floating photocatalytic device based on TiO_2 photocatalyst immobilized over an ad hoc synthesized ter-polymer. The developed methylmethacrylate, α-methylstyrene and perfluoroctyl methacrylate ter-polymer is characterized by a highly porous morphology and inherent hydrophobicity, which enable a stable buoyancy. Furthermore, the support, displaying the characteristic optical transparency and oxygen permeability of PMMA, was engineered to possess enhanced thermal stability, mechanical resistance and photostability, in order to promote the device durability. The compatibility of the inorganic top coating was enhanced by a series of strategies: (1) an ad hoc polymer casting method leading to reorganization of the hydrophobic chains and dual wetting features of the opposite film sides; (2) the corona treatment of the polymer surface aimed at increasing hydrophilicity and creating surface pitting to bolster adhesion of the inorganic coating; (3) the deposition of an intermediate SiO_2 layer, which improves the adhesion of the TiO_2 top layer and protects the polymer support from the radical species generated by photocatalytic oxidation. The adopted TiO_2 layer contains commercial nanoparticles with high photocatalytic activity bound together by a titania sol promoting adhesion. The final device showed promising results in photocatalytic degradation tests of both water and gas phase pollutants, also in recycle tests. Tests were carried out under both UV and simulated solar irradiation, with either front or back irradiation, confirming the ability of the device to work also when capsized. Future work will further optimize the TiO_2 amount to boost the photocatalytic activity of the device by both increasing the nanoparticles content in the top layer and using fibers and sponge architectures as an alternative to polymer films.

Supplementary Materials: The following are available online at http://www.mdpi.com/2073-4344/8/11/568/s1, Figure S1. ^{1}H NMR spectrum of POMA monomer, Figure Figure S2. ^{1}H NMR spectrum of MMA_ α-methylstyrene_POMA polymer, Figure S3. O_2 transfer with respect to time of the synthesized MMA_α-methylstyrene_POMA, Figure S4. FT-IR spectra of MMA_α-methylstyrene_POMA collected before and after the UV stability test at the air (I) and PTFE (II) side, Figure S5. Ethanol disappearance and acetaldehyde and CO_2 formation during the photocatalytic test under UV irradiation, Figure S6. Cross-sectional SEM images after UV irradiation for over 15 hours in working conditions.

Author Contributions: Conceptualization, V.S., D.M., S.A.; methodology, L.T., V.S., L.R.; formal analysis and data curation, L.R., V.S., D.M.; writing—original draft preparation, D.M., V.S., L.R.; writing—review and editing, D.M., S.A., V.S., L.R.; resources, M.A.O., H.F., S.A.; supervision, M.A.O., H.F., S.A.

Funding: This research received no external funding.

Acknowledgments: The authors wish to thank Stefano Farris and Riccardo Rampazzo of the Dipartimento di Scienze per gli Alimenti, la Nutrizione e l'Ambiente (Defens) at the Università degli Studi di Milano, for assistance during corona treatment and gas permeability measurements.

Conflicts of Interest: The authors declare no conflict of interest.

References

1. Crippa, M.; Bianchi, A.; Cristofori, D.; D'Arienzo, M.; Merletti, F.; Morazzoni, F.; Scotti, R.; Simonutti, R. High dielectric constant rutile–polystyrene composite with enhanced percolative threshold. *J. Mater. Chem. C* **2013**, *1*, 484–492. [CrossRef]

2. Zhang, S.; Cao, J.; Shang, Y.; Wang, L.; He, X.; Li, J.; Zhao, P.; Wang, Y. Nanocomposite polymer membrane derived from nano TiO_2-PMMA and glass fiber nonwoven: High thermal endurance and cycle stability in lithium ion battery applications. *J. Mater. Chem. A* **2015**, *3*, 17697–17703. [CrossRef]

3. Li, W.; Li, H.; Zhang, Y.-M. Preparation and investigation of PVDF/PMMA/TiO2 composite film. *J. Mater. Sci.* **2009**, *44*, 2977–2984. [CrossRef]

4. Motaung, T.E.; Luyt, A.S.; Bondioli, F.; Messori, M.; Saladino, M.L.; Spinella, A.; Nasillo, G.; Caponetti, E. PMMA–titania nanocomposites: Properties and thermal degradation behaviour. *Polym. Degrad. Stab.* **2012**, *97*, 1325–1333. [CrossRef]

5. Pan, B.; Pan, B.; Zhang, W.; Lv, L.; Zhang, Q.; Zheng, S. Development of polymeric and polymer-based hybrid adsorbents for pollutants removal from waters. *Chem. Eng. J.* **2009**, *151*, 19–29. [CrossRef]

6. Wang, M.; Yang, G.; Jin, P.; Tang, H.; Wang, H.; Chen, Y. Highly hydrophilic poly(vinylidene fluoride)/meso-titania hybrid mesoporous membrane for photocatalytic membrane reactor in water. *Sci. Rep.* **2016**, *6*, 19148. [CrossRef] [PubMed]

7. Díez-Pascual, A.M.; Díez-Vicente, A.L. Nano-TiO_2 Reinforced PEEK/PEI Blends as Biomaterials for Load-Bearing Implant Applications. *ACS Appl. Mater. Interfaces* **2015**, *7*, 5561–5573. [CrossRef] [PubMed]

8. Salabat, A.; Mirhoseini, F. Applications of a new type of poly(methyl methacrylate)/TiO_2 nanocomposite as an antibacterial agent and a reducing photocatalyst. *Photochem. Photobiol. Sci.* **2015**, *14*, 1637–1643. [CrossRef] [PubMed]

9. Ravirajan, P.; Haque, S.A.; Durrant, J.R.; Bradley, D.D.C.; Nelson, J. The Effect of Polymer Optoelectronic Properties on the Performance of Multilayer Hybrid Polymer/TiO_2 Solar Cells. *Adv. Funct. Mater.* **2005**, *15*, 609–618. [CrossRef]

10. Lee, H.J.; Leventis, H.C.; Haque, S.A.; Torres, T.; Grätzel, M.; Nazeeruddin, M.K. Panchromatic response composed of hybrid visible-light absorbing polymers and near-IR absorbing dyes for nanocrystalline TiO_2-based solid-state solar cells. *J. Power Sources* **2011**, *196*, 596–599. [CrossRef]

11. Wen, R.; Guo, J.; Zhao, C.; Liu, Y. Nanocomposite Capacitors with Significantly Enhanced Energy Density and Breakdown Strength Utilizing a Small Loading of Monolayer Titania. *Adv. Mater. Interfaces* **2018**, *5*, 1701088. [CrossRef]

12. Cao, J.; Wang, L.; He, X.; Fang, M.; Gao, J.; Li, J.; Deng, L.; Chen, H.; Tian, G.; Wang, J.; et al. In situ prepared nano-crystalline TiO_2–poly(methyl methacrylate) hybrid enhanced composite polymer electrolyte for Li-ion batteries. *J. Mater. Chem. A* **2013**, *1*, 5955–5961. [CrossRef]

13. Chen, Y.-H.; Liu, Y.-Y.; Lin, R.-H.; Yen, F.-S. Photocatalytic degradation of p-phenylenediamine with TiO_2-coated magnetic PMMA microspheres in an aqueous solution. *J. Hazard. Mater.* **2009**, *163*, 973–981. [CrossRef] [PubMed]

14. Chen, Y.-H.; Chen, L.-L.; Shang, N.-C. Photocatalytic degradation of dimethyl phthalate in an aqueous solution with Pt-doped TiO_2-coated magnetic PMMA microspheres. *J. Hazard. Mater.* **2009**, *172*, 20–29. [CrossRef] [PubMed]

15. Xu, Y.; Wen, W.; Wu, J.-M. Titania nanowires functionalized polyester fabrics with enhanced photocatalytic and antibacterial performances. *J. Hazard. Mater.* **2018**, *343*, 285–297. [CrossRef] [PubMed]

16. Zhou, X.; Shao, C.; Yang, S.; Li, X.; Guo, X.; Wang, X.; Li, X.; Liu, Y. Heterojunction of g-C3N4/BiOI Immobilized on Flexible Electrospun Polyacrylonitrile Nanofibers: Facile Preparation and Enhanced Visible Photocatalytic Activity for Floating Photocatalysis. *ACS Sustain. Chem. Eng.* **2018**, *6*, 2316–2323. [CrossRef]

17. Krýsa, J.; Waldner, G.; Měšťánková, H.; Jirkovský, J.; Grabner, G. Photocatalytic degradation of model organic pollutants on an immobilized particulate TiO$_2$ layer. *Appl. Catal. B Environ.* **2006**, *64*, 290–301. [CrossRef]

18. Tu, W.; Lin, Y.-P.; Bai, R. Enhanced performance in phenol removal from aqueous solutions by a buoyant composite photocatalyst prepared with a two-layered configuration on polypropylene substrate. *J. Environ. Chem. Eng.* **2016**, *4*, 230–239. [CrossRef]

19. Ni, L.; Li, Y.; Zhang, C.; Li, L.; Zhang, W.; Wang, D. Novel floating photocatalysts based on polyurethane composite foams modified with silver/titanium dioxide/graphene ternary nanoparticles for the visible-light-mediated remediation of diesel-polluted surface water. *J. Appl. Polym. Sci.* **2016**, *133*. [CrossRef]

20. Han, H.; Bai, R. Highly effective buoyant photocatalyst prepared with a novel layered-TiO$_2$ configuration on polypropylene fabric and the degradation performance for methyl orange dye under UV–Vis and Vis lights. *Sep. Purif. Technol.* **2010**, *73*, 142–150. [CrossRef]

21. Xing, Z.; Zhang, J.; Cui, J.; Yin, J.; Zhao, T.; Kuang, J.; Xiu, Z.; Wan, N.; Zhou, W. Recent advances in floating TiO$_2$-based photocatalysts for environmental application. *Appl. Catal. B Environ.* **2018**, *225*, 452–467. [CrossRef]

22. Yang, M.; Di, Z.; Lee, J.-K. Facile control of surface wettability in TiO$_2$/poly(methyl methacrylate) composite films. *J. Colloid Interface Sci.* **2012**, *368*, 603–607. [CrossRef] [PubMed]

23. Soliveri, G.; Sabatini, V.; Farina, H.; Ortenzi, M.A.; Meroni, D.; Colombo, A. Double side self-cleaning polymeric materials: The hydrophobic and photoactive approach. *Colloids Surf. A Physicochem. Eng. Asp.* **2015**, *483*, 285–291. [CrossRef]

24. Ma, J.Z.; Hu, J.; Zhang, Z.J. Polyacrylate/silica nanocomposite materials prepared by sol-gel process. *Eur. Polym. J.* **2007**, *43*, 4169–4177. [CrossRef]

25. McLaren, A.C.; McLaren, S.G.; Hickmon, M.K. Sucrose, xylitol, and erythritol increase PMMA permeability for depot antibiotics. *Clin. Orthop. Relat. Res.* **2007**, *461*, 60–63. [CrossRef] [PubMed]

26. Sabatini, V.; Farina, H.; Montarsolo, A.; Ardizzone, S.; Ortenzi, M.A. Novel Synthetic Approach to Tune the Surface Properties of Polymeric Films: Ionic Exchange Reaction between Sulfonated Polyarylethersulfones and Ionic Liquids. *Polym. Plast. Technol. Eng.* **2017**, *56*, 296–309. [CrossRef]

27. Sabatini, V.; Cattò, C.; Cappelletti, G.; Cappitelli, F.; Antenucci, S.; Farina, H.; Ortenzi, M.A.; Camazzola, S.; Di Silvestro, G. Protective features, durability and biodegration study of acrylic and methacrylic fluorinated polymer coatings for marble protection. *Prog. Org. Coat.* **2018**, *114*, 47–57. [CrossRef]

28. Cappelletti, G.; Ardizzone, S.; Meroni, D.; Soliveri, G.; Ceotto, M.; Biaggi, C.; Benaglia, M.; Raimondi, L. Wettability of bare and fluorinated silanes: A combined approach based on surface free energy evaluations and dipole moment calculations. *J. Colloid Interface Sci.* **2013**, *389*, 284–291. [CrossRef] [PubMed]

29. Huang, W.; Kim, J.-B.; Bruening, M.L.; Baker, G.L. Functionalization of Surfaces by Water-Accelerated Atom-Transfer Radical Polymerization of Hydroxyethyl Methacrylate and Subsequent Derivatization. *Macromolecules* **2002**, *35*, 1175–1179. [CrossRef]

30. Chiantore, O.; Trossarelli, L.; Lazzari, M. Photooxidative degradation of acrylic and methacrylic polymers. *Polymer* **2000**, *41*, 1657–1668. [CrossRef]

31. McNeill, I.C.; Sadeghi, S.M.T. Thermal stability and degradation mechanisms of poly(acrylic acid) and its salts: Part 1—Poly(acrylic acid). *Polym. Degrad. Stab.* **1990**, *29*, 233–246. [CrossRef]

32. Zuev, V.V.; Bertini, F.; Audisio, G. Investigation on the thermal degradation of acrylic polymers with fluorinated side-chains. *Polym. Degrad. Stab.* **2006**, *91*, 512–516. [CrossRef]

33. Carradò, A.; Sokolova, O.; Donnio, B.; Palkowski, H. Influence of corona treatment on adhesion and mechanical properties in metal/polymer/metal systems. *J. Appl. Polym. Sci.* **2011**, *120*, 3709–3715. [CrossRef]

34. Soliveri, G.; Piffori, V.; Panzarasa, G.; Ardizzone, S.; Cappelletti, G.; Meroni, D.; Sparnacci, K.; Falciola, L. Self-cleaning properties in engineered sensors for dopamine electroanalytical detection. *Analyst* **2015**, *140*, 1486–1494. [CrossRef] [PubMed]

35. Arabatzis, I.; Antonaraki, S.; Stergiopoulos, T.; Hiskia, A.; Papaconstantinou, E.; Bernard, M.; Falaras, P. Preparation, characterization and photocatalytic activity of nanocrystalline thin film TiO_2 catalysts towards 3,5-dichlorophenol degradation. *J. Photochem. Photobiol. A Chem.* **2002**, *149*, 237–245. [CrossRef]

36. Pifferi, V.; Rimoldi, L.; Meroni, D.; Segrado, F.; Soliveri, G.; Ardizzone, S.; Falciola, L. Electrochemical characterization of insulating silica-modified electrodes: Transport properties and physicochemical features. *Electrochem. Commun.* **2017**, *81*, 102–105. [CrossRef]

37. Antonello, A.; Soliveri, G.; Meroni, D.; Cappelletti, G.; Ardizzone, S. Photocatalytic remediation of indoor pollution by transparent TiO_2 films. *Catal. Today* **2014**, *230*, 35–40. [CrossRef]

38. Piera, E.; Ayllón, J.A.; Doménech, X.; Peral, J. TiO_2 deactivation during gas-phase photocatalytic oxidation of ethanol. *Catal. Today* **2002**, *76*, 259–270. [CrossRef]

39. U.S. Food and Drug Administration. *Summary Report on Antimicrobials Sold or Distributed for Use in Food-Producing Animals*; U.S. Food and Drug Administration: Washington, DC, USA, 2016.

40. Rimoldi, L.; Pargoletti, E.; Meroni, D.; Falletta, E.; Cerrato, G.; Turco, F.; Cappelletti, G. Concurrent role of metal (Sn, Zn) and N species in enhancing the photocatalytic activity of TiO_2 under solar light. *Catal. Today* **2018**, *313*, 40–46. [CrossRef]

41. Palominos, R.A.; Mondaca, M.A.; Giraldo, A.; Peñuela, G.; Pérez-Moya, M.; Mansilla, H.D. Photocatalytic oxidation of the antibiotic tetracycline on TiO_2 and ZnO suspensions. *Catal. Today* **2009**, *144*, 100–105. [CrossRef]

42. Wei, R.; Ge, F.; Huang, S.; Chen, M.; Wang, R. Occurrence of veterinary antibiotics in animal wastewater and surface water around farms in Jiangsu Province, China. *Chemosphere* **2011**, *82*, 1408–1414. [CrossRef] [PubMed]

43. Calamari, D.; Zuccato, E.; Castiglioni, S.; Bagnati, R.; Fanelli, R. Strategic Survey of Therapeutic Drugs in the Rivers Po and Lambro in Northern Italy. *Environ. Sci. Technol.* **2003**, *37*, 1241–1248. [CrossRef]

44. Czekalski, N.; Berthold, T.; Caucci, S.; Egli, A.; Bürgmann, H. Increased Levels of Multiresistant Bacteria and Resistance Genes after Wastewater Treatment and Their Dissemination into Lake Geneva, Switzerland. *Front. Microbiol.* **2012**, *3*, 1–18. [CrossRef] [PubMed]

45. Addamo, M.; Augugliaro, V.; Di Paola, A.; García-López, E.; Loddo, V.; Marcì, G.; Palmisano, L. Removal of drugs in aqueous systems by photoassisted degradation. *J. Appl. Electrochem.* **2005**, *35*, 765–774. [CrossRef]

46. Magalhães, F.; Lago, R.M. Floating photocatalysts based on TiO_2 grafted on expanded polystyrene beads for the solar degradation of dyes. *Sol. Energy* **2009**, *83*, 1521–1526. [CrossRef]

47. Magalhães, F.; Moura, F.C.C.; Lago, R.M. TiO_2/LDPE composites: A new floating photocatalyst for solar degradation of organic contaminants. *Desalination* **2011**, *276*, 266–271. [CrossRef]

48. Maino, G.; Meroni, D.; Pifferi, V.; Falciola, L.; Soliveri, G.; Cappelletti, G.; Ardizzone, S. Electrochemically assisted deposition of transparent, mechanically robust TiO_2 films for advanced applications. *J. Nanopart. Res.* **2013**, *15*, 2087. [CrossRef]

49. Rimoldi, L.; Meroni, D.; Cappelletti, G.; Ardizzone, S. Green and low cost tetracycline degradation processes by nanometric and immobilized TiO_2 systems. *Catal. Today* **2017**, *281*, 38–44. [CrossRef]

50. Reyes, C.; Fernández, J.; Freer, J.; Mondaca, M.A.; Zaror, C.; Malato, S.; Mansilla, H.D. Degradation and inactivation of tetracycline by TiO_2 photocatalysis. *J. Photochem. Photobiol. A Chem.* **2006**, *184*, 141–146. [CrossRef]

MDPI

St. Alban-Anlage 66

4052 Basel

Switzerland

Tel. +41 61 683 77 34

Fax +41 61 302 89 18

www.mdpi.com

Catalysts Editorial Office

E-mail: catalysts@mdpi.com

www.mdpi.com/journal/catalysts